W9-BAL-537

y intercept of least-squares line

$$\hat{\beta}_0 = \bar{y} - \hat{\beta}_1 \bar{x}$$

Student's *t* (single mean)

$$t = \frac{\bar{y} - \mu_0}{s/\sqrt{n}}$$

Student's *t* (comparing two means)

$$t = \frac{(\bar{y}_1 - \bar{y}_2) - D_0}{s\sqrt{\frac{1}{n_1} + \frac{1}{n_2}}}$$

pooled estimate of σ^2

$$s^2 = \frac{\sum_{i=1}^{n_1} (y_i - \bar{y}_1)^2 + \sum_{i=1}^{n_2} (y_i - \bar{y}_2)^2}{n_1 + n_2 - 2}$$

correlation coefficient

$$r = \frac{n\sum_{i=1}^{n} x_i y_i - \left(\sum_{i=1}^{n} x_i\right)\left(\sum_{i=1}^{n} y_i\right)}{\sqrt{\left[n\sum_{i=1}^{n} x_i^2 - \left(\sum_{i=1}^{n} x_i\right)^2\right]\left[n\sum_{i=1}^{n} y_i^2 - \left(\sum_{i=1}^{n} y_i\right)^2\right]}}$$

Mann-Whitney *U*

$$U = n_1 n_2 + \frac{n_1(n+1)}{2} - T_A$$

and

$$U = n_1 n_2 + \frac{n_2(n_2+1)}{2} - T_B$$

chi-square statistic for contingency tables

$$X^2 = \sum_{i=1}^{k} \frac{(n_i - np_i)^2}{np_i}$$

Spearman's rank correlation coefficient

$$r_s = 1 - \frac{6\sum_{i=1}^{n} d_i^2}{n(n^2 - 1)}$$

INTRODUCTION TO PROBABILITY AND STATISTICS

Fifth Edition

William Mendenhall
University of Florida

Duxbury Press
North Scituate, Massachusetts

INTRODUCTION TO PROBABILITY AND STATISTICS

Fifth Edition

Duxbury Press
A Division of Wadsworth Publishing Company, Inc.

© 1979 by Wadsworth Publishing Company, Inc., Belmont, California 94002.
All rights reserved. No part of this book may be reproduced, stored in a retrieval system, or transcribed, in any form or by any means, electronic, mechanical, photocopying, recording, or otherwise, without the prior written permission of the publisher, Duxbury Press, a division of Wadsworth Publishing Company, Inc., Belmont, California.

Introduction to Probability and Statistics, Fifth Edition was edited and prepared for composition by Carol Beal. Interior design was provided by Duxbury Press. The cover was designed by Oliver Kline.

Library of Congress Cataloging in Publication Data

Mendenhall, William.
Introduction to probability and statistics.
First ed. published in 1963 under title: Introduction to statistics.
Includes bibliographies and index.
1. Mathematical statistics. 2. Probabilities.
I. Title.
QA276.M425 1979 519.2 78-15037
ISBN 0-87872-189-4

Printed in the United States of America
2 3 4 5 6 7 8 9 83 82 81 80 79

CONTENTS

13 The Analysis of Variance 421

14 Nonparametric Statistics 475

15 A Summary and Conclusion 519

PREFACE

The teaching objective of this fifth edition of *Introduction to Probability and Statistics* is to present a connected introduction to statistics which presents statistical inference, *the objective of statistics,* as its theme. To achieve this objective, the fifth edition retains the features of earlier editions, slightly modified, with some important additions. These changes include the following:

1. A statement of chapter objectives precedes each chapter. This addition will give the student a clear picture of each chapter's learning objectives and assist in creating a cohesive presentation of subject.
2. Highlighted tips on problem solving have been inserted at pertinent points in chapters to aid students in the diagnosis and the solution of specific types of problems.
3. Special sections emphasizing assumptions have been added at the end of the methods chapters.
4. The definition of the sample variance, Chapter 3, has been changed, indicating division of the sum of squares of deviations by $(n - 1)$. The relevance of this change to the statement of Tchebysheff's Theorem is mentioned in a footnote.
5. To present greater flexibility in the depth of presentation of probability, combinatorial mathematics (Chapter 4) has been moved to the end of the chapter and identified as an optional section. The first part of the chapter can be used by instructors who wish to present a brief introduction to the basic concepts of probability, moving rapidly into statistical inference. An instructor who wishes to present a more in-depth coverage of probability can insert the optional section at its traditional location.
6. The definition of random sampling has been moved forward in the text, to the end of Chapter 5, so that the term can be used in a discussion of

practical sampling situations associated with the binomial probability distribution, Chapter 6.
7. The order of presentation of the material on the normal probability distribution has been changed, permitting a student to become more familiar with the properties of the distribution (areas, etc.) before exposure to the Central Limit Theorem.
8. A section containing a realistic case study is added to Chapter 10; it demonstrates a practical application of regression analysis and, at the same time, introduces the student to a typical computer regression analysis printout.
9. Chapter 14, nonparametric statistics, has been revised, bringing the general discussion more in tune with recent developments in that area.
10. Like the fourth edition, many exercises, drawn from news articles or scientific literature, have been added to motivate and to add relevancy to the material.
11. Many less obvious changes have been made to clarify explanations, figures, etc., in response to instructor and student reviews. These changes, made in each edition, play an important role in the evolution of a clear presentation of the material.

As stated in the preface for earlier editions, the intent of the original text, and all subsequent editions, is to provide a cohesive, connected presentation of statistics that identifies inference as its objective and stresses the relevance of statistics in learning about the world in which we live.

To accomplish this goal, we think that the student should understand how statistical inference-making procedures work and why they function better than our own intuitive inference-making machinery. Thus, we believe that an elementary but clear understanding of the role of probability in making inferences and assessing their reliabilities is essential to an intelligent use of statistical methods. Consequently, a strong effort has been made to weld the theories of probability and inference throughout the book.

The student should understand the importance of each chapter in the text and how each relates to the objective of the course, statistical inference. Too often, students see statistics as the computation of means and standard deviations, as probability, as tests of hypotheses and estimation, but they do not see these topics as connected and forming a coherent picture of statistics. To create this connectivity, we attempt to explain at the beginning of each chapter its relevance to statistical inference and its relation to preceding chapters.

The student should also see that the same statistical problems occur in business, biology, engineering, and the social sciences and should realize that the basic concepts of statistics and inferential methods apply in all fields. Only the frequency of occurrence of specific types of problems varies from one field to another. Thus, I think that a first course in statistics should stress the unity of concepts, rather than being field-oriented. Specialized courses can follow in the student's field of application.

Three distinct points of emphasis in the text are a result of this philosophy. The objective of statistics—inference—is stated in Chapter 1, and the introduction to every subsequent chapter identifies chapter objectives and explains the role of the chapter material in making inferences and measuring their reliabilities. Second, not only is probability presented in a very elementary (but thorough) manner, based on the concept of the sample space, but it is employed at the earliest opportunity (Chapter 6) in two very practical inferential problems, as well as in subsequent discussions when the complexity of the probabilistic problem is within the grasp of the student. Finally, a large number of examples and exercises with practical application in a variety of fields indicate the broad applicability of statistical methodology.

The chapters on the analysis of enumerative data, important considerations in the design of experiments, the analysis of variance, and nonparametric statistics provide flexibility in the construction of an introductory course. Students in business and in the social and biological sciences at the University of Florida take a one-quarter course based on the first ten chapters of the text. This is followed by specialized courses in each student's field of application; or the student can take a second quarter covering the remainder of this text. For example, business majors follow the introductory course with a second-semester course on index numbers and time-series analysis. A course oriented toward the social sciences might utilize some combination of Chapters 11, 12, 13, and 14; an elementary course in the theory of probability, with some emphasis on inference, might include Chapters 1 through 8 and Chapter 14.

Exercises are graduated in difficulty so that all students can solve some of the exercises, a substantial number can solve most, and a few of the best students are challenged to solve all without error. Symbols are used to identify areas of application in business, industry, and the sciences. A thorough knowledge of definitions and concepts is essential before the students attempt the exercises. Unannounced quizzes are very effective in helping the student learn the new language of statistics.

A semiprogrammed study guide to help students who have individual difficulties with the subject matter is available. This study guide, now in its fifth edition, benefits from considerable user experience.

The number of colleagues and friends who have given kind assistance in helping me write this text has increased over time. William Miller and the late Paul Benson of Bucknell University helped in developing and testing the first edition and made comments concerning its revision. Much credit should go to the numerous reviewers who aided in the writing of the second, third, and fourth editions and thereby contributed to the quality of the current edition. These include John T. Webster, William Brelsford, Robert Crovelli, Douglas Chapman, Arthur Coladarci, the late Paul Meyer, Joyce Curry, and Frank Deane. I thank the reviewers of this edition for both comments and encouragement, in particular, Roy E. Myers, Pennsylvania State University, P. V. Rao and Dennis D. Wackerly, University of Florida, and John N. Quiring, Grand Valley State College.

Barbara Beaver, University of California at Riverside, prepared the solutions manual for the text. She and James Bolognese, Merck Inc., provided many helpful comments that were used in acquiring solutions to exercises. I would also like to express my appreciation to the staff of Duxbury Press, particularly Carol Beal. Thanks are also due to E. J. Pearson, A. Hald, W. H. Beyer, and D. L. Burkholder, and to R. A. Wilcox and C. W. Dunnett, for their kind permission to use tables reprinted in Appendix II. I am indebted to the typists who have given their best and endured the worst: Florence Valentine, who typed the original draft of the manuscript; Angie Anastasia and Gay Midelis, who prepared the second revision; Mary Jackson and Catherine Kennedy, who typed the third edition; Ellen Evans, who typed the fourth; and Carol Rozear and others who typed drafts of the current edition. Finally, as before, I acknowledge the assistance of my family and their partnership in this writing endeavor.

WILLIAM MENDENHALL

CHAPTER OBJECTIVES

GENERAL OBJECTIVE The purpose of this chapter is to identify the nature of statistics, its objective, and how it plays an important role in the sciences, in industry, and, ultimately, in our daily lives.

SPECIFIC OBJECTIVES

1. To answer the question, "What is statistics?" *Sections 1.1, 1.2, 1.3*
2. To identify statistical inference as the objective of modern statistics. *Section 1.3*
3. To identify the contributions that statistics can make to inference making based on sample data. *Section 1.4*
4. To define the basic words and concepts used in statistics. *Sections 1.1 to 1.5*

1 WHAT IS STATISTICS?

1.1

ILLUSTRATIVE STATISTICAL PROBLEMS

What is statistics? How does it function? How does it help to solve certain practical problems? Rather than attempt a definition at this point, let us examine several problems that might come to the attention of the statistician. From these examples we can then select the essential elements of a statistical problem.

In predicting the outcome of a national election, pollsters interview a predetermined number of people throughout the country and record their preference. On the basis of this information a prediction is constructed. Similar problems are encountered in market research (What fraction of smokers prefer cigarette brand *A*?); in sociology (What fraction of rural homes have electricity?); in industry (What fraction of items purchased, or produced, are defective?).

The yield (production) of a chemical plant is dependent upon many factors. By observing these factors and the yield over a period of time, we can construct a prediction equation relating yield to the observed factors. How do we find a good prediction equation? If the equation is used to predict yield, the prediction will rarely equal the true yield; that is, the prediction will almost always be in error. Can we place a limit on the prediction error? Which factors are the most important in predicting yield? One encounters similar problems in the fields of education, sociology, psychology, and the physical sciences. For instance, colleges wish to predict the grade-point average of college students based upon information obtained *prior* to college entrance. Here, we might relate grade-point average to rank in high school class, the type of high school, college board examination scores, socioeconomic status, or any number of similar factors.

In addition to prediction, the statistician is concerned with decision making based upon observed data. Consider the problems of determining the effectiveness of a new cold vaccine. For simplification, let us assume that 10 people have received the new cold vaccine and are observed over a winter season. Of these 10, 8 survive the winter without acquiring a cold. Is the vaccine effective?

Two physically different teaching techniques are used to present a subject to two groups of students of comparable ability. At the end of the instructional period, a measure of achievement is obtained for each group. On the basis of this information, we ask: Do the data present sufficient evidence to indicate that one method produces, on the average, higher student achievement?

Consider the inspection of purchased items in a manufacturing plant. On the basis of such an inspection, each lot of incoming goods must be either accepted or rejected and returned to the supplier. The inspection might involve drawing a sample of 10 items from each lot and recording the number of defectives. The decision to accept or reject the lot could then be based upon the number of defective items observed.

A company manufacturing complex electronic equipment produces some systems that function properly but also some that, for unknown reasons, do not. What makes good systems good and bad systems bad? In attempting to answer this question, we might make certain internal measurements on a system in order to find important factors which differentiate between an acceptable and an unacceptable product. From a sample of good and bad systems, data could then be collected that might shed light on the fundamental design or on production variables affecting system quality.

1.2

THE POPULATION AND THE SAMPLE

The examples we have cited are varied in nature and complexity, but each involves prediction or decision making. In addition, each of these examples involves sampling. A specified number of items (objects or bits of information)—a sample—is drawn from a much larger body of data which we call the population. The pollster draws a sample of opinion (those interviewed) from the statistical population, which is the set of opinions corresponding to all the eligible voters in the country. In predicting the fraction of smokers who prefer cigarette brand *A*, we assume that those interviewed yield a representative sample of the population of all smokers. The sample for the cold vaccine experiment consists of observations made on the 10 individuals receiving the vaccine. The sample is presumably representative of data pertinent to a much larger body of people—the population—who could have received the vaccine.

Which is of primary interest, the sample or the population? In all the examples given above, we are primarily interested in the population. We cannot interview all the people in the United States, hence we must predict their behavior on the basis of information obtained from a representative sample. Similarly, it is practically impossible to give all possible users a cold vaccine. The manufacturer of the drug is interested in its effectiveness in preventing colds in the purchasing public (the population) and must predict this effectiveness from information extracted from the sample. Hence, the sample may be of immediate interest, but we are primarily interested in describing the population from which the sample is drawn.

DEFINITION

A population is the set representing all measurements of interest to the sample collector.

DEFINITION

A sample is a subset of measurements selected from the population of interest.

As used by most people, the word "sample" has two meanings. It can refer to the set of objects on which measurements are to be taken or it can refer to the objects themselves. A similar double use could be made of the word "population."

For example, you read in the newspapers that a Gallup Poll was based on a sample of 1823 people. In this use of the word "sample," the objects selected for the sample are clearly people. Presumably each person is interviewed on a particular question, and that person's response represents a single item of data. The collection of data corresponding to the people represents a sample of data.

In a study of sample survey methods, it is important to distinguish between the objects measured and the measurements themselves. To experimenters the objects measured are called "experimental units." The sample survey statistician calls them "elements of the sample."

To avoid a proliferation of terms, we will use the word "sample" in its everyday meaning. Most of the time we will be referring to the set of measurements made on the experimental units (elements of the sample). If occasionally we use the term in referring to a collection of experimental units, the context of the discussion will clarify the meaning.

1.3 THE ESSENTIAL ELEMENTS OF A STATISTICAL PROBLEM

The objective of statistics and the essential elements of a statistical problem are now vaguely apparent.

> The objective of statistics is to make inferences (predictions, decisions) about a population based upon information contained in a sample.

How will we achieve this objective? We will find that every statistical problem contains five elements. The first and foremost of these is a clear specification of the question to be answered and of the population of data that is related to the question.

The second element of a statistical problem is deciding how the sample will be selected. This is called the design of the experiment or sampling procedure. This element is important because data cost money and time. In fact, it is not unusual for an experiment or a statistical survey to cost $50,000 to $500,000, and the costs of many biological or technological experiments can run into the millions. And what do these experiments and surveys produce? Numbers on a sheet of paper or, in brief, information. So planning the experiment is important. Including too many observations in the sample is often costly and wasteful; including too few is also unsatisfactory. But most important, you will learn that the method used to collect the sample will often affect the amount of information per observation. A good sampling design can sometimes reduce the costs of data collection to one-tenth or as little as one-hundredth of the cost of another sampling design.

The third element of a statistical problem involves the analysis of the sample data. No matter how much information the data contain about the practical question, you must use an appropriate method of data analysis to extract the desired information from the data.

The fourth element of a statistical problem is using the sample data to make an inference about the population. As you will subsequently learn, many different procedures can be employed to make an estimate or decision about some characteristic of a population or to predict the value of some member of the population. For example, ten different methods might be available to estimate human response to a new drug but one procedure might be much more accurate than another. Therefore, you will wish to employ the best inference-making procedure when you use sample data to make an estimate or decision about a population or a prediction about some member of a population.

The final element of a statistical problem identifies what is perhaps the most important contribution of statistics to inference making. It answers the question, "How good is the inference?" To illustrate, suppose you manage a small manufacturing concern. You arrange for an agency to conduct a statistical survey for you and it estimates that your company's product will gain 34% of the market this year. How much faith can you place in this estimate? You will quickly discern that you are lacking some important information. Of what value is the estimate without a measure of its reliability? Is the estimate accurate to within 1%, 5%, or 20%? Is it reliable enough to be used in setting production goals? As you will subsequently learn, statistical estimation, decision-making, and prediction procedures enable you to calculate a measure of goodness for every inference. Consequently, in a practical inference-making situation, every inference should be accompanied by a measure which tells you how much faith you can place in the inference.

To summarize, a statistical problem involves the following:

1. A clear definition of the objective of the experiment and the pertinent population.
2. The design of the experiment or sampling procedure.
3. The collection and analysis of data.
4. The procedure for making inferences about the population based upon sample information.
5. The provision of a measure of goodness (reliability) for the inference.

It is extremely important to note that the steps in the solution of a statistical problem are sequential; that is, you must identify the population of interest and plan how you will collect the data before you can collect and analyze it. And all these operations must precede the ultimate goal, making inferences about the population based on information contained in the sample. These steps—carefully identifying the pertinent population and designing the experiment or sampling procedure—often are omitted. The experimenter may select the sample from the wrong population or may plan the data collection in a manner that intuitively seems reasonable or logical but that may be an extremely poor plan from a statistical point of view. The resulting data may be difficult or impossible to analyze, may contain little or no pertinent information, or, inadvertently, the sample may not be representative of the population of interest. This means that every experimenter should be knowledgeable in the statistical design of experiments and (or) sample surveys or should consult an applied statistician for the appropriate design *before* the data are collected.

1.4

THE ROLE OF STATISTICS AND THE STATISTICIAN

We may now ask: What is the role of the statistician in attaining the objective in what we have described as statistical problems? People have been making observations and collecting data for centuries. Furthermore, they have been using the data as a basis for prediction and decision making completely unaided by statistics. What, then, do statisticians and statistics have to offer?

Statistics is an area of science concerned with the extraction of information from numerical data and its use in making inferences about a population from which the data are obtained. In some respects the statistician quantifies information and studies various designs and sampling procedures, searching for the procedure that yields a specified amount of information in a given situation at a minimum cost. Therefore, one major contribution of statistics and statisticians is in designing experiments and surveys, thereby reducing their cost and size. The second major contribution is in the inference making itself. The statistician studies various inferential procedures, looking for the best predictor or decision-making process for a given situation. Even more important, the statistician provides information concerning the goodness of an inferential procedure. When we predict, we would like to know something about the error in our prediction. If we make a decision, we want to know the chance that our decision is correct. Our built-in individual prediction and decision-making systems do not provide immediate answers to these important questions. They could be evaluated only by observation over a long period of time. In contrast, statistical procedures do provide answers to these questions.

1.5

SUMMARY

Statistics is an area of science concerned with the design of experiments or sampling procedures, the analysis of data, and the making of inferences about a population of measurements from information contained in a sample. The statistician is concerned with developing and using procedures for design, analysis, and inference making which will provide the best inference at a minimum cost. In addition to making the best inference, the statistician is concerned with providing a quantitative measure of the goodness of the inference-making procedure.

A careful identification of the target population and the design of the sampling procedure are often omitted but are essential steps in drawing inferences from experimental data. Poorly designed sampling procedures will often produce data that are of little or no value (although this may not be obvious to the experimenter). And after you make an inference, cast a critical eye upon it. Be sure you acquire a measure of its reliability.

1.6
A NOTE TO THE READER

We have stated the objective of statistics and, hopefully, have answered the question: What is statistics? The remainder of this text is devoted to the development of the basic concepts involved in statistical methodology. In other words, we wish to explain how statistical techniques actually work and why.

As you proceed through this text, you may wonder why we are devoting so much space to the making of inferences about populations from sample data and so little to the design of experiments. The reason is that the principles of good design do not become apparent until you know how to make inferences. As a consequence, your first contact with a design that may greatly increase the information in an experiment is in Section 9.5. An elementary discussion of experimental design follows in Chapter 12. The purpose of this text is to introduce you to the concepts and some of the elementary methods of statistics. For assistance in designing real-life experiments or surveys, you will want to consult a professional statistician.

Statistics is a very heavy user of applied mathematics. Most of the fundamental rules (called theorems in mathematics) are developed and based upon a knowledge of the calculus or higher mathematics. Inasmuch as this is meant to be an introductory text, we omit proofs except where they can be easily derived. Where concepts or theorems can be shown to be intuitively reasonable, we shall attempt to give a logical explanation. Hence, we shall attempt to convince you with the aid of examples and intuitive arguments rather than with rigorous mathematical derivations.

You should refer occasionally to Chapter 1 and review the objective of statistics and the elements of a statistical problem. Each of the following chapters should, in some way, be directed toward answering the questions posed here. Each is essential to completing the overall picture of statistics.

REFERENCES

Careers in Statistics, American Statistical Association and the Institute of Mathematical Statistics, 1974.

Tanur, J. M., et al., eds., *Statistics: A Guide to the Unknown*. San Francisco: Holden-Day, 1972.

EXERCISES

1.1 Suppose that you wish to obtain an estimate of the gasoline consumption (miles per gallon) of the Ford Fiesta. Describe the population that would likely be of interest to you and from which you would like to select your sample.

1.2 You are a candidate for your state legislature and you wish to survey voter attitudes regarding your chances of winning. Identify the population of interest to you and from which you would like to select your sample. How will this population be dependent upon time?

1.3 A medical researcher wishes to estimate the survival time of a patient after the onset of a particular type of cancer and after a particular regimen of radiotherapy. Identify the population of interest to the medical researcher. Can you perceive some problems in sampling this population?

1.4 An educational researcher wishes to evaluate the effectiveness of a new method for teaching reading to deaf students. Achievement at the end of a period of teaching is to be measured by a student's score on a reading test. Discuss the population (or populations) that might be of interest to the researcher.

EXPERIENCES WITH REAL DATA

Students frequently confuse experimental units with sample data and samples with populations. Let us examine these terms in a realistic setting by viewing the summaries of one or more research investigations.

Visit the research library of your college or university and select a research journal in a field of study of interest to you. For example, a student majoring in the biological sciences might choose *Ecology, The Journal of General Microbiology, The Australian Journal of Experimental Biology and Medical Science,* or one of the many other biological research journals. Similar research journals are available in the social sciences, physical sciences, or in the various fields of business. Once the research journal has been selected, pick an article that deals with experimentation, either in the laboratory or a social science experiment involving a sample survey. Even though the article may not contain the actual data collected in the experiment, discuss the research objective and the experimental procedure that was employed to generate the data. Identify the experimental units, the sample, and, particularly, the population. Presumably the researcher had some question that he or she wished to answer or a research hypothesis to confirm. Explain how statistical inference about the population will answer the researcher's questions or yield information concerning the research hypothesis.

CHAPTER OBJECTIVES

GENERAL OBJECTIVE Because functional notation and summation notation are used throughout this text, the objective of this chapter is to review (or introduce) these two important mathematical notations.

SPECIFIC OBJECTIVES

1. To develop an understanding of functional notation. This notation will be used in the subsequent discussion of summation notation. *Section 2.2*
2. To develop an understanding of the meaning of summation symbols. *Sections 2.3, 2.4*
3. To show how summation notation is used to provide instructions for the summation of numbers. Summation notation will be used in the summing of sample data in Chapter 3 and subsequent chapters. *Section 2.4*
4. To present three useful theorems that can be used to find simplified expressions for formulas that appear in Chapter 3 and subsequent chapters. *Section 2.5*

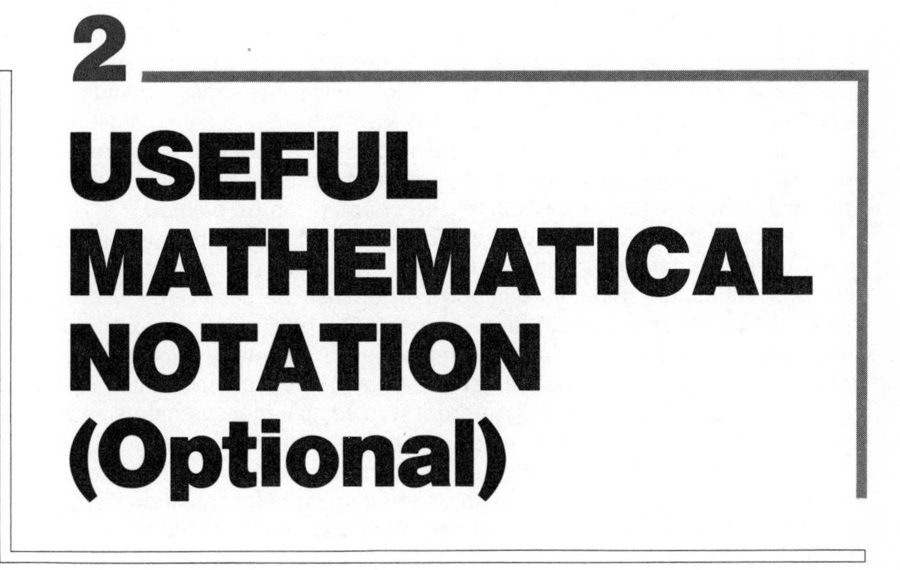

2 USEFUL MATHEMATICAL NOTATION (Optional)

2.1

INTRODUCTION

As you will learn subsequently, the statistical analysis of data will require numerous arithmetical operations which are too cumbersome to describe in words. The solution to this difficulty is to express the operations in terms of one or more formulas. Then, to analyze a given set of sample data, you substitute the sample measurements into the appropriate set of formulas.

One of the most common operations in a statistical analysis of data is the process of summation (addition). Consequently, we need a symbol to instruct the reader to sum the sample measurements or perhaps a set of numbers computed from the sample measurements. This symbol, called summation notation, will be very familiar to some readers and entirely new to others.

If you are unfamiliar with the notation or need a review, it is suggested that you read this chapter and work the exercises at the ends of the sections. If you are familiar with summation notation, the chapter can be omitted.

Understanding summation notation requires that you be familiar with two topics from elementary algebra: functional notation and the notion of a sequence. Functional notation is important because we shall often wish to sum some function of the sample measurements (to give a simple example, their squares). Thus the function of the sample measurements to be summed will be given as a formula in functional notation. Sequences are important because we must order (identify by arranging in order) the sample measurements so that we can give instructions on which measurements are to be summed.

At first glance this chapter may seem more mathematical than necessary. Please bear with us. You will need an understanding of this notation if you are to understand the formulas that will appear in subsequent chapters.

2.2

FUNCTIONAL NOTATION

Let us consider two sets* of elements (objects, numbers, or anything we want to use) and their relation to one another. Let the symbol x represent an element of the first set and y an element of the second. One could specify a number of different rules defining relationships between x and y. For instance, if our elements are people, we have rules for determining whether x and y, two persons, are first cousins. Or, suppose that x and y are integers taking values 1, 2, 3, 4,. . . . For some reason we might wish to say that x and y are "related" if $x = y$.

Now let us direct our attention to a specific relationship useful in mathematics called a functional relation between x and y.

*A set is a collection of specific things.

DEFINITION

A function consists of two sets of elements and a defined correspondence between an element of the first set, x, and an element of the second set, y, such that for each element x there corresponds one, and only one, element y.

A function may be exhibited as a collection of ordered pairs of elements, written (x, y) where x represents an element from the first set and y represents an element from the second. In fact, a function is often defined as a collection of ordered pairs of elements with the property stated in the definition.

Most often, x and y will be variables taking numerical values. The two sets of elements would represent all the possible numerical values that x and y might take, and the rule defining the correspondence between them would be an equation. For example, if we state that x and y are real numbers and that

$$y = x + 2,$$

then y is a function of x. Assigning a value to x (that is, choosing an element of the set of all real numbers), there corresponds one, and only one, value of y. When $x = 1$, $y = 3$. When $x = -4$, $y = -2$; and so on.

The area, A, of a circle is related to the radius, r, by the formula

$$A = \pi r^2.$$

Note that if we assign a value to r, the value of A can be determined from the formula. Hence A is a function of r. Likewise, the circumference of a circle, C, is a function of the radius.

As a third example of a functional relation, consider a classroom containing 20 students. Let x represent a specific body in the classroom and y represent a name. Is y a function of x? To answer the question, we examine x and y in the light of the definition. We note that each body has a name y attached to it. Hence when x is specified, y will be uniquely determined. According to our definition, y is a function of x. Note that the defining rule for functional relations does not require that x and y take numerical values. We shall encounter an important functional relation of this type in Chapter 4.

Having defined a functional relation, we may now turn to functional notation.

Mathematical writing frequently uses the same phrases or refers to a specific object many times in the course of a discussion. Unnecessary repetition wastes the reader's time, takes up valuable space, and is cumbersome to the writer. Hence the mathematician resorts to mathematical symbolism which is in some respects a type of mathematical shorthand. Rather than state "The area of a circle, A, is a function of the radius, r," the mathematician would write $A(r)$ or $A = f(r)$. The expression

$$y = f(x)$$

tells us that y is a function of the variable appearing in parentheses, namely x. Note that this expression does not tell us the specific functional relation that exists between y and x.

Consider the function of x,

$$y = 3x + 2.$$

In functional notation, this would be written as

$$f(x) = 3x + 2.$$

It is understood that $y = f(x)$.

Functional notation is especially advantageous when we wish to indicate the value of the function y when x takes a specific value, say $x = 2$. For the example above, we see that when $x = 2$,

$$\begin{aligned} y &= 3x + 2 \\ &= 3(2) + 2 \\ &= 8. \end{aligned}$$

Rather than write this, we use the simpler notation

$$f(2) = 8.$$

Similarly, $f(5)$ would be the value of the function when $x = 5$:

$$\begin{aligned} f(5) &= 3(5) + 2 \\ &= 15 + 2 \\ &= 17. \end{aligned}$$

Thus $f(5) = 17$.

To find the value of the function when x equals any value, say $x = c$, we substitute the value of c for x in the equation and obtain

$$f(c) = 3c + 2.$$

Formulas for probabilities in later chapters will be expressed in functional notation.

EXAMPLE 2.1

Given a function of x,

$$g(x) = \frac{1}{x} + 3x \qquad \text{where } x \neq 0.$$

(Note that x cannot equal 0 because then $1/x$ is undefined.)

(a) Find $g(4)$.

$$\begin{aligned} g(4) &= \frac{1}{4} + 3(4) \\ &= .25 + 12 \\ &= 12.25. \end{aligned}$$

(b) Find $g\left(\frac{1}{a}\right)$.

$$g\left(\frac{1}{a}\right) = \frac{1}{(1/a)} + 3\left(\frac{1}{a}\right)$$
$$= a + \frac{3}{a}$$
$$= \frac{a^2 + 3}{a}.$$

EXAMPLE 2.2

Let $p(y) = (1 - a)a^y$. Find $p(2)$, $p(3)$, and $p(0)$.

$p(2) = (1 - a)a^2$;
$p(3) = (1 - a)a^3$;
$p(0) = (1 - a)a^0 = 1 - a$, since $a^0 = 1$.

EXAMPLE 2.3

Let $f(y) = 4$. Find $f(2)$ and $f(3)$.

$f(2) = 4$;
$f(3) = 4$.

Note that $f(y)$ always equals 4, regardless of the value of y. Hence when a value of y is assigned, the value of the function is determined and will equal 4.

EXERCISES

2.1 If $f(x) = x^2 - 3x + 1$, find $f(0)$, $f(1)$, and $f(2)$.

2.2 If $p(y) = (1/2)^y$, find $p(0)$, $p(1)$, and $p(4)$.

2.3 If $f(y) = y^2 + 3$, find $f(0)$ and $f(3)$.

2.4 If $f(y) = y^3$, find $f(1)$, $f(2)$, and $f(3)$.

2.5 If $f(y) = y$, find $f(2)$, $f(-3)$, and $f(-1)$.

2.6 If $g(y) = (y - 1)^2$, find $g(1)$, $g(2)$, and $g(3)$.

2.3

NUMERICAL SEQUENCES

In statistics we shall be concerned with samples consisting of sets of measurements. There will be a first measurement, a second, and so on. Introducing the notion of a mathematical sequence at this point will provide us with a simple notation for discussion of data and will, at the same time, supply our first practical application of functional notation. This will, in turn, be used in the summation notation introduced in Section 2.4.

A set of objects, $a_1, a_2, a_3, a_4, \ldots$, ordered in the sense that we can identify the first member of the set a_1, the second a_2, etc., is called a sequence. Most often, $a_1, a_2, a_3, \ldots$, called elements of the sequence, are numbers but this is not a requirement. For example, the numbers

$$1, 5, 4, 8, 7, 11, 10, \ldots$$

form a sequence moving from left to right. Note that the elements are ordered only in their position in the sequence and need not be ordered in magnitude. Likewise, in some card games, we are interested in a sequence of cards; for example,

10, jack, queen, king, ace.

Although nonnumerical sequences are of interest, we shall be concerned solely with numerical sequences. Specifically, data obtained in a sample from a population will be regarded as a sequence of measurements.

Since sequences will form an important part of subsequent discussions, let us turn to a shortcut method of writing sequences that utilize functional notation. Inasmuch as the elements of a sequence are ordered in position in the sequence, it would seem natural to attempt to write a formula for a typical element of the sequence as a function of its position. For example, consider the sequence

$$3, 4, 5, 6, 7, \ldots,$$

where each element in the sequence is one greater than the preceding element. Let y be a position variable for the sequence so that y can take values 1, 2, 3, 4, for positions 1, 2, 3, 4, . . . , respectively. Then we might write a formula for the element in position y as

$$f(y) = y + 2.$$

Thus the first element in the sequence would be in position $y = 1$, and $f(1)$ would equal

$$f(1) = 1 + 2 = 3.$$

Likewise, the second element would be in position $y = 2$, and the second element of the sequence would be

$$f(2) = 2 + 2 = 4.$$

A brief check convinces us that this formula works for all elements of the sequence. Note that finding a proper formula (function) is a matter of trial and error and requires a bit of practice.

EXAMPLE 2.4

Given the sequence

$$1, 4, 9, 16, 25, \ldots,$$

find a formula that expresses a typical element in terms of a position variable y.

We note that each element is the square of the position variable; hence

$$f(y) = y^2.$$

EXAMPLE 2.5

The formula for the typical element of the following sequence is not so obvious:

$$0, 3, 8, 15, 24, 35, \ldots .$$

The typical element would be

$$f(y) = y^2 - 1.$$

Readers of mathematical writings (and this includes the author) prefer consistency in the use of mathematical notation. Unfortunately, this is not always practical. Writers are limited by the number of symbols available and also by the desire to make their notation consistent with some other texts on the subject. Hence x and y are very often used in referring to a variable, but we could just as well use i, j, k, or z. For instance, suppose that we wish to refer to a set of measurements and denote the measurements as a variable y. We might write this sequence of measurements as

$$y_1, y_2, y_3, y_4, y_5, \ldots ,$$

using a subscript to denote a particular element in the sequence. We are now forced to choose a new variable to denote the position of an element in the sequence (since y has been used). Suppose that we use the letter i. Then we could write a typical element as

$$f(i) = y_i.$$

Note that, as previously, $f(1) = y_1$, $f(2) = y_2$, etc.

EXERCISES

2.7 If y is the position variable for a sequence and the formula for a typical element is $y^2 + 1$, give the first three elements of the sequence.

2.8 Given the sequence

$$2, 3, 4, 5, 6, \ldots ,$$

give a formula for a typical element of the sequence as a function of y, the position variable.

2.4

SUMMATION NOTATION

As we shall observe in Chapter 3, in analyzing statistical data we shall often be working with sums of numbers and will need a simple notation for indicating a sum. For instance, consider the sequence of numbers

$$1, 2, 3, 4, 5, \ldots$$

and suppose that we wish to discuss the sum of the squares of the first four numbers of the sequence. Using summation notation, this would be written as

$$\sum_{y=1}^{4} y^2.$$

Interpretation of the summation notation is relatively easy. The Greek letter Σ (capital sigma), corresponding to "S" in the English alphabet (the first letter in the word "Sum"), tells us to sum elements of a sequence. A typical element of the sequence is given to the right of the summation symbol, and the position variable, called the variable of summation, is shown beneath. For our example, y^2 is a typical element, y is the variable of summation, and the implied sequence is

$$1, 4, 9, 16, 25, 36, \ldots .$$

Which elements of the sequence should appear in the sum? The position of the first element in the sum is indicated below the summation sign, the last above. The sum would include all elements proceeding *in order* from the first to last. In our example, the sum would include the sum of the elements commencing with the first and ending with the fourth:

$$\begin{aligned}\sum_{y=1}^{4} y^2 &= (1)^2 + (2)^2 + (3)^2 + (4)^2 \\ &= 1 + 4 + 9 + 16 \\ &= 30.\end{aligned}$$

EXAMPLE 2.6

$$\begin{aligned}\sum_{y=2}^{4} (y - 1) &= (2 - 1) + (3 - 1) + (4 - 1) \\ &= 6.\end{aligned}$$

EXAMPLE 2.7

$$\begin{aligned}\sum_{x=2}^{5} 3x &= 3(2) + 3(3) + 3(4) + 3(5) \\ &= 42.\end{aligned}$$

We emphasize that the typical element is a function only of the variable of summation. All other symbols are regarded as constants.

EXAMPLE 2.8

$$\sum_{i=1}^{3} (y_i - a) = (y_1 - a) + (y_2 - a) + (y_3 - a).$$

In this example, note that i is the variable of summation and that it appears as a subscript in the typical element.

EXAMPLE 2.9

$$\sum_{i=1}^{2} (x - i + 1) = (x - 1 + 1) + (x - 2 + 1)$$
$$= 2x - 1.$$

EXAMPLE 2.10

$$\sum_{y=2}^{4} y - 1 = (2 + 3 + 4) - 1$$
$$= 8.$$

Note the difference between Example 2.6 and Example 2.10. The quantity $(y - 1)$ is the typical element in Example 2.6, while y is the typical element in Example 2.10.

EXAMPLE 2.11

Suppose the measurements 122, 129, and 124 represent the blood pressure measurements on 3 twenty-year-old females. Let y_1 represent the measurement 122 and, similarly, let $y_2 = 129$ and $y_3 = 124$. Find

$$\sum_{i=1}^{3} y_i \quad \text{and} \quad \sum_{i=1}^{3} y_i^2.$$

Solution You can see that the expression

$$\sum_{i=1}^{3} y_i = y_1 + y_2 + y_3$$

provides a compact way of writing the sum of the sample measurements. Then,

$$\sum_{i=1}^{3} y_i = y_1 + y_2 + y_3 = 122 + 129 + 124 = 375.$$

Similarly,

$$\sum_{i=1}^{3} y_i^2$$

is a compact way of writing the sum of squares of the sample measurements. Thus,

$$\sum_{i=1}^{3} y_i^2 = y_1^2 + y_2^2 + y_3^2 = (122)^2 + (129)^2 + (124)^2$$
$$= 14{,}884 + 16{,}641 + 15{,}376 = 46{,}901.$$

EXERCISES

2.9 Evaluate the following summations.

(a) $\sum_{y=0}^{5} (y - 4)$ (b) $\sum_{y=2}^{6} (y^2 - 5)$

(c) $\sum_{i=1}^{4} (y_i - 2)$ (d) $\sum_{i=1}^{3} (y + 2i)$

2.10 If y is the position variable for a sequence and the formula for a typical element is $y^2 - 1$:

(a) Write the first four elements of the sequence.

(b) Use summation notation to write an expression for the sum of the first four terms of the sequence.

(c) Find the sum, part (b).

2.11 If y is the position variable for a sequence and the formula for a typical element is $(y - 3)^2$:

(a) Write the first five elements of the sequence.

(b) Use summation notation to write an expression for the sum of the first five terms of the sequence.

(c) Find the sum, part (b).

2.12 Consider the set of the opinions of all persons at your college or university regarding the desirability of a tax on gas-guzzling automobiles. Let a person favoring the tax be represented as a 1 and a person opposed as a 0 (for the sake of simplicity, assume that everyone has an opinion). As an ultimate objective, we wish to sample this set of opinions in order to estimate the proportion of students favoring the tax.

(a) Describe the population of interest.

(b) Suppose the first five measurements in the sample are $y_1 = 1$, $y_2 = 1$, $y_3 = 0$, $y_4 = 0$, and $y_5 = 1$. Use summation notation to write expressions for the sum and the sum of the squares of these five measurements.

(c) Find the sum and the sum of the squares of the five measurements.

(d) Suppose that the complete sample contains $n = 100$ measurements which are represented as $y_1, y_2, \ldots, y_{99}, y_{100}$ and that 73 of these favor the tax. Find

$$\sum_{i=1}^{100} y_i, \quad \sum_{i=1}^{100} y_i^2, \quad \text{and} \quad \left(\sum_{i=1}^{100} y_i\right)^2.$$

2.13 To estimate weekly loss due to theft, a clothing store recorded the total dollar loss over a period of 10 weeks. The losses, recorded to the nearest ten dollars, were

$$y_1 = 360,\ y_2 = 430,\ y_3 = 210,\ y_4 = 320,\ y_5 = 550,$$
$$y_6 = 170,\ y_7 = 240,\ y_8 = 370,\ y_9 = 280,\ y_{10} = 290.$$

(a) Describe the population of interest to the store manager.

(b) Find $\sum_{i=1}^{10} y_i$.

(c) Find $\sum_{i=2}^{4} y_i, \quad \sum_{i=2}^{4} y_i^2, \quad \text{and} \quad \left(\sum_{i=1}^{3} y_i\right)^2.$

2.14 For $y_i = 2i^2 - 1$, find:

(a) $\sum_{i=1}^{5} y_i$ (b) $\sum_{i=1}^{4} y_i^2$ (c) $\left(\sum_{i=1}^{4} y_i\right)^2$ (d) $\sum_{i=1}^{5} xy_i$.

2.5 USEFUL THEOREMS RELATING TO SUMS

Consider the summation

$$\sum_{y=1}^{3} 5.$$

The typical element is 5 and it does not change. The sequence is, therefore,

$$5, 5, 5, 5, \ldots,$$

and

$$\sum_{y=1}^{3} 5 = 5 + 5 + 5$$
$$= 15.$$

THEOREM 2.1

Let c be a constant (an element that does not involve the variable of summation) and y be the variable of summation. Then,

$$\sum_{y=1}^{n} c = nc.$$

Proof

$$\sum_{y=1}^{n} c = c + c + c + \cdots + c,$$

where the sum involves n elements. Then,

$$\sum_{y=1}^{n} c = nc.$$

EXAMPLE 2.12

$$\sum_{y=1}^{4} 3a = 4(3a) = 12a.$$

(Note that y is the variable of summation and, therefore, that a is a constant.)

EXAMPLE 2.13

$$\sum_{i=1}^{3} (3x - 5) = 3(3x - 5).$$

(Note that i is the variable of summation and, therefore, that x is a constant.)

EXAMPLE 2.14

$$\sum_{y=4}^{10} 3a = \sum_{y=1}^{10} 3a - \sum_{y=1}^{3} 3a = 10(3a) - 3(3a) = 21a.$$

(Note that since the sum of the elements does not commence with $y = 1$, it can be written as the difference between the sum of the elements from 1 to 10 and the sum from 1 to 3.)

A second theorem is illustrated using Example 2.7. We note that 3 is a common factor in each term. Therefore,

$$\begin{aligned}\sum_{x=2}^{5} 3x &= 3(2) + 3(3) + 3(4) + 3(5) \\ &= 3(2 + 3 + 4 + 5) \\ &= 3 \sum_{x=2}^{5} x.\end{aligned}$$

Thus it would appear that the summation of a constant times a variable is equal to the constant times the summation of the variable.

THEOREM 2.2

Let c be a constant. Then

$$\sum_{i=1}^{n} cy_i = c \sum_{i=1}^{n} y_i.$$

Proof

$$\begin{aligned}\sum_{i=1}^{n} cy_i &= cy_1 + cy_2 + cy_3 + \cdots + cy_n \\ &= c(y_1 + y_2 + \cdots + y_n) \\ &= c \sum_{i=1}^{n} y_i.\end{aligned}$$

THEOREM 2.3

$$\sum_{i=1}^{n} (x_i + y_i + z_i) = \sum_{i=1}^{n} x_i + \sum_{i=1}^{n} y_i + \sum_{i=1}^{n} z_i.$$

Proof

$$\sum_{i=1}^{n} (x_i + y_i + z_i) = x_1 + y_1 + z_1 + x_2 + y_2 + z_2 + x_3 + y_3 + z_3 + \cdots + x_n + y_n + z_n.$$

Regrouping, we have

$$\begin{aligned}\sum_{i=1}^{n} (x_i + y_i + z_i) &= (x_1 + x_2 + \cdots + x_n) + (y_1 + y_2 + \cdots + y_n) + (z_1 + z_2 + \cdots + z_n)\\ &= \sum_{i=1}^{n} x_i + \sum_{i=1}^{n} y_i + \sum_{i=1}^{n} z_i.\end{aligned}$$

In words we would say that the summation of a typical element which is itself a sum of a number of terms is equal to the sum of the summations of the terms.

Theorems 2.1, 2.2, and 2.3 can be used jointly to simplify summations. Consider the following examples:

EXAMPLE 2.15

$$\begin{aligned}\sum_{x=1}^{3} (x^2 + ax + 5) &= \sum_{x=1}^{3} x^2 + \sum_{x=1}^{3} ax + \sum_{x=1}^{3} 5\\ &= \sum_{x=1}^{3} x^2 + a \sum_{x=1}^{3} x + 3(5)\\ &= (1 + 4 + 9) + a(1 + 2 + 3) + 15\\ &= 6a + 29.\end{aligned}$$

EXAMPLE 2.16

$$\begin{aligned}\sum_{i=1}^{4} (x^2 + 3i) &= \sum_{i=1}^{4} x^2 + \sum_{i=1}^{4} 3i\\ &= 4x^2 + 3 \sum_{i=1}^{4} i\\ &= 4x^2 + 3(1 + 2 + 3 + 4)\\ &= 4x^2 + 30.\end{aligned}$$

EXAMPLE 2.17

Suppose you have a set of sample measurements, $y_1, y_2, y_3, \ldots, y_n$, and that you subtract a constant, say c, from each measurement; then the sum of $y_1 - c, y_2 - c, \ldots, y_n - c$, is

$$\sum_{i=1}^{n} (y_i - c) = \sum_{i=1}^{n} y_i - \sum_{i=1}^{n} c = \sum_{i=1}^{n} y_i - nc.$$

EXERCISES

Utilize Theorems 2.1, 2.2, and 2.3 to simplify and evaluate the following summations.

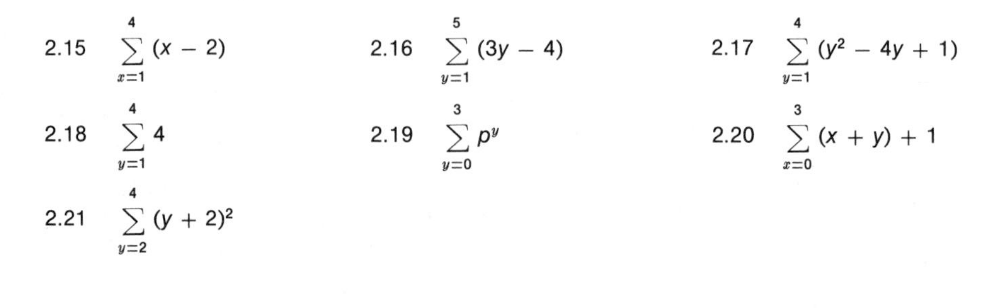

2.15 $\sum_{x=1}^{4} (x - 2)$

2.16 $\sum_{y=1}^{5} (3y - 4)$

2.17 $\sum_{y=1}^{4} (y^2 - 4y + 1)$

2.18 $\sum_{y=1}^{4} 4$

2.19 $\sum_{y=0}^{3} p^y$

2.20 $\sum_{x=0}^{3} (x + y) + 1$

2.21 $\sum_{y=2}^{4} (y + 2)^2$

2.6
SUMMARY

Two types of mathematical notations have been presented, functional notation and summation notation. The former is used in summation notation to express the typical element as a function of the variable of summation. Formulas will also be presented in functional notation in subsequent chapters. Summation notation will be employed in Chapter 3 and succeeding chapters.

SUPPLEMENTARY EXERCISES

[Throughout, starred (*) exercises are optional.]

2.22 If y is the position variable for a sequence and the formula for a typical element is $y^2 + y - 1$:

(a) Write the first four elements of the sequence.

(b) Use summation notation to write an expression for the sum of the first four terms of the sequence.

(c) Find the sum, part (b).

2.23 Given the sequence

$$0, 1, 4, 9, 16, \ldots,$$

(a) Write a formula for the element in the yth position, $y = 1, 2, 3, 4, 5, \ldots$. [*Hint:* Note that the element in the third position is $(3 - 1)^2$.]
(b) Write an expression for the sum of the first five terms of the sequence.
(c) Find the sum, part (b).

2.24 If y is the position variable for a sequence and the formula for a typical element is $(c + y)$:
(a) Give the first three elements of the sequence.
(b) Write an expression for the sum of the first three elements of the sequence.
(c) Find the sum, part (b).

2.25 Given $f(y) = 4y + 3$, find
(a) $f(0)$ (b) $f(1)$ (c) $f(2)$ (d) $f(-1)$
(e) $f(-2)$ (f) $f(a^2)$ (g) $f(-a)$ (h) $f(1 - y)$

2.26 Given $f(y) = (y - 2)^2$, find
(a) $f(2)$ (b) $f(-3)$ (c) $f(-1)$
(d) $f(x)$ (e) $f(a - 1)$

2.27 Given $f(x) = x^2 - x + 1$, find
(a) $f(-2)$ (b) $f(a + b)$

2.28 If $g(y) = (y^2 + 1)/(y + 1)$, find
(a) $g(1)$ (b) $g(-1)$ (c) $g(4)$ (d) $g(-a)$

2.29 If $p(x) = (1 - a)^x$, find $p(0)$ and $p(1)$.

2.30 If $f(x) = 3x^2 - 3x + 1$ and $g(x) = x - 3$, find
(a) $f(1/2)$ (b) $g(-3)$ (c) $f(1/x)$

2.31 If $h(x) = 2$, find
(a) $h(0)$ (b) $h(1)$ (c) $h(2)$

2.32 Suppose that you have a sample of five measurements, where $y_1 = 3.1$, $y_2 = 2.0$, $y_3 = 4.4$, $y_4 = 2.5$, and $y_5 = 2.1$. Find

$$\sum_{i=1}^{5} y_i, \quad \sum_{i=1}^{5} y_i^2, \quad \text{and} \quad \left(\sum_{i=1}^{5} y_i\right)^2$$

2.33 Refer to Exercise 2.32. What is the effect on the sum of the measurements of subtracting 3.1 from each measurement? Use the summation theorems to simplify the expression for

$$\sum_{i=1}^{5} (y_i - 3.1)$$

Utilize Theorems 2.1, 2.2, and 2.3 to simplify and evaluate the following summations.

2.34 $\sum_{y=1}^{3} y^3$ 2.35 $\sum_{y=1}^{3} 7$ 2.36 $\sum_{x=2}^{3} (1 + 2x + x^2)$

2.37 $\sum_{i=1}^{5} (x^2 + 2i)$ 2.38 $\sum_{y=0}^{5} (x^2 + y^2)$ 2.39 $\sum_{x=0}^{2} (x^3 + 2ix)$

2.40 $\sum_{x=1}^{4} (x + xy^2)$ 2.41 $\sum_{i=1}^{2} (y_i - i)$ 2.42 $\sum_{i=1}^{n} (y_i - a)$

2.43 $\sum_{i=1}^{n} (y_i - a)^2$

Using the accompanying set of measurements, calculate the sums in Exercises 2.44 through 2.49.

i	1	2	3	4	5	6	7	8	9	10	11	12	13
y_i	3	12	10	−6	0	11	2	−9	−5	8	−7	4	−5

2.44 $\sum_{i=1}^{13} 2y_i$ 2.45 $\sum_{i=1}^{13} (2y_i - 5)$ 2.46 $\sum_{i=3}^{10} y_i^2$

2.47 $\sum_{i=1}^{10} (y_i^2 + y_i)$ 2.48 $\sum_{i=1}^{13} (y_i - 2)^2$ 2.49 $\sum_{i=1}^{13} y_i^2 - \frac{1}{13}\left(\sum_{i=1}^{13} y_i\right)^2$

Verify the following identities. Each of them is a shortcut formula that will be used in later chapters. The symbols $\bar{x}$ and $\bar{y}$ appearing in these identities have the following definitions:

$$\bar{x} = \frac{\sum_{i=1}^{n} x_i}{n}; \qquad \bar{y} = \frac{\sum_{i=1}^{n} y_i}{n}.$$

*2.50 $$\sum_{i=1}^{n} (y_i - \bar{y})^2 = \sum_{i=1}^{n} y_i^2 - \frac{\left(\sum_{i=1}^{n} y_i\right)^2}{n}.$$

*2.51 $$\sum_{i=1}^{n} (x_i - \bar{x})(y_i - \bar{y}) = \sum_{i=1}^{n} x_i y_i - \frac{\left(\sum_{i=1}^{n} x_i\right)\left(\sum_{i=1}^{n} y_i\right)}{n}.$$

*2.52 $$\frac{\sum_{i=1}^{n} (y_i - \bar{y})^2}{n - 1} = \frac{1}{n - 1}\left\{\sum_{i=1}^{n} y_i^2 - \frac{1}{n}\left(\sum_{i=1}^{n} y_i\right)^2\right\}.$$

*2.53 $$\frac{\sum_{i=1}^{n} (x_i - \bar{x})(y_i - \bar{y})}{\sum_{i=1}^{n} (x_i - \bar{x})^2} = \frac{n \sum_{i=1}^{n} x_i y_i - \left(\sum_{i=1}^{n} x_i\right)\left(\sum_{i=1}^{n} y_i\right)}{n \sum_{i=1}^{n} x_i^2 - \left(\sum_{i=1}^{n} x_i\right)^2}.$$

2.54 Suppose the numbers 5, 1, 6, 2, 1, 3 represent a random sample of $n = 6$ measurements from a population. Let $y_1 = 5$, $y_2 = 1$, $y_3 = 6$, $y_4 = 2$, $y_5 = 1$, and $y_6 = 3$.

(a) Find $\sum_{i=1}^{n} y_i$. (b) Find $\sum_{i=1}^{n} y_i^2$.

2.55 Use the results of Exercise 2.54 to calculate the quantity indicated in Exercise 2.50.

2.56 A random sampling by a department store of $n = 7$ monthly inventory losses gave the following measurements (in thousands of dollars): 3.9, 3.8, 4.6, 3.2, 4.3, 4.1, and 3.8. Let $y_1 = 3.9$, $y_2 = 3.8$, . . . , $y_7 = 3.8$.

(a) Find $\sum_{i=1}^{n} y_i$. (b) Find $\sum_{i=1}^{n} y_i^2$.

2.57 Use the results of Exercise 2.56 to calculate the quantity indicated in Exercise 2.52.

CHAPTER OBJECTIVES

GENERAL OBJECTIVES Methods for describing sets of data are needed so that (1) you will better understand data sets and thus more easily construct theories about the phenomena they represent and (2) you will be able to phrase an inference (make descriptive statements) about a population based on sample data. Consequently, the objective of this chapter is to find a single compact method for describing a set of data.

SPECIFIC OBJECTIVES

1. To develop the notion of a relative frequency distribution (or relative frequency histogram), a very good graphical method for describing a set of data. As you will subsequently see, a model for the relative frequency distribution for a population of data, a probability distribution (Chapter 5), will be the tool employed in making inferences about populations based on sample data. *Section 3.2*
2. To present important numerical descriptive methods, particularly measures of central tendency and variation. *Sections 3.3, 3.4, 3.5*
3. To provide you with a method for interpreting a standard deviation so that you can construct a mental picture of the frequency distribution for a set of data. *Section 3.6*
4. To present an easy way to calculate the standard deviation of a set of measurements. *Sections 3.7, 3.8*

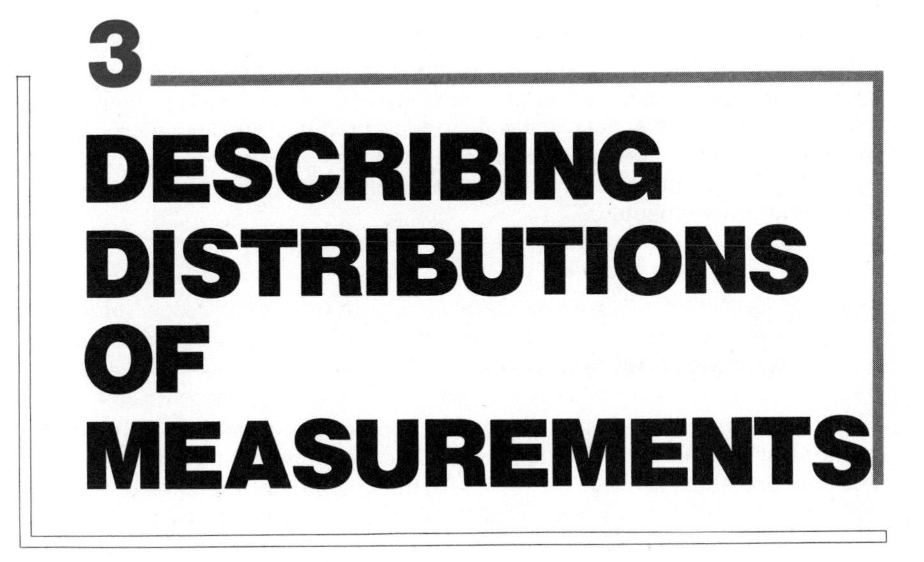

3 DESCRIBING DISTRIBUTIONS OF MEASUREMENTS

3.1

INTRODUCTION

After a brief detour in Chapter 2, we return to the main objective of our study—making inferences about a large body of data, the population, based upon information contained in a sample. A most peculiar difficulty arises: How will the inferences be phrased? How do we describe a set of measurements, whether they be the sample or the population? If the population were before us, how could we describe this large set of measurements?

Numerous texts have been devoted to the methods of descriptive statistics—that is, the methods of describing sets of numerical data. Essentially, these methods can be categorized as graphical methods and numerical methods. In this text we shall restrict our discussion to a few graphical and numerical methods which are useful not only for descriptive purposes but also for statistical inference. The reader interested in descriptive statistics should refer to the references at the end of the chapter.

3.2

A GRAPHICAL METHOD

It would seem natural to introduce appropriate graphical and numerical methods of describing sets of data through consideration of a set of real data. The data presented in Table 3.1 represent the grade-point averages of 30 Bucknell University freshmen recorded at the end of the freshman year.

A cursory examination of the data indicates that the lowest grade-point average in the sample is 1.4, the largest 2.9. How are the other 28 measurements distributed? Do most lie near 1.4, near 2.9, or are they evenly distributed over the interval from 1.4 to 2.9? To answer this question, we divide the interval into an arbitrary number of subintervals of equal length, the number depending upon the amount of data available. (As a rule of thumb, the number of subintervals chosen would range from 5 to 20, the larger the amount of data available, the more subintervals employed.) For instance we might use the subintervals 1.35 to 1.55, 1.55 to 1.75, 1.75 to 1.95, etc. Note that the points dividing the subintervals have been chosen so that it is impossible for a measurement to fall on the point of division, thus eliminating any ambiguity regarding the disposition of a particular measurement. The subintervals, called classes in statistical language, form cells or pockets similar to the pockets of a billiard table. We wish to determine the manner in which the measurements are distributed among the pockets, or classes. A tally of the data from Table 3.1 is presented in Table 3.2.

Each of the 30 measurements falls in one of eight classes which, for purposes of identification, we shall number. The identification number appears in the first column of Table 3.2 and the corresponding class boundaries are given in the second column. The third column of the table is used for the tally, a mark entered opposite the appropriate class for each measurement falling in the class. For example, 3 of the 30 measurements fall in class 1, three in class 2, three in class 3, seven in class 4, etc. The number of measurements falling in a particular

TABLE 3.1

Grade-point averages of 30 Bucknell University freshmen

1.5	2.6	1.4	2.0	1.4
1.8	2.1	2.6	2.0	1.6
2.4	2.5	2.2	2.0	1.9
2.2	2.0	1.9	2.5	2.9
2.1	2.3	2.0	2.2	2.4
2.2	2.3	1.7	2.2	1.6

class, say class i, is called the class frequency and is designated by the symbol f_i. The class frequency is given in the fourth column of Table 3.2. The last column of this table presents the fraction of the total number of measurements falling in each class. We call this the relative frequency. If we let n represent the total number of measurements, for instance, in our example $n = 30$, then the relative frequency for the ith class would equal f_i divided by n:

$$\text{relative frequency} = \frac{f_i}{n}.$$

The resulting tabulation can be presented graphically in the form of a frequency histogram (Figure 3.1). Rectangles are constructed over each class interval, their height being proportional to the number of measurements (class frequency) falling in each class interval. Viewing the frequency histogram, we see at a glance the manner in which the grade-point averages are distributed over the interval.

It is often more convenient to modify the frequency histogram by plotting class relative frequency rather than class frequency. A relative frequency histogram is presented in Figure 3.2. Statisticians rarely make a distinction between the frequency histogram and the relative frequency histogram and refer to either as a frequency histogram or simply a histogram. If corresponding values of

TABLE 3.2

Tabulation of relative frequencies for a histogram

Class i	Class Boundaries	Tally	Class Frequency f_i	Relative Frequency
1	1.35–1.55	111	3	3/30
2	1.55–1.75	111	3	3/30
3	1.75–1.95	111	3	3/30
4	1.95–2.15	~~1111~~ 11	7	7/30
5	2.15–2.35	~~1111~~ 11	7	7/30
6	2.35–2.55	1111	4	4/30
7	2.55–2.75	11	2	2/30
8	2.75–2.95	1	1	1/30
			Total $n = 30$	1

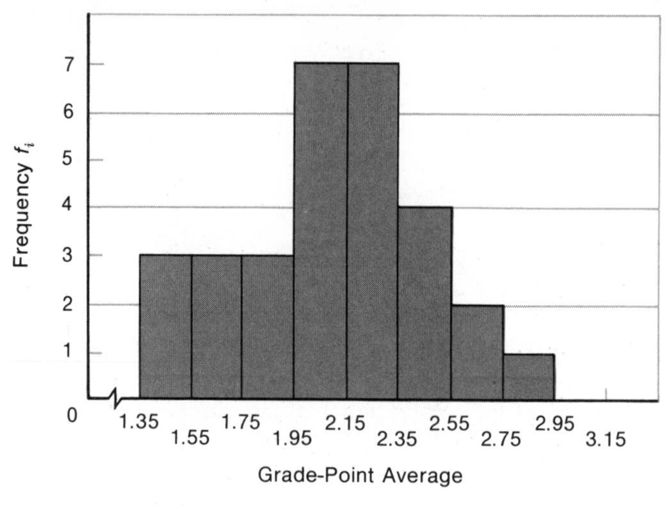

FIGURE 3.1
Frequency histogram

frequency and relative frequency are marked along the vertical axes of the graphs, then the frequency and the relative frequency histograms are identical (see Figures 3.1 and 3.2).

Although we were interested in describing the set of $n = 30$ measurements, we are much more interested in the population from which the sample was drawn. We might view the 30 grade-point averages as a representative sample drawn from the population of grade-point averages of the freshmen currently in

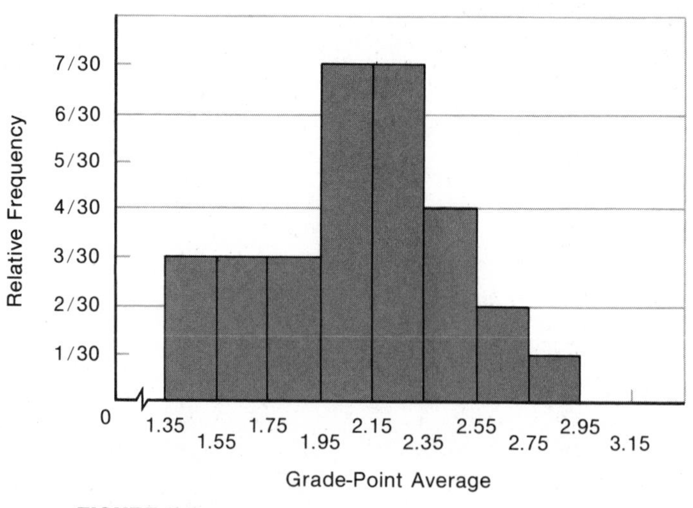

FIGURE 3.2
Relative frequency histogram

attendance at Bucknell University. Or, if we are interested in the academic achievement of freshmen college students in general, we might consider our sample representative of the achievement of the population of freshmen attending Bucknell *or colleges similar to Bucknell*. In either case, if we had the grade-point averages for the entire population, we could construct a population relative frequency histogram.

Let us consider the relative frequency histogram for the sample in greater detail. What fraction of the students attained grade-point averages equal to 2.2 or better? Checking the relative frequency histogram, we see that the fraction would involve all classes to the right of 2.15. Using Table 3.2, we see that 14 students achieved grade-point averages equal to or greater than 2.2. Hence the fraction is 14/30 or approximately 47 percent. We note that this is also the percentage of the total area of the histogram, Figure 3.2, lying to the right of 2.15. Suppose that we were to write each grade-point average on a piece of paper, place the 30 slips of paper in a hat, mix them, and then draw one paper from the hat. What is the chance that this paper would contain a grade-point average equal to or greater than 2.2? Since 14 of the 30 slips are marked with numbers equal to or greater than 2.2, we would say that we have 14 chances out of 30. Or, we might say that the probability is 14/30. You have undoubtedly encountered the word "probability" in ordinary conversation and we are content to defer definition and discussion of its significance until Chapter 4.

Let us now direct our attention to the population from which the sample was drawn. What fraction of students in the population attained a grade-point average equal to or greater than 2.2? If we possessed the relative frequency histogram for the population, we could give the exact answer to this question by calculating the fraction of total area lying to the right of 2.15. Unfortunately, since we do not have such a histogram, we are forced to make an inference. We must estimate the true population fraction, basing our estimate upon information contained in the sample. Our estimate would likely be 14/30 or 47 percent. Suppose that we wish to state the chance or probability that a student drawn from the population would have a grade-point average equal to or greater than 2.2. Without knowledge of the population relative frequency histogram, we would infer that the population histogram is similar to the sample histogram and that approximately 14/30 of the measurements in the population would be equal to or greater than 2.2. Naturally, this estimate would be subject to error. We shall examine the magnitude of this estimation error in Chapter 8.

The relative frequency histogram is often called a frequency distribution because it shows the manner in which the data are distributed along the abscissa of the graph. We note that the rectangles constructed above each class are subject to two interpretations. They represent the fraction of observations falling in a given class. Also, if a measurement is drawn from the data, a particular class relative frequency is also the chance or probability that the measurement will fall in that class. The most significant feature of the sample frequency histogram is that it provides information on the population frequency histogram which describes the population. We would expect the two frequency histograms, sample and population, to be similar. Such is the case. The degree of resemblance will increase as more and more data are added to the sample. If the sample were enlarged to include the entire population, the sample and population would be synonymous and the histograms would be identical.

In the preceding discussion we showed you how to construct a frequency distribution for the grade-point average data, Table 3.1, and we explained how such a distribution could be interpreted. Before concluding this topic, we will summarize the principles that you should employ in constructing a frequency distribution for a set of data.

Principles to Employ in Constructing a Frequency Distribution

1. Determine the number of classes. It is usually best to have from 5 to 20 classes. The larger the amount of data available, the more classes that should be employed. If the number of classes is too small, we might be concealing important characteristics of the data by grouping. If we have too many classes, empty classes may result and the distribution would be meaningless. The number of classes should be determined from the amount of data present and the uniformity of the data. A small sample would require fewer classes.
2. Find the range and determine the class width. As a general rule for finding the class width, divide the difference between the largest and smallest measurement by the number of classes desired and add enough to the quotient to arrive at a convenient figure for class width. All classes should be of equal width. This allows us to make uniform comparisons of the class frequencies.
3. Locate the class boundaries. To locate the class boundaries, commence with the lowest class so that you include the smallest measurement. Then add the remaining classes. Class boundaries should be chosen so that it is impossible for a measurement to fall on a boundary.

EXERCISES

3.1 In order to decide on the number of service counters needed for stores to be built in the future, a supermarket chain wished to obtain information on the length of time (in minutes) required to service customers. To obtain information on the distribution of customer service times, a sample of 1000 customer's service times were recorded. Sixty of these are shown in the accompanying tabulation.

3.6	1.9	2.1	.3	.8	.2
1.0	1.4	1.8	1.6	1.1	1.8
.3	1.1	.5	1.2	.6	1.1
.8	1.7	1.4	.2	1.3	3.1
.4	2.3	1.8	4.5	.9	.7
.6	2.8	2.5	1.1	.4	1.2
.4	1.3	.8	1.3	1.1	1.2
.8	1.0	.9	.7	3.1	1.7
1.1	2.2	1.6	1.9	5.2	.5
1.8	.3	1.1	.6	.7	.6

(a) Construct a relative frequency histogram for the data.
(b) What fraction of the service times are less than or equal to one minute?

3.2 The length of time (in months) between the onset of a particular illness and its recurrence was recorded for $n = 50$ patients. The times of recurrence are as follows:

2.1	4.4	2.7	32.3	9.9
9.0	2.0	6.6	3.9	1.6
14.7	9.6	16.7	7.4	8.2
19.2	6.9	4.3	3.3	1.2
4.1	18.4	.2	6.1	13.5
7.4	.2	8.3	.3	1.3
14.1	1.0	2.4	2.4	18.0
8.7	24.0	1.4	8.2	5.8
1.6	3.5	11.4	18.0	26.7
3.7	12.6	23.1	5.6	.4

(a) Construct a relative frequency histogram for the data.
(b) Give the fraction of recurrence times less than or equal to 10.

3.3 Twenty-eight applicants interested in working for the Food Stamp program took an examination designed to measure their aptitude for social work. The following test scores were obtained:

79	97	86	76
93	87	98	68
84	88	81	91
86	87	70	94
77	92	66	85
63	68	98	88
46	72	59	79

Construct a relative frequency histogram for the test scores. (Use subintervals of width 9 beginning at 44.5.)

3.3 NUMERICAL DESCRIPTIVE METHODS

Graphical methods are extremely useful in conveying a rapid general description of collected data and in presenting data. This supports, in many respects, the saying that a picture is worth a thousand words. There are, however, limitations to the use of graphical techniques for describing and analyzing data. For instance, suppose that we wish to discuss our data before a group of people and have no method of describing the data other than verbally. Unable to present the histogram visually, we would be forced to use other descriptive measures which would convey to the listeners a mental picture of the histogram. A second and not so obvious limitation of the histogram and other graphical techniques is that they are difficult to use for purposes of statistical inference. Presumably we use the sample histogram to make inferences about the shape and position of the population histogram which describes the population and is unknown to us. Our inference is based upon the correct assumption that some degree of similarity will exist between the two histograms, but we are then faced with the problem of measuring the degree of similarity. We know when two figures are identical, but this situation will not likely occur in practice. Hence, if the sample and population histograms differ, how can we measure the degree of difference or, expressing it positively, the degree of similarity? To be more specific, we might wonder about the degree of similarity between the histogram, Figure 3.2, and the frequency histogram for the population of grade-point averages from which the sample was drawn. Although these difficulties are not insurmountable, we prefer to seek other descriptive measures which readily lend themselves for use as predictors of the shape of the population frequency distribution.

The limitations of the graphical method of describing data can be overcome by the use of numerical descriptive measures. Thus we would like to use the sample data to calculate a set of numbers which will convey to the statistician a good mental picture of the frequency distribution and which will be useful in making inferences concerning the population.

3.4 MEASURES OF CENTRAL TENDENCY

In constructing a mental picture of the frequency distribution for a set of measurements, we would likely envision a histogram similar to that shown in Figure 3.2 for the data on grade-point averages. One of the first descriptive measures of interest would be a measure of central tendency, that is, a measure of the center of the distribution. We note that the grade-point data ranged from a low of 1.4 to a high of 2.9, the center of the histogram being located in the vicinity of 2.1. Let us now consider some definite rules for locating the center of a distribution of data.

One of the most common and useful measures of central tendency is the arithmetic average of a set of measurements. This is also often referred to as the arithmetic mean or, simply, the mean of a set of measurements.

DEFINITION

The arithmetic mean of a set of n measurements $y_1, y_2, y_3, \ldots, y_n$ is equal to the sum of the measurements divided by n.

Recall that we are always concerned with both the sample and the population, each of which possesses a mean. In order to distinguish between the two, we shall use the symbol $\bar{y}$ for the mean of the sample and μ (the Greek letter mu) for the mean of the population. Since the n sample measurements can be denoted by the symbols $y_1, y_2, y_3, \ldots, y_n$, a formula for the sample mean would be

Sample Mean

$$\bar{y} = \frac{\sum_{i=1}^{n} y_i}{n},$$

and this can be used as a measure of central tendency for the sample.

EXAMPLE 3.1

Find the mean of the set of measurements 2, 9, 11, 5, 6.

$$\bar{y} = \frac{\sum_{i=1}^{n} y_i}{n} = \frac{2 + 9 + 11 + 5 + 6}{5} = 6.6.$$

Even more important than locating the center of a set of sample measurements, $\bar{y}$ will be employed as an estimator (predictor) of the value of the unknown population mean μ.

For example, the mean of the data, Table 3.1, is equal to

$$\bar{y} = \frac{\sum_{i=1}^{n} y_i}{n} = \frac{62.5}{30} = 2.08.$$

Note that this falls approximately in the center of the set of measurements. The mean of the entire population of grade-point averages, μ, is unknown to us but if we were to estimate its value, our estimate of μ would be 2.08.

A second measure of central tendency is the median.

DEFINITION

The median of a set of n measurements $y_1, y_2, y_3, \ldots, y_n$ is defined to be the value of y that falls in the middle when the measurements are arranged in order of magnitude.

EXAMPLE 3.2

Find the median for the set of five measurements

9, 2, 7, 11, 14.

Solution Arranging the measurements in order of magnitude, 2, 7, 9, 11, 14, we would choose 9 as the median. If the number of measurements is even, we choose the median as the value of y halfway between the two middle measurements.

EXAMPLE 3.3

Find the median for the set of measurements

9, 2, 7, 11, 14, 6.

Solution Arranged in order of magnitude, 2, 6, 7, 9, 11, 14, we would choose the median halfway between 7 and 9, which is 8.

Our rule for locating the median may seem a bit arbitrary for the case where we have an even number of measurements, but recall that we calculate the sample median either for descriptive purposes or as an estimator of the population median. If it is used for descriptive purposes, we may be as arbitrary as we please. If it is used as an estimator of the population median, "the proof of the pudding is in the eating." A rule for locating the sample median is poor or good depending upon whether it tends to give a poor or good estimate of the population median.

You will note that we have not specified a symbol for the population median. This is because most of the common methods of statistical inference suitable for an elementary course in statistics are based upon the use of the sample mean rather than the median. We say this, being wholly aware of the popularity of the median in the social sciences, but point out that it is used more often for descriptive purposes than for statistical inference. We also note that other measures of central tendency exist which have practical application in certain situations, but limitations of time and space forbid their discussion here. As we proceed in this text we will use the sample mean, exclusively, as a measure of central tendency.

3.5

MEASURES OF VARIABILITY

Having located the center of a distribution of data, our next step is to provide a measure of the variability or dispersion of the data. Consider the two distributions shown in Figure 3.3. Both distributions are located with a center of $y = 4$, but there is a vast difference in the variability of the measurements about the mean for the two distributions. Most of the measurements in Figure 3.3(a) vary from 3 to 5, while in Figure 3.3(b) they vary from 0 to 8. Variation is a very important characteristic of data. For example, if we are manufacturing bolts,

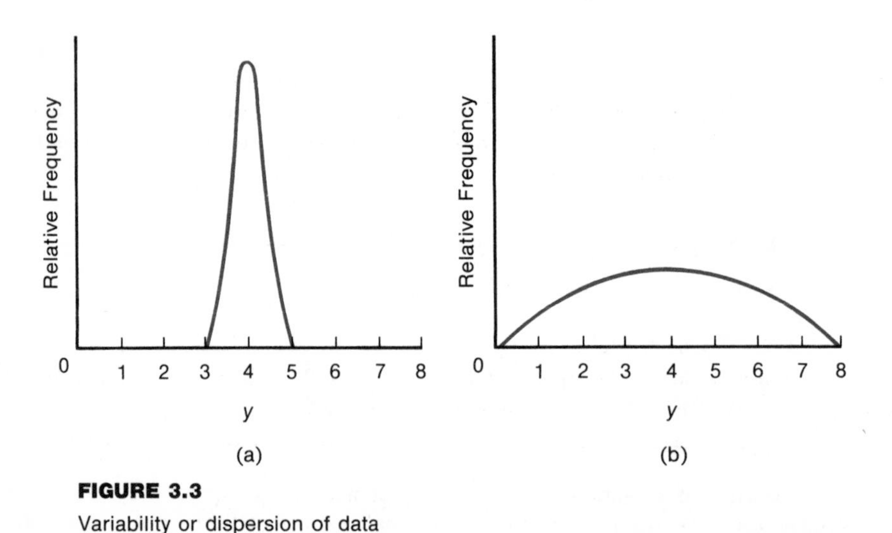

FIGURE 3.3
Variability or dispersion of data

excessive variation in the bolt diameter would imply a high percentage of defective product. On the other hand, if we are using an examination to discriminate between good and poor accountants, we would be most unhappy if the examination always produced test grades with little variation, as this would make discrimination very difficult indeed. In addition to the practical importance of variation in data, it is obvious that a measure of this characteristic is necessary to the construction of the mental image of the frequency distribution. Numerous measures of variability exist, and we shall discuss a few of the most important.

The simplest measure of variation is the range.

DEFINITION

The range of a set of n measurements $y_1, y_2, y_3, \ldots, y_n$ is defined to be the difference between the largest and smallest measurement.

For our grade-point data, Table 3.1, we note that the measurements varied from 1.4 to 2.9. Hence the range is equal to $(2.9 - 1.4) = 1.5$.

Unfortunately, the range is not completely satisfactory as a measure of variation. Consider the two distributions of Figure 3.4. Both distributions have the same range, but the data of Figure 3.4(b) are more variable than the data of Figure 3.4(a). To overcome this limitation of the range, we introduce quartiles and percentiles. Remember that if we specify an interval along the y-axis of our histogram, the percentage of area under the histogram lying above the interval is equal to the percentage of the total number of measurements falling in that interval. Since the median is the middle measurement when the data are arranged in order of magnitude, the median would be the value of y such that half the area of the histogram would lie to its left, half to the right. Similarly, we might define quartiles as values of y that divide the area of the histogram into quarters.

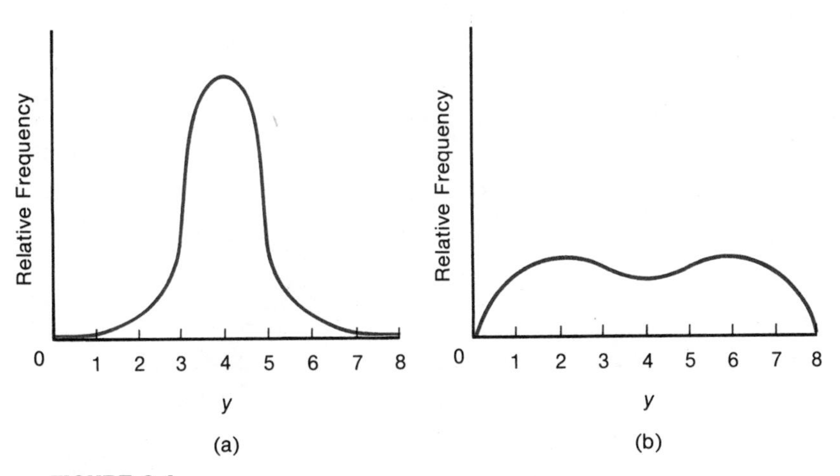

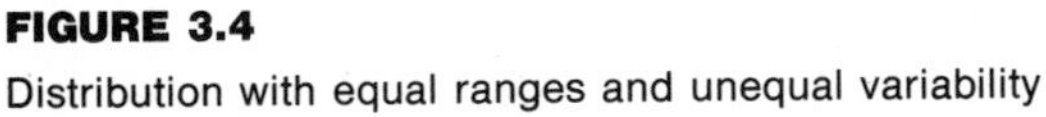

FIGURE 3.4

Distribution with equal ranges and unequal variability

DEFINITION

Let $y_1, y_2, \ldots, y_n$ be a set of n measurements arranged in order of magnitude. The lower quartile (first quartile) is the value of y that exceeds 1/4 of the measurements and is less than the remaining 3/4. The second quartile is the median. The upper quartile (third quartile) is the value of y that exceeds 3/4 of the measurements and is less than 1/4.

Locating the lower quartile on a histogram, Figure 3.5, we note that 1/4 of the area lies to the left of the lower quartile, 3/4 to the right. The upper quartile is the value of y such that 3/4 of the area lies to the left, 1/4 to the right.

For some applications, particularly when you have large quantities of data, it is preferable to use percentiles.

DEFINITION

Let $y_1, y_2, \ldots, y_n$ be a set of n measurements arranged in order of magnitude. The pth percentile is the value of y such that p percent of the measurements are less than that value of y and $(100 - p)$ percent are greater.

For example, the ninetieth percentile for a set of data would be the value of y that exceeds 90 percent of the measurements and is less than 10 percent. Just as in the case of quartiles, 90 percent of the area of the histogram would lie to the left of the ninetieth percentile.

The range possesses simplicity in that it can be expressed as a single number. Quartiles and percentiles, on the other hand, provide more information about data location and variation, but several numbers must be given to provide an adequate description. Can we find a measure of variability expressible as a single number but more sensitive than the range?

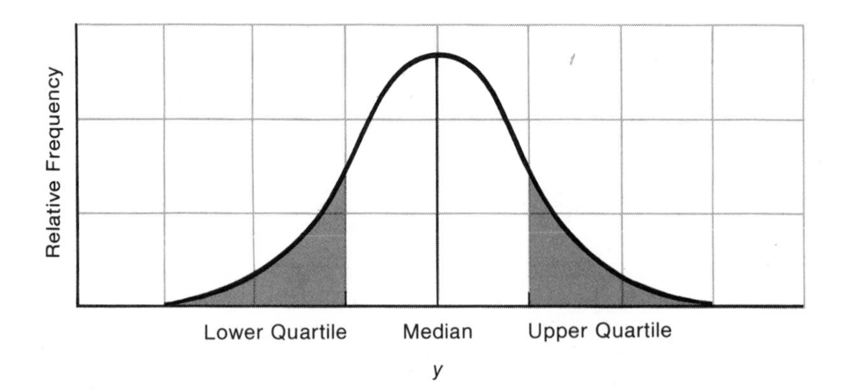

FIGURE 3.5

Location of quartiles

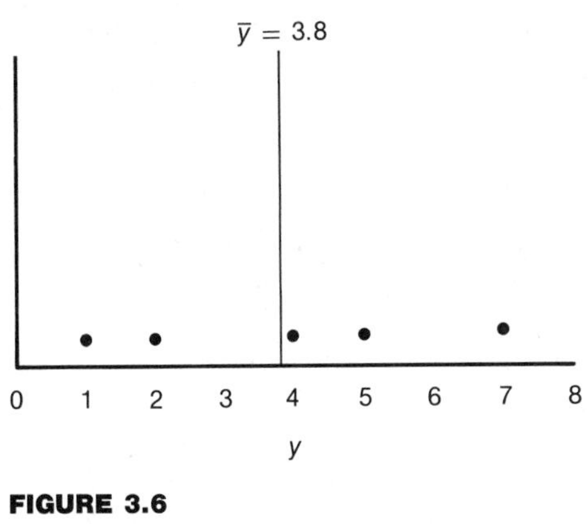

FIGURE 3.6
Dot diagram

Consider, as an example, the set of measurements 5, 7, 1, 2, 4. We can depict these data graphically, as in Figure 3.6, by showing the measurements as dots falling along the y-axis. Figure 3.6 is called a dot diagram.

Calculating the mean as the measure of central tendency, we obtain

$$\bar{y} = \frac{\sum_{i=1}^{n} y_i}{n} = \frac{19}{5} = 3.8$$

and locate it on the dot diagram. We can now view variability in terms of distance between each dot (measurement) and the mean, $\bar{y}$. If the distances are large, we can say that the data are more variable than if the distances are small. Being more explicit, we shall define the deviation of a measurement from its mean to be the quantity $(y_i - \bar{y})$. Note that measurements to the right of the mean represent positive deviations, and those to the left, negative deviations. The values of y and the deviations for our example are shown in columns 1 and 2 of Table 3.3.

TABLE 3.3
Computation of $\sum_{i=1}^{n} (y_i - \bar{y})^2$

y_i	$y_i - \bar{y}$	$(y_i - \bar{y})^2$
5	1.2	1.44
7	3.2	10.24
1	−2.8	7.84
2	−1.8	3.24
4	.2	.04
$\sum_{i=1}^{5} y_i = 19$	0	22.80

If we now agree that deviations contain information on variation, our next step is to construct a formula based upon the deviations which will provide a good measure of variation. As a first possibility we might choose the average of the deviations. Unfortunately, this will not work because some of the deviations are positive, some are negative, and the sum of the deviations is always equal to zero (unless round-off errors have been introduced into the calculations). This can be shown using the summation theorems of Chapter 2. Given n measurements $y_1, y_2, \ldots, y_n$,

$$\sum_{i=1}^{n} (y_i - \bar{y}) = \sum_{i=1}^{n} y_i - \sum_{i=1}^{n} \bar{y}.$$

But since $\bar{y}$ is a constant (i.e., not a function of the variable of summation, i),

$$\sum_{i=1}^{n} \bar{y} = n\bar{y}.$$

Then,

$$\begin{aligned}
\sum_{i=1}^{n} (y_i - \bar{y}) &= \sum_{i=1}^{n} y_i - \sum_{i=1}^{n} \bar{y} \\
&= \sum_{i=1}^{n} y_i - n\bar{y} \\
&= \sum_{i=1}^{n} y_i - n\left(\frac{\sum_{i=1}^{n} y_i}{n}\right) \\
&= \sum_{i=1}^{n} y_i - \sum_{i=1}^{n} y_i = 0.
\end{aligned}$$

Note that the deviations, the second column of Table 3.3, sum to zero.

You will readily observe an easy solution to this problem. Why not calculate the average of the absolute values of the deviations? This method has, in fact, been employed as a measure of variability but it is difficult to interpret and it tends to be unsatisfactory for purposes of statistical inference. We prefer overcoming the difficulty caused by the sign of the deviations by working with the sum of their squares,

$$\sum_{i=1}^{n} (y_i - \bar{y})^2.$$

For a fixed number of measurements, when this quantity is large, the data will be more variable than when it is small.

DEFINITION

The variance of a population of N measurements $y_1, y_2, \ldots, y_N$ is defined to be the average of the square of the deviations of the measurements about their mean μ. The population variance is

denoted by σ^2 (σ is the lowercase Greek letter sigma) and is given by the formula

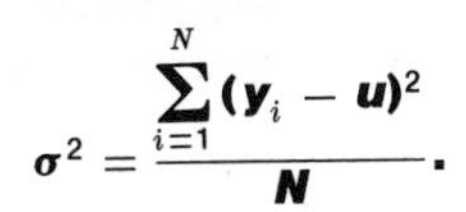

$$\sigma^2 = \frac{\sum_{i=1}^{N} (y_i - u)^2}{N}.$$

Note that we use N to denote the number of measurements in the population and n to denote the number of measurements in the sample.

In defining the variance of the sample measurements, we will modify our averaging procedure, dividing the sum of squares of deviations by $(n - 1)$ rather than n.

DEFINITION

The variance of a sample of n measurements, $y_1, y_2, \ldots, y_n$, is defined to be the sum of the squared deviations of the measurements about their mean $\bar{y}$ divided by $(n - 1)$. The sample variance is denoted by s^2 and is given by the formula

$$s^2 = \frac{\sum_{i=1}^{n} (y_i - \bar{y})^2}{n - 1}.$$

This is done because, ultimately, we will want to use the sample variance s^2 to estimate the population variance σ^2. Dividing the sum of squares of deviations by n produces estimates that tend to underestimate σ^2. Division by $(n - 1)$ eliminates this difficulty.

For example, we can calculate the variance for the set of $n = 5$ measurements presented in Table 3.3. The square of the deviation of each measurement is recorded in the third column of Table 3.3. Adding, we obtain

$$\sum_{i=1}^{5} (y_i - \bar{y})^2 = 22.80.$$

The sample variance would equal

$$s^2 = \frac{\sum_{i=1}^{5} (y_i - \bar{y})^2}{n - 1} = \frac{22.80}{5 - 1} = 5.70.$$

At this point, you might be understandably disappointed with the practical significance attached to the variance as a measure of variability. Large variances imply a large amount of variation, but this only permits comparison of several sets of data. When we attempt to say something specific concerning a single set of data, we are at a loss. For example, what can be said about the variability of a set of data with a variance equal to 100? This question cannot be answered with the facts at hand. We shall remedy this situation by introducing a new definition and, in Section 3.6, a theorem and a rule.

DEFINITION

The standard deviation of a set of n measurements, $y_1, y_2, y_3, \ldots, y_n$, is equal to the positive square root of the variance.

The variance is measured in terms of the square of the original units of measurement. Thus, if the original measurements were in inches, the variance would be expressed in square inches. Taking the positive square root of the variance, we obtain the standard deviation, which returns our measure of variability to the original units of measurement. The sample standard deviation is denoted by the symbol s and the population standard deviation by the symbol σ.

Sample Standard Deviation

$$s = \sqrt{s^2} = \sqrt{\frac{\sum_{i=1}^{n}(y_i - \bar{y})^2}{n - 1}}$$

Having defined the standard deviation, you might wonder why we bothered to define the variance in the first place. Actually, both the variance and the standard deviation play an important role in statistics, a fact that you must accept on faith at this stage of our discussion.

3.6 ON THE PRACTICAL SIGNIFICANCE OF THE STANDARD DEVIATION

We now introduce an interesting and useful theorem developed by the Russian mathematician, Tchebysheff. Proof of the theorem is not difficult, but we omit it from our discussion.

THEOREM 3.1

Tchebysheff's Theorem: Given a number k greater than or equal to 1 and a set of n measurements $y_1, y_2, \ldots, y_n$, at least $(1 - 1/k^2)$ of the measurements will lie within k standard deviations of their mean.

Tchebysheff's Theorem applies to any set of measurements, and for purposes of illustration we could refer to either the sample or the population. We shall use the notation appropriate for populations, but the reader should realize

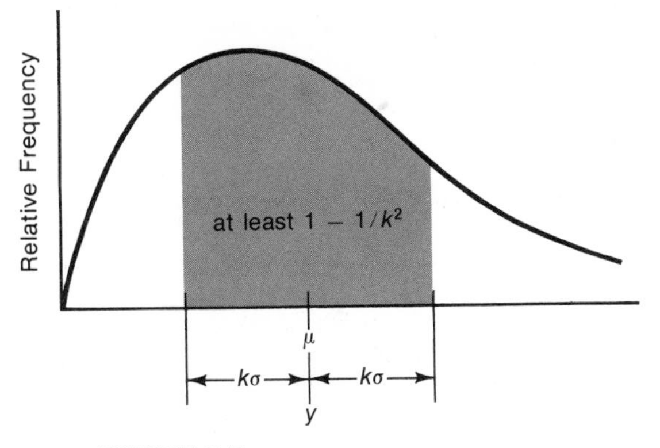

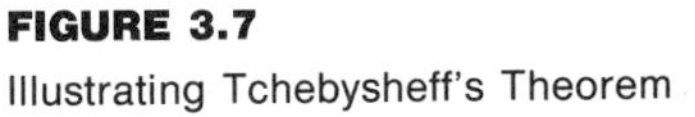

FIGURE 3.7
Illustrating Tchebysheff's Theorem

that we could just as easily use $\bar{y}$ and s, the mean and standard deviation for the sample.*

The idea involved in Tchebysheff's Theorem is illustrated in Figure 3.7. An interval is constructed by measuring a distance of $k\sigma$ on either side of the mean μ. Note that the theorem is true for any number we wish to choose for k as long as it is greater than or equal to 1. Then, computing the fraction $1 - 1/k^2$, we see that Tchebysheff's Theorem states that *at least* that fraction of the total number, n, of measurements will lie in the constructed interval.

Let us choose a few numerical values for k and compute $1 - 1/k^2$ (see Table 3.4). When $k = 1$, the theorem states that at least $1 - 1/(1)^2 = 0$ of the measurements lie in the interval $\mu - \sigma$ to $\mu + \sigma$, a most unhelpful and uninformative result. However, when $k = 2$, we observe that at least $1 - 1/(2)^2 = 3/4$ of the measurements will lie in the interval $\mu - 2\sigma$ to $\mu + 2\sigma$. At least 8/9 of the measurements will lie within three standard deviations of the mean, i.e., in the interval $\mu - 3\sigma$ to $\mu + 3\sigma$. Let us consider an example where we will use the mean and standard deviation (or variance) to construct a mental image of the distribution of measurements from which the mean and standard deviation were obtained.

*The proof of Tchebysheff's Theorem for a finite number of measurements is based on a variance defined as

$$s'^2 = \frac{\sum_{i=1}^{n} (y_i - \bar{y})^2}{n},$$

i.e., with a divisor of n rather than the $(n - 1)$ used in s^2. Since s^2 will always be larger than s'^2, because $n > (n - 1)$, Tchebysheff's Theorem will always hold when s is used to form the intervals about $\bar{y}$. In any case, there will be very little numerical difference in the values of s and s' for moderate to large values of n.

TABLE 3.4
Illustrative values of $1 - 1/k^2$

k	$1 - 1/k^2$
1	0
2	3/4
3	8/9

EXAMPLE 3.4

The mean and variance of a sample of $n = 25$ measurements are 75 and 100, respectively. Use Tchebysheff's Theorem to describe the distribution of measurements.

Solution We are given $\bar{y} = 75$ and $s^2 = 100$. The standard deviation is $s = \sqrt{100} = 10$. The distribution of measurements is centered about $\bar{y} = 75$, and Tchebysheff's Theorem states:

a. *At least* 3/4 of the 25 measurements lie in the interval $\bar{y} \pm 2s = 75 \pm 2(10)$, that is, 55 to 95.
b. *At least* 8/9 of the measurements lie in the interval $\bar{y} \pm 3s = 75 \pm 3(10)$, that is, 45 to 105.

We emphasize the "at least" in Tchebysheff's Theorem because the theorem is very conservative, applying to *any* distribution of measurements. In most situations, the fraction of measurements falling in the specified interval will exceed $1 - 1/k^2$.

We now state a rule that describes accurately the variability of a particular bell-shaped distribution and describes reasonably well the variability of other mound-shaped distributions of data. The frequent occurrence of mound-shaped and bell-shaped distributions of data in nature and hence the applicability of our rule leads us to call it the Empirical Rule.

The Empirical Rule: Given a distribution of measurements that is approximately bell-shaped (see Figure 3.8), the interval

1. $\mu \pm \sigma$ will contain approximately 68 percent of the measurements.
2. $\mu \pm 2\sigma$ will contain approximately 95 percent of the measurements.
3. $\mu \pm 3\sigma$ will contain all or almost all of the measurements.

The bell-shaped distribution, Figure 3.8, is commonly known as the normal distribution and will be discussed in detail in Chapter 7. The point we wish to make here is that the Empirical Rule is extremely useful and provides an excellent description of variation for many types of data.

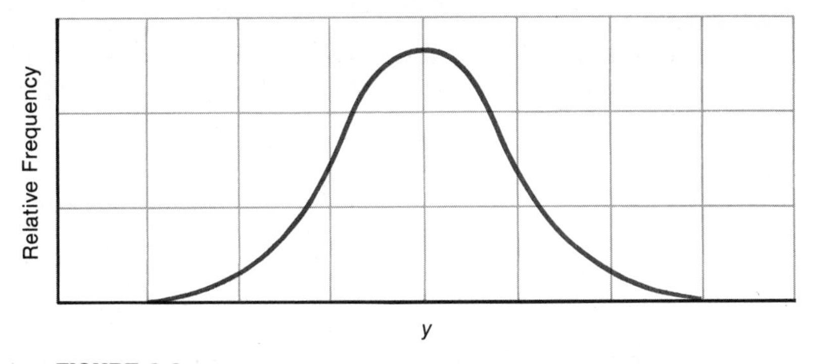

FIGURE 3.8

The normal distribution

EXAMPLE 3.5

A time study was conducted to determine the length of time necessary to perform a specified operation in a manufacturing plant. The length of time necessary to complete the operation was measured for each of $n = 40$ workmen. The mean and standard deviation were found to equal 12.8 and 1.7, respectively. Describe the data.

Solution To describe the data, we calculate the intervals

$$\bar{y} \pm s \;\; = 12.8 \pm 1.7, \text{ or } 11.1 \text{ to } 14.5;$$
$$\bar{y} \pm 2s = 12.8 \pm 2(1.7), \text{ or } 9.4 \text{ to } 16.2;$$
$$\bar{y} \pm 3s = 12.8 \pm 3(1.7), \text{ or } 7.7 \text{ to } 17.9.$$

Although we have no prior information on the distribution of the data, there is a very good chance that it will be mound-shaped and that the Empirical Rule will provide a good description of the data. According to the Empirical Rule, we would expect approximately 68 percent of the measurements to fall in the interval 11.1 to 14.5, 95 percent in the interval 9.4 to 16.2, and all or almost all in the interval 7.7 to 17.9.

If we doubt that the distribution of measurements is mound-shaped or wish, for some reason, to be conservative, we can apply Tchebysheff's Theorem and be absolutely certain of our statements. Tchebysheff's Theorem would tell us that at least 3/4 of the measurements fell in the interval 9.4 to 16.2 and at least 8/9 in the interval 7.7 to 17.9.

How well does the Empirical Rule apply to the grade-point data of Table 3.1? We will show, in Section 3.7, that the mean and standard deviation for the $n = 30$ measurements is $\bar{y} = 2.08$ and $s = .37$. The appropriate intervals were calculated and the number of measurements falling in each interval recorded. The results are shown in Table 3.5 with k in the first column and the interval $\bar{y} \pm ks$ in the second column, using $\bar{y} = 2.08$ and $s = .37$. The frequency or number of measurements falling in each interval is given in the third column and the relative frequency in the fourth column. Note that the observed relative

TABLE 3.5

Frequency of measurements lying within k standard deviations of the mean for the data, Table 3.1

k	Interval $\bar{y} \pm ks$	Frequency in Interval	Relative Frequency
1	1.71–2.45	19	.63
2	1.34–2.82	29	.97
3	.97–3.19	30	1.0

frequencies are quite close to the relative frequencies specified in the Empirical Rule.

Another way to see how well the Empirical Rule and Tchebysheff's Theorem apply to the grade-point data is to mark off the intervals $\bar{y} \pm s$, $\bar{y} \pm 2s$, and $\bar{y} \pm 3s$ on the relative frequency histogram for the data. This is shown in Figure 3.9. Now recall that the area under the histogram over an interval is proportional to the number of measurements falling in the interval and visually observe the proportion of the area above the interval, $\bar{y} \pm s$. You will observe that this proportion is near the .68 specified by the Empirical Rule. Similarly, you will note that almost all of the area lies above the interval $\bar{y} \pm 2s$. Clearly both the Empirical Rule and Tchebysheff's Theorem, using $\bar{y}$ and s, provide a good description for the grade-point data.

To conclude, note that Tchebysheff's Theorem is a fact that can be proved mathematically and it applies to any set of data. It gives a *lower* bound to the

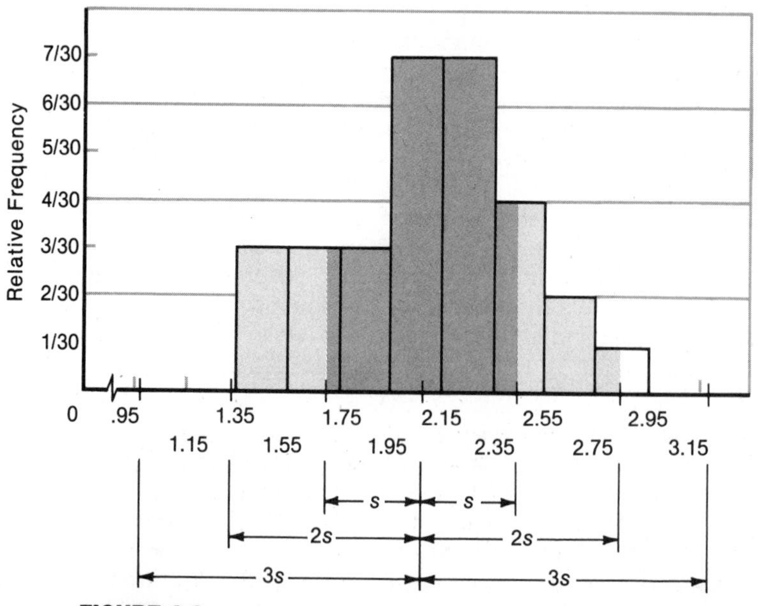

FIGURE 3.9

Histogram for the grade-point data (Figure 3.1) with intervals $\bar{y} \pm s$, $\bar{y} \pm 2s$, and $\bar{y} \pm 3s$ superimposed

fraction of measurements to be found in an interval $\bar{y} \pm ks$, where k is some number greater than or equal to 1. In contrast, the Empirical Rule is an arbitrary statement about the behavior of data, a "rule of thumb." Although the percentages contained in the rule come from the area under the normal curve (Figure 3.8), the same percentages hold approximately for distributions with varying shapes as long as they tend to be roughly mound-shaped (the data tend to pile up near the center of the distribution). We have shown this to be true for the grade-point data. If you need further convincing, calculate $\bar{y}$ and s for a set of data of your choosing and check the fraction of measurements falling in the intervals $\bar{y} \pm s$, $\bar{y} \pm 2s$, and $\bar{y} \pm 3s$. We think you will find that the observed relative frequencies are reasonably close to the values specified in the Empirical Rule.

EXERCISES

3.4 Given the $n = 5$ measurements, 2, 4, 0, 3, 1:
(a) Find $\bar{y}$.
(b) Find s.

3.5 Is your breath rate normal? Actually, there is no standard breath rate for humans. It can vary from as low as 4 breaths per minute to as high as 70 or 75 for a person engaged in strenuous exercise. Suppose that the resting breath rates for college-age students possess a relative frequency distribution that is mound-shaped with a mean equal to 12 and a standard deviation of 2.3 breaths per minute. What fraction of all students would possess breath rates in the interval
(a) 9.7 to 14.3 breaths per minute?
(b) 7.4 to 16.6 breaths per minute?
(c) more than 18.9 or less than 5.1 breaths per minute?

3.6 An analytical chemist wished to determine the number of moles of cupric ions in a given volume of solution by electrolysis. The solution was partitioned into $n = 30$ portions of .200 milliliter each. Each of the $n = 30$ unknown portions was tested. The average number of moles of cupric ions for the $n = 30$ portions was found to be .17 mole; the standard deviation was .01 mole. Describe the distribution of the measurements for the $n = 30$ portions of the solution:
(a) Use Tchebysheff's Theorem.
(b) Use the Empirical Rule. (Would you expect the Empirical Rule to be suitable for describing these data?)
(c) Suppose that the chemist had only employed $n = 4$ portions of the solution for the experiment and obtained the readings .15, .19, .17, and .15. Would the Empirical Rule be suitable for describing the $n = 4$ measurements? Why?

3.7 According to the Environmental Protection Agency (EPA), chloroform, which in its gaseous form is suspected of being a cancer-causing agent, is present in small quantities in all of the country's 240,000 public water sources. If the mean and standard deviation of the amounts of chloroform present in the water sources are 34 and 53 micrograms per liter, respectively, describe the distribution for the population of all public water sources.

3.8 A profile of the incomes of families in the United States in 1976 was provided in an October 1976 release by the U.S. Census Bureau. Some of the statistics, in the form of percentiles, are shown in the accompanying table.

Percentile	Family Income ($)
7	4,000
14	6,000
22	8,000
30	10,000
38	12,000
50	15,000
69	20,000
82	25,000
98	50,000

(a) Explain what is meant by the statement that the 82nd percentile is $25,000.
(b) Use these percentiles to draw a rough sketch of the relative frequency distribution of family incomes in 1976.

3.9 Trichinosis, a disease derived from improperly cooked pork products (main source: pork sausage), may be on the increase. Two hundred eighty-four (284) cases of trichinosis, including one fatality, were reported to the Communicable Disease Center in 1975. This number was 2 1/2 times higher than the mean number of cases reported during the previous five years and it represents the highest annual incidence since 1961.* From your knowledge about data variation, do you think these data confirm an increase in the per capita rate of incidence of trichinosis? Explain. (*Note:* This exercise is only intended to stimulate discussion. More clear-cut decisions will be derived from more complete sets of data in later chapters.)

3.7

A SHORT METHOD FOR CALCULATING THE VARIANCE

The calculation of the variance and standard deviation of a set of measurements is no small task regardless of the method employed, but it is particularly tedious if one proceeds, according to the definition, by calculating each deviation individually as shown in Table 3.3. We shall use the data of Table 3.3 to illustrate a shorter method of calculation. The tabulations are presented in Table 3.6 in two columns, the first containing the individual measurements and the second containing the squares of the measurements.

We now calculate

$$\sum_{i=1}^{n} y_i^2 - \frac{\left(\sum_{i=1}^{n} y_i\right)^2}{n} = 95 - \frac{(19)^2}{5}$$

$$= 95 - \frac{361}{5} = 95 - 72.2$$

$$= 22.8$$

and notice that it is exactly equal to the sum of squares of the deviations, $\sum_{i=1}^{n} (y_i - \bar{y})^2$, given in the third column of Table 3.3.

* *Source:* "VM 1362 Trichinosis," *Veterinary Medicine Newsletter,* Florida Cooperative Extension Service (August 1977), p. 6.

TABLE 3.6
Table for simplified calculations of $\sum_{i=1}^{n}(y_i - \bar{y})^2$

y_i	y_i^2
5	25
7	49
1	1
2	4
4	16
19	95

This is no accident. We will show that the sum of squares of the deviations is always equal to

$$\sum_{i=1}^{n}(y_i - \bar{y})^2 = \sum_{i=1}^{n} y_i^2 - \frac{\left(\sum_{i=1}^{n} y_i\right)^2}{n}.$$

The proof is obtained by using the summation theorems, Chapter 2:

$$\begin{aligned}\sum_{i=1}^{n}(y_i - \bar{y})^2 &= \sum_{i=1}^{n}(y_i^2 - 2\bar{y}y_i + \bar{y}^2)\\ &= \sum_{i=1}^{n} y_i^2 - 2\bar{y}\sum_{i=1}^{n} y_i + \sum_{i=1}^{n}\bar{y}^2\\ &= \sum_{i=1}^{n} y_i^2 - \frac{2\sum_{i=1}^{n} y_i}{n}\sum_{i=1}^{n} y_i + n\bar{y}^2\\ &= \sum_{i=1}^{n} y_i^2 - \frac{2}{n}\left(\sum_{i=1}^{n} y_i\right)^2 + n\left(\frac{\sum_{i=1}^{n} y_i}{n}\right)^2\end{aligned}$$

or

$$\sum_{i=1}^{n}(y_i - \bar{y})^2 = \sum_{i=1}^{n} y_i^2 - \frac{\left(\sum_{i=1}^{n} y_i\right)^2}{n}.$$

We call this formula the shortcut method of calculating the sum of squares of deviations needed in the formula for the variance and standard deviation. Comparatively speaking, it is short because it eliminates all the subtractions required for calculating the individual deviations. A second and not so obvious advantage is that it tends to give better computational accuracy than the method utilizing the deviations. The beginning statistics student frequently finds the variance that he has calculated at odds with the answer in the text. This is usually caused by rounding off decimal numbers in the computations. We suggest that rounding off be held at a minimum since it may seriously affect the results of computation of the variance. A third advantage is that the shortcut method is especially suitable for use with many electronic calculators, some of which accumulate $\sum_{i=1}^{n} y_i$ and $\sum_{i=1}^{n} y_i^2$ simultaneously.

Before leaving this topic, we will calculate the standard deviation for the $n = 30$ grade-point averages, Table 3.1. You can verify the following:

$$\sum_{i=1}^{n} y_i = 62.5,$$

$$\sum_{i=1}^{n} y_i^2 = 134.19.$$

Using the shortcut formula,

$$\begin{aligned}\sum_{i=1}^{n} (y_i - \bar{y})^2 &= \sum_{i=1}^{n} y_i^2 - \frac{\left(\sum_{i=1}^{n} y_i\right)^2}{n} \\ &= 134.19 - \frac{(62.5)^2}{30} = 134.19 - 130.21 \\ &= 3.98.\end{aligned}$$

It follows that the standard deviation is

$$s = \sqrt{\frac{\sum_{i=1}^{n}(y_i - \bar{y})^2}{n-1}} = \sqrt{\frac{3.98}{29}} = .37.$$

EXAMPLE 3.6

Calculate $\bar{y}$ and s for the measurements 85, 70, 60, 90, 81.

Solution

y_i	y_i^2
85	7,225
70	4,900
60	3,600
90	8,100
81	6,561
386	30,386

$$\bar{y} = \frac{386}{5} = 77.2.$$

$$\begin{aligned}\sum_{i=1}^{n} (y_i - \bar{y})^2 &= \sum_{i=1}^{n} y_i^2 - \frac{\left(\sum_{i=1}^{n} y_i\right)^2}{n} \\ &= 30{,}386 - \frac{(386)^2}{5} \\ &= 30{,}386 - 29{,}799.2 \\ &= 586.8.\end{aligned}$$

$$s = \sqrt{\frac{\sum_{i=1}^{n}(y_i - \bar{y})^2}{n-1}} = \sqrt{\frac{586.8}{4}} = \sqrt{146.7}$$
$$= 12.1.$$

3.8
A CHECK ON THE CALCULATION OF *s*

Tchebysheff's Theorem and the Empirical Rule can be used to detect gross errors in the calculation of s. Thus, we know that at least 3/4 or, in the case of a mound-shaped distribution, nearly 95 percent of a set of measurements will lie within two standard deviations of their mean. Consequently, most of the sample measurements will lie in the interval $\bar{y} \pm 2s$, and the range will approximately equal $4s$. This is, of course, a very rough approximation but from it we can acquire a useful check that will detect large errors in the calculation of s.

Letting R equal the range,

$$R \approx 4s,$$

then s is approximately equal to $R/4$; that is,

$$s \approx \frac{R}{4}.$$

The computed value of s using the shortcut formula should be of roughly the same order as the approximation.

EXAMPLE 3.7

Use the approximation above to check the calculation of s for Example 3.6.

Solution The range of the five measurements is

$$R = 90 - 60 = 30.$$

Then,

$$s \approx \frac{R}{4} = \frac{30}{4} = 7.5.$$

This is of the same order as the calculated value, $s = 12.1$.

You should note that the range approximation is not intended to provide an accurate value for s. Rather, its purpose is to detect gross errors in calculating such as the failure to divide the sum of squares of deviations by $(n - 1)$ or the failure to take the square root of s^2. Both of these errors yield solutions that are many times larger than the range approximation to s.

EXAMPLE 3.8

Use the range approximation to determine an approximate value for the standard deviation for the data of Table 3.1.

Solution The range is $R = 2.9 - 1.4 = 1.5$. Then

$$s \approx \frac{R}{4} = \frac{1.5}{4} = .375.$$

We have shown that $s = .37$ for the data of Table 3.1. The approximation is very close to the actual value of s.

Tips on Problem Solving

1. Always use the shortcut formula when calculating $\sum_{i=1}^{n} (y_i - \bar{y})^2$. This will help reduce rounding errors.
2. Be careful about rounding numbers. Carry your calculations of $\sum_{i=1}^{n} (y_i - \bar{y})^2$ to six significant figures.
3. Always find the range of the set of measurements (this is easy) and then calculate the range approximation to s as a check on your calculations.

EXERCISES

3.10 Given $n = 5$ measurements, 4, 1, 1, 3, 5:
(a) Calculate $\bar{y}$.
(b) Use the shortcut formula, Section 3.7, to calculate s^2 and s.

3.11 Given $n = 8$ measurements, 0, 1, 1, 3, 1, 2, 2, 1:
(a) Calculate $\bar{y}$.
(b) Use the shortcut formula, Section 3.7, to calculate s^2 and s.

3.12 The length of time required for an automobile driver to respond to a particular emergency situation was recorded for $n = 10$ drivers. The times, in seconds, were .5, .8, 1.1, .7, .6, .9, .7, .8, .7, .8.
(a) Scan the data and use the procedure of Section 3.8 to find an approximate value for s. Use this value to check your calculations in part (b).
(b) Calculate the sample mean, $\bar{y}$, and the standard deviation, s. Compare with part (a).

3.13 Most purchasers of new automobiles would agree that EPA ratings for mileage per gallon (mpg) of gasoline for new cars are suspiciously high and consequently are of limited value to consumers. A record of the actual mileages per gallon for 12 fill-ups of a 1975 VW Rabbit (automatic transmission, mostly in-town driving) are as follows:

21.2	24.8	22.7
24.4	22.3	23.4
23.7	24.3	21.8
25.1	25.2	24.9

(a) Describe the population from which this sample was selected. Are the sample measurements representative of the population of mpg ratings for all 1975 VW Rabbits equipped with automatic transmissions?

(b) Use the mean of the data to estimate the population mean mpg for the 1975 VW Rabbit described above.

(c) As you will learn in Chapters 8 and 9, the sample standard deviation will be used in assessing the accuracy of the estimate, part (b). Use the range to obtain an approximate value for s. Then calculate s and use the range approximation as a check on gross errors in your calculations.

3.14 To estimate the amount of lumber in a tract of timber, an owner decided to count the number of trees with diameters exceeding 12 inches in randomly selected 50×50-feet squares. Seventy 50×50 squares were randomly selected from the tract and the number of trees (with diameters in excess of 12 inches) were counted for each. The data are as follows:

7	8	7	10	4	8	6
9	6	4	9	10	9	8
3	9	5	9	9	8	7
10	2	7	4	8	5	10
9	6	8	8	8	7	8
6	11	9	11	7	7	11
10	8	8	5	9	9	8
8	9	10	7	7	7	5
8	7	9	9	6	8	9
5	8	8	7	9	13	8

(a) Construct a relative frequency histogram to describe this data.

(b) Calculate the sample mean, $\bar{y}$, as an estimate of μ, the mean number of timber trees for all 50×50-feet squares in the tract.

(c) Calculate s for the data. Construct the intervals $\bar{y} \pm s$, $\bar{y} \pm 2s$, and $\bar{y} \pm 3s$. Count the percentage of squares falling in each of the three intervals and compare with the corresponding percentages given by the Empirical Rule and Tchebysheff's Theorem.

3.15 Refer to times of recurrence data, Exercise 3.2.

(a) Find the range.

(b) Use the procedure of Section 3.8 to find an approximate value for s.

(c) Compute s for the data and compare with your answer for part (b).

3.16 Refer to the times of recurrence data, Exercise 3.2. Examine the data and count the number of observations falling in the intervals $\bar{y} \pm s$, $\bar{y} \pm 2s$, and $\bar{y} \pm 3s$. (Use the value of s computed in Exercise 3.15.)

(a) Do the fractions falling in these intervals agree with Tchebysheff's Theorem? the Empirical Rule?

(b) Why might the Empirical Rule be unsuitable for describing these data?

3.17 Suppose that some measurements occur more than once and that the data, y_1, y_2, . . . , y_k, are arranged in a frequency table as shown here.

Observations	Frequency f_i
y_1	f_1
y_2	f_2
$\vdots$	$\vdots$
y_k	f_k
	n

Then,

$$\bar{y} = \frac{\sum_{i=1}^{k} y_i f_i}{n} \qquad \text{where} \qquad n = \sum_{i=1}^{k} f_i$$

and

$$\sum_{i=1}^{n} (y_i - \bar{y})^2 = \sum_{i=1}^{k} y_i^2 f_i - \frac{\left\{\sum_{i=1}^{k} y_i f_i\right\}^2}{n}$$

Although these formulas for grouped data are primarily of value when you have a large number of measurements, demonstrate their use for the sample 1, 0, 0, 1, 3, 1, 3, 2, 3, 0, 0, 1, 1, 3, 2.

(a) Calculate $\bar{y}$ and $\sum_{i=1}^{n} (y_i - \bar{y})^2$ directly using the formulas for ungrouped data.

(b) The frequency table for the $n = 15$ measurements is shown below.

y	$f(y)$
0	4
1	5
2	2
3	4
	$n = 15$

Calculate $\bar{y}$ and $\sum_{i=1}^{n} (y_i - \bar{y})^2$ using the formulas for grouped data. Compare with your answers to part (a).

3.18 To illustrate the utility of the Empirical Rule, consider a distribution that is heavily skewed to the right, as shown in the accompanying figure.

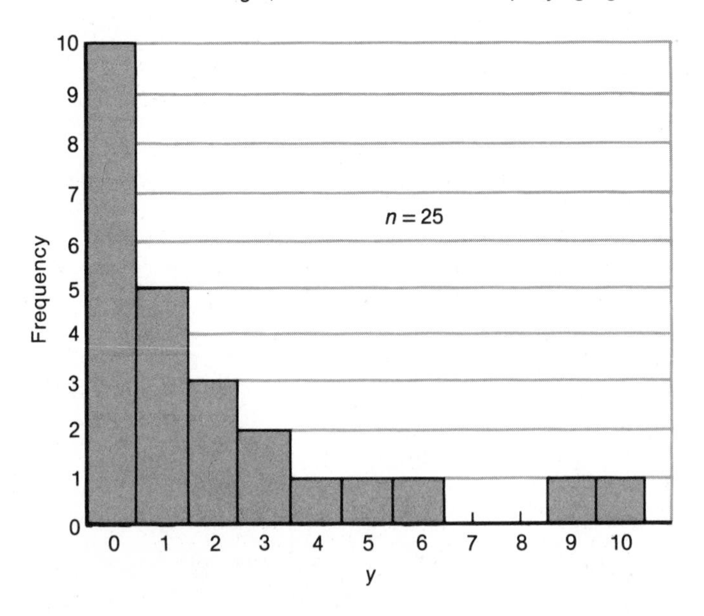

(a) Calculate $\bar{y}$ and s for the data shown. (Note: there are ten 0s, five 1s, etc.)
(b) Construct the intervals $\bar{y} \pm s$, $\bar{y} \pm 2s$, and $\bar{y} \pm 3s$ and locate them on the frequency distribution.
(c) Calculate the proportion of the $n = 25$ measurements falling in each of the three intervals. Compare with Tchebysheff's Theorem and the Empirical Rule. Note that while the proportion falling in the interval $\bar{y} \pm s$ does not agree closely with the Empirical Rule, the proportions falling in the intervals $\bar{y} \pm 2s$ and $\bar{y} \pm 3s$ agree very well. Many times this is true, even for non-mound-shaped distributions of data.

3.9

THE EFFECT OF CODING ON $\bar{y}$ AND s^2 (OPTIONAL)

Data are frequently coded to simplify the calculation of $\bar{y}$ and s^2. Thus, it is easier to calculate the mean and variance of the set of measurements $(-.1, .2, .1, 0, .2)$, than for (99.9, 100.2, 100.1, 100.0, 100.2). The first set was obtained by subtracting 100 from each measurement of the second set. Similarly, we might wish to simplify a set of measurements by multiplying or dividing by a constant. Thus, it is easier to work with the set (3, 1, 4, 6, 4, 2) than with (.003, .001, .004, .006, .004, .002). The first set was obtained by multiplying each element of the second set by 1,000.

Data are coded by performing one or both of two operations. Thus one may subtract (or add) a constant, c, to each measurement, multiply (or divide) each measurement by a constant, k, or do both. The objective of coding is to obtain data for which we can more easily determine the mean and variance. Then we wish to use these quantities to find $\bar{y}$ and s^2, the mean and variance, of the original data.

Let $y_1, y_2, \ldots, y_n$ be the original measurements with mean and variance, $\bar{y}$ and s^2. Similarly, let $\bar{x}$ and s_x^2 be the mean and variance of the coded data. How are these quantities related for the two operations of coding? Four theorems answer this question. Proof of the theorems will be left as exercises for the interested reader.

First consider subtracting a constant from each measurement. Thus the coded variable is $x = y - c$. The following two theorems relate $\bar{x}$ and s_x^2 to $\bar{y}$ and s^2.

THEOREM 3.2

Let $y_1, y_2, \ldots, y_n$ be n measurements and let $x_i = y_i - c$, $i = 1, 2, \ldots, n$. Then,

$$\bar{x} = \bar{y} - c.$$

The implication of Theorem 3.2 is that the difference in means between the uncoded and coded data, $\bar{y}$ and $\bar{x}$, will equal the quantity c. Thus, one can compute $\bar{x}$ for the coded data and obtain

$$\bar{y} = \bar{x} + c.$$

THEOREM 3.3

Let $y_1, y_2, \ldots, y_n$ be n measurements and let $x_i = y_i - c, i = 1, 2, \ldots, n$. Then,

$$s_x^2 = s^2.$$

That the variances of the coded and uncoded data should be equal is quite reasonable. A variance is a measure of the variability of a set of measurements. Adding or subtracting the same constant to each measurement would not change the distance between any pair of observations and thus would not change the variability of the data.

EXAMPLE 3.9

Find $\bar{y}$ and s^2 for the sample 99.9, 100.2, 100.1, 100.0, 100.2, by coding $x_i = y_i - 100, i = 1, 2, \ldots, 5$.

Solution The coded data are $-.1, .2, .1, 0, .2$. Then,

$$\sum_{i=1}^{5} x_i = .4,$$

$$\sum_{i=1}^{5} x_i^2 = .1,$$

$$\bar{x} = \frac{\sum_{i=1}^{n} x_i}{n} = \frac{.4}{5} = .08,$$

and

$$s_x^2 = \frac{\sum_{i=1}^{n} x_i^2 - \frac{\left(\sum_{i=1}^{n} x_i\right)^2}{n}}{n-1} = \frac{.1 - \frac{(.4)^2}{5}}{4} = .017.$$

Then, by Theorem 3.2,

$$\bar{y} = \bar{x} + c = .08 + 100 = 100.08.$$

By Theorem 3.3,

$$s^2 = s_x^2 = .017.$$

The second method of coding is multiplying each measurement by a constant, k. The following theorems relate the coded and uncoded means and variances.

THEOREM 3.4

Let $y_1, y_2, \ldots, y_n$ be n measurements and let $x_i = ky_i$, $i = 1, 2, \ldots, n$, where $k \neq 0$. Then,

$$\bar{x} = k\bar{y} \quad \text{and} \quad \bar{y} = \frac{\bar{x}}{k}.$$

THEOREM 3.5

Let $y_1, y_2, \ldots, y_n$ be n measurements and let $x_i = ky_i$, $i = 1, 2, \ldots, n$, where $k \neq 0$. Then,

$$s_x^2 = k^2 s^2 \quad \text{and} \quad s^2 = \frac{1}{k^2} s_x^2.$$

The effect of multiplying each observation by a constant, k, on the relation between $\bar{x}$ and $\bar{y}$ was perhaps predictable. Thus, the coded mean, $\bar{x}$, is k times as large as $\bar{y}$. In agreement with this relation, Theorem 3.5 states that $s_x^2 = k^2 s^2$ or, equivalently, $s_x = ks$. Thus, like the relation between the means, the standard deviation of x is k times the standard deviation for y.

EXAMPLE 3.10

Calculate $\bar{y}$ and s^2 for the sample .003, .001, .004, .006, .004, .002 by coding $x_i = 1{,}000y_i$, $i = 1, 2, \ldots, 6$.

Solution The coded data are 3, 1, 4, 6, 4, 2. Then,

$$\sum_{i=1}^{6} x_i = 20, \qquad \sum_{i=1}^{6} x_i^2 = 82,$$

$$\bar{x} = \frac{\sum_{i=1}^{n} x_i}{n} = \frac{20}{6} = 3.33,$$

and

$$s_x^2 = \frac{\sum_{i=1}^{n} x_i^2 - \frac{\left(\sum_{i=1}^{n} x_i\right)^2}{n}}{n-1} = \frac{82 - \frac{(20)^2}{6}}{5} = 3.07.$$

Then, by Theorem 3.4,

$$\bar{y} = \frac{\bar{x}}{k} = \frac{3.33}{1{,}000} = .00333.$$

By Theorem 3.5,

$$s^2 = \frac{s_x^2}{k^2} = \frac{3.07}{(1{,}000)^2} = .00000307.$$

Both methods of coding are frequently applied to the same set of data. The effects of these operations are given in Theorem 3.6:

THEOREM 3.6

Let $y_1, y_2, \ldots, y_n$ be n measurements and let

$$x_i = k(y_i - c).$$

Then,

$$\bar{x} = k(\bar{y} - c) \qquad \text{or} \qquad \bar{y} = \frac{\bar{x}}{k} + c$$

and

$$s_x^2 = k^2 s^2 \qquad \text{or} \qquad s^2 = \frac{s_x^2}{k^2}.$$

EXAMPLE 3.11

Suppose that both methods of coding are applied to the data 99.9, 100.2, 100.1, 100.0, 100.2. Subtract $c = 100$ from each observation and multiply by $k = 10$. Find $\bar{y}$ and s^2.

Solution $x_i = 10(y_i - 100)$, $i = 1, 2, \ldots, 5$, and the coded data are -1, 2, 1, 0, 2. Then,

$$\sum_{i=1}^{5} x_i = 4, \qquad \sum_{i=1}^{5} x_i^2 = 10,$$

$$\bar{x} = \frac{\sum_{i=1}^{n} x_i}{n} = \frac{4}{5} = .8,$$

and

$$s_x^2 = \frac{\sum_{i=1}^{n} x_i^2 - \frac{\left(\sum_{i=1}^{n} x_i\right)^2}{n}}{n-1} = \frac{10 - \frac{(4)^2}{5}}{4} = 1.7.$$

Applying Theorem 3.6,

$$\bar{y} = \frac{\bar{x}}{k} + c = \frac{.8}{10} + 100 = 100.08,$$

$$s^2 = \frac{s_x^2}{k^2} = \frac{1.7}{(10)^2} = .017.$$

EXERCISES

3.19 Given the sample measurements 3,005, 2,998, 3,000, 3,003, 2,999, code by subtracting a suitable constant. Then use Theorems 3.2 and 3.3 to find the sample mean and variance. Note the reduction in computing effort achieved by the coding.

3.20 Code the sample data .020, .021, .019, .020, .021, .018, by multiplying by 1,000. Use Theorems 3.4 and 3.5 to find the sample mean and variance.

3.21 Refer to Exercise 3.20. Code by subtracting .020 from each observation and then multiplying by 1,000. Use Theorem 3.6 to find the sample mean and variance.

3.22 A retail oil dealer knows that a particular customer's use during the winter months is an average of 105 gallons per week with a standard deviation of 12 gallons. Use Theorems 3.4 and 3.5 to find the mean and standard deviation of the customer's fuel oil consumption per day.

3.23 Simplify the calculation of $\bar{y}$ and s^2 and s for Exercise 3.14 by subtracting 5 from each count before performing the calculations.

3.10
SUMMARY AND COMMENTS

Methods for describing sets of measurements fall into one of two categories, graphical methods and numerical methods. The relative frequency histogram is an extremely useful graphical method for characterizing a set of measurements. Numerical descriptive measures are numbers that attempt to create a mental image of the frequency histogram (or frequency distribution). We have restricted the discussion to measures of central tendency and variation, the most useful of which are the mean and standard deviation. While the mean possesses intuitive descriptive significance, the standard deviation is significant only when used in conjunction with Tchebysheff's Theorem and the Empirical Rule. The objective of sampling is the description of making inferences about the population from which the sample was obtained. This is accomplished by using the sample mean, $\bar{y}$, and the quantity, s^2, as estimators of the population mean, μ, and variance, σ^2.

Many descriptive methods and numerical measures have been presented in this chapter, but these constitute only a small percentage of those which might have been discussed. In addition, many special computational techniques usually found in elementary texts have been omitted. This was necessitated by the limited time available in an elementary course and because the advent and common use of electronic computers have minimized the importance of special computational formulas. But, more important, the inclusion of such techniques would tend to detract from and obscure the main objective of modern statistics and this text—statistical inference.

REFERENCES

Barr, A. J., J. H. Goodnight, J. P. Sall, and J. T. Helwig, *A User's Guide to SAS 76*. Raleigh, N.C.: SAS Institute, Inc., 1976.

Brown, M. B., ed., *Biomedical Computer Programs*. Berkeley, Calif.: University of California, 1977.

Freund, J. E., *Modern Elementary Statistics*, 4th ed. Englewood Cliffs, N.J.: Prentice-Hall, Inc., 1973.

Hoel, P. G., *Elementary Statistics*, 4th ed. New York: John Wiley & Sons, Inc., 1976.

Nie, N., C. H. Hull, J. G. Jenkins, K. Steinbrenner, and D. H. Bent, *Statistical Package for the Social Sciences*, 2nd ed. New York: McGraw-Hill Book Company, 1975.

SUPPLEMENTARY EXERCISES

[Starred (*) exercises are optional.]

3.24 Conduct the following experiment: toss 10 coins and record y, the number of heads observed. Repeat this process $n = 50$ times, thus providing 50 values of y. Construct a relative frequency histogram for these measurements.

3.25 The following measurements represent the grade-point average of 25 college freshmen:

2.6	1.8	2.6	3.7	1.9
2.1	2.7	3.0	2.4	2.3
3.1	2.6	2.6	2.5	2.7
2.7	2.9	3.4	1.9	2.3
3.3	2.2	3.5	3.0	2.5

Construct a relative frequency histogram for these data.

3.26 Given the set of $n = 6$ measurements 5, 7, 2, 1, 3, 0, calculate $\bar{y}$, s^2, and s.

3.27 Given the set of $n = 7$ measurements 3, 0, 1, 5, 3, 0, 4, calculate $\bar{y}$, s^2, and s.

3.28 Calculate $\bar{y}$, s^2, and s for the data in Exercise 3.24.

3.29 Calculate $\bar{y}$, s^2, and s for the data in Exercise 3.25.

3.30 Refer to the histogram constructed in Exercise 3.24 and find the fraction of measurements lying in the interval $\bar{y} \pm 2s$. (Use the results of Exercise 3.28.) Are the results consistent with Tchebysheff's Theorem? Is the frequency histogram of Exercise 3.24 relatively mound-shaped? Does the Empirical Rule adequately describe the variability of the data in Exercise 3.24?

3.31 Repeat the instructions of Exercise 3.30; use the interval $\bar{y} \pm s$.

3.32 Refer to the grade-point data in Exercise 3.25. Find the fraction of measurements falling in the intervals $\bar{y} \pm s$ and $\bar{y} \pm 2s$. Do these results agree with Tchebysheff's Theorem and the Empirical Rule?

3.33 In contrast to aptitude tests, which are predictive measures of what one can accomplish with training, achievement tests tell what an individual can do at the time of the test. Mathematics achievement test scores for 400 students were found to have a mean and a variance equal to 600 and 4900, respectively. If the distribution of test scores was bell-shaped, approximately how many of the scores would fall in the interval 530 to 670? Approximately how many scores would be expected to fall in the interval 460 to 740?

3.34 The following five measurements represent the amount of milk, in ounces, discharged by an automatic milk dispenser: 7.1, 7.2, 7.0, 7.3, 7.0. Calculate $\bar{y}$, s^2, and s.

3.35 Refer to Exercise 3.34. Calculate the fraction of measurements in the intervals $\bar{y} \pm s$ and $\bar{y} \pm 2s$. Do these results agree with Tchebysheff's Theorem? Why is the Empirical Rule inappropriate for describing the variability of these data?

3.36 Use the range approximation for s (Section 3.8) to check the calculated value of s for the data in Exercise 3.25.

3.37 Petroleum pollution in seas and oceans stimulates the growth of some types of bacteria. A count of petroleumlytic microorganisms (bacteria per 100 milliliters) in 10 portions of seawater gave the following readings: 49, 70, 54, 67, 59, 40, 61, 69, 71, 52.
(a) Observe the data and guess the value for s by use of the range approximation.
(b) Calculate $\bar{y}$ and s and compare with the range approximation of part (a).

3.38 Why do statisticians prefer to divide the sum of squares of deviations of the sample measurements by $(n - 1)$ rather than n when estimating a population variance σ^2?

3.39 After 20 generations of selective breeding for high locomotor activity, a strain of mice is found to have a mean of 23 revolutions per minute and a standard deviation of 4.5 revolutions per minute in an activity wheel. If the distribution of activity scores is bell-shaped and if the experimenter regards high activity as more than 18.5 revolutions per minute, what percentage would have activity scores between 18.5 and 32 revolutions per minute?

3.40 The percentage of city telephone subscribers who use unlisted numbers is on the increase. The distribution of the percentage of unlisted numbers in cities possesses a mean and standard deviation that are near 14 and 6 percent, respectively. If you were to pick a city at random, is it likely that the percentage of unlisted numbers will exceed 20 percent? Explain. Within what limits would you expect the percentage to fall?

3.41 A machining operation produces bolts with an average diameter of .51 inch and a standard deviation of .01 inch. If the distribution of bolt diameters is approximately normal, what fraction of total production would possess diameters falling in the interval .49 to .53 inch?

3.42 Refer to Exercise 3.41. Suppose that the bolt specification required a diameter equal to $.5 \pm .02$ inch. Bolts not satisfying this requirement are considered defective. If the machining operation functioned as described in Exercise 3.41, what fraction of total production would result in defective bolts?

3.43 An hour examination in statistics produced the following scores:

83	91	88	82	81
84	62	93	50	63
96	68	73	80	97
95	91	38	82	72
88	83	78	93	69
78	91	87	84	91

Construct a relative frequency histogram for the examination scores.

3.44 Compute $\bar{y}$, s^2, and s for the data in Exercise 3.43.

3.45 Use the range approximation for s (Section 3.8) to check the calculated value of s for the data in Exercise 3.43.

3.46 Refer to Exercise 3.43. Find the fraction of scores in the intervals $\bar{y} \pm s$ and $\bar{y} \pm 2s$ and compare with Tchebysheff's Theorem and the Empirical Rule.

3.47 Find the range for the data in Exercise 3.43. Find the ratio of range to s. If one possessed a large amount of data having a bell-shaped distribution, the range would be expected to equal how many standard deviations?

3.48 Find the ratio of the range to s for the data in Exercise 3.24.

3.49 The range approximation for s can be improved if it is known that the sample is drawn from a bell-shaped distribution of data. Thus, the calculated s should not differ substantially from the range divided by the appropriate ratio given in the following table:

Number of measurements	5	10	25
Expected ratio of range to s	2.5	3	4

For the data in Exercise 3.27, estimate s as suggested. Compare this estimate with the calculated s.

3.50 Repeat the instructions of Exercise 3.49; use the data given in Exercises 3.26 and 3.34.

3.51 From the following data, a student calculated s to be .263. On what grounds might we doubt his accuracy? What is the correct value (nearest hundredth)?

17.2	17.1	17.0	17.1	16.9
17.0	17.1	17.0	17.3	17.2
17.1	17.0	17.1	16.9	17.0
17.1	17.3	17.2	17.4	17.1

3.52 A set of 128 examination scores in history produced a mean and standard deviation equal to 92 percent and 10 percent, respectively. Would you expect the relative frequency distribution for these scores to be bell-shaped? Why? (Note that a score cannot exceed 100 percent.) What proportion of the 128 scores would you expect to find between 60 and 100 percent?

3.53 Given the set of $n = 6$ measurements, 0, 1, 2, 0, 3, 2, find $\bar{y}$ and s^2.

3.54 If a dairy truck delivers an average of 210 bottles of milk a day with a standard deviation of 20 bottles, approximately what portion of the time could the driver anticipate delivering between 190 and 250 bottles a day? Assume a mound-shaped distribution of deliveries.

3.55 An industrial concern uses an employee screening test with average score μ and standard deviation $\sigma = 10$. Assume that the test-score distribution is approximately bell-shaped and that a score of 65 qualifies an applicant for further consideration. What is the value of μ such that approximately 2.5 percent of the applicants qualify for further consideration?

3.56 The intelligence quotient (IQ) expresses intelligence as a ratio of the mental age to the chronological age, multiplied by 100. Thus the average (when mental age equals chronological age) is 100. Construct a relative frequency histogram for the following IQ scores:

100	103	99	101	100	120	109	82
101	112	95	118	118	89	114	113
92	137	130	94	87	93	111	96
93	98	101	96	84	86	89	90

3.57 Refer to Exercise 3.56.

(a) Calculate $\bar{y}$ and s.

(b) Use the range approximation for s to check your calculations.

(c) Find the number of scores in the intervals $\bar{y} \pm s$, $\bar{y} \pm 2s$, and $\bar{y} \pm 3s$. Compare the proportions of measurements in these intervals with the proportions specified by Tchebysheff's Theorem and the Empirical Rule.

3.58 Attendances at a high school's basketball games were recorded and found to have a sample mean and variance of 420 and 25, respectively. Calculate $\bar{y} \pm s$, $\bar{y} \pm 2s$, $\bar{y} \pm 3s$, and state the approximate fraction of measurements we would expect to fall in these intervals according to the Empirical Rule:

$\bar{y} \pm s$ __________ fraction __________
$\bar{y} \pm 2s$ __________ fraction __________
$\bar{y} \pm 3s$ __________ fraction __________

3.59 The mean duration of television commercials on a given network is 75 seconds, with a standard deviation of 20 seconds. Assuming that duration times are approximately normally distributed:

(a) What is the approximate probability that a commercial will last less than 35 seconds?

(b) What is the approximate probability that a commercial will last longer than 55 seconds?

3.60 A random sample of 100 foxes was examined by a team of veterinarians to determine the prevalence of a particular type of parasite. Counting the number of parasites per fox, the veterinarians found that 69 foxes possessed no parasites, 17 possessed one, etc. The following is a frequency tabulation of the data:

Number of parasites, y	0	1	2	3	4	5	6	7	8
Number of foxes, f	69	17	6	3	1	2	1	0	1

(a) Construct a relative frequency histogram for y, the number of parasites per fox.

(b) Calculate $\bar{y}$ and s for the sample.

(c) What fraction of the parasite counts fall within two standard deviations of the mean? three? Do these results agree with Tchebysheff's Theorem? the Empirical Rule?

3.61 Consider a population consisting of the number of teachers per college at small two-year colleges. Suppose that the number of teachers per college has an average $\mu = 175$ and a standard deviation $\sigma = 15$.

(a) Use Tchebysheff's Theorem to make a statement about the percentage of colleges that have between 145 and 205 teachers.

(b) Assume that the population is normally distributed. What fraction of the colleges have more than 190 teachers?

3.62 In the terminology of a psychologist, acquisition refers to the stage during which a new response is learned and is gradually strengthened. The following data, obtained from a learning experiment, represent rates of acquisition as measured by the number of trials it took subjects to successfully perform a memory task.

12	10	16	7	18
13	14	20	9	23
8	13	14	6	19
6	11	15	10	16

Computing s for these data, the experimenter obtained a value of 8.90. Use the range approximation for s as a check on this computation. Do these estimated and computed values for s appear to differ excessively?

3.63 A recent study by the Highway Loss Data Institute reported that the average loss payment per insurance claim by automobile owners during the first half of 1974 was \$495 with a standard deviation of \$75.* Assume that the distribution of loss payment per claim is mound-shaped.

**Source: Money,* September 1974.

(a) Describe the distribution of loss payments during this period.

(b) Approximately what fraction of the loss payments exceeded $570 during this period?

*3.64 Code the data given in Exercise 3.34 by subtracting 7 from each observation. Calculate the sample mean and variance for the coded data and then use Theorems 3.2 and 3.3 to find $\bar{y}$ and s^2.

*3.65 Code the data given in Exercise 3.34 by subtracting 7 from each observation and then multiplying by 10. Calculate $\bar{x}$ and s_x^2 for the coded data and use Theorem 3.6 to find $\bar{y}$ and s^2.

*3.66 Code the data given in Exercise 3.25 by subtracting 2.5 from each observation. Use Theorems 3.2 and 3.3 to find $\bar{y}$ and s^2.

*3.67 Code the data given in Exercise 3.25 by subtracting 2.5 from each observation and then multiplying by 10. Calculate $\bar{x}$ and s_x^2 for the coded data and use Theorem 3.6 to find $\bar{y}$ and s^2.

*3.68 Code the data given in Exercise 3.51 by subtracting 17 from each observation. Use Theorems 3.2 and 3.3 to find $\bar{y}$ and s^2.

*3.69 Code the data given in Exercise 3.51 by subtracting 17 from each observation and then multiplying by 10. Calculate $\bar{x}$ and s_x^2 for the coded data and use Theorem 3.6 to find $\bar{y}$ and s^2.

*3.70 Use the summation theorems, 2.1, 2.2, and 2.3, to prove Theorems 3.2, 3.3, 3.4, and 3.5.

*3.71 Use Theorems 2.1, 2.2, and 2.3 to prove Theorem 3.6.

EXPERIENCES WITH REAL DATA

1. Select an area of your undergraduate major that utilizes experimental data. Typical data sources would be chemistry, biology, psychology, geology, or physics laboratories. Or you might seek data contained in the social science or business journals. Either by experimentation or by use of a professional journal, select a sample of at least $n = 25$ observations on some random variable.

(a) Define the population from which your sample was drawn.

(b) Construct a relative frequency histogram for the data.

(c) Calculate $\bar{y}$ and s for the data.

(d) Do the data appear to be mound-shaped and thereby satisfy the requirements of the Empirical Rule?

(e) What fraction of the observations lie within two standard deviations of $\bar{y}$? three? Do these results agree with Tchebysheff's Theorem? the Empirical Rule?

2. Examine the population of all telephone subscribers in your community (see your local telephone directory). Randomly select a page from the directory and repeat the process until you have selected $n = 30$ pages. Let y be the number of subscribers per page and find y for each of the $n = 30$ pages.

(a) Construct a relative frequency histogram for your data.

(b) Calculate $\bar{y}$ and s.

(c) Find the fraction of observations in the intervals $\bar{y} \pm s$ and $\bar{y} \pm 2s$. Does the Empirical Rule provide a satisfactory description of the variability of the data?

(d) Knowing the number of pages, N, you could estimate the total number of subscribers by multiplying N by the estimated average number of subscribers per page, $\bar{y}$. Find the estimate and compare with the actual number of subscribers (the telephone company should have this figure).

Calculation of $\bar{y}$ and s^2 can most easily be accomplished on an electronic desk calculator that is programmed to simultaneously accumulate y and y^2. That is, as you enter the data, the calculator accumulates the sum of the y values and y^2 values. When the last measurement has been entered, the calculator will give the values, $\sum_{i=1}^{n} y_i$ and $\sum_{i=1}^{n} y_i^2$, that are needed to calculate $\bar{y}$ and s^2.

For large data sets, you might wish to use packaged programs and an electronic computer to perform the calculations. Useful packaged programs are available in the Biomed (Biomedical Programs), the SAS (Statistical Analysis Systems), and the SPSS (Statistical Package for the Social Sciences) program libraries (see the References).

CHAPTER OBJECTIVES

GENERAL OBJECTIVES The objective of this chapter is to develop an understanding of the basic concepts of probability. These concepts will be used in subsequent chapters to make inferences and to evaluate the reliability of inferences about populations based on sample data.

SPECIFIC OBJECTIVES

1. To motivate a study of probability by giving a simple example of the use of probability in making a decision based on sample data. *Section 4.1*
2. To present a model for the repetition of an experiment, and to explain how the model provides a straightforward mechanism for calculating the probability of an experimental outcome. *Sections 4.2, 4.3*
3. To present a method for representing an event as a composition of two or more other events and then to give two probability laws that you can use to find the probability of the composition. *Sections 4.4, 4.5, 4.6*
4. To introduce the ideas of numerical events and random variables. *Section 4.7*
5. Optional topics (not needed in subsequent chapters) include Bayes' Rule and some counting rules that are sometimes useful in calculating the probability of an event. *Sections 4.8, 4.9*

4 PROBABILITY

4.1

INTRODUCTION

Probability and statistics are related in a most curious way. In essence, probability is the vehicle that enables the statistician to use information in a sample to make inferences or describe the population from which the sample was obtained. We illustrate this relationship with a simple example.

Consider a balanced die with its familiar six faces. By balanced we mean that the chance of observing any one of the six sides on a single toss of the die is just as likely as any other. Tossing the die might be viewed as an experiment that could conceivably be repeated a very large number of times, thus generating a population of numbers where the measurements, y, would be either 1, 2, 3, 4, 5, or 6. Assume that the population is so large that each value of y occurs with equal frequency. Note that we do not actually generate the population; rather, it exists conceptually. Now let us toss the die once and observe the value of y. This one measurement represents a sample of $n = 1$ drawn from the population. What is the probability that the sample value of y will equal 2? Knowing the structure of the population, we realize that each value of y has an equal chance of occurring and hence the probability that $y = 2$ is $1/6$. This example illustrates the type of problem considered in probability theory. The population is assumed to be known and we are concerned with calculating the probability of observing a particular sample. Exactly the opposite is true in statistical problems, where we assume the population unknown, the sample known, and we wish to make inferences about the population. Thus probability reasons from the population to the sample; statistics acts in reverse, moving from the sample to the population.

To illustrate how probability is used in statistical inference, consider the following example. Suppose that a die is tossed $n = 10$ times and the number of dots, y, appearing is recorded after each toss. This represents a sample of $n = 10$ measurements drawn from a much larger body of tosses, the population, which could be generated if we wished. Suppose that all 10 measurements resulted in $y = 1$. We wish to use this information to make an inference concerning the population of tosses; specifically, we wish to infer that the die is or is not balanced. Having observed 10 tosses, each resulting in $y = 1$, we would be somewhat suspicious of the die and would likely reject the theory that the die was balanced. We reason as follows: If the die were balanced as we hypothesize, observing 10 identical measurements is most improbable. Hence, either we observed a rare event or else our hypothesis is false. We would likely be inclined to the latter conclusion. Notice that the decision was based upon the probability of observing the sample, assuming our theory to be true.

The above illustration emphasizes the importance of probability in making statistical inferences. In the following discussion of the theory of probability, we shall assume the population known and calculate the probability of drawing various samples. In doing so, we are really choosing a model for a physical situation because the actual composition of a population is rarely known in practice. Thus the probabilist models a physical situation (the population) with probability much as the sculptor models with clay.

Chapter 4 is divided into segments. The first, Sections 4.2 and 4.3, provides a probabilistic model for the repetition of an experiment and, at the same time, presents a clear-cut method for calculating the probability of an observed event. We call this the sample point approach. The second segment, Sections 4.4 through 4.6, presents the methodology for a second method for calculating the probability of an event, which we call the event composition approach. Acquiring an understanding of the probabilistic model and the two methods for calculating the probability of an event are the major objectives of Chapter 4.

4.2

THE SAMPLE SPACE

Data are obtained either by observation of uncontrolled events in nature or by controlled experimentation in the laboratory. To simplify our terminology, we seek a word that will apply to either method of data collection and hence define the term experiment.

DEFINITION

An experiment is the process by which an observation (or measurement) is obtained.

Note that the observation need not be numerical. Typical examples of experiments are:

1. Recording a test grade.
2. Making a measurement of daily rainfall.
3. Interviewing a voter to obtain his preference prior to an election.
4. Inspecting a light bulb to determine whether it is a defective or an acceptable product.
5. Tossing a coin and observing the face that appears.

A population of measurements results when the experiment is repeated many times. For instance, we might be interested in the length of life of television tubes produced in a plant during the month of June. Testing a single tube to failure and measuring length of life would represent a single experiment, while repetition of the experiment for all tubes produced during this period would generate the entire population. A sample would represent the results of some small group of experiments selected from the population.

Let us now direct our attention to a careful analysis of an experiment and the construction of a mathematical model for a population. A by-product of our development will be a systematic and direct approach to the solution of probability problems.

We commence by noting that each experiment may result in one or more outcomes which we will call events and denote by capital letters. Consider the following experiment.

EXAMPLE 4.1

Experiment: Toss a die and observe the number appearing on the upper face. Some events would be:

1. Event A: observe an odd number.
2. Event B: observe a number less than 4.
3. Event E_1: observe a 1.
4. Event E_2: observe a 2.
5. Event E_3: observe a 3.
6. Event E_4: observe a 4.
7. Event E_5: observe a 5.
8. Event E_6: observe a 6.

The events detailed above do not represent a complete listing of all possible events associated with the experiment but suffice to illustrate a point. The reader will readily note a difference between events A and B and events E_1, E_2, E_3, E_4, E_5, and E_6. Event A will occur if event E_1, E_3, or E_5 occurs, that is, if we observe a 1, 3, or 5. Thus A could be decomposed into a collection of simpler events, namely, E_1, E_3, and E_5. Likewise, event B will occur if E_1, E_2, or E_3 occur and could be viewed as a collection of smaller or simpler events. In contrast, we note that it is impossible to decompose events $E_1, E_2, E_3, \ldots, E_6$. Events $E_1, E_2, \ldots, E_6$ are called simple events and A and B are compound events.

DEFINITION

An event that cannot be decomposed is called a simple event. Simple events will be denoted by the symbol *E* with a subscript.

The events $E_1, E_2, \ldots, E_6$ represent a complete listing of all simple events associated with the experiment, Example 4.1. An interesting property of simple events is readily apparent. An experiment will result in one and only one of the simple events. For instance, if a die is tossed, we will observe either a 1, 2, 3, 4, 5, or 6, but we cannot possibly observe more than one of the simple events at the same time. Hence a list of simple events provides a breakdown of all possible indecomposable outcomes of the experiment. For purposes of illustration, consider the following examples:

EXAMPLE 4.2

Experiment: Toss a coin. Simple events:

E_1: observe a head,
E_2: observe a tail.

EXAMPLE 4.3

Experiment: Toss two coins. Simple events:

Event	Coin 1	Coin 2
E_1	Head	Head
E_2	Head	Tail
E_3	Tail	Head
E_4	Tail	Tail

It would be extremely convenient if we were able to construct a model for an experiment that could be portrayed graphically. We do this by creating a correspondence between simple events and a set of points. To each simple event we assign a point, called a sample point.* Thus the symbol E_i will now be associated with either simple event E_i or its corresponding sample point. The resulting diagram is called a Venn diagram.

Example 4.1 may be viewed symbolically in terms of the Venn diagram shown in Figure 4.1. Six sample points are shown corresponding to the six possible simple events enumerated in Example 4.1. Likewise, a Venn diagram for the two-coin-toss experiment, Example 4.3, would represent an experiment in which there are four sample points.

DEFINITION

The set of all sample points for the experiment is called the sample space and is represented by the symbol *S*. We say that *S* is the totality of all sample points.

What is an event in terms of the sample points? We recall that event *A*, Example 4.1, occurred if any one of the simple events E_1, E_3, or E_5 occurred. That is, we observe *A*, an odd number, if we observe either a 1, 3, or 5. Event *B*, a number less than 4, occurs if E_1, E_2, or E_3 occurs. Thus, if we designate the

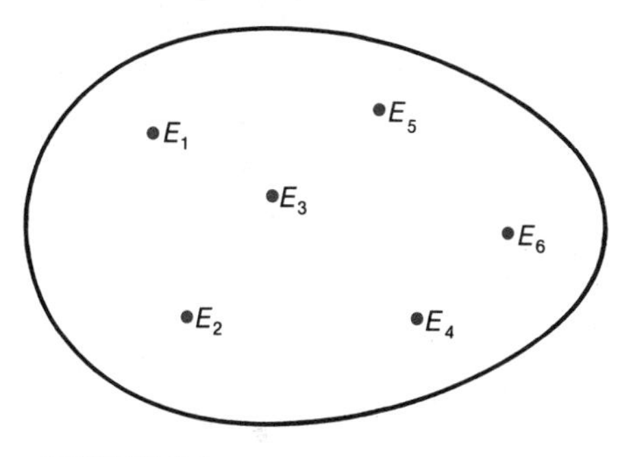

FIGURE 4.1
Venn diagram for die tossing

*The preceding terminology and the use of sample points to represent simple events is consistent with W. Feller's treatment of this subject in his text on probability. Feller's introductory text on probability is considered to be a classic.

sample points associated with an event (those for which the event occurs), the event is as clearly defined as if we had presented a verbal description of it. The event "observe E_1, E_3, or E_5" is obviously the same as the event "observe an odd number." Both represent event A.

DEFINITION

An event is a specific collection of sample points.

To decide whether a sample point is included in a particular event, check to see if the occurrence of the sample point implies occurrence of the event. If it does, that sample point is included in the event. For example, in the die-tossing experiment, the sample point E_1 is in the event A, "observe an odd number," because if E_1 occurs, then A will occur.

Keep in mind that the foregoing discussion refers to the outcome of a single experiment and that the performance of the experiment will result in the occurrence of one and only one sample point. A particular event will occur if any sample point in the event occurs.

An event could be represented on the Venn diagram by encircling the sample points in that event. Events A and B for the die-tossing problem are shown in Figure 4.2. Note that points E_1 and E_3 are in both events A and B and that both A and B occur if either E_1 or E_3 occurs.

4.3 THE PROBABILITY OF AN EVENT

Populations of observations are obtained by repeating an experiment a very large number of times. Some fraction of this very large number of experiments will result in E_1, another fraction in E_2, etc. From a practical point of view, we think of the fraction of the population resulting in an event A as the probability of A.

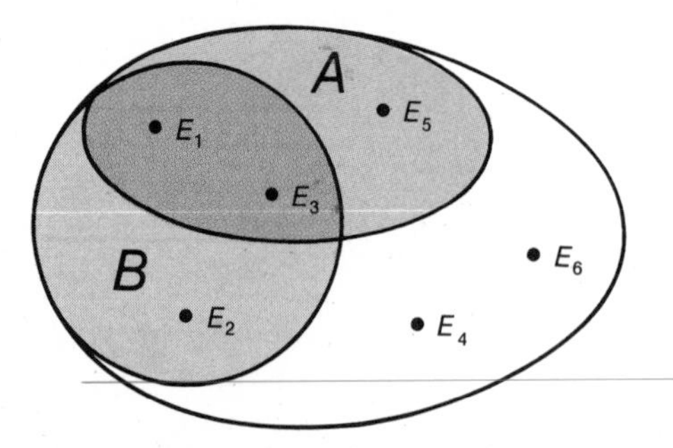

FIGURE 4.2

Events A and B for die tossing

Putting it another way, if an experiment is repeated a large number of times, N, and A is observed n times, the probability of A is

$$P(A) = \frac{n}{N}.$$

This practical interpretation of the meaning of probability, a view held by most laypersons, might properly be called the relative frequency concept of probability.

In practice, the exact composition of the population is rarely known and hence the exact values of the probabilities for various events are unknown. Mathematically speaking, we ignore this aspect of the problem and take the probabilities as given, hence providing a model for a real population. For instance, we would assume that for a large population of die tosses, Example 4.1, the numbers 1, 2, 3, 4, 5, and 6 should appear with approximately the same relative frequency and therefore that

$$P(E_1) = P(E_2) = \cdots = P(E_6) = 1/6.$$

That is, we assume that the die is perfectly balanced. Is there such a thing as a perfectly balanced die? Probably not, but we would be inclined to think that the probability of the sample points would be so near 1/6 that our assumption is quite valid for practical purposes and provides a good model for die tossing.

We complete our model for the population by adding the following:

To each point in the sample space we assign a number called the probability of E_i, denoted by the symbol $P(E_i)$, such that:

1. $0 \leq P(E_i) \leq 1$, for all i.
2. $\sum_S P(E_i) = 1$.

The two requirements placed upon the probabilities of the sample points are necessary in order that the model conform to our relative frequency concept of probability. Thus we require that a probability be greater than or equal to 0 and less than or equal to 1 and that the sum of the probabilities over the entire sample space, S, be equal to 1. Furthermore, from a practical point of view, we would choose the $P(E_i)$ in a realistic way so that they agree with the observed relative frequency of occurrence of the sample points.

Keeping in mind that a particular event is a specific collection of sample points, we can now state a simple rule for the probability of any event, say event A.

DEFINITION

The probability of an event *A* is equal to the sum of the probabilities of the sample points in *A*.

Note that the definition agrees with our intuitive concept of probability.

EXAMPLE 4.4

Calculate the probability of the event A for the die-tossing experiment, Example 4.1.

Solution Event A, "observe an odd number," includes sample points E_1, E_3, E_5. Hence

$$\begin{aligned} P(A) &= P(E_1) + P(E_3) + P(E_5) \\ &= 1/6 + 1/6 + 1/6 \\ &= 1/2. \end{aligned}$$

EXAMPLE 4.5

Calculate the probability of observing exactly one head in a toss of two coins.

Solution Construct the sample space letting H represent a head, T a tail.

Event	First Coin	Second Coin	$P(E_i)$
E_1	H	H	1/4
E_2	H	T	1/4
E_3	T	H	1/4
E_4	T	T	1/4

It would seem reasonable to assign a probability of 1/4 to each of the sample points. We are interested in:

Event A: observe exactly one head.

Sample points E_2 and E_3 are in A. Hence

$$\begin{aligned} P(A) &= P(E_2) + P(E_3) \\ &= 1/4 + 1/4 \\ &= 1/2. \end{aligned}$$

EXAMPLE 4.6

Consider the following experiment involving two urns. Urn 1 contains two white balls and one black ball. Urn 2 contains one white ball. A ball is drawn from urn 1 and placed in urn 2. Then a ball is drawn from urn 2. What is the probability that the ball drawn from urn 2 will be white? See Figure 4.3.

Solution The problem is easily solved once we have listed the sample points. For convenience, number the white balls 1, 2, and 3, with ball 3 residing in urn 2. The sample points are listed below using W_i to represent the ith white ball and B to represent the black ball. For example, E_1 is the event that white ball 1 is drawn from urn 1, placed in urn 2, and then is drawn from urn 2.

	Drawn From:		
Event	Urn 1	Urn 2	$P(E_i)$
E_1	W_1	W_1	1/6
E_2	W_1	W_3	1/6
E_3	W_2	W_2	1/6
E_4	W_2	W_3	1/6
E_5	B	B	1/6
E_6	B	W_3	1/6

Event A, drawing a white ball from urn 2, occurs if E_1, E_2, E_3, E_4, or E_6 occurs. Once again, it would seem reasonable to assume the sample points equally likely and to assign a probability of 1/6 to each of the six sample points. Hence

$$\begin{aligned} P(A) &= P(E_1) + P(E_2) + P(E_3) + P(E_4) + P(E_6) \\ &= 1/6 + 1/6 + 1/6 + 1/6 + 1/6 \\ &= 5/6. \end{aligned}$$

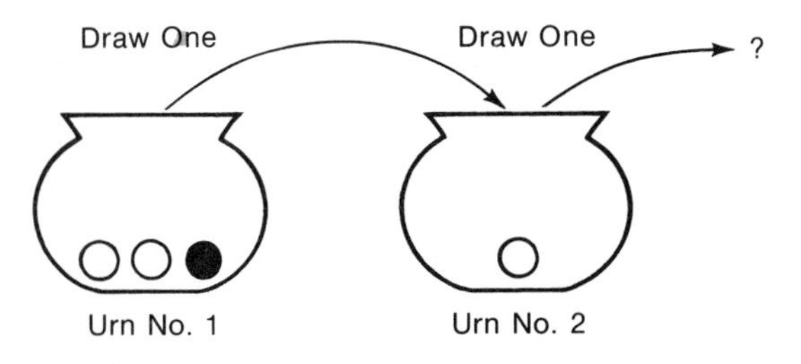

FIGURE 4.3
Representation of the experiment in Example 4.6

Note that the sample space provides a probabilistic model for a population that possesses, in addition to elegance, a great deal of utility. The model provides us with a simple, logical, and direct method for calculating the probability of an event or, if you like, the probability of a sample drawn from a theoretical population. The student who has had prior experience with probability problems will recognize the advantage of a systematic procedure for their solution. The disadvantages soon become apparent. Listing the sample points can be quite tedious and one must be certain that none has been omitted. Since the total number of sample points in S may run into the millions, it is very convenient to have counting rules to simplify the counting of sample points. Although not necessary for an understanding of the basic concepts of probability, counting rules are included in Section 4.9 (optional). If you wish to develop an ability to solve complex probability problems using the sample point approach, you should move directly to Section 4.9.

Tips on Problem Solving: Calculating the probability of an event: the sample point approach.

1. Use the following steps for calculating the probability of an event by summing the probabilities of the sample points:
   (a) Define the experiment.
   (b) Identify a typical simple event. List the simple events associated with the experiment and test each to make certain that they cannot be decomposed. This defines the sample space, S.
   (c) Assign reasonable probabilities to the sample points in S, making certain that $0 \leq P(E_i) \leq 1$ and $\sum_S P(E_i) = 1$.
   (d) Define the event of interest, A, as a specific collection of sample points. (A sample point is in A if A occurs when the sample point occurs. Test *all* sample points in S to locate those in A.)
   (e) Find $P(A)$ by summing the probabilities of sample points in A.
2. When the sample points are equiprobable, the sum of the probabilities of the sample points in A, step (e), can be acquired by counting the points in A and multiplying by the probability per sample point.
3. Calculating the probability of an event by using the five-step procedure described in part 1 is systematic and will lead to the correct solution if all the steps are correctly followed. Major sources of error are:
   (a) failing to define the experiment clearly [step (a)],
   (b) failing to specify simple events [step (b)],
   (c) failing to list all the simple events,
   (d) failing to assign valid probabilities to the sample points.

EXERCISES

4.1 An experiment involves tossing a single die. Specify the sample points in the events:

A: observe a 4.
B: observe an even number.
C: observe a number less than 3.
D: observe both A and B.
E: observe either A or B or both.
F: observe both A and C.

Calculate the probabilities of the events D, E, and F by summing the probabilities of the appropriate sample points.

4.2 According to *Webster's New Collegiate Dictionary,* a divining rod is "a forked rod believed to indicate [divine] the presence of water or minerals by dipping downward when held over a vein." To test the claims of success by a divining rod expert, four cans are buried in the ground, two empty and two filled with water. The expert will use the divining rod to test each of the four cans and will decide which two contain water.

(a) Define the experiment.
(b) List the sample points in S.
(c) If the rod is completely useless in locating water, what is the probability that the expert correctly identifies (by guessing) the two cans containing water?

4.3 Refer to Exercise 4.2. Suppose the experiment was conducted using five cans, three empty and two containing water. Answer parts (a), (b), and (c) of Exercise 4.2.

4.4 Patients arriving at a hospital outpatient clinic can select one of three stations for service. Suppose that physicians are randomly assigned to the stations and that the patients have no station preference. Three patients arrive at the clinic and their selection of stations is observed.
(a) List the sample points for the experiment.
(b) Let A be the event that each station receives a patient. List the sample points in A.
(c) Make a reasonable assignment of probabilities to the sample points and find $P(A)$.

4.5 A tea taster is required to taste and rank three varieties of tea, A, B, and C, according to the taster's preference.
(a) Define the experiment.
(b) List the sample points in S.
(c) If the taster had no ability to distinguish a difference in taste between teas, what is the probability that the taster will rank tea type A as best? As the least desirable?

4.6 The odds are two to one that when A and B play racquetball, A wins. Suppose A and B play three matches and that the winners of the matches are recorded. Using the letters A and B to denote the winner of each match, the eight sample points are listed in the accompanying table. As you will subsequently learn, under certain conditions,

Winner of Match			Sample Point	
1	2	3	i	$P(E_i)$
A	A	A	1	8/27
A	A	B	2	4/27
A	B	A	3	4/27
A	B	B	4	2/27
B	A	A	5	4/27
B	A	B	6	?
B	B	A	7	2/27
B	B	B	8	1/27

it is reasonable to assume that the sample point probabilities are as listed in the table. Using these probabilities, find the following:
(a) $P(E_6)$.
(b) The probability that A wins at least two of the three matches.

4.7 An investor has the option of investing in three of five recommended stocks. Unknown to the investor, only two will show a substantial profit within the next five years. If the investor selects the three stocks at random (giving every combination of three stocks an equal chance of selection), what is the probability that the investor selects the two profitable stocks?

4.8 Two dice are tossed. What is the probability that the sum of the numbers showing on the dice is equal to 7? 11?

4.9 Four union men, two from a minority group, are assigned to four distinctly different one-man jobs.
(a) Define the experiment.
(b) List the sample points in S.
(c) If the assignment to the job is unbiased, that is, if any one ordering of assignments is as probable as any other, what is the probability that the two men from the minority group are assigned to the two least desirable jobs?

4.10 Two city commissioners are to be selected from a total of five to form a subcommittee to study the city's traffic problems.
(a) Define the experiment.
(b) List the sample points in S.
(c) If all possible pairs of commissioners have an equal probability of selection, what is the probability that commissioners Jones and Smith will be selected?

4.11 A popular test to control the quality of brand-name food products is obtained by presenting three specimens identical in appearance to each member of a panel of tasters. In each case, two of the specimens are from batches of stock known to possess the desired taste, while the third specimen is from the latest batch. Each panelist is told to select the specimen that is different from the other two. Suppose that there are four tasters (designated T_1, T_2, T_3, and T_4) on the panel. Define the following events as specific collections of sample points:
(a) The sample space S.
(b) The event A that T_1, T_2, and T_3 are each "successful" in identifying the specimen from the latest batch.
(c) The event B that exactly three of the four tasters are successful.
(d) The event C that at least three of the four tasters are successful.
(e) Assign reasonable probabilities to the sample points and find the probabilities of events A, B, and C.

4.4

COMPOUND EVENTS

Most events of interest in practical situations are compound events that require enumeration of a large number of sample points. Actually, we find a second approach available for calculating the probability of events which obviates the listing of sample points and is therefore much less tedious and time consuming. It is based upon the classification of events, event relations, and two probability laws which will be discussed in this section and in Sections 4.5 and 4.6, respectively. We call it the "event composition approach" for finding the probability of an event.

Compound events, as the name suggests, are formed by some composition of two or more events. Composition takes place in one of two ways, or in some combination of the two, namely, a union or an intersection.

DEFINITION

Let A and B be two events in a sample space, S. The union of A and B is defined to be the event containing all sample points in A or B or both. We denote the union of A and B by the symbol $(A \cup B)$.

Defined in ordinary terms, a union is the event that *either* event A or event B or both A and B occur. For instance, in Example 4.1,

A: E_1, E_3, E_5,
B: E_1, E_2, E_3.

The union $(A \cup B)$ would be the collection of points E_1, E_2, E_3, and E_5. This is shown diagrammatically as the shaded area in Figure 4.4.

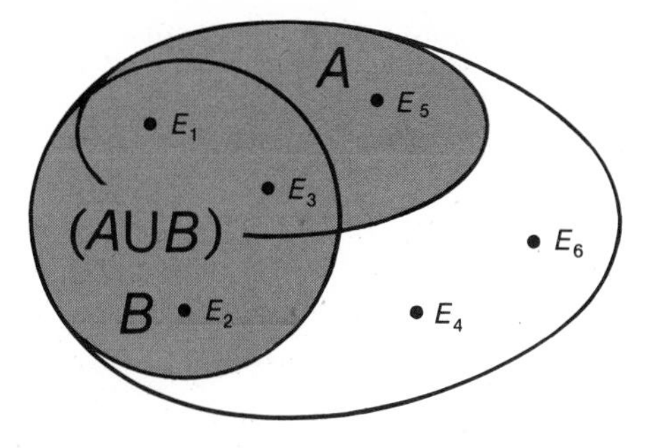

FIGURE 4.4
Event ($A \cup B$) in Example 4.1

DEFINITION

Let *A* and *B* be two events in a sample space, *S*. The intersection of *A* and *B* is the event composed of all sample points that are in both *A* and *B*. An intersection of events *A* and *B* is represented by the symbol *AB*. (Some authors use $A \cap B$.)

The intersection AB is the event that *both* A and B occur. It would appear in a Venn diagram as the overlapping area between A and B. The intersection AB for Example 4.1 would be the event consisting of points E_1 and E_3. If either E_1 or E_3 occurs, both A and B occur. This is shown diagrammatically as the shaded area in Figure 4.5.

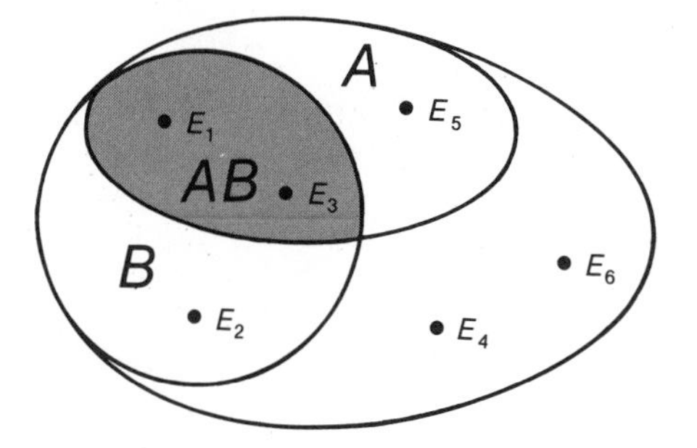

FIGURE 4.5
Intersection AB

EXAMPLE 4.7

Refer to the experiment, Example 4.3, where two coins are tossed, and define:

Event A: at least one head,
Event B: at least one tail.

Define events A, B, AB, and ($A \cup B$) as collections of sample points.

Solution Recall that the sample points for this experiment were:

E_1: *HH* (head on first coin, head on second),
E_2: *HT*,
E_3: *TH*,
E_4: *TT*.

The occurrence of sample points E_1, E_2, and E_3 implies and hence defines event *A*. The other events could similarly be defined:

Event *B*: E_2, E_3, E_4,
Event *AB*: E_2, E_3,
Event $(A \cup B)$: E_1, E_2, E_3, E_4.

Note that $(A \cup B) = S$, the sample space, and is thus certain to occur.

4.5
EVENT RELATIONS

We shall define three relations between events: complementary, independent, and mutually exclusive events. You will have many occasions to inquire whether two or more events bear a particular relationship to one another. The test for each relationship is inherent in the definition, as we shall illustrate, and their use in the calculation of the probability of an event will become apparent in Section 4.6.

DEFINITION

The complement of an event *A* is the collection of all sample points in *S* and not in *A*. The complement of *A* is denoted by the symbol $\bar{A}$.

Since

$$\sum_S P(E_i) = 1,$$

$$P(A) + P(\bar{A}) = 1,$$

and

$$P(A) = 1 - P(\bar{A}),$$

which is a useful relation for obtaining $P(A)$ when $P(\bar{A})$ is known or easily calculated.

Two events are often related in such a way that the probability of occurrence of one depends upon whether the second has or has not occurred. For instance, suppose that one experiment consists in observing the weather on a specific day. Let *A* be the event "observe rain" and *B* be the event "observe an overcast sky." Events *A* and *B* are obviously related. The probability of rain, $P(A)$, is not the same as the probability of rain given prior information that the day is cloudy. The probability of A, $P(A)$, would be the fraction of the entire population of observations that result in rain. Now let us look only at the subpopulation of observations that result in B, a cloudy day, and the fraction of these which result in A. This fraction, called the conditional probability of A given B, may equal $P(A)$, but we would expect the chance of rain, given that the day is cloudy, to be larger. The conditional probability of A, given that B has occurred, is denoted as

$$P(A \mid B),$$

where the vertical bar in the parentheses is read "given" and events appearing to the right of the bar are the events that have occurred.

We shall define the conditional probabilities of B given A and A given B as follows.

DEFINITION

$$P(B \mid A) = \frac{P(AB)}{P(A)}$$

and

$$P(A \mid B) = \frac{P(AB)}{P(B)}.$$

You can see that this definition of conditional probability is consistent with the relative frequency concept of probability by attaching some numbers to the probabilities in our weather example. Recall that A denotes rain on a given day; B denotes the day is cloudy. Now suppose that 10 percent of all days are rainy and cloudy [that is, $P(AB) = .10$] and 30 percent of all days are cloudy [$P(B) = .30$].

The situation that we have just described is graphically portrayed in Figure 4.6. Each sample point in event B, denoted by the large egg-shaped area, is associated with a single cloudy day. Since 30 percent of all days will be cloudy, we can regard this area as .3. Ten percent of all days, or 1/3 of all cloudy days, will also be rainy. These days are included in the color-shaded event, AB. Hence, if a single day is selected from the set of all days representing the population,

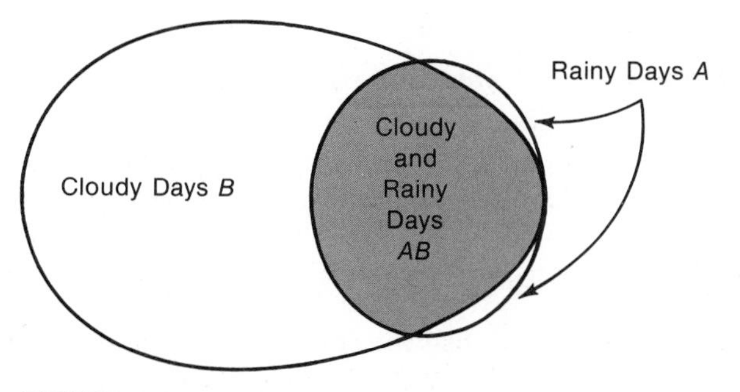

FIGURE 4.6
Events A and B

what is the probability that we will select a rainy day, given that we know the day is cloudy? That is, what is $P(A \mid B)$?

Since we already know that the day is cloudy, we know that the sample point to be selected must fall in event B (Figure 4.6). One-third of these days will result in rain. Hence the probability that we will select a rainy day is

$$P(A \mid B) = 1/3.$$

You can see that this result agrees with our definition for $P(A \mid B)$. That is,

$$P(A \mid B) = \frac{P(AB)}{P(B)} = \frac{.10}{.30} = 1/3.$$

EXAMPLE 4.8

Calculate $P(A \mid B)$ for the die-tossing experiment described in Example 4.1.

Solution Given that B has occurred, we are concerned only with sample points E_1, E_2, and E_3, which occur with equal frequency. Of these, E_1 and E_3 imply event A. Hence

$$P(A \mid B) = 2/3.$$

Or, we could obtain $P(A \mid B)$ by substituting into the equation

$$P(A \mid B) = \frac{P(AB)}{P(B)} = \frac{1/3}{1/2} = 2/3.$$

Note that $P(A \mid B) = 2/3$ while $P(A) = 1/2$, indicating that A and B are dependent upon each other.

DEFINITION

Two events, *A* and *B*, are said to be independent if either

$$\mathbf{P(A \mid B) = P(A)}$$

or

$$\mathbf{P(B \mid A) = P(B).}$$

Otherwise, the events are said to be dependent.

Translating this definition into words, two events are independent if the occurrence or nonoccurrence of one of the events does not change the probability of the occurrence of the other event. Note that if $P(A \mid B) = P(A)$, then $P(B \mid A)$ will also equal $P(B)$. Similarly, if $P(A \mid B)$ and $P(A)$ are unequal, then $P(B \mid A)$ and $P(B)$ will be unequal.

A third useful event relation was observed but not specifically defined in our discussion of simple events. Recall that an experiment could result in one and only one simple event. No two could occur at exactly the same time. Two events, A and B, are said to be mutually exclusive, if, when one occurs, it excludes the possibility of occurrence of the other. Another way to say this is to state that the intersection, AB, will contain no sample points. It would then follow that $P(AB) = 0$.

DEFINITION

Two events, A and B, are said to be mutually exclusive if the event AB contains no sample points.

Mutually exclusive events have no overlapping area in a Venn diagram (see Figure 4.7).

EXAMPLE 4.9

Refer to the die-tossing experiment in Example 4.1. Are events A and B mutually exclusive? Are they complementary? Are they independent?

Solution

Event A: E_1, E_3, E_5,
Event B: E_1, E_2, E_3.

The event AB is the set of sample points in both A and B. You can see that AB includes points E_1 and E_3. Therefore, A and B are not mutually exclusive. They are not complementary because B is not the set of all points in S which are not in A. The test for independence lies in the definition. That is, we shall check to see if $P(A|B) = P(A)$. From Example 4.8, $P(A|B) = 2/3$. Then since $P(A) = 1/2$, $P(A|B) \neq P(A)$ and, by definition, events A and B are dependent.

EXAMPLE 4.10

Given two mutually exclusive events, A and B, with $P(A)$ and $P(B)$ not equal to zero, are A and B independent events?

Solution Since A and B are mutually exclusive, if A occurs, B cannot occur and vice versa. Then $P(B|A) = 0$. But $P(B)$ was said to be greater than zero. Hence $P(B|A)$ is not equal to $P(B)$, and according to the definition the events are dependent.

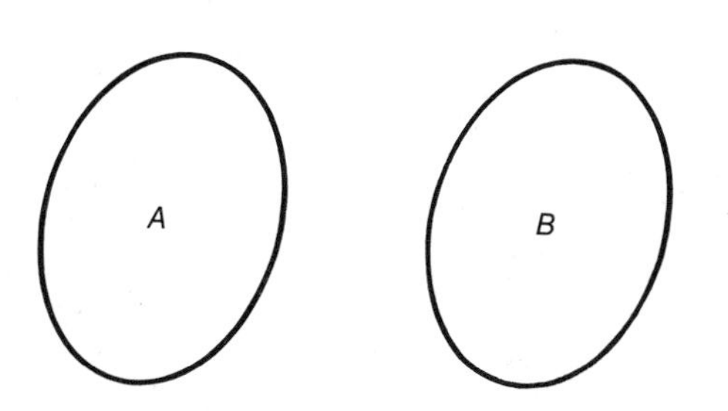

FIGURE 4.7
Mutually exclusive events

4.6

TWO PROBABILITY LAWS AND THEIR USE

As previously stated, a second approach to the solution of probability problems is based upon the classification of compound events, event relations, and two probability laws which we now state and illustrate. The "laws" can be simply stated and taken as fact as long as they are consistent with our model and with reality. The first is called the Additive Law of Probability and applies to unions.

> **The Additive Law of Probability:** The probability of a union $(A \cup B)$ is equal to
>
> $$P(A \cup B) = P(A) + P(B) - P(AB).$$
>
> If A and B are mutually exclusive, $P(AB) = 0$ and
>
> $$P(A \cup B) = P(A) + P(B).$$

The Additive Law conforms to reality and our model. As you will note from Figure 4.8, the sum, $P(A) + P(B)$, contains the sum of the probabilities of all sample points in $(A \cup B)$ but includes a double counting of the probabilities of all points in the intersection, AB. Subtracting $P(AB)$ gives the correct result.

The second law of probability is called the Multiplicative Law and applies to intersections.

> **The Multiplicative Law of Probability:** Given two events, A and B, the probability of the intersection, AB, is
>
> $$\begin{aligned} P(AB) &= P(A)P(B \mid A) \\ &= P(B)P(A \mid B). \end{aligned}$$
>
> If A and B are independent, $P(AB) = P(A)P(B)$.

The Multiplicative Law follows from the definition of conditional probability.

The use of the probability laws for calculating the probability of a compound event is less direct than the listing of sample points and requires a bit of

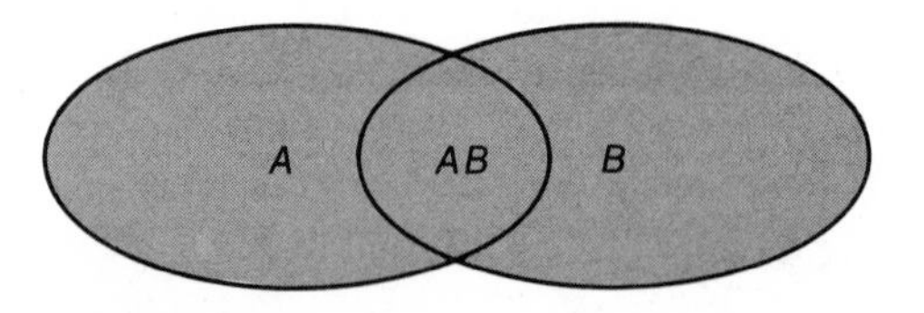

FIGURE 4.8

The union of two events, A and B ($A \cup B$ is shaded)

experience and ingenuity. The approach involves the expression of the event of interest as a union or intersection (or combination of both) of two or more events whose probabilities are known or easily calculated. This can often be done in many ways. The trick is to find the right combination, a task requiring no little amount of creativity in some cases. The usefulness of event relations is now apparent. If the event of interest is expressed as a union of mutually exclusive events, the probabilities of the intersections will equal zero. If they are independent, we can use the unconditional probabilities to calculate the probability of an intersection. Examples 4.11 through 4.16 illustrate the use of the probability laws and the technique described above.

EXAMPLE 4.11

Calculate $P(AB)$ and $P(A \cup B)$ for Example 4.1.

Solution Recall that $P(A) = P(B) = 1/2$ and $P(A \mid B) = 2/3$. Then,

$$\begin{aligned} P(AB) &= P(B)P(A \mid B) \\ &= (1/2)(2/3) \\ &= 1/3, \end{aligned}$$

$$\begin{aligned} P(A \cup B) &= P(A) + P(B) - P(AB) \\ &= 1/2 + 1/2 - 1/3 \\ &= 2/3. \end{aligned}$$

You will notice that these solutions agree with those obtained using the sample point approach.

EXAMPLE 4.12

Consider the experiment in which two coins are tossed. Let A be the event that the toss results in at least one head. Find $P(A)$.

Solution $\bar{A}$ is the collection of sample points implying the event "two tails." Because $\bar{A}$ is the complement of A,

$$P(A) = 1 - P(\bar{A}).$$

The event $\bar{A}$ will occur if both of two independent events occur, "tail on the first coin," T_1, and "tail on the second coin," T_2. Then $\bar{A}$ is the intersection of T_1 and T_2, or

$$\bar{A} = T_1T_2.$$

Applying the Multiplicative Law and noting that T_1 and T_2 are independent events,

$$P(\bar{A}) = P(T_1)P(T_2) = (1/2)(1/2) = 1/4.$$

Then, $P(A) = 1 - P(\bar{A}) = 1 - 1/4 = 3/4$.

EXAMPLE 4.13

Refer to Example 4.12 and find the probability of exactly one head.

Solution Event B, exactly one head, is the union of two mutually exclusive events, B_1 and B_2, where

B_1: head on the first coin, tail on the second,
B_2: head on the second coin, tail on the first.

Then B_1 and B_2 are both intersections of independent events:

$$B_1 = H_1T_2,$$
$$B_2 = H_2T_1.$$

Applying the Multiplicative Law for independent events,

$$\begin{aligned} P(B_1) &= P(H_1)P(T_2) \\ &= (1/2)(1/2) = 1/4, \\ P(B_2) &= P(H_2)P(T_1) \\ &= (1/2)(1/2) = 1/4. \end{aligned}$$

Then, $P(B) = P(B_1) + P(B_2)$, because B_1 and B_2 are mutually exclusive events, and

$$P(B) = 1/4 + 1/4 = 1/2.$$

You will note that this problem was solved in Example 4.5 using the sample point approach.

EXAMPLE 4.14

A city council contains eight members, of which two are local contractors. If two councilmen are selected at random to fill vacancies on the zoning committee, what is the probability that both of the contractors will be selected?

Solution Let D_1 be the event that the first councilman selected is a contractor and, correspondingly, let D_2 be the event that the second is a contractor. The event that both councilmen are contractors is A. Then A will occur if both D_1 and D_2 occur or, equivalently, A is the intersection of D_1 and D_2:

$$A = D_1D_2.$$

Applying the Multiplicative Law,

$$P(A) = P(D_1)P(D_2|D_1).$$

The probability of selecting a contractor on the first draw is 2/8 or 1/4. Similarly, the probability that the second councilman will be a contractor, given that the first was a contractor, is 1/7. Then,

$$P(A) = P(D_1)P(D_2|D_1) = (1/4)(1/7) = 1/28.$$

(*Note:* This solution can be easily obtained using the sample point approach.)

EXAMPLE 4.15

Two cards are drawn from a deck of 52 cards. Calculate the probability that the draw will include an ace and a ten.

Solution Event A: draw an ace and a ten. Then $A = B \cup C$, where
B: draw the ace on the first draw and the ten on the second,
C: draw the ten on the first draw and the ace on the second.

Note that B and C were chosen so as to be mutually exclusive and also intersections of events with known probabilities. Thus,

$$B = B_1B_2 \quad \text{and} \quad C = C_1C_2,$$

where

B_1: draw an ace on the first draw,
B_2: draw a ten on the second draw,
C_1: draw a ten on the first draw,
C_2: draw an ace on the second draw.

Applying the Multiplicative Law,

$$\begin{aligned} P(B_1B_2) &= P(B_1)P(B_2 | B_1) \\ &= (4/52)(4/51) \end{aligned}$$

and

$$P(C_1C_2) = (4/52)(4/51).$$

Then, applying the Additive Law,

$$\begin{aligned} P(A) &= P(B) + P(C) \\ &= (4/52)(4/51) + (4/52)(4/51) \\ &= \frac{8}{663}. \end{aligned}$$

The student is cautioned to check each composition carefully to be certain that it is actually equal to the event of interest.

EXAMPLE 4.16

Use the laws of probability to solve the "two-urns" problem, Example 4.6.

Solution Define the event of interest as:

Event A: ball drawn from urn 2 is white.

Then,

$$A = (B \cup C),$$

where

B: draw a white ball from urn 1 and a white ball from urn 2,
C: draw a black ball from urn 1 and a white ball from urn 2.

Note that B and C were chosen to be mutually exclusive and that both are intersections. Thus

$$B = B_1A$$

and

$$C = C_1A,$$

where

B_1: draw a white ball from urn 1,
C_1: draw a black ball from urn 1.

Then,

$$P(A) = P(B \cup C)$$
$$= P(B) + P(C) - P(BC).$$

Since B and C are mutually exclusive, $P(BC) = 0$. Then,

$$P(A) = P(B) + P(C)$$
$$= P(B_1A) + P(C_1A).$$

Applying the Multiplicative Law,

$$P(B_1A) = P(B_1)P(A \mid B_1)$$
$$= (2/3)(1)$$
$$= 2/3,$$
$$P(C_1A) = P(C_1)P(A \mid C_1)$$
$$= (1/3)(1/2)$$
$$= 1/6.$$

Substituting,

$$P(A) = P(B_1A) + P(C_1A)$$
$$= 2/3 + 1/6$$
$$= 5/6.$$

Example 4.16 shows you that a problem often can be solved using either the sample point approach (Example 4.6) or the event composition approach. Because the number of sample points was small, the sample point approach was the easier method for solving this problem. If the number of balls in the urn had been large, it would have been difficult to enumerate the large number of sample points. Then the event composition approach should have been used. The method of solution would be identical to the solution for Example 4.16. The difficulty of the solution would not depend upon the number of balls in the urn.

Tips on Problem Solving: Calculating the probability of an event: event composition approach.

1. Use the following steps for calculating the probability of an event using the event composition approach:
   (a) Define the experiment.
   (b) Clearly visualize the nature of the sample points. Identify a few to clarify your thinking.
   (c) Write an equation expressing the event of interest, say A, as a composition of two or more events using either or both of the two forms of composition (unions and intersections). Note that this equates point sets. Make certain that the event implied by the composition and event A represent the same set of sample points.

(d) Apply the Additive and Multiplicative Laws of Probability to step (c) and find $P(A)$.

2. Be careful with step (c). You often can form many compositions that will be equivalent to event A. The trick is to form a composition in which all the probabilities appearing in step (d) will be known. Thus, you must visualize the results of step (d) for any composition and select the one for which the component probabilities are known.
3. Always write down letters to represent events described in an exercise. Then write down the probabilities that are given and assign them to events. Identify the probability that is requested in the exercise. This may help you to arrive at the appropriate event composition.

EXERCISES

4.12 A certain genetic characteristic occurs in mice with probability equal to .2. If three mice are randomly selected from a large number of unrelated litters, what is the probability that all three mice possess the genetic characteristic?

4.13 An experiment generates a sample space containing eight simple events $E_1, \ldots, E_8$ with $P(E_i) = 1/8$, $i = 1, \ldots, 8$. The events A and B are defined as:

A: $E_1, E_4, E_6,$
B: $E_3, E_4, E_5, E_6, E_7.$

Find the following:
(a) $P(A)$.
(b) $P(\bar{A})$.
(c) $P(A \cup B)$.
(d) $P(AB)$.
(e) $P(A|B)$.
(f) Are events A and B mutually exclusive? Why?
(g) Are events A and B independent? Why?

4.14 Television commercials are designed to appeal to the most likely viewing audience of the sponsored program. However, S. Ward, in "Children's Reactions to Commercials,"* notes that children often possess a very low understanding of commercials, even for those designed to appeal especially to children. Ward's studies show that the percentages of children understanding TV commercials for different age groups are as given in the accompanying table.

	Age		
	5–7	8–10	11–12
Don't understand	55%	40%	15%
Understand	45%	60%	85%

* *Journal of Advertising Research,* April 1972.

An advertising agent has shown a television commercial to a 6-year-old and another to a 9-year-old child in a laboratory experiment to test their understanding of the commercials.

(a) What is the probability that the message of the commercial is understood by the 6-year-old child?

(b) What is the probability that both children demonstrate an understanding of the TV commercials?

(c) What is the probability that one or the other, or both, children demonstrate an understanding of the TV commercials?

4.15 A survey of consumers in a particular community showed that 10 percent were dissatisfied with plumbing jobs done in their homes. Fifty percent of the complaints dealt with plumber *A*. If plumber *A* does 40 percent of the plumbing jobs in the town, what is the probability that you will obtain:

(a) An unsatisfactory plumbing job, given that the plumber was *A*?

(b) A satisfactory plumbing job, given that the plumber was *A*?

4.16 A lie detector will show a positive reading (indicate a lie) 10 percent of the time when a person is telling the truth and 95 percent of the time when the person is lying. If two people are suspects in a one-man crime and if (for certain) one is guilty:

(a) What is the probability that the detector shows a positive reading for both suspects?

(b) What is the probability that the detector shows a positive reading for the guilty suspect and a negative reading for the innocent?

(c) What is the probability that the detector is completely wrong, that is, that it gives a positive reading for the innocent suspect and a negative reading for the guilty?

(d) What is the probability that it gives a position reading for either or both of the two suspects?

4.17 According to a recent study, the number of people susceptible to Chinese restaurant syndrome, a reaction linked to the food seasoner, monosodium glutamate, has been greatly exaggerated. A particular researcher reported that in a study of approximately 600 diners, only 3 to 5 percent reported the sensations commonly associated with the syndrome. Suppose we take 5 percent to be in fact correct. If three unrelated people dine in a particular Chinese restaurant:

(a) What is the probability that all three will experience the syndrome?

(b) What is the probability that only one of the three will experience it?

(c) What is the probability that at least one of the three will experience it?

(d) And if all three actually experienced the syndrome, what would you infer? (*Note:* We are not asking for a statistical inference. We only want your reaction to this event and the reasoning behind your statements.)

4.18 A survey of people in a given region showed that 20 percent were smokers. The probability of death due to lung cancer, given that a person smoked, was roughly ten times the probability of death due to lung cancer, given that a person did not smoke. If the probability of death due to lung cancer in the region is .006, what is the probability of death due to lung cancer given that a person is a smoker?

4.19 Experience has shown that 50 percent of the time, a particular union–management contract negotiation had led to a contract settlement within a 2-week period, 60 percent of the time the union strike fund has been adequate to support a strike, and 30 percent of the time both conditions have been satisfied. What is the probability of a contract settlement given that you know that the union strike fund is adequate to support a strike? Is settlement of a contract within a 2-week period dependent on whether the union strike fund is adequate to support a strike?

4.20 Suppose it is known that at a particular company, the probability of remaining with the company 10 years or more is 1/6. A man and a woman start work at the company on the same day.

(a) What is the probability that the man will work there less than 10 years?
(b) What is the probability that both the man and woman will work there less than 10 years? Assume that they are unrelated and hence their lengths of service are independent of each other.
(c) What is the probability that one or the other, or both, will work longer than 10 years?

4.21 Suppose that the probability of exposure to the flu during an epidemic is .6. Experience has shown that a serum is 80 percent successful in preventing an inoculated person who is exposed to the flu from acquiring it. A person not inoculated faces a probability of .90 of acquiring the flu if exposed to it. Two persons, one inoculated and one not, are capable of performing a highly specialized task in a business. Assume that they are not at the same location, are not in contact with the same people, and cannot expose each other. What is the probability that at least one will get the flu?

4.22 Two people enter a room and their birthdays (ignoring years) are recorded.
(a) Identify the nature of the sample points in S.
(b) How many sample points are in S?
(c) Identify the sample points in

A: both persons have the same birthday.

(d) Find $P(A)$.
(e) Find $P(\bar{A})$.

4.23 If n people enter a room, find the probability that

A: none of the persons have the same birthday,
B: at least two of the persons have the same birthday.

Solve for:
(a) $n = 3$.
(b) $n = 4$.
[*Note:* Surprisingly, $P(B)$ increases rapidly as n increases. For example, for $n = 20$, $P(B) = .411$; for $n = 40$, $P(B) = .891$.]

4.7

RANDOM VARIABLES

Observations generated by an experiment fall into one of two categories, quantitative or qualitative. For example, the daily production in a manufacturing plant would be a quantitative or numerically measurable observation, while weather descriptions, such as rainy, cloudy, sunny, would be qualitative. Statisticians are concerned with both quantitative and qualitative data, although the former are perhaps more common.

In some instances it is possible to convert qualitative data to quantitative by assigning a numerical value to each category to form a scale. Industrial production is often scaled according to first grade, second grade, etc. Cigarette tobaccos are graded and foods ranked according to preference by persons employed as taste testers.

In this text we shall be concerned primarily with quantitative observations such as the grade-point data discussed in Example 3.1. The events of interest associated with the experiment are the values that the data may take and are, therefore, numerical events. Suppose that the variable measured in the experiment is denoted by the symbol y. Recalling that an event is a collection of sample points, it would seem that certain sample points would be associated with one numerical event, say $y = 2$, another set associated with $y = 3$, etc., covering all possible values of y. Such is, in fact, the case. The variable, y, is called a random variable because the value of y observed for a particular experiment is a chance (or random) event associated with a probabilistic model, the sample space.

Note that each sample point implies one and only one value of y, although many sample points may imply the same value of y. For example, consider the experiment that consists of tossing two coins. Suppose that we are interested in y = number of heads. The random variable, y, can take three values, $y = 0$, $y = 1$, or $y = 2$. The sample points for the experiment are

E_1: HH,
E_2: HT,
E_3: TH,
E_4: TT.

The numerical event $y = 0$ includes sample point E_4; $y = 1$ includes sample points E_2 and E_3; and $y = 2$ includes sample point E_1. Note that one and only one value of y corresponds to each sample point, but the converse is not true. Two sample points, E_2 and E_3, are associated with the same value of y. Two conclusions may be drawn from this example. We note that the relationship between the random variable, y, and the sample points satisfies the definition of a functional relation presented in Chapter 2. Choose any point in the sample space, S, and there corresponds one and only one value of y. Hence, we choose the following definition for a random variable.

DEFINITION

A random variable is a numerical valued function defined over a sample space.

We also note that the numerical events associated with the random variable, y, are mutually exclusive events. The outcome of a single experiment can result in one and only one value of y. At this point we have almost completed our probabilistic model, the theoretical frequency distribution for a population of numerical measurements discussed in Chapter 3.

It would seem appropriate to recapitulate. Recalling Chapter 3, populations are described by frequency distributions. Areas under the frequency distribution are associated with the fraction of measurements in the population falling in a particular interval, or they may be interpreted as probabilities. The purpose of Chapter 4 was to construct a theoretical model for the frequency distribution, or probability distribution as it is called in probability theory, for the population. Preceding sections of this chapter provide both the model and the machinery for achieving this result. To complete the picture, we need to calculate the probabilities associated with each value of y. These probabilities, presented in the form of a table or a formula, are called the probability distribution for the random variable, y. In reality, the probability distribution is the theoretical frequency distribution for the population. Random variables and probability distributions form the topic of Chapter 5.

4.8

BAYES' RULE (OPTIONAL)

We most frequently wish to find the conditional probability of an event A, given that an event B has occurred at a prior point in time. Thus, we might wish to know the probability of rain tomorrow given that it has rained during the preceding seven days. Or, we might wish to know the probability of drawing two aces from a deck of cards given that the deck contains only two aces. Hence, we assume that some state of nature exists, and we wish to calculate the probability of some event that will occur in the future.

Equally interesting is the probability that a particular state of nature exists given that a certain sample is observed. If the first two cards in a five-card poker hand yield a pair, what is the probability that the hand is a full house? A similar but more interesting problem occurs in X-raying people for tuberculosis. A positive X-ray result may or may not imply that the person tested has tuberculosis. Given that the X-ray result for a certain patient is positive, what is the probability that he has tuberculosis? The answer to this probabilistic question suggests an approach to statistical inference. If the probability is high, we will infer that the patient has tuberculosis. This section presents an application of the Multiplicative Law and a formula derived by the probabilist, Thomas Bayes.

Consider an experiment that involves the selection of a sample from one of k populations, call them $H_1, H_2, \ldots, H_k$. The sample is observed, but it is not known from which population the sample was selected. Suppose that the sample results in an event A. Then the problem is to determine the population from which the sample was selected. This inference will be based on the conditional probabilities, $P(H_i|A)$, $i = 1, 2, \ldots, k$.

To find the probability that the sample was selected from population i given that event A was observed, $P(H_i|A)$, $i = 1, 2, \ldots, k$, note that A could have been observed if the sample were selected from population 1, population 2, or any one of the k populations, $H_1, H_2, \ldots, H_k$. The probability that population i was selected *and* that event A occurred is the intersection of the events H_i and A, or (AH_i). These events, $(AH_1), (AH_2), \ldots, (AH_k)$, are mutually exclusive and hence

$$P(A) = P(AH_1) + P(AH_2) + \cdots + P(AH_k).$$

Then the probability that the sample came from population i is

$$P(H_i|A) = \frac{P(AH_i)}{P(A)} = \frac{P(H_i)P(A|H_i)}{\sum_{J=1}^{k} P(AH_J)} = \frac{P(H_i)P(A|H_i)}{\sum_{J=1}^{k} P(H_J)P(A|H_J)}.$$

This expression for $P(H_i|A)$ is known as Bayes' Rule for the probability of causes. As you can see, it follows easily from the definition of conditional probability.

Finding $P(H_i | A)$ requires knowledge of the probabilities, $P(H_i)$ and $P(A | H_i)$, $i = 1, 2, \ldots, k$. One can often determine the probability of an event if the population is known and hence can find $P(A|H_i)$, but $P(H_i)$, $i = 1, 2, \ldots, k$, are the probabilities that certain states of nature exist and are either unknown or difficult to ascertain. This set of probabilities $P(H_i)$ is called the prior probability distribution because it gives the distribution of the states of nature prior to conducting the experiment.

In the absence of knowledge concerning the values of $P(H_i)$, Bayes suggested that these probabilities should be taken as equal. That is, he assumed the populations over which his experiment was defined to be equiprobable and, therefore, assigned equal probabilities to the $P(H_i)$, $i = 1, 2, \ldots, k$. We would not agree that the assignment of equal prior probabilities to the populations $H_1, H_2, \ldots, H_k$ is logical, but the procedure for selecting one of the set $H_1, H_2, \ldots, H_k$, given the observation of event A, is certainly intriguing. In many instances the experimenter does have some notion—sometimes vague, sometimes exact—of the prior probabilities for $H_1, H_2, \ldots, H_k$. Then, inferring the population—that is, which of the set $H_1, H_2, \ldots, H_k$ is the true population—can be achieved by finding $P(H_i | A)$ and selecting H as the population that gives the highest probability, $P(H_i | A)$.

EXAMPLE 4.17

An individual is selected at random from a community in which 1 percent are afflicted with tuberculosis and is X-rayed to detect the presence of the disease. A positive X-ray indication of the disease can occur when a nontubercular person is tested. Let T denote the event that the person selected is tubercular and let E indicate a positive X-ray result. The probability of a positive X-ray result, given that the person selected is tubercular, is $P(E | T) = .90$. The corresponding probability for a nontubercular person is $P(E | \bar{T}) = .01$. What is the probability that a person will be tubercular given that his X-ray result is positive?

Solution The population H_1 and H_2 corresponds to the two sets of people, those who have tuberculosis and those who do not. (Thus H_1 is equivalent to event T, H_2 to $\bar{T}$.) The prior probabilities are $P(T) = .01$ and $P(\bar{T}) = .99$. The event E, a positive X-ray result, can occur either if the person does have tuberculosis or if he does not. These two mutually exclusive events are the intersections (ET) and $(E\bar{T})$. The probability of E is the probability of the union of (ET) and $(E\bar{T})$, or

$$\begin{aligned} P(E) &= P(ET) + P(E\bar{T}) \\ &= P(T)P(E | T) + P(\bar{T})P(E | \bar{T}) \\ &= (.01)(.90) + (.99)(.01) \\ &= .0189. \end{aligned}$$

Then the probability that a person has tuberculosis, given a positive X-ray result, is

$$P(T | E) = \frac{P(ET)}{P(E)} = \frac{P(T)P(E | T)}{P(E)} = \frac{(.01)(.90)}{.0189} = .476.$$

You can see how Bayes' Rule could be used to make an inference about the population from which the X-rayed person (Example 4.17) was selected. We have shown that the probability that a person has tuberculosis, given a positive X-ray result, is .476. Similarly, we could compute $P(\bar{T} \mid E)$, the probability that a person does not have tuberculosis, given a positive X-ray result, to be $1 - P(T \mid E)$, or .524. Now suppose that you randomly select a sample of one person and find that the person's X-ray is positive. What do you infer about the population from which the person was selected? Tubercular or not? Because $P(\bar{T} \mid E)$ is slightly larger than $P(T \mid E)$, we think that you would be inclined to believe that the person is not tubercular. Thus Bayes' Rule was used to make an inference about a population based on an observed sample. Naturally, in a practical situation, additional tests would be conducted to verify whether tuberculosis was or was not present.

Another way to view Bayes' Rule is to view it as a method of incorporating the information from sample observations to adjust the probability of some event. For example, if we had no information on the result of a person's X-ray examination, we would regard the probability that he or she is tubercular as .01 (because 1 percent of all people are afflicted with tuberculosis). However, if we are given the added information that the person's X-ray was positive, the probability that he or she is tubercular is

$$P(T \mid E) = .476.$$

Thus, based on the X-ray information, we have adjusted the probability that the person is tubercular from a small probability, .01, to a rather high probability, .476.

EXERCISES

4.24 As items come to the end of a production line, an inspector chooses which items are to go through a complete inspection. Ten percent of all items produced are defective. Sixty percent of all defective items go through a complete inspection, and 20 percent of all good items go through a complete inspection. Given that an item is completely inspected, what is the probability it is defective?

4.25 Medical case histories indicate that many different illnesses may produce identical symptoms. Suppose that a particular set of symptoms, which we will denote as event H, occur only when any one of three illnesses, A, B, or C, occur (for the sake of simplicity, we will assume that illnesses A, B, and C are mutually exclusive). Studies show that the probabilities of getting the three illnesses are

$$P(A) = .01,$$
$$P(B) = .005,$$
$$P(C) = .02.$$

The probabilities of developing the symptoms, H, given a specific illness are

$$P(H \mid A) = .90,$$
$$P(H \mid B) = .95,$$
$$P(H \mid C) = .75.$$

Assuming that an ill person shows the symptoms H, what is the probability that the person has illness A?

4.26 A student answers a multiple-choice examination question that possesses four possible answers. Suppose that the probability that the student knows the answer to the question is .8 and the probability that the student will guess is .2. Assume that if the student guesses, the probability of selecting the correct answer is .25. If the student correctly answers a question, what is the probability that the student really knew the correct answer?

4.27 Suppose that 5 percent of all people filing the long income tax form seek deductions which they know to be illegal and an additional 2 percent will incorrectly list deductions because of lack of knowledge of the income tax regulations. Of the 5 percent guilty of cheating, 80 percent will deny knowledge of the error if confronted by an investigator. If a filer of the long form is confronted with an unwarranted deduction and he denies knowledge of the error, what is the probability that he is guilty?

4.28 A particular football team is known to run 30 percent of its plays to the left, 70 percent to the right. A linebacker on an opposing team notes that the right guard shifts his stance most of the time (80 percent) when plays go to the right and that he uses a balanced stance the remainder of the time. When plays go to the left, the guard takes a balanced stance 90 percent of the time and the shift stance the remaining 10 percent of the time. On a particular play, the linebacker notes that the guard takes a balanced stance. What is the probability that the play will go to the left?

4.9

RESULTS USEFUL IN COUNTING SAMPLE POINTS (OPTIONAL)

The preceding sections cover the basic concepts of probability and provide a background that will help you to understand the role that probability plays in making inferences. To introduce the concept of a sample space, we used only examples and exercises for which the total number of sample points in the sample space was small. This allowed you to list the sample points in S, to identify the sample points in the event of interest, and then to calculate its probability. Although these examples and exercises were adequate for our learning objective, most real-life problems involve many more sample points. Consequently, we include this optional section for the student who wishes to improve his or her problem-solving ability (*Note:* Developing the ability to solve problems requires practice.)

We will present three theorems that fall in the realm of combinatorial mathematics and which can be of assistance in solving probability problems that involve a large number of sample points. For example, suppose that you are interested in the probability of an event A and you know that the sample points in S are equiprobable. Then

$$P(A) = \frac{n_A}{N},$$

where

n_A = number of sample points in A,
N = number of sample points in S.

Often we can use counting rules to find the values of n_A and N and thereby eliminate the necessity of listing the sample points in S.

The first theorem, which we call the *mn* rule, is as follows.

THEOREM 4.1

With m elements $a_1, a_2, a_3, \ldots, a_m$ and n elements $b_1, b_2, \ldots, b_n$ it is possible to form mn pairs that contain one element from each group.

Proof Verification of the theorem can be seen by observing the rectangular table in Figure 4.9. There will be one square in the table for each a_i, b_j combination—a total of mn squares.

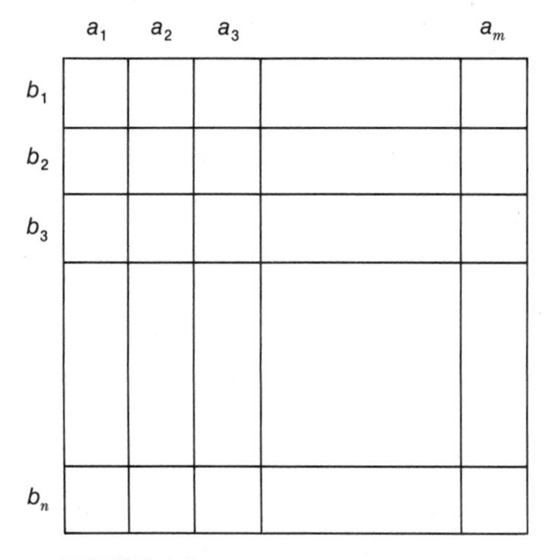

FIGURE 4.9

Table that indicates the number of pairs (a_i, b_j)

To illustrate, suppose that four companies have job openings in each of three areas, sales, manufacturing, and personnel. How many job opportunities are available to you? You can see that you have two sets of "things," companies (four) and types of jobs (three). Therefore, as shown in Figure 4.10, there are three jobs for each of the four companies, or (4) (3) = 12 possible pairings of companys and jobs. This example illustrates a use of the *mn* rule.

EXAMPLE 4.18

Two dice are tossed. How many sample points are associated with the experiment?

Solution The first die can fall in one of six ways; that is, $m = 6$. Likewise, the second die can fall in $n = 6$ ways. Since an outcome of this experiment

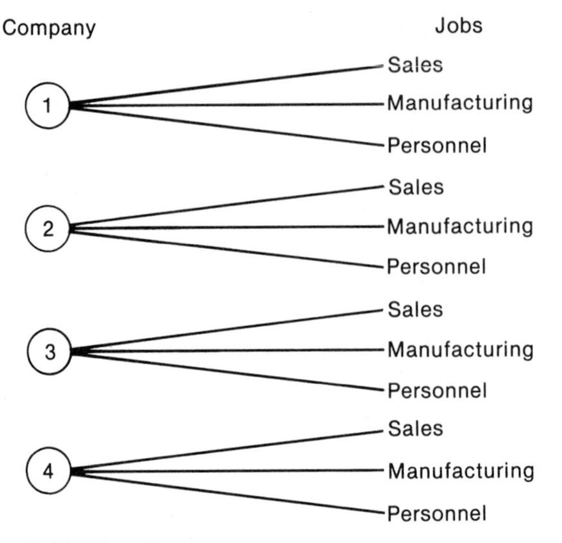

FIGURE 4.10
Company-job combinations

involves a pairing of the numbers showing on the faces of the two dice, the total number, N, of sample points is

$$N = mn = 6(6) = 36.$$

EXAMPLE 4.19

How many sample points are associated with the experiment, Example 4.6?

Solution A ball can be chosen from urn 1 in one of $m = 3$ ways. After one of these ways has been chosen, a ball may be drawn from urn 2 in $n = 2$ ways. The total number of sample points is

$$N = mn = 3(2) = 6.$$

As noted, the *mn* rule gives the number of pairs you can form in selecting one object from each of two groups. The rule can be extended to apply to triplets formed by selecting one object from each of three groups, quadruplets formed by selecting one object from each of four groups, etc. The application to triplets is shown in Figure 4.11. If you have m elements in the first group, n in the second, and t in the third, the total number of triplets that you can form, taking one object from each group, is equal to mnt, the number of branchings shown in Figure 4.11.

EXAMPLE 4.20

How many sample points are in the sample when three coins are tossed?

Solution Each coin can land in one of two ways. Hence

$$N = 2(2)(2) = 8.$$

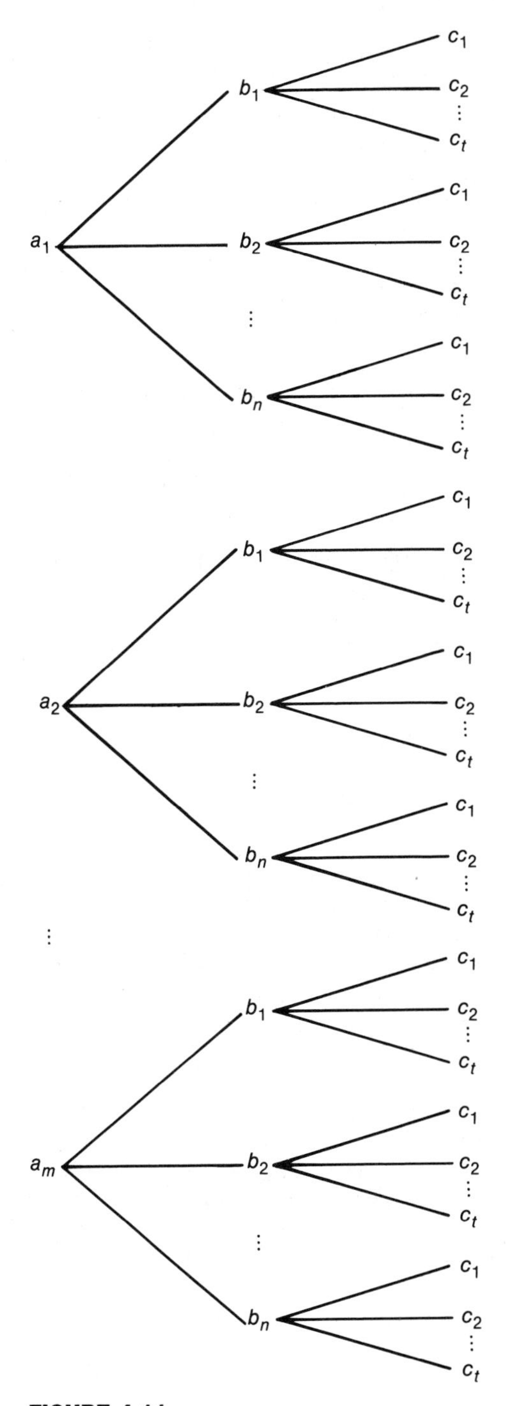

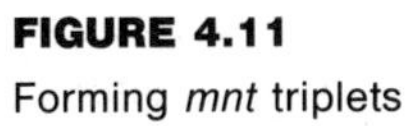

FIGURE 4.11
Forming *mnt* triplets

EXAMPLE 4.21

A truck driver can take three routes in going from City A to City B, four from City B to City C, and three from City C to City D. If in going from A to D the driver must proceed from A to B to C to D, how many possible A-to-D routes are available to the driver?

Solution Let

$$m = \text{number of routes from } A \text{ to } B = 3,$$
$$n = \text{number of routes from } B \text{ to } C = 4,$$
$$t = \text{number of routes from } C \text{ to } D = 3.$$

Then the total number of ways that you can construct a complete route, taking one subroute from each of the three groups (A to B), (B to C), and (C to D), is

$$mnt = (3)(4)(3) = 36.$$

A second useful mathematical result is associated with orderings or permutations. For instance, suppose that we have three books, b_1, b_2, and b_3. In how many ways can the books be arranged on a shelf, taking them two at a time? We enumerate, listing all combinations of two in the first column and a reordering of each in the second column:

Combinations of Two	Reordering of Combinations
b_1b_2	b_2b_1
b_1b_3	b_3b_1
b_2b_3	b_3b_2

The number of permutations is six, a result easily obtained from the mn rule. The first book can be chosen in $m = 3$ ways, and, once selected, the second book can be chosen in $n = 2$ ways. The result is $mn = 6$.

In how many ways can three books be arranged on a shelf taking three at a time? Enumerating, we obtain

$$b_1b_2b_3 \qquad b_2b_1b_3 \qquad b_3b_1b_2$$
$$b_1b_3b_2 \qquad b_2b_3b_1 \qquad b_3b_2b_1$$

a total of six. This, again, could be obtained easily by the extension of the mn rule. The first book can be chosen and placed in $m = 3$ ways. After choosing the first, the second can be chosen in $n = 2$ ways, and finally the third in $t = 1$ way. Hence the total number of ways is

$$N = mnt = 3 \cdot 2 \cdot 1 = 6.$$

DEFINITION

An ordered arrangement of r distinct objects is called a permutation. The number of ways of ordering n distinct (different) objects taken r at a time will be designated by the symbol P^n_r.

THEOREM 4.2

$$P_r^n = n(n-1)(n-2)\cdots(n-r+1).$$

Proof We are concerned with the number of ways of filling r positions with n distinct objects. Applying the extension of the mn rule, the first object can be chosen in one of n ways. After choosing the first, the second can be chosen in $(n-1)$ ways, the third in $(n-2)$ ways, and the rth in $(n-r+1)$ ways. Hence the total number of ways is

$$P_r^n = n(n-1)(n-2)\cdots(n-r+1).$$

Expressed in terms of factorials,

$$P_r^n = \frac{n!}{(n-r)!}.$$

[You should recall that $n! = n(n-1)(n-2)\cdots 3\cdot 2\cdot 1$ and $0! = 1$. Thus $4! = 4\cdot 3\cdot 2\cdot 1 = 24$.]

EXAMPLE 4.22

Three lottery tickets are drawn from a total of 50. Assume that order is of importance. How many sample points are associated with the experiment?

Solution The total number of sample points is

$$P_3^{50} = \frac{50!}{47!} = 50(49)(48) = 117{,}600.$$

EXAMPLE 4.23

A piece of equipment is composed of five parts which may be assembled in any order. A test is to be conducted to determine the length of time necessary for each order of assembly. If each order is to be tested once, how many tests must be conducted?

Solution The total number of tests would equal

$$P_5^5 = \frac{5!}{0!} = 5(4)(3)(2)(1) = 120.$$

The enumeration of the permutations of books in the previous discussion was performed in a systematic manner, first writing the combinations of n books taken r at a time and then writing the rearrangements of each combination. In many situations, ordering is unimportant and we are interested solely in the number of possible combinations. For instance, suppose that an experiment involves the selection of 5 men, a committee, from a total of 20 candidates. Then the simple events associated with this experiment correspond to the different combinations of men selected from the group of 20. How many simple events (different combinations) are associated with this experiment? Since order in a single selection is unimportant, permutations are irrelevant. Thus we are interested in the number of combinations of $n = 20$ things taken $r = 5$ at a time.

DEFINITION

The number of combinations of *n* objects taken *r* at a time will be denoted by the symbol, C_r^n. [*Note:* Some authors prefer the symbol $\binom{n}{r}$.]

THEOREM 4.3

$$C_r^n = \frac{P_r^n}{r!} = \frac{n!}{r!(n-r)!}.$$

Proof The number of combinations of n objects taken r at a time is apparently related to the number of permutations since it was used in enumerating P_r^n. The relationship can be developed using the mn rule. (We shall use the symbols a and b instead of m and n since n is used in P_r^n and C_r^n.)

Let $a = C_r^n$ and let b equal the number of ways of rearranging each combination once chosen, or P_r^r. Then,

$$\begin{aligned} P_r^n &= ab \\ &= (C_r^n)(b). \end{aligned}$$

Note that $b = P_r^r = r!$ Therefore,

$$P_r^n = C_r^n(r!)$$

or

$$C_r^n = \frac{P_r^n}{r!} = \frac{n!}{r!(n-r)!}.$$

EXAMPLE 4.24

A radio tube may be purchased from five suppliers. In how many ways can three suppliers be chosen from the five?

Solution

$$C_3^5 = \frac{5!}{3!2!} = \frac{(5)(4)}{2} = 10.$$

The following example illustrates the use of the counting rules in the solution of a probability problem.

EXAMPLE 4.25

Five manufacturers, of varying but unknown quality, produce a certain type of electronic tube. If we were to select three manufacturers at random, what is the chance that the selection would contain exactly two of the best three?

Solution Without enumerating the sample points, we would likely agree that each point, that is, any combination of three, would be assigned equal probability. If N points are in S, then each point receives probability

$$P(E_i) = \frac{1}{N}.$$

Let n be the number of points in which two of the best three manufacturers are selected. Then the probability of including two of the best three manufacturers in a selection of three is

$$P = \frac{n}{N}.$$

Our problem is to use the counting rules to find n and N.

Since order within a selection is unimportant and is unrecorded, each selection is a combination and hence

$$N = C_3^5 = \frac{5!}{3!2!} = 10.$$

Determination of n is more difficult, but it can be obtained using the *mn* rule. Let a be the number of ways of selecting exactly two from the best three, or

$$C_2^3 = \frac{3!}{2!1!} = 3,$$

and let b be the number of ways of choosing the remaining manufacturer from the two poorest, or

$$C_1^2 = \frac{2!}{1!1!} = 2.$$

Then the total number of ways of choosing two of the best three in a selection of three is $n = ab = 6$.

Hence the probability, P, is equal to

$$P = \frac{6}{10}.$$

Many other counting rules are available in addition to the three presented in this section. If you are interested in this topic, you should consult one of the many texts on combinatorial mathematics.

Tips on Problem Solving: Many students have difficulty deciding which (if any) of the three counting rules to apply in a given problem. The following tips may help.

1. Look at the problem and note whether a simple event is formed by:
   (a) selecting elements from each of *two* (*or more*) sets (a situation which suggests the use of the *mn* rule),
   (b) or selecting r elements from a *single* set of n elements (a situation which suggests the use of either combinations or permutations).
2. If the situation is 1(b), you must decide whether you should use combi-

nations or permutations. If every different ordering of the elements in the group of r leads to a different simple event, then use permutations. If ordering does not produce a new simple event, then use combinations. So, your diagnostic thought process should perform the checks indicated in the decision tree shown in the accompanying diagram.

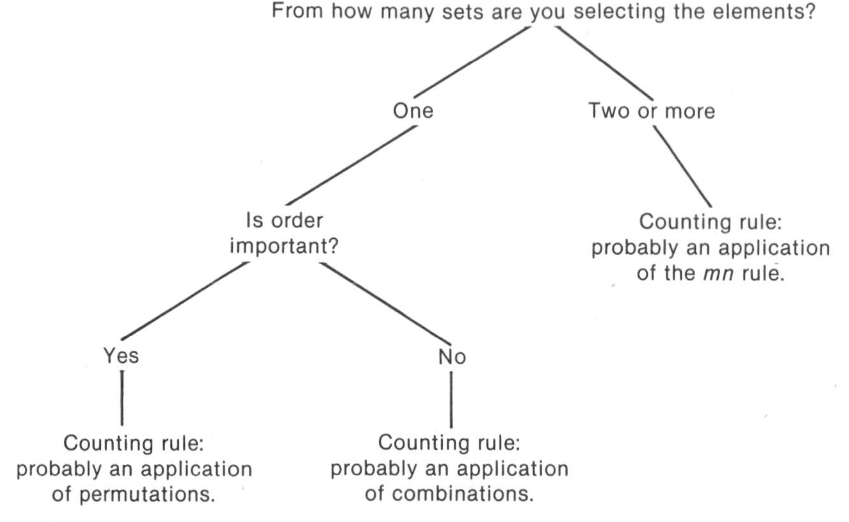

3. If you have difficulty visualizing the appropriate counting rule (or rules) to use for a problem involving a large number of sample points, construct a miniature version of the problem so that you can manually count them. This may help you to see how the more complex version can be solved.

EXERCISES

4.29 A salesperson in New York is preparing an itinerary to visit six major cities. The distance traveled, and hence the cost of the trip, will depend upon which city is visited first, second, . . . , sixth. How many different itineraries (and hence trip costs) are possible?

4.30 A company wishes to fill three vice-presidential positions, sales, manufacturing, and finance, from among 24 lower-echelon company managers. How many different options do they have for filling the positions?

4.31 An experiment consists in assigning 10 workmen to 10 different jobs. In how many different ways can the 10 men be assigned to the 10 jobs?

4.32 Refer to Exercise 4.31. Suppose there are only four different jobs available for the 10 men. In how many different ways can four men be selected from the 10 and assigned to the four jobs?

4.33 A study is to be conducted to determine the attitudes of nurses in a hospital to various administrative procedures that are currently employed. If a sample of 10 nurses is to be selected from a total of 90, how many different samples could be selected? (Note that order within a sample is unimportant.)

4.34 Buyers of television sets are offered a choice of one of three different styles. How many different outcomes could result if one customer makes a selection? two customers? ten?

4.35 If five cards are to be selected, one after the other, in sequence from a 52-card deck, each card being replaced in the deck before the next draw, how many different selections are possible?

4.36 Refer to Exercise 4.35. Suppose that the five cards are drawn from the 52-card deck, simultaneously and without replacement. How many different hands could be selected?

4.37 The following actual case occurred in the city of Gainesville, Florida, in 1976. The eight-member Human Relations Advisory Board considered the complaint of a woman who claimed discrimination, based upon her sex, on the part of a local surveying company. The Board, composed of five women and three men, voted 5–3 in favor of the plaintiff, the five women voting in favor of the plaintiff, the three men against. The attorney representing the company appealed the Board's decision by claiming sex bias on the part of the Board members. If the vote in favor of the plaintiff was 5–3 and the Board members were not sex biased, what is the probability that the vote would split along sex lines (five women for, three men against)?

4.38 Ten thousand tickets, each worth $5.00, are issued for a lottery.
(a) If you purchase two tickets, how many different pairs of tickets are available?
(b) Suppose that the lottery operators choose four tickets from the 10,000 and that each is judged a winner. What is the probability that both of your tickets will be winners?

4.39 For each of twenty questions on a multiple-choice test, a student can choose one of five possible answers.
(a) How many completely different sets of answers are possible for the test?
(b) If a person guessed on all of the questions, what is the probability that all of the questions would be answered correctly?

4.40 In a psychological learning experiment, a mouse is given the option of choosing one of five paths, two of which are expected to be more attractive to the mouse than the others.
(a) If two mice are chosen for the experiment, how many different simple events are associated with the experiment?
(b) If no learning has occurred and it is equally likely that a mouse would choose any one of the paths, what is the probability that both mice would choose one of the two "attractive" paths?

4.41 Refer to Exercise 4.40. Suppose that 10 mice were employed in the experiment, that no learning had occurred, and hence that it is equally likely that a mouse would choose any one of the paths.
(a) What is the probability that all 10 mice would choose one of the "attractive" paths?
(b) Noting the result of part (a), do you think learning occurred or do you think the observed outcome occurred purely due to chance?

4.42 A lineup of 10 men is conducted to test the ability of a witness to identify three burglary suspects. Suppose that the three burglary suspects who committed the crime are in the lineup. If the witness is actually unable to identify the suspects but feels compelled to make a choice, what is the probability that the three guilty men are selected by chance? What is the probability that the witness selects three innocent men?

4.43 The sizes of wildlife populations are often estimated using the capture–recapture method. For example, suppose that a forest contains an unknown number, N, of deer. A sample of K is caught, marked, and released. Then the capture process is

repeated and the number of marked animals in the sample is observed. The reasoning is that the proportion of marked animals in the second sample must be related to N, the size of the deer population. To illustrate how N is related to the recapture probabilities (we shall use small numbers to simplify our calculations), suppose that 4 deer were originally captured, marked, and released. After a short period of time, five deer were again captured and it was found that one was marked. Calculate the probability of recapturing exactly one marked deer if the size of the deer population was equal to:
(a) $N = 8$.
(b) $N = 10$.
(c) $N = 15$.
(d) $N = 20$.
(e) $N = 25$.
Graph the recapture probabilities as a function of N.

4.44 To understand how a single elimination tennis tournament works, we will illustrate the matchings for $n = 8$ players. The players are assigned to eight positions. Player No. 1 plays No. 2, No. 3 plays No. 4, . . . , and player No. 7 plays No. 8. Then the winner of the No. 1-No. 2 pair plays the winner of No. 3-No. 4 and the winner of the No. 5-No. 6 pair plays the winner of No. 7-No. 8 in the semifinal round. The winners of these two matches meet in the final match to decide the winner. Jones and Smith, the two best players, can definitely defeat any of the other players. If the eight players are randomly assigned to the eight starting positions, what is the probability that Jones and Smith meet in the finals? (Does it seem as though the answer should be 1/2?) (*Hint:* You can obtain insight into this problem by solving the problem for a single elimination tournament containing $n = 4$ players. This will enable you to list the simple events.)

4.10 SUMMARY

The theories of both probability and statistics are concerned with samples drawn from populations. Probability assumes the population known and calculates the probability of observing a particular sample. Statistics assumes the sample to be known and, with the aid of probability, attempts to describe the frequency distribution of the population that is unknown. Chapter 4 is directed toward the construction of a model, the sample space, for the frequency distribution of a population of observations. The theoretical frequencies, representing probabilities of events, can be obtained by one of two methods:

1. The summation of the probabilities of the sample points in the event of interest.
2. The joint use of event composition (compound events) and the laws of probability.

REFERENCES

Cramer, H., *The Elements of Probability Theory and Some of Its Applications*. New York: John Wiley & Sons, Inc., 1955; Stockholm: Almqvist and Wiksell, 1954.

Feller, W., *An Introduction to Probability Theory and Its Applications,* Vol. 1, 3rd ed. New York: John Wiley & Sons, Inc., 1968.

Meyer, P. L., *Introductory Probability and Statistical Applications,* 2nd ed. Reading, Mass.: Addison-Wesley Publishing Company, Inc., 1970.

Parzen, E., *Modern Probability Theory and Its Applications.* New York: John Wiley & Sons, Inc., 1960.

Riordan, J., *An Introduction to Combinatorial Analysis.* New York: John Wiley & Sons, Inc., 1958.

Scheaffer, R. L., and W. Mendenhall, *Introduction to Probability: Theory and Applications.* North Scituate, Mass.: Duxbury Press, 1975.

SUPPLEMENTARY EXERCISES

[Starred (*) exercises are optional]

4.45 Two cold tablets are accidentally placed in a box containing two aspirin tablets. The four tablets are identical in appearance. One tablet is selected at random from the box and is swallowed by patient A. A tablet is then selected at random from the three remaining tablets and is swallowed by patient B. Define the following events as specific collections of sample points:
(a) The sample space S.
(b) The event A that patient A obtained a cold tablet.
(c) The event B that exactly one of the two patients obtained a cold tablet.
(d) The event C that neither patient obtained a cold tablet.

4.46 Two dice are tossed. Let A be the event that the first die shows an odd number and let B be the event that the second die shows a number greater than 2. Find $P(A)$ and $P(B)$. (Solve by listing the sample points.) Find the probability that both A and B will occur. Find the probability that either A or B or both occur.

4.47 A coin is tossed four times and the outcome is recorded for each toss.
(a) List the sample points for the experiment.
(b) Let A be the event that the experiment yields exactly three heads. List the sample points in A.
(c) Make a reasonable assignment of probabilities to the sample points and find $P(A)$.

4.48 A retailer sells two styles of high-priced high-fidelity consoles that experience indicates are in equal demand. (Fifty percent of all potential customers prefer style 1, 50 percent favor style 2.) If he stocks four of each, what is the probability that the first four customers seeking a console all purchase the same style?
(a) Define the experiment.
(b) List the sample points.
(c) Define the event of interest, A, as a specific collection of sample points.
(d) Assign probabilities to the sample points and find $P(A)$.
(Note that this example is illustrative of a very important problem associated with product inventory.)

4.49 A boxcar contains seven complex electronic systems. Unknown to the purchaser, three are defective. Two are selected out of the seven for thorough testing and then classified as defective or nondefective.
(a) List the sample points for this experiment.
(b) Let A be the event that the selection includes no defectives. List the sample points in A.
(c) Assign probabilities to the sample points and find $P(A)$.

4.50 A die is tossed two times; what is the probability that the sum of the numbers observed will be greater than 9?

4.51 If three coins are tossed, what is the probability of getting exactly two heads? at least two heads?

4.52 Suppose that two of the six sparkplugs on a six-cylinder automobile engine require replacement. If the mechanic removes two plugs at random, what is the probability that he will select the two defective plugs? At least one of the two defective plugs? Solve by listing the sample points.

4.53 A pair of dice is thrown. What is the probability that the sum equals 2? Equals 7?

4.54 Two seeds are chosen from a packet containing seven seeds, two that will produce blue flowers, three that will produce white, and two that will produce red. What is the probability that both seeds produce flowers of the same color? (Solve by listing the sample points.)

4.55 Toss a die and a coin. If event A is the occurrence of a head and an even number and event B is the occurrence of a head and a 1, find $P(A)$, $P(B)$, $P(AB)$, and $P(A \cup B)$. (Solve by listing the sample points.)

4.56 Refer to Exercise 4.46. Let C be the event that the sum of the numbers appearing on the dice is even. Find $P(C)$, $P(AC)$, $P(BC)$, $P(A \cup C)$, $P(B \cup C)$, $P(ABC)$, and $P(A \cup B \cup C)$.

4.57 A man takes either a bus or the subway to work with probabilities .3 and .7, respectively. When he does take the bus, he is late 30 percent of the days. When he takes the subway, he is late 20 percent of the days. If the man is late for work on a particular day, what is the probability that he took the bus?

4.58 A supermarket is having a sale of unlabeled cans. Included in the sale are 200 cans of corn, 300 cans of beets, and 500 cans of peaches. What is the probability that the first housewife will get a can of vegetables? of corn? Are these two events independent? mutually exclusive? Given that she draws a can of vegetables, what is the probability that it will be corn?

4.59 Refer to Exercise 4.45. Find each of the following by summing probabilities of sample points: $P(A)$, $P(B)$, $P(AB)$, $P(A \cup B)$, $P(C)$, $P(AC)$, and $P(A \cup C)$.

4.60 Refer to Exercise 4.11. Suppose that the latest batch does actually possess the desired taste so that the three specimens for each taster are identical. Use the Multiplicative Law of Probability to find the probabilities attached to the sample points. Find each of the following by summing probabilities of sample points: $P(A)$, $P(B)$, $P(AB)$, $P(A \cup B)$, $P(C)$, $P(AC)$, and $P(A \cup C)$.

4.61 Suppose that independent events A and B have nonzero probabilities. Show that A and B cannot be mutually exclusive.

4.62 Refer to Exercise 4.46 and find $P(A|B)$ and $P(B|A)$. Then calculate $P(AB)$ and $P(A \cup B)$; use the laws of probability.

4.63 Refer to Exercise 4.55 and calculate $P(AB)$ and $P(A \cup B)$; use the laws of probability.

4.64 A salesman figures that the probability of his consummating a sale during the first contact with a client is .4 but improves to .55 on the second contact if the client did not buy during the first contact. Suppose that the salesman will make one and only one callback to any client. If the salesman contacts a client, calculate:
(a) The probability that the client will buy.
(b) The probability that the client will not buy.

*4.65 Two dice are tossed. Use the mn rule to count the total number of sample points in the sample space, S.

*4.66 Three coins are tossed. Use the mn rule to count the total number of sample points in S.

*4.67 Refer to the experiment in Exercise 4.55, and use the mn rule to count the number of points in S.

* 4.68 How many different telephone numbers of five digits can be formed if the first digit must be a 3 or a 4?

* 4.69 Prove that $C_r^n = C_{n-r}^n$.

* 4.70 Use combinatorial methods to count the number of sample points in the sample space defined for Exercise 4.54. Solve Exercise 4.54 by using combinatorial methods.

* 4.71 A piece of equipment can be assembled in three operations, which may be arranged in any sequence.
  (a) Give the total number of ways that the equipment can be assembled.
  (b) Comparative tests are to be conducted to determine the best assembly procedure. If each assembly procedure is to be tested and compared with every other procedure exactly once, how many tests must be conducted?

* 4.72 Use combinatorial methods to count the number of sample points in the sample space for the experiment described in Exercise 4.52. Solve Exercise 4.52 by using combinatorial methods.

* 4.73 A chemist wishes to observe the effect of temperature, pressure, and the amount of catalyst on the yield of a particular chemical in a chemical reaction. If the experimenter chooses to use two levels of temperature, three of pressure, and two of catalyst, how many experiments must be conducted in order to run each temperature–pressure–catalyst combination exactly once?

* 4.74 How many four-digit numbers can be formed from the digits 5, 6, 7, 8, 9 if each digit can be used only once? If the digits may be repeated?

* 4.75 How many ways can the letters in the word "charm" be arranged? in the word "church"?

* 4.76 If there are 18 players on a baseball team, how many different 9-man teams can be organized if each player can play any position?

* 4.77 An airline has six flights from New York to California and seven flights from California to Hawaii per day. How many different flight arrangements can the airline offer from New York to Hawaii?

* 4.78 A man is in the process of buying a new car. He has a choice of 3 engine makes, 7 body styles, and 14 colors. How many different cars does he have to choose from?

* 4.79 Solve Example 4.14 using the sample point approach.

* 4.80 A personnel director for a corporation has hired 10 new engineers. If three (distinctly different) positions are open at a Cleveland plant, in how many ways can he fill the positions?

* 4.81 Five cards are drawn from an ordinary 52-card bridge deck. Given two events A and B as follows:

  A: all five cards are spades,
  B: the five cards include an ace, king, queen, jack, and ten, all of the *same* suit.

  (a) Define the event AB.
  (b) Define the event $A \cup B$.
  (c) Give the probability of the event A.
  (d) Give the probability of the event B.
  (e) Give the probability of the event AB.
  (f) Give the probability of the event $(A \cup B)$.
  [Solve parts (c) to (e) by summing probabilities of the sample points.]

4.82 A certain article is visually inspected successively by two different inspectors. When a defective article comes through, the probability that it gets by the first inspector is .1. Of those that do get past the first inspector, the second inspector will "miss" 5 of 10. What fraction of the defectives will get by both inspectors?

* 4.83 Eight colleges join to form a basketball league. If each team is required to play every other team twice during the season, answer the following questions:
(a) What is the total number of league games that will be played?
(b) How many games will be played in which the contestants are two of the best three teams?
(c) For any given game, what is the probability that two of the best three teams will be playing?

4.84 Two ambulances are kept in readiness for emergencies. Owing to the demand on their time as well as the chance of mechanical failure, the probability that a specific ambulance will be available when needed is 9/10. The availability of one ambulance is independent of the other.
(a) In the event of a catastrophe, what is the probability that both ambulances will be available?
(b) What is the probability that neither will be available?
(c) If an ambulance is needed in an emergency, what is the probability that it will be available?

4.85 The failure rate for a guided missile control system is 1 in 1,000. Suppose that a duplicate but completely independent control system is installed in each missile so that if the first fails, the second can still take over. The reliability of a missile is the probability that it does not fail. What is the reliability of the modified missile?

4.86 Consider the following fictitious problem. Suppose that it is known that at a particular supermarket, the probability of waiting 5 minutes or longer for checkout at the cashier's counter is .2. On a given day, a man and his wife decide to shop individually at the market, each checking out separately at different cashier counters. If they both reach cashier counters at the same time, answer the following questions:
(a) What is the probability that the man will wait less than 5 minutes for checkout?
(b) What is the probability that both the man and his wife will be checked out in less than 5 minutes? Assume that the checkout times for the two are independent events.
(c) What is the probability that one or the other, or both, will wait 5 minutes or more?

4.87 A basketball player hits on 60 percent of his shots from the floor. What is the probability that he makes exactly two of his next three shots?

4.88 A quality-control plan calls for accepting a large lot of crankshaft bearings if a sample of seven is drawn and none is defective. What is the probability of accepting the lot if none in the lot is defective? If 1/10 are defective? If 1/2 are defective?

4.89 How many ways may a student choose two physics courses from six being given at different times?

4.90 It is said that only 40 percent of all people in a community favor the development of a mass-transit system. If four citizens are selected at random from the community, what is the probability that all four favor the mass-transit system? That none favor the mass-transit system?

4.91 The weatherman forecasts rain with probability .6 today and .4 tomorrow. Experience has shown that in this particular locale, it rains one day in four and the probability of rain on two successive days is .15. If it is raining outside when we hear his forecast (as so often happens), what is the probability of rain tomorrow?

4.92 A rat in a maze may go left, right, or straight ahead as he comes to each of four intersections in the maze. What is the probability that the rat correctly threads the maze on his first try if there is only one correct path?

* 4.93 A monkey is given 12 blocks—3 shaped like squares, 3 like rectangles, 3 like triangles, and 3 like circles. If he draws 3 of each kind in order—say, 3 triangles, then 3 squares, etc.—would you suspect that the monkey associates identically shaped figures? Calculate the probability of this event.

4.94 A research physician compared the effectiveness of two blood-pressure drugs, A and B, by administering the two drugs to each of four pairs of identical twins. Drug A was given to one member of a pair, drug B to the other. If, in fact, there is no difference in the effect of the drugs, what is the probability that the drop in the blood-pressure reading for drug A would exceed the corresponding reading for drug B for all four pairs of twins? Suppose that drug B created a greater drop in blood pressure than drug A for each of the four pairs of twins. Do you think this provides sufficient evidence to indicate that drug B is more effective in lowering blood pressure than drug A?

*4.95 Refer to Exercise 4.81. Find $P(B|A)$. Then calculate $P(AB)$ and $P(A \cup B)$ by the use of the laws of probability.

*4.96 An electronic fuse is produced by five production lines in a manufacturing operation. The fuses are costly, are quite reliable, and are shipped to suppliers in 100-unit lots. Because testing is destructive, most buyers of the fuses test only a small number before deciding to accept or reject lots of incoming fuses. All five production lines produce fuses at the same rate and normally produce only 2 percent defective fuses which are randomly dispersed in the output. Unfortunately, production line 1 suffered mechanical difficulty and produced 5 percent defectives during the month of March. This situation became known to the manufacturer after the fuses had been shipped. A customer received a lot produced in March and tested three fuses. One failed. What is the probability that the lot came from one of the four other lines?

4.97 To reduce the cost of detecting a disease, blood tests are conducted on a pooled sample of blood collected from a group of n people. If no indication of the disease is present in the pooled blood sample (as is usually the case), none have the disease. If analysis of the pooled blood sample indicates that the disease is present, each individual must submit to a blood test. The individual tests are conducted in sequence. If among a group of five people, one person possesses the disease, what is the probability that it requires six blood tests (including the pooled test) to detect the single diseased person? If two people possess the disease, what is the probability that it requires six tests to locate both diseased people?

4.98 Two light bulbs are selected at random from a pack containing 3 good bulbs and 2 defective ones. Let y be the number of defective bulbs selected. Find the probability that $y = 1$.

*4.99 A student prepares for an exam by studying a list of 10 problems. He can solve a certain 6 of these. For the exam, the instructor selects 5 questions at random from the list of 10. What is the probability that the student can solve all 5 problems on the exam?

*4.100 Five homes and three stores have applied for telephones and only one three-party line is available. Parties are assigned to the available three-party line at random. Let the events A and B be defined as follows:

A: the line contains 2 homes and 1 store,
B: the line contains at least 1 home.

(a) Find $P(A)$.
(b) Find $P(B)$.

*4.101 A housewife is asked to rank four brands (A, B, C, D) of common household cleaner according to her preference, No. 1 being the one she prefers best, etc. She really has *no* preference among the four brands. Hence, any ordering is equally likely to occur.
(a) What is the probability that brand A is ranked first?
(b) What is the probability that C is first and D is second in the rankings?
(c) What is the probability that A is ranked *either first or second*?

4.102 A manufacturer is considering the purchase of transistors to be used in the production of an electronic system. The transistors can be purchased from any of four suppliers and it is assumed that these four products vary in quality. The manufacturer wishes to select two of the four in such a way that he is fairly certain of including at least one of the two best suppliers. Although he plans an experimental program to assist in making the choice, you are asked the following: If the choice were based on no information and the two suppliers were randomly chosen from the four, what is the probability that the selection would include at least one of the two best?

* 4.103 An employer plans to interview 10 men for possible employment. Two people are to be hired. Five of the men are affiliated with fraternities and five are not.
(a) In how many ways could the employer select two men, disregarding fraternal affiliations?
(b) In how many ways could he select two fraternity men?
(c) Assuming that the employer has no preference regarding fraternal affiliations and that the men have equal qualifications, what is the probability that two fraternity men will be hired?

4.104 An accident victim will die unless he or she receives in the next 10 minutes an amount of type A Rh-positive blood which can be supplied by a single donor. It requires 2 minutes to "type" a prospective donor's blood and 2 minutes to complete the transfer of blood. A large number of untyped donors are available and 40 percent of them have type A Rh-positive blood. What is the probability that the accident victim will be saved if there is only one blood-typing kit available?

4.105 Each of two packages of six flashlight batteries contains exactly two inoperable batteries. If two batteries are selected from each package, what is the probability that all four batteries will function?

4.106 Refer to Exercise 4.105. Suppose that two batteries were randomly selected from package 1 and mixed with those in package 2. Then, two were randomly drawn from the eight in package 2. What is the probability that both will function?

* 4.107 An assembler of electric fans uses motors from two sources. Company A supplies 90 percent of the motors and company B supplies the other 10 percent of the motors. Suppose it is known that 5 percent of the motors supplied by company A are defective while 3 percent of the motors supplied by company B are defective. An assembled fan is found to have a defective motor. What is the probability that this motor was supplied by company B?

4.108 How many times should a coin be tossed in order that the probability of observing at least one head be equal to or greater than .9?

4.109 An oil prospector will drill a succession of holes in a given area to find a productive well. The probability that he is successful on a given trial is .2.
(a) What is the probability that the third hole drilled is the first that locates a productive well?
(b) If his total resources allow the drilling of no more than three holes, what is the probability that he locates a productive well?

4.110 Suppose that two defective refrigerators have been included in a shipment of six refrigerators. The buyer begins to test the six refrigerators one at a time.
(a) What is the probability that the last defective refrigerator is found on the fourth test?
(b) What is the probability that no more than four refrigerators need be tested before both of the defective refrigerators are located?
(c) Given that one of the two defective refrigerators has been located in the first two tests, what is the probability that the remaining refrigerator is found in the third or fourth test?

4.111 Two men each toss a coin and obtain a "match." That is, both coins are either heads or tails. If the process is repeated three times, answer the following questions:
(a) What is the probability of three matches?

(b) What is the probability that all six tosses (three for each man) result in tails?
(c) Coin tossing provides a model for many practical experiments. Suppose that the coin tosses represented the answers given by two students for three specific true–false questions on an examination. If the two students gave three matches for answers, would the low probability found in (a) suggest collusion?

* 4.112 Seventy percent of all cattle are treated by an injected vaccine to combat a serious disease. The probability of recovery from the disease is 1 in 20 if untreated and 1 in 5 if treated. If an infected cow has recovered, what is the probability that the cow received the preventive vaccine?

* 4.113 In a sixteen-player single elimination tennis tournament, the players are assigned to sixteen positions. The player in the first position (call this player No. 1), plays player No. 2, player No. 3 plays No. 4, . . . , and player No. 15 plays No. 16. In the second round of the tournament, the winner of the first pair (No. 1 and No. 2) plays the winner of the second pair (No. 3 and No. 4), and so on. The second-round winners play the pair adjacent to them in the two semifinal matches and the winners of these matches play in the final match. If the sixteen players are randomly assigned to the sixteen positions and if the two best players will defeat any of the remaining fourteen, what is the probability that they will play each other in the final match? What is the probability for a 32-player single elimination match? (This exercise is an extension of Exercise 4.44.)

EXPERIENCES WITH REAL DATA

Although the theory of probability can be employed to find the probability of most events, it is often easier to acquire an approximate value for a probability by simulating the experiment either manually or on an electronic computer. That is, we construct a model of the population and then randomly sample over and over again. If the event of interest, say event A, occurs n times in a long series of N trials, we calculate the approximate probability of event A to be

$$P(A) \approx \frac{n}{N}$$

For example, suppose that 50 percent of all restaurants in a city are unsanitary as judged by city and state health regulations. If 10 restaurants are selected at random from this population, what is the probability that 7 or more will be judged unsanitary? Then the experiment consists of tossing 10 coins and event A is defined to be the observation of 7 or more unsanitary restaurants in the sample of 10.

Although we will find a way to solve this problem in Chapter 6 by use of the theory of probability, we can easily obtain an approximate answer by a coin-tossing simulation. Tossing a single coin once is analogous to selecting a single restaurant from the population described above. The observation of a head could correspond to an unsanitary restaurant, a tail to a sanitary one. Flipping the coin 10 times (or tossing 10 coins once) would be equivalent to sampling 10 restaurants from the population.

Select 10 coins, mix thoroughly, toss, and record the number of heads (unsanitary restaurants) in the sample. Record whether y is 7 or larger. Repeat this process $N = 100$ times and count the number of times y is 7 or larger. This will give you the number of times, n, your event of interest was observed. Then calculate an approximate value for the probability of observing 7 or more unsanitary restaurants in a sample of 10, as

$$P = P(7 \text{ or more}) \approx \frac{n}{N} = \frac{n}{100}$$

This approximation may be poor because $N = 100$ repetitions of the experiment is not large enough to acquire an accurate approximation to $P(A)$.

To obtain a more accurate approximation to $P(A)$, combine the data collected by all the members of your class and use these data to approximate $P(A)$. This value of P should be close to the exact value of $P(A)$, which is .172.

To see how the experimentally obtained approximations for P vary, collect the approximations from each member of your class and describe them using a relative frequency histogram. Notice how they cluster about the exact value, $P(A) = .172$. Although we only have one approximation based on the larger value of N for the combined data, you can see that this value will tend to fall closer to .172 than the approximations based on $N = 100$ repetitions of the experiment.

The simulation study described above could be conducted for populations where the proportion of "unsanitary restaurants" is something other than .5, say p. This would be equivalent to flipping an unbalanced coin where the probability of a head was equal to p. For example, you could use a die for a simulation study, where $p = 1/6, 1/3, 1/2, 2/3$, or $5/6$. And, an electronic computer can be programmed to "toss an unbalanced coin" using any value of p that you choose. If the simulation is conducted on a computer, a very large number of repetitions of the experiment can be conducted in a matter of minutes, thereby giving a very accurate approximation to $P(A)$.

CHAPTER OBJECTIVES

GENERAL OBJECTIVES In Chapter 4 we gave an example to illustrate how probability is used in making inferences about a population based on information contained in a sample, and then we presented the concepts of probability that help us find the probability of experimental outcomes. Because most samples are measurements on random variables, we need to be able to find the probabilities associated with such measurements. The objective of this chapter is to distinguish between two types of random variables and to explain how to find the probability that a random variable will assume specific values.

SPECIFIC OBJECTIVES

1. To identify the role that random variables play in making inferences. *Sections 5.1, 5.2*
2. To identify the two types of random variables—discrete and continuous—and to present probability models appropriate for each. *Sections 5.3, 5.4, 5.5*
3. To define the expected value of a random variable and to identify this quantity as the mean of its probability distribution. *Section 5.6*
4. To define and identify the variance of a random variable as an expectation and to explain how the expected value and variance of a random variable can be used to describe its probability distribution. *Section 5.6*
5. To note the relationship between random variables, samples, and statistical inference and to define what is meant by random sampling. We will use random sampling and introduce the topic of statistical inference as a by-product of Chapter 6. *Section 5.7*

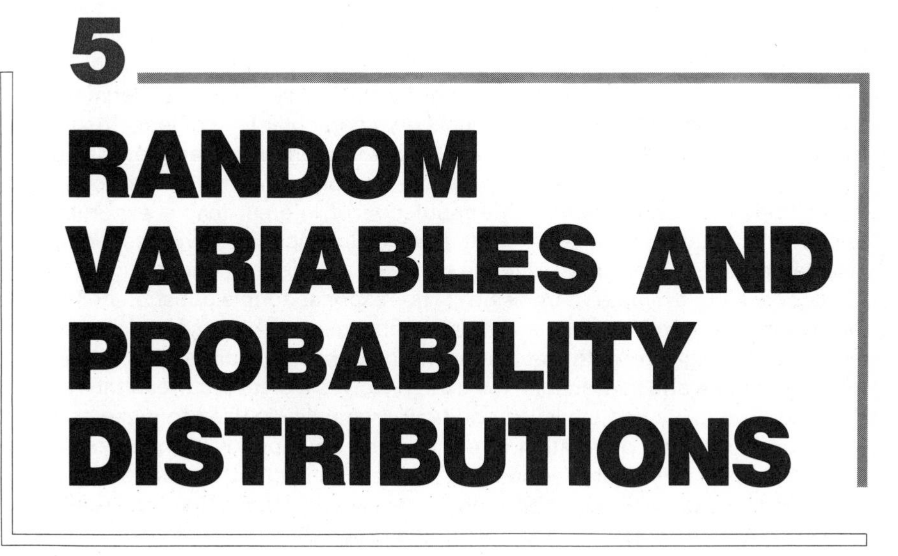

5 RANDOM VARIABLES AND PROBABILITY DISTRIBUTIONS

5.1

IN RETROSPECT

As we enter Chapter 5 you may experience a difficulty familiar to most beginning students. The broad picture of statistics, its objective and how it works, becomes lost in a maze of detail. Unfortunately, because of the nature of the subject, the sketch developed in introductory texts is completed only at the end of the book. The student possesses an ever-growing fragment of the picture as he or she progresses from chapter to chapter, but does not have the meaningful whole until the end. We would attempt to alleviate this problem in two ways. First, we suggest that you construct the complete picture for one or two examples for ready reference as we proceed through the text. Visualize an experiment—for instance, counting the number of insects on a particular type of leaf. Visualize a repetition of the experiment and hence the generation of the sample. Obviously, the sample has been drawn for a purpose—to obtain information concerning a larger body of measurements, the population of interest to the experimenter. Specify a population parameter of interest, for example, the average number of insects per leaf, μ, and then estimate or make a decision concerning its value. How good is the estimate or decision? The response to this question concludes the picture but unfortunately is not easily obtained at this stage. As a second example you might refer to the die-throwing problem described in Section 4.1. Although simple and impractical, the example gives a clear picture of the relation between probability and statistics and how probability is used to make an inference concerning the population of die throws. Equipped with these examples, you will more readily relate various subject areas to each other and to the whole.

Second, as we proceed through the text we will attempt to give a reason for the study of each subject area and to refocus your attention on an overall picture of statistics, much as we are doing in this section. Some may view such repetition as wasteful, but we think it not only desirable but, for most readers, essential.

5.2

RANDOM VARIABLES: HOW THEY RELATE TO STATISTICAL INFERENCE

Recall that an experiment very often yields a numerical measurement that varies from sample point to sample point. For this reason, we defined a random variable (Section 4.10) as a numerical valued function defined over a sample space. In other words, a variable y is a random variable if the value that y assumes is a chance or random numerical event. The daily closing price of an industrial stock is a numerical event. Observing the number of defects on a piece of new furniture or recording the grade-point average of a particular student are other examples of experiments that yield random numerical events. The population associated with the experiments results when the experiment is repeated a number of times and a relatively large body of data is obtained. As previously noted, we never actually measure each member of the population, but we can certainly conceive of doing so. In lieu of this, we wish to obtain a small set of these measurements, called the sample, and use the information in the sample to infer the nature of the population.

We have stated that a measurement obtained from an experiment results in a specific value of the random variable of interest and represents a measurement drawn from a population. How can a single measurement or a larger sample of, say, n measurements be used to make inferences about the population of interest? With the die-tossing example, Section 4.1, firmly in mind, we would suggest that we calculate the probability of the observed numerical event—the sample—for a large set of possible populations and choose the one that gives the highest probability of observing the sample. We would like to think that the method of inference described above appears reasonable and intuitively appealing to you. Note that it is not claimed to be the best, however "best" might be defined. We only suggest that it is intuitively reasonable. (In defense of the procedure, we might add that it is the basis for one of the more important methods for the statistical estimation of the parameters of a population and can be shown to provide "good" inferences in many situations.) For those who fail to grasp the argument, we defer further discussion of inference until Chapter 6. At this point it is sufficient to note that the procedure requires a knowledge of the probability associated with each value of the random variable. In other words, we require the probability distribution for the random variable, a distribution which represents the theoretical frequency histogram for the population of numerical measurements. The theory of probability presented in Chapter 4 provides the mechanism for calculating these probabilities for many random variables.

5.3
A CLASSIFICATION OF RANDOM VARIABLES

All the experiments described in Chapter 4 had one characteristic in common: their sample spaces contained a finite or at least countable number of sample points.* This enabled us to assign probabilities to the sample points so that the sum of their respective probabilities was equal to 1. Not all experiments possess this characteristic. Measuring the length of time for a sprinter to run the 100-yard dash is an experiment that can yield an infinitely large number of simple events (and sample points) that cannot be counted. Theoretically, with measuring equipment of perfect accuracy, we could associate each possible time measurement with a unique point contained in a line interval. Thus each of the infinitely large number of points on a line interval is a possible time for a sprinter and represents a sample point for the experiment. Since each time point represents a sample point for the experiment, the resulting sample space contains an infinitely large (and uncountable) number of sample points. Can we apportion nonzero

*Countable means that you can associate the values that the random variable can assume with the integers 1, 2, 3, 4, . . . (that is, you can count them).

probabilities to an infinite number of points and, at the same time, satisfy the requirement that the probabilities of the sample points sum to 1? Oddly enough, the answer is yes, but only when the infinite number of points can be counted. An example of this situation will be given in Section 5.4. In all other cases, the answer is no. This leads us to define two different types of random variables and probability models for each.

Random variables are classified as one of two types: *discrete* or *continuous*.

DEFINITION

A discrete random variable is one that can assume a countable number of values.

A discrete random variable is easily identified by examining the number of the values it may assume. If the number of values that the random variable may assume can be counted, it must be discrete.

Typical examples of discrete random variables are:

1. The number of defective bolts in a sample of 10 drawn from industrial production.
2. The number of rural electrified homes in a township.
3. The number of malfunctions of an airplane engine over a period of time.
4. The number of people in the waiting line in a doctor's office.

DEFINITION

A continuous random variable is one that can assume the infinitely large number of values corresponding to the points on a line interval.

The word "continuous," an adjective, means proceeding without interruption. It, in itself, provides the key for identifying continuous random variables. Look for a measurement with a set of values that form points on a line with no interruptions or intervening spaces between them.

Typical examples of continuous random variables are:

1. The height of a human.
2. The length of life of a human cell.
3. The amount of sugar in an orange.
4. The length of time required to complete an assembly operation in a manufacturing process.

The distinction between discrete and continuous random variables is an important one since different probability models are required for each. Accordingly, the probability distribution for discrete and continuous random variables will be discussed in Sections 5.4 and 5.5, respectively.

EXERCISES

5.1 Identify the following as discrete or continuous random variables:
(a) The height of water in a dam.
(b) The amount of money awarded a plaintiff by a court in a damage suit.
(c) The number of people waiting for treatment at a hospital emergency room.
(d) The total points scored in a football game.
(e) The number of claims received by an insurance company during a day.

5.2 Identify the following as discrete or continuous random variables.
(a) The number of incoming telephone calls at a telephone switchboard during a 5-minute interval.
(b) The amount of rainfall in Gainesville, Florida, for a 1-week period.
(c) The length of time for an automobile driver to respond when faced with an impending collision.
(d) The number of aircraft near-collisions observed by an air controller over a 24-hour period.
(e) The bacteria count per cubic centimeter in your drinking water.

5.3 Identify the following as discrete or continuous random variables:
(a) The increase in length of life achieved by a cancer patient as a result of surgery.
(b) The tensile breaking strength, in pounds per square inch, of 1-inch-diameter steel cable.
(c) The number of deer killed per year in a state wildlife preserve.
(d) The number of overdue accounts in a department store at a particular point in time.
(e) Your blood pressure.

5.4

PROBABILITY DISTRIBUTIONS FOR DISCRETE RANDOM VARIABLES

The probability distribution for a discrete random variable is a formula, table, or graph that provides the probability associated with each value of the random variable. Since each value of the variable y is a numerical event, we may apply the methods of Chapter 4 to obtain the appropriate probabilities. It is interesting to note that the events cannot overlap because one and only one value of y is assigned to each sample point, and hence that the values of y represent mutually exclusive numerical events. Summing $p(y)$ over all values of y would equal the sum of the probabilities of all sample points and, hence, equal 1. We may therefore state two requirements for a probability distribution,

1. $0 \leq p(y) \leq 1$.

2. $\sum_{\text{all } y} p(y) = 1$.

EXAMPLE 5.1

Consider an experiment that consists of tossing two coins and let y equal the number of heads observed. Find the probability distribution for y.

Solution The sample points for this experiment with their respective probabilities are as follows:

Sample Point	Coin 1	Coin 2	$P(E_i)$	y
E_1	H	H	1/4	2
E_2	H	T	1/4	1
E_3	T	H	1/4	1
E_4	T	T	1/4	0

Because sample point E_1 is associated with the simple event, "observe a head on coin 1 and a head on coin 2," we assign it the value $y = 2$. Similarly, we assign $y = 1$ to point E_2, etc. The probability of each value of y may be calculated by adding the probabilities of the sample points in that numerical event. The numerical event $y = 0$ contains one sample point, E_4; $y = 1$ contains two sample points, E_2 and E_3; and $y = 2$ contains one point, E_1. The values of y with respective probabilities are given in Table 5.1. Observe that $\sum_{y=0}^{2} p(y) = 1$.

Noting that $p(y)$ is a function of y, according to the definition of a functional relation given in Section 2.2, we might look for a simple mathematical equation to replace Table 5.1. Although it is not particularly obvious, a casual check will confirm that the equation

$$p(y) = \frac{C_y^2}{4}$$

provides exactly the same information as does Table 5.1, that is,

$$p(0) = \frac{C_0^2}{4} = 1/4,$$

$$p(1) = \frac{C_1^2}{4} = 1/2,$$

and $$p(2) = \frac{C_2^2}{4} = 1/4.$$

TABLE 5.1

Probability distribution for y (y = number of heads)

y	Sample Points in y	$p(y)$
0	E_4	1/4
1	E_2, E_3	1/2
2	E_1	1/4
		$\sum_{y=0}^{2} p(y) = 1$

The origin of our particular equation is not essential at this point. Let us accept the fact that it *is* a representation of the probability distribution for y, as can be verified, and defer further discussion on this point until Chapter 6.

We have presented $p(y)$ in tabular form and as a formula. We complete the discussion of this example by giving $p(y)$ in graphical form, that is, in the form of a frequency histogram discussed in Section 3.2. The histogram for the random variable y would contain three classes, corresponding to $y = 0$, $y = 1$, and $y = 2$. Since $p(0) = 1/4$, the theoretical relative frequency for $y = 0$ is $1/4$; $p(1) = 1/2$ and hence the theoretical frequency for $y = 1$ is $1/2$, etc. The histogram is given in Figure 5.1.

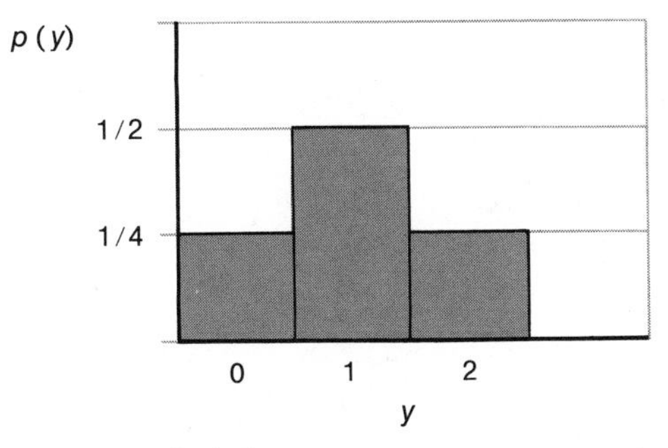

FIGURE 5.1
Probability histogram showing $p(y)$ for Example 5.1

If you were to draw a sample from this population—that is, if you were to throw two balanced coins, say $n = 100$ times, and each time record the number of heads observed, y, and then construct a histogram using the 100 measurements on y—you would find that the histogram for your sample would appear very similar to that for $p(y)$, Figure 5.1. If you were to repeat the experiment $n = 1{,}000$ times, the similarity would be much more pronounced.

EXAMPLE 5.2

Let y equal the number observed on the throw of a single balanced die. Find $p(y)$.

Solution The sample points for this experiment are given in Table 5.2. You would assign $y = 1$ to E_1, $y = 2$ to E_2, etc. Since each value of y contains only one sample point, $p(y)$, the probability distribution for y would appear as shown in the fifth column of Table 5.2. You can also see that

$$p(y) = 1/6$$

gives the probability distribution as a formula. The corresponding histogram is given in Figure 5.2.

TABLE 5.2
Tossing a die: probability distribution for y

Sample Point	Number on Upper Face	$P(E_i)$	y	$p(y)$
E_1	1	1/6	1	1/6
E_2	2	1/6	2	1/6
E_3	3	1/6	3	1/6
E_4	4	1/6	4	1/6
E_5	5	1/6	5	1/6
E_6	6	1/6	6	1/6

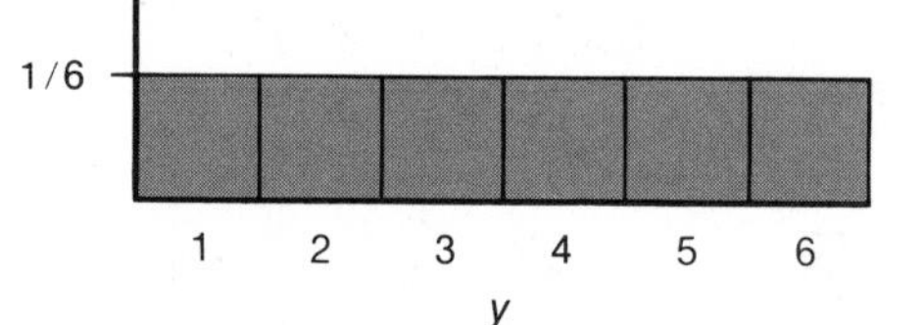

FIGURE 5.2
Probability histogram for $p(y) = 1/6$ in Example 5.2

EXAMPLE 5.3

As a final example, we shall consider a random variable that may assume a countable infinity of values. The experiment consists of tossing a coin until the first head appears. Let y equal the number of tosses. Find the probability distribution for y.

Solution The sample points, infinite in number, are given below. Let the symbol *TTTH* represent the results of the first four tosses proceeding from left to right, with *H* representing a head and *T* a tail.

E_1: *H*
E_2: *TH*
E_3: *TTH*
E_4: *TTTH*
E_5: *TTTTH*
⋮ ⋮
etc. etc.

You can see that E_{65} would be the sample point associated with the event that each of the first 64 tosses resulted in a tail and the 65th resulted in a head. Conceivably, the experiment might never end. Let us now calculate the probability of each sample point and assign values of y to each of these points. The probability of E_1, a head on the first toss, is 1/2. The probability of E_2, a tail and then a head, is an intersection of two independent events and can be obtained by use of the Multiplicative Law of Probability. Hence, $P(E_2) = (1/2)(1/2)$. Likewise, $P(E_3) = (1/2)(1/2)(1/2) = (1/2)^3$ and

TABLE 5.3
Tossing a coin until the first head appears (y = number of tosses)

Sample Point	$P(E_i)$	y	$p(y)$
E_1	1/2	1	1/2
E_2	1/4	2	1/4
E_3	1/8	3	1/8
E_4	1/16	4	1/16
E_5	1/32	5	1/32
⋮	⋮	⋮	⋮
etc.			

it would follow that $P(E_{50}) = (1/2)^{50}$. The appropriate probability distribution is given in the fourth column of Table 5.3, or by the equation $p(y) = (1/2)^y$. The frequency histogram is shown in Figure 5.3. Will $p(y)$ satisfy the second requirement of a probability distribution; that is, will

$$\sum_{y=1}^{\infty} p(y) = 1?$$

Summing, we obtain

$$\sum_{y=1}^{\infty} p(y) = p(1) + p(2) + p(3) + \cdots$$
$$= 1/2 + 1/4 + 1/8 + \cdots.$$

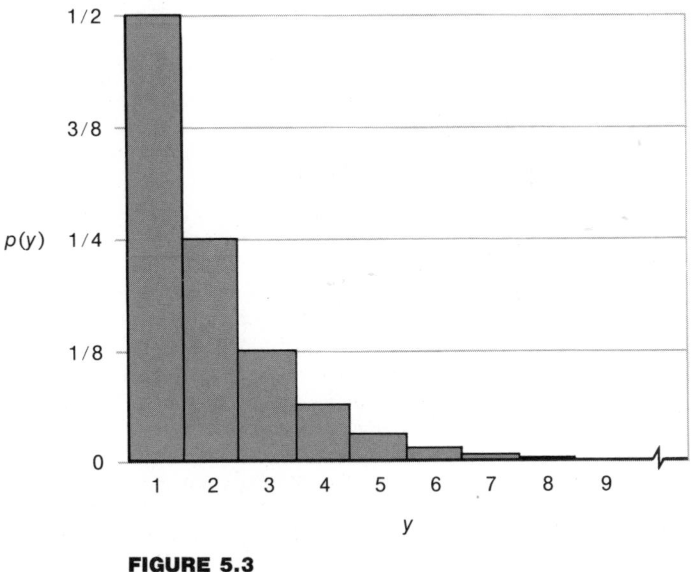

FIGURE 5.3
$p(y)$ for Example 5.3

The reader familiar with high school algebra will recognize this as the sum of an infinite geometric progression with common ratio equal to 1/2. A college algebra book or mathematical handbook will verify that this sum is equal to (or, correctly, approaches as a limit) 1. Thus we observe a sample space that contains an infinite number of sample points, each of which is assigned a positive (nonzero) probability, and the sum of the probabilities over the infinite number of sample points is equal to 1. The probability distribution for y is shown in Figure 5.3.

Tips on Problem Solving: To find the probability distribution for a discrete random variable, construct a table listing each value that the random variable y can assume. Then calculate $p(y)$ for each value of y.

EXERCISES

5.4 A salesman figures that each contact results in a sale with probability .2. During a given day he contacts two prospective clients. Calculate the probability distribution for y, the number of clients who sign a sales contract (assume that the contacts represent independent events). Construct a probability histogram for $p(y)$.

5.5 A key ring contains four office keys that are identical in appearance. Only one will open your office door. Suppose you randomly select one key and try it. If it does not fit, you randomly select one of the three remaining keys. If that does not fit, you randomly select one of the last two. Let y equal the number of keys you try before you find the one that opens the door ($y = 1, 2, 3, 4$). Find the probability distribution for y. Construct a probability histogram for $p(y)$.

5.6 A piece of electronic equipment contains six transistors, two of which are defective. Three transistors are selected at random, removed from the piece of equipment, and inspected. Let y equal the number of defectives observed, where $y = 0, 1,$ or 2. Find the probability distribution for y. Express your results graphically as a probability histogram.

5.7 Simulate the experiment described in Exercise 5.6 by marking six marbles (or coins, or cards) so that two represent defectives and four represent nondefectives. Place the marbles in a hat, mix, draw three, and record y, the number of "defectives observed." Replace the marbles and repeat the process until a total of $n = 100$ observations on y have been recorded. Construct a relative frequency histogram for this sample and compare it with the population probability distribution, Exercise 5.6.

5.8 In order to verify the accuracy of their financial accounts, companies utilize auditors on a regular basis to verify accounting entries. Suppose that the company's employees make erroneous entries 5 percent of the time. If an auditor randomly checks three entries:

(a) Find the probability distribution for y, the number of errors detected by the auditor.

(b) Construct a probability histogram for $p(y)$.

(c) Find the probability that the auditor will detect more than one error.

5.9 Past experience has shown that on the average only 1 in 10 wells drilled hits oil. Let

y be the number of drillings until the first success (oil is struck). Assume that the drillings represent independent events.

(a) Find $p(1)$, $p(2)$, and $p(3)$.
(b) Give a formula for $p(y)$.
(c) Graph $p(y)$.

5.5

PROBABILITY DISTRIBUTIONS FOR CONTINUOUS RANDOM VARIABLES

Recall that we cannot assign probabilities to the sample points associated with a continuous random variable and that a completely different population model is required. We invite you to refocus your attention on the discussion of the relative frequency histogram, Section 3.2, and specifically to the histogram for the thirty student grade-point averages, Figure 3.2. If more and more measurements were obtained, we might reduce the width of the class interval. The outline of the histogram would change slightly, for the most part becoming less and less irregular. When the number of measurements becomes very large and the intervals very small, the relative frequency would appear, for all practical purposes, as a smooth curve. (See Figure 5.4.)

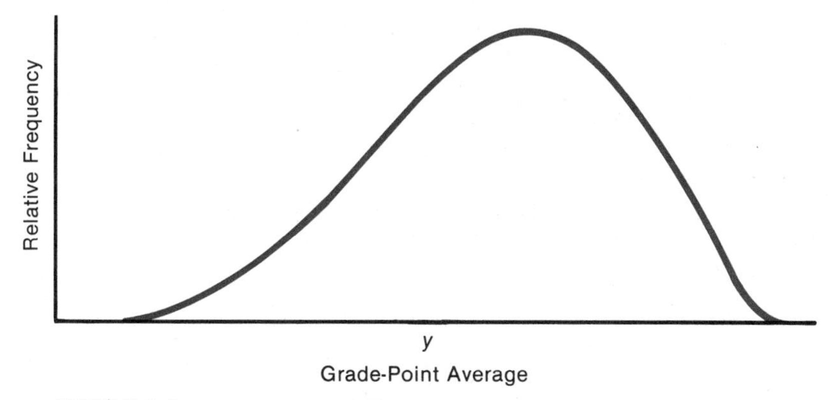

FIGURE 5.4

A relative frequency histogram for a population

The relative frequency associated with a particular class in the population is the fraction of measurements in the population falling in that interval and, also, is the probability of drawing a measurement in that class. If the total area under the relative frequency histogram were adjusted to equal 1, then areas under the frequency curve would correspond to probabilities. Indeed, this was the basis for the application of the Empirical Rule in Chapter 3.

Let us now construct a model for the probability distribution for a continuous random variable. Assume that the random variable, y, may take values on a real line as in Figure 5.4. We will then distribute one unit of probability along the line much as a person might distribute a handful of sand, each measurement in the population corresponding to a single grain. The probability, grains of sand or measurements, will pile up in certain places, and the result will be a probability distribution, which might appear like the one shown in Figure 5.5. The depth or density of probability, which varies with y, may be represented by a mathematical equation $f(y)$, called the probability distribution (or the probability density function) for the random variable y. The function, $f(y)$, represented graphically in Figure 5.5, provides a mathematical model for the population relative frequency histogram which exists in reality. The density function, $f(y)$, is defined so that the total area under the curve is equal to 1 and therefore that the area lying above a given interval will equal the probability that y will fall in that interval. Thus, the probability that $a < y < b$ (a is less than y and y is less than b) is equal to the shaded area under the density function between the two points a and b.

Two puzzling questions remain. How do we choose the model, that is, the probability distribution, $f(y)$, appropriate for a given physical situation? And then, how can we calculate the area under the curve corresponding to a given interval? The first question is the more difficult because the answer will depend upon the type of measurement involved. To grasp the reasoning better, let us consider the model for the discrete random variable. Using *good judgment,* we assigned realistic probabilities to the sample points associated with the experiment and then, applying the techniques of Chapter 4, derived the probability distribution for the random variable of interest. A similar approach is available for some continuous random variables but it is, unfortunately, beyond the scope of this course. The important thing to note, however, is that good judgment was required in assigning the probabilities to the sample points and that, although inexact, the probabilities were sufficiently accurate for practical purposes. Thus they provided a model for reality. Similarly, for continuous random variables, we utilize all available information and then use our best judgment in choosing the model, $f(y)$. In some cases the approximate distribution may be derived using the theory of mathematical statistics. In others, we may rely on the frequency histograms for samples drawn from the population or histograms for data drawn from similar populations. The form of the sample histogram will often suggest a reasonable choice for $f(y)$.

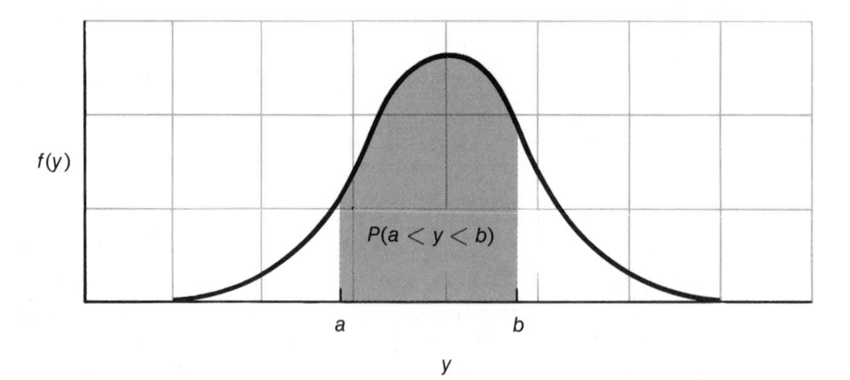

FIGURE 5.5
Probability distribution, $f(y)$

Once the model, $f(y)$, has been chosen, we may calculate areas under the curve by using the integral calculus or by applying rectangular approximations, or, as a last resort, we could make an approximation of the area by visual inspection. We will leave this problem to the mathematician. The areas associated with most useful probability distributions have been calculated and appear in tabular form. (See Appendix II.)

We conclude with a comment to the "doubting Thomas" concerned because of the lack of agreement between $f(y)$ and the true relative frequency curve for the population. Can useful results be obtained from an inaccurate population model? We need only look to the physicist, the chemist, and the engineer. The equations, formulas, and various numerical expressions used in all the sciences are simply mathematical models which, fortunately, provide good approximations to reality. The engineer uses his equations to determine the size and location of members in a bridge or an airplane wing and is concerned only that the resulting structure perform the task for which it was designed. Similarly, in statistics the attainment of the end result is the criterion by which we measure the worth of a statistical technique. Does the technique provide good inferences, that is, predictions or decisions, concerning the population from which the sample was drawn? Many statistical techniques possess this desirable property, a fact that can be shown experimentally and confirmed by the many extremely useful applications of statistics in industry and in the physical and social sciences.

5.6

MATHEMATICAL EXPECTATION

The probability distribution, described in Sections 5.4 and 5.5, provides a model for the theoretical frequency distribution of a random variable and hence must possess a mean, variance, standard deviation, and other descriptive measures associated with the theoretical population that it represents. Recalling that both the mean and variance are averages (Sections 3.4 and 3.5), we shall confine our attention to the problem of calculating the mean value of a random variable defined over a theoretical population. This mean is called the expected value of the random variable.

The method for calculating the population mean or expected value of a random variable can be more easily understood by considering an example. Let y equal the number of heads observed in the toss of two coins. For convenience, we give $p(y)$:

y	$p(y)$
0	1/4
1	1/2
2	1/4

Let us suppose that the experiment is repeated a large number of times, say $n = 4{,}000{,}000$ times. Intuitively, we would expect to observe approximately

1 million zeros, 2 million ones, and 1 million twos. Then the average value of y would equal

$$\frac{\text{sum of measurements}}{n} = \frac{1{,}000{,}000(0) + 2{,}000{,}000(1) + 1{,}000{,}000(2)}{4{,}000{,}000}$$

$$= \frac{1{,}000{,}000(0)}{4{,}000{,}000} + \frac{2{,}000{,}000(1)}{4{,}000{,}000} + \frac{1{,}000{,}000(2)}{4{,}000{,}000}$$

$$= (1/4)(0) + (1/2)(1) + (1/4)(2).$$

Note that the first term in this sum is equal to $(0)p(0)$, the second is equal to $(1)p(1)$, and the third is equal to $(2)p(2)$. The average value of y is then equal to

$$\sum_{y=0}^{2} yp(y) = 1.$$

You will observe that this result was not an accident and that it would be intuitively reasonable to define the expected value of y for a discrete random variable as follows.

DEFINITION

Let y be a discrete random variable with probability distribution $p(y)$ and let $E(y)$ represent the expected value of y. Then

$$E(y) = \sum_{y} yp(y),$$

where the elements are summed over all values of the random variable y.

Note that if $p(y)$ is an accurate description of the relative frequencies for a real population of data, then $E(y) = \mu$, the mean of the population. We shall assume this to be true and let $E(y)$ be synonymous with μ. That is, we shall assume that

$$\mu = E(y)$$

The method for calculating the expected value of y for a continuous random variable is rather similar from an intuitive point of view, but, in practice, it involves the use of the calculus and is therefore beyond the scope of this text.

EXAMPLE 5.4

Consider the random variable y representing the number observed on the toss of a single die. The probability distribution for y is given in Example 5.2. Then the expected value of y would be

$$E(y) = \sum_{y=1}^{6} yp(y) = (1)p(1) + (2)p(2) + \cdots + (6)p(6)$$

$$= (1)(1/6) + (2)(1/6) + \cdots + (6)(1/6)$$

$$= \frac{1}{6}\sum_{y=1}^{6} y = \frac{21}{6} = 3.5.$$

Note that this value, $\mu = E(y) = 3.5$, exactly locates the center of the probability distribution, Figure 5.2.

EXAMPLE 5.5

Eight thousand tickets are to be sold at $1.00 each in a lottery conducted to benefit the local fire company. The prize is a $3,000.00 automobile. If Ed Smith has purchased two tickets, what is his expected gain?

Solution Smith's gain, y, may take one of two values. Either he will lose $2.00 (that is, his gain will be −$2.00) or will win $2,998.00 with probabilities 7,998/8,000 and 2/8,000, respectively. The probability distribution for the gain, y, is as follows:

y	$p(y)$
−$2.00	$\frac{7,998}{8,000}$
$2,998.00	$\frac{2}{8,000}$

The expected gain will be

$$\begin{aligned} E(y) &= \sum_y yp(y) \\ &= (-\$2.00)\left(\frac{7,998}{8,000}\right) + (\$2,998.00)\left(\frac{2}{8,000}\right) \\ &= -\$1.25. \end{aligned}$$

Recall that the expected value of y is the average of the theoretical population that would result if the lottery were repeated an infinitely large number of times. If this were done, Smith's average or expected gain per lottery would be a loss of $1.25.

EXAMPLE 5.6

Determine the yearly premium for a $1,000.00 insurance policy covering an event which, over a long period of time, has occurred at the rate of 2 times in 100. Let y equal the yearly financial gain to the insurance company resulting from the sale of the policy and let C equal the unknown yearly premium. We will calculate the value of C such that the expected gain, $E(y)$, will equal zero. Then C is the premium required to break even. To this the company would add administrative costs and profit.

Solution The first step in the solution is to determine the values which the gain, y, may take and then to determine $p(y)$. If the event does not occur during the year, the insurance company will gain the premium of $y = C$ dollars. If the event does occur, the gain will be negative. That is, the company will lose $1,000 less the premium of C dollars already collected. Then $y = -(1,000 - C)$ dollars. The probabilities associated with these two values of y are 98/100 and 2/100, respectively. The probability distribution for the gain would be:

y = gain	$p(y)$
C	$\frac{98}{100}$
$-(1{,}000 - C)$	$\frac{2}{100}$

Since we want the insurance premium C such that, in the long run (for many similar policies), the mean gain will equal 0, we will set the expected value of y equal to zero and solve for C. Then

$$E(y) = \sum_y yp(y)$$

$$= C\left(\frac{98}{100}\right) + [-(1{,}000 - C)]\left(\frac{2}{100}\right) = 0$$

or

$$\frac{98}{100}C + \left(\frac{2}{100}\right)C - 20 = 0.$$

Solving this equation for C, we obtain

$$C = \$20.$$

Concluding, if the insurance company were to charge a yearly premium of \$20, the average gain calculated for a large number of similar policies would equal zero. The actual premium would equal \$20 plus administrative costs and profit.

Just as we used numerical descriptive measures to describe a relative frequency distribution (Chapter 3), we wish to use the mean and variance (ultimately, the standard deviation) of a random variable to describe its probability distribution. Knowing μ and σ, we could use Tchebysheff's Theorem or the Empirical Rule to describe $p(y)$. The variance σ^2 of a random variable is defined to be the mean value of the square of the deviation of y from its mean. That is,

$$\sigma^2 = E[(y - \mu)^2].$$

This leads us to the problem of finding the mean value of a function of a random variable, namely the expected value of $(y - \mu)^2$.

There are many other situations where we might only know the probability distribution of a random variable y but want to know the mean value of some function of y. For example, if a life insurance company has issued a fixed number of policies in a given year, then the profit per year P is a function of y, the number of policyholders who die. Now suppose we only know the equation relating P to y and the probability distribution for y. We want to know the mean value of P. It is this problem that we now consider—finding the expected value of a function of a random variable.

The rule for finding the expected value of a function of y, say $g(y)$, can be obtained by considering a simple example, the coin-tossing problem, Example 5.1, where y equals the number of heads observed when tossing two coins. Suppose that we wish to find the expected value of $g(y) = y^2$.

In accordance with the definition of a function, Section 2.2, only one value of the function corresponds to each value of y. Hence the quantity y^2 would represent a numerical event that varies over the sample space with only one value of y^2 being assigned to each sample point. Without proceeding further, it is reasonably clear that any function of a random variable, y, will itself be a random variable. The probability distribution for y, with associated values of y^2, is as follows:

y	y^2	$p(y)$
0	0	1/4
1	1	1/2
2	4	1/4

Repeating the experiment a large number of times, say $n = 4{,}000{,}000$, we would expect approximately 1,000,000 values of $y^2 = 0$, 2,000,000 values of $y^2 = 1$, and 1,000,000 values of $y^2 = 4$. The average value for y^2 would be

$$\frac{\text{sum of measurements}}{n} = \frac{1{,}000{,}000(0) + 2{,}000{,}000(1) + 1{,}000{,}000(4)}{4{,}000{,}000}$$

$$= (1/4)(0) + (1/2)(1) + (1/4)(4)$$

$$= \sum_{y=0}^{2} y^2 p(y).$$

You may obtain the expected value for other functions of y and other random variables using the technique employed above. The results will agree with the following definition.

DEFINITION

Let y be a discrete random variable with probability distribution $p(y)$, and let $g(y)$ be a numerical valued function of y. Then the expected value of $g(y)$ is

$$E[g(y)] = \sum_{y} g(y)p(y).$$

As noted earlier, the variance of a random variable, y, the expected value of $(y - \mu)^2$, is of particular interest to us because it, along with μ, can be used to describe the probability distribution for y.

DEFINITION

Let y be a discrete random variable with probability distribution $p(y)$; then the variance* of y is

$$\sigma^2 = E[(y - \mu)^2] = \sum_y (y - \mu)^2 p(y).$$

We conclude with two examples.

EXAMPLE 5.7

Find the variance, σ^2, for the population associated with Example 5.1, the tossing of two coins. The expected value of y, μ, was shown to equal 1.

Solution The variance is equal to the expected value of $(y - \mu)^2$, or

$$\begin{aligned}\sigma^2 = E[(y - \mu)^2] &= \sum_y (y - \mu)^2 p(y) \\ &= (0 - 1)^2 p(0) + (1 - 1)^2 p(1) + (2 - 1)^2 p(2) \\ &= 1(1/4) + 0(1/2) + 1(1/4) \\ &= 1/2.\end{aligned}$$

Calculating σ^2 can be facilitated by using the following table. Note that σ^2 is the total of column 4.

y	$(y - \mu)^2$	$p(y)$	$(y - \mu)^2 p(y)$
0	1	1/4	1/4
1	0	1/2	0
2	1	1/4	1/4
			1/2

EXAMPLE 5.8

Let y be a random variable with probability distribution given by the following table:

y	$p(y)$
−1	.05
0	.10
1	.40
2	.20
3	.10
4	.10
5	.05

*Three theorems, directly analogous to the three summation theorems of Section 2.5, exist for expectations. These theorems can be used to show that

$$\sigma^2 = \sum_y (y - \mu)^2 p(y) = \sum_y y^2 p(y) - \mu^2 = E(y^2) - \mu^2,$$

a result that is analogous to the shortcut formula for the sum of squares of deviations, Chapter 3. See Exercises 5.52 and 5.53.

Find μ, σ^2, and σ. Graph $p(y)$ and locate the interval, $\mu \pm 2\sigma$, on the graph. What is the probability that y will fall in the interval $\mu \pm 2\sigma$?

Solution

$$\mu = E(y) = \sum_{y=-1}^{5} yp(y)$$
$$= (-1)(.05) + (0)(.10) + (1)(.40) + \cdots + (4)(.10) + (5)(.05)$$
$$= 1.70$$

$$\sigma^2 = E[(y - \mu)^2] = \sum_{y=-1}^{5} (y - \mu)^2 p(y)$$
$$= (-1 - 1.7)^2(.05) + (0 - 1.7)^2(.10) + \cdots + (5 - 1.7)^2(.05)$$
$$= 2.11$$

and

$$\sigma = \sqrt{\sigma^2} = \sqrt{2.11} = 1.45$$

The interval $\mu \pm 2\sigma$ is $1.70 \pm (2)(1.45)$ or -1.20 to 4.60.

The graph of $p(y)$ and the interval, $\mu \pm 2\sigma$, are shown in Figure 5.6. You can see that $y = -1, 0, 1, 2, 3, 4$ fall in the interval. Therefore,

$$P[\mu - 2\sigma < y < \mu + 2\sigma] = p(-1) + p(1) + p(2) + \cdots + p(4)$$
$$= (.05) + (.10) + (.40) + (.20) + (.10) + (.10)$$
$$= .95$$

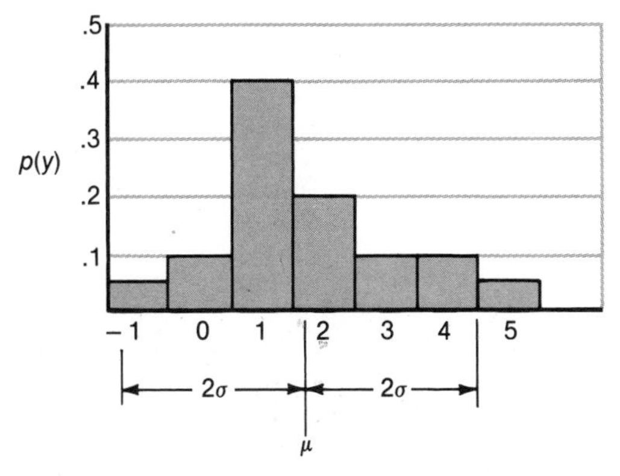

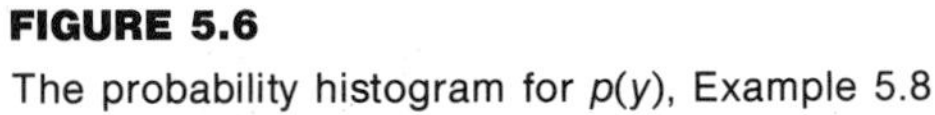

FIGURE 5.6

The probability histogram for $p(y)$, Example 5.8

Tips on Problem Solving: To find the expected value of a discrete random variable y, construct a table containing three columns, the first for y and the second for $p(y)$. Then multiply each y value by its corresponding probability and enter in the third column. The sum of this third column, the sum of $yp(y)$, will give you the expected value of y.

To find the expected value of a function $g(y)$ of a discrete random variable y, start with a table containing four columns, the first for y, the second for $g(y)$, the third for $p(y)$, and the fourth for the cross products, $g(y)p(y)$. First calculate the value of $g(y)$ for each value of y and enter in column 2. Then obtain the cross product $g(y)p(y)$ for each value of y and enter in column 4. The sum of column 4 will give the expected value of $g(y)$.

EXERCISES

5.10 Let y be a random variable with probability distribution given by the following table:

y	$p(y)$
0	1/8
1	1/4
2	1/2
3	1/8

Find the expected value and variance of y. Construct a graph of the probability distribution. Do μ and σ permit a rough description of $p(y)$?

5.11 Let y represent the number of times a housewife visits a grocery store in a 1-week period. Assume that the following is the probability distribution of y.

y	$p(y)$
0	.1
1	.5
2	.3
3	.1

Find the expected value of y. This is the average number of times a housewife visits the store.

5.12 A manufacturing representative is considering the option of taking out an insurance policy to cover possible losses incurred by marketing a new product. If the product is a complete failure, the representative feels that a loss of $80,000 would be incurred; if it is only moderately successful, a loss of $25,000 would be incurred. Insurance actuaries have determined from market surveys and other available information that the probabilities that the product will be a failure or only moderately successful are .01 and .05, respectively. Assuming that the manufacturing representative would be willing to ignore all other possible losses, what premium should the insurance company charge for the policy in order to break even?

5.13 A manufacturing company ships its product in two different sizes of truck trailers, an $8 \times 10 \times 30$ and an $8 \times 10 \times 40$. If 30 percent of its shipments are made using the 30-foot trailer and 70 percent using the 40-foot trailer, find the mean volume shipped per trailer load (assume that the trailers are always full).

5.14 The number N of residential homes that a fire company can serve depends on the distance r (in city blocks) that a fire engine can cover in a specified (fixed) period of time. If we assume that N is proportional to the area of a circle r blocks from the firehouse, then

$$N = C\pi r^2,$$

where C is a constant, $\pi = 3.1416 \ldots$, and r, a random variable, is the number of blocks that a fire engine can move in the specified time interval. For a particular fire company, $C = 8$, the probability distribution for r is as shown in the accompanying table, and $p(r) = 0$, $r \leq 20$, and $r \geq 27$.

r	21	22	23	24	25	26
$p(r)$	.05	.20	.30	.25	.15	.05

Find the expected value of N, the number of homes that the fire department can serve.

5.15 The probability distribution for a random variable y is as shown in the table.

y	0	1	2	3	4	5
$p(y)$	.05	.3	.3	.2	.1	.05

(a) Find $E(y)$.
(b) Find σ^2.
(c) Sketch $p(y)$ and locate the interval $\mu \pm 2\sigma$ on the graph.
(d) Find the probability that y falls in the interval $\mu \pm 2\sigma$.

5.7

RANDOM SAMPLING

Since probability distributions are theoretical models for population relative frequency distributions, samples selected from populations can be viewed as observations on random variables. As noted in earlier chapters, the way a sample is selected from a population affects the quantity of information in the sample (the topic of Chapter 12) as well as the probabilities of observing particular samples. Since the probability of observed sample outcomes is the basis for inference making, a topic which is introduced in Chapters 6, 7, and 8, it is important that we define one of the most commonly employed sampling procedures at this point. It is called simple random sampling.

Simple random sampling gives every different sample in the population an equal chance of being selected. To illustrate, suppose that we wish to select a sample of $n = 2$ from a population containing $N = 4$ elements (we are choosing a small value of N to simplify our discussion). If the four elements are identified by the symbols, y_1, y_2, y_3, and y_4, then there are six different samples that could be selected from the population, namely,

Sample	Observations in Sample
1	y_1, y_2
2	y_1, y_3
3	y_1, y_4
4	y_2, y_3
5	y_2, y_4
6	y_3, y_4

If the sample was selected in such a way that each of these six samples had an equal chance of selection (probability equal to 1/6), the sample would be called a simple random (or, simply, "random") sample.

It can be shown* that the number of ways of selecting $n = 2$ elements from a set of $N = 4$, denoted by the symbol C_2^4, is

$$C_2^4 = \frac{4!}{2!2!} = \frac{4 \cdot 3 \cdot 2 \cdot 1}{(2 \cdot 1)(2 \cdot 1)} = 6.$$

[*Note:* The symbol $n!$ (read "n factorial") is used to denote the product, $n(n-1)(n-2) \cdots 3 \cdot 2 \cdot 1$. Thus $6! = 6 \cdot 5 \cdot 4 \cdot 3 \cdot 2 \cdot 1$. The quantity $0!$ is defined to be equal to 1.] In general, the number of ways of selecting n elements from a set of N is

$$C_n^N = \frac{N!}{n!(N-n)!}.$$

For example, if you wish to conduct an opinion poll of 5,000 people based on a sample of $n = 100$, there are

$$C_{100}^{5,000}$$

different combinations of people who could be selected in the sample. If the sampling is conducted in such a way that each of these combinations has an equal probability of being selected, then the sample is called a simple random sample.

DEFINITION

Let N and n represent the numbers of elements in the population and sample, respectively. If the sampling is conducted in such a way that each of the C_n^N samples has an equal probability of being selected, the sampling is said to be random and the result is said to be a simple random sample.

Perfect random sampling is difficult to achieve in practice. If the population is not too large, we might write each of the N numbers on a poker chip, mix the total, and select a sample of n chips. The numbers on the poker chips would specify the measurements to appear in the sample. Other techniques are available when the population is large. One of these, the use of a random-number table, is discussed in Section 12.3.

In many situations, the population is conceptual, as in an observation made during a laboratory experiment. Here the population is envisioned to be the infinitely large number of measurements obtained when the experiment is repeated over and over again. If we wish a sample of $n = 10$ measurements from this population, we repeat the experiment 10 times and hope that the results represent, to a reasonable degree of approximation, a random sample.

Although the primary purpose of this discussion was to clarify the meaning of a random sample, we would like to mention that some sampling techniques are

*A derivation of this result along with examples and applications is given in optional Section 4.9. An understanding of the derivation is not essential to our discussion.

partly systematic and partly random. For instance, if we wish to determine the voting preference of the nation in a presidential election, we would not likely choose a random sample from the population of voters. Just due to a pure chance, all the voters appearing in the sample might be drawn from a single city, say San Francisco, which might not be at all representative of the population. We would prefer a random selection of voters from smaller political districts, perhaps states, allotting a specified number to each state. The information from the randomly selected subsamples drawn from the respective states would be combined to form a prediction concerning the entire population of voters in the country. The purpose of systematic sampling, as in the design of experiments in general, is to obtain a maximum of information for a fixed sample size. This, we recall, was one of the five elements of a statistical problem discussed in Chapter 1.

EXERCISES

5.16 Evaluate:
(a) 5! (b) C_3^6. (c) C_0^7.

5.17 How many different samples of $n = 2$ elements can be selected from $N = 5$? List them.

5.18 How many different samples of $n = 10$ could be selected from a population containing $N = 100$ elements?

5.8

SUMMARY

Random variables, representing numerical events defined over a sample space, may be classified as discrete or continuous random variables depending upon whether the number of sample points in the sample space is or is not countable. The theoretical population frequency distribution for the discrete random variable is called a probability distribution and often may be derived using the techniques of Chapter 4. The model for the frequency distribution for a continuous random variable is a mathematical function, $f(y)$, called a probability distribution or probability density function. This function, usually a smooth curve, is defined over a line interval and is chosen such that the total area under the curve is equal to 1. The probabilities associated with a continuous random variable are given as areas under the probability distribution, $f(y)$.

A mathematical expectation is the average of a random variable calculated for the theoretical population defined by its probability distribution.

As noted in earlier discussions, inferences about a population will be based on the probability of an observed sample and this probability will depend upon how the sample was selected from the population. The probability distribution for a random variable plays a role in inference making because it enables us to calculate the probability that a single observation, randomly selected from the population, will assume one or more specific values. The most common method for sampling more than one observation from a population is called simple random (or simply "random") sampling. A simple random sample is selected from a population in such a way that every possible different sample has an equal probability of selection. Simple random sampling will be used in Chapter 6 and in subsequent chapters in this text.

REFERENCES

Freund, J. E., *Mathematical Statistics,* 2nd ed. Englewood Cliffs, N.J.: Prentice-Hall, Inc., 1971.

Mendenhall, W., and R. L. Scheaffer, *Mathematical Statistics with Applications.* North Scituate, Mass.: Duxbury Press, 1973.

Mood, A. M., et al., *Introduction to the Theory of Statistics,* 3rd ed. New York: McGraw-Hill Book Company, 1974.

Mosteller, F., R. E. K. Rourke, and G. B. Thomas, Jr., *Probability with Statistical Applications,* 2nd ed. Reading, Mass.: Addison-Wesley Publishing Company, Inc., 1970.

SUPPLEMENTARY EXERCISES

Starred (*) exercises are optional.

5.19 Toss three coins and let y equal the number of heads observed. Calculate the probability distribution for y. Construct a probability histogram for the distribution. Find the expected value and variance of y.

5.20 Given the discrete random variable y and its probability distribution, $p(y)$:

y	$p(y)$
0	1/8
2	1/4
3	1/2
4	1/8

(a) Find $\mu = E(y)$.
(b) Find $E(y^2)$.
(c) Find σ^2 and σ.

5.21 Identify the following as continuous or discrete random variables:
(a) The number of homicides in Detroit during a 1-month period.
(b) The length of time between arrivals at an outpatient clinic.
(c) The number of typing errors on a page of typing.
(d) The number of defective light bulbs in a packet containing four bulbs.
(e) The time required to finish an examination.

5.22 Given the following probability distribution for the random variable y:

y	0	1	3	4
$p(y)$		2/10	4/10	3/10

(a) What is $p(0)$?
(b) Find μ, the expected value of y.
(c) Find σ^2.

5.23 Refer to Exercise 5.19. Using the probability histogram, find the fraction of the total population lying within two standard deviations of the mean. Compare with Tchebysheff's Theorem.

5.24 Refer to Exercise 5.6. Find μ, the expected value of y, for the theoretical population by using the probability distribution obtained in Exercise 5.6. Find the sample mean, $\bar{y}$, for the $n = 100$ measurements generated in Exercise 5.7. Does $\bar{y}$ provide a good estimate of μ?

5.25 Find the population variance, σ^2, for Exercise 5.6 and the sample variance, s^2, for Exercise 5.7. Compare them.

5.26 Find the variance of y, the number of dots observed on the throw of a single die. (Refer to Example 5.2.) Use the probability histogram to compute the fraction of the population lying within two standard deviations of the mean.

5.27 Draw a sample of $n = 50$ measurements from the die-throw population of Example 5.2 by tossing a die 50 times and recording y after each toss. Calculate $\bar{y}$ and s^2 for the sample. Compare $\bar{y}$ with the expected value of y, Example 5.4, and s^2 with the variance of y obtained in Exercise 5.26. Do $\bar{y}$ and s^2 provide good estimates of μ and σ^2?

5.28 Use the probability distribution obtained in Exercise 5.6 to find the fraction of the total population of measurements that lies within two standard deviations of the mean. Compare with Tchebysheff's Theorem. Repeat for the sample in Exercise 5.7.

5.29 A personnel director wishes to select three out of five candidates interviewed. Although all five candidates appear to be equally desirable, there likely exists a "best" candidate, a second best, etc. Refer to the three, and unknown, best candidates as "successes" and let y equal the number of "successes" appearing in the personnel director's selection; that is, $y = 1$, 2, or 3. Find the probability distribution for y. Construct a probability histogram for y. What is the probability that he selects at least two of the three best?

5.30 Refer to Exercise 5.29. Find the expected value μ and variance σ^2 for y.

5.31 A heavy-equipment salesman can contact either one or two customers per day with probability 1/3 and 2/3, respectively. Each contact will result in either no sale or a \$50,000 sale with probability 9/10 and 1/10, respectively. What is the expected value of his daily sales?

5.32 A county containing a large number of rural homes is thought to have 60 percent insured against a fire. Four rural home owners are chosen at random from the entire population and y are found to be insured against a fire. Find the probability distribution for y. What is the probability that at least three of the four will be insured?

5.33 A fire-detection device utilizes three temperature-sensitive cells acting independently of each other in such a manner that any one or more may actuate the alarm. Each cell possesses a probability of $p = .8$ of actuating the alarm when the temperature reaches 100 degrees or more. Let y equal the number of cells actuating the alarm when the temperature reaches 100 degrees. Find the probability distribution for y. Find the probability that the alarm will function when the temperature reaches 100 degrees.

5.34 Find the expected value and variance for the random variable y defined in Exercise 5.33.

5.35 If you toss a pair of dice, the sum S of the numbers of dots appearing on the upper faces of the dice can assume the value of an integer in the interval, $2 \leq S \leq 12$.
(a) Find the probability distribution for S.
(b) Find $E(S)$.
(c) Find σ^2.

(d) Graph $p(S)$ and locate the interval, $\mu \pm 2\sigma$.
(e) What is the probability that S will assume a value in the interval, $\mu \pm 2\sigma$?

* 5.36 Express the probability distribution, Exercise 5.6, as a formula. (The distribution is known to statisticians as a hypergeometric probability distribution.)

5.37 A potential customer for a $20,000 fire insurance policy possesses a home in an area, which, according to experience, may sustain a total loss in a given year with probability of .001 and a 50 percent loss with probability .01. Ignoring all other partial losses, what premium should the insurance company charge for a yearly policy in order to break even?

5.38 The manufacturer of a low-calorie dairy drink wishes to compare the taste appeal of a new formula (formula *B*) with that of the standard formula (formula *A*). Each of four judges is given three glasses in random order, two containing formula *A* and the other containing formula *B*. Each judge is asked to state which glass he most enjoyed. Suppose that the two formulas are equally attractive. Let y be the number of judges stating a preference for the new formula.
(a) Find the probability function for y.
(b) What is the probability that at least three of the four judges state a preference for the new formula?
(c) Find the expected value of y. (d) Find the variance of y.

5.39 A fisherman has probability 1/3 of catching a fish before a given worm must be replaced. He never uses the same worm to catch more than one fish. Suppose he has three worms in his bait can. Let y be the number of fish he catches before running out of worms.
(a) Find the probability function for y.
(b) What is the probability that the fisherman catches more than one fish?
(c) What is the expected number of fish caught?
(d) What is the variance of the number of fish caught?

5.40 Continuing Exercise 5.39, we shall assume that if a fish is caught, it is (with probability 1/2) too small and must be released. If a fish is caught and it is not too small, imagine that it weighs 1 pound. Let w be the total weight of the fish actually kept.
(a) Find the probability distribution for w.
(b) What is the probability that the total weight of the fish retained exceeds 1 pound?
(c) What is the expected weight of the fish retained?
(d) Find the variance of w.

5.41 The probability distribution for y, the number of new automobile sales per day at a small distributorship, is as shown in the table.

y	$p(y)$
0	.10
1	.20
2	.40
3	.15
4	.10
5	.03
6	.01
7	.01

(a) Find the expected number of sales per day.
(b) Find σ^2 and σ.
(c) What is the probability that y will fall in the interval, $\mu \pm 2\sigma$?
(d) If three days are chosen at random, what is the probability that $y \geq 3$ for all three days?

5.42 To test the willow-wand theory of water dowsing, each of 10 men is asked to explore given terrain and specify a single spot at which water can be found. Exactly 5 of these men will use the willow wand. Let y be the number out of the 5 willow-wand users who are successful in locating water. Suppose that exactly 4 men out of the 10 are successful in locating water and that the willow-wand theory is a fraud.
(a) Find the probability distribution for y.
(b) Find the probability that at least 3 of the 4 successful men used the willow wand.
(c) What is the expected number of successful willow-wand users?
(d) Find the variance of y.

5.43 An investment can result in one of three outcomes: a \$7,000 gain, a \$4,000 gain, or a \$10,000 loss with probabilities .55, .20, and .25, respectively. Find the expected gain for a potential investor.

5.44 Accident records collected by an automobile insurance company give the following information. The probability that an insured driver has an automobile accident is .15. If an accident has occurred, the damage to the vehicle amounts to 20 percent of its market value with probability .80, 60 percent of its market value with probability .12, and a total loss with probability .08. What premium should the company charge on a \$4,000 car so that the expected gain by the company is zero?

* 5.45 The Poisson probability distribution provides a good model for count data when y, the count, represents the number of "rare events" observed in a given unit of time, distance, area, or volume. The number of bacteria per small volume of fluid and the number of traffic accidents at a given intersection during a given period of time possess, approximately, a Poisson probability distribution which is given by the formula

$$p(y) = \frac{\mu^y e^{-\mu}}{y!},$$

where $y = 0, 1, 2, 3, \ldots$; $e = 2.718 \ldots$; and μ is the population mean. If the average number of accidents over a specified section of highway is two per week:
(a) Construct a probability histogram for y. (Use Table 2, Appendix II, to evaluate $e^{-\mu}$.)
(b) Calculate the probability that no accidents will occur during a given week.
(c) It can be shown that the standard deviation of a Poisson random variable is

$$\sigma = \sqrt{\mu}.$$

Construct the interval $\mu \pm 2\sigma$ and locate it on the graph, part (a). What is the probability that y will fall in the interval $\mu \pm 2\sigma$?

* 5.46 Suppose that a random system of police patrol is devised so that a patrolman may visit a given location on his beat $y = 0, 1, 2, 3, \ldots$ times per half-hour period and that the system is arranged so that he visits each location on an average of once per time period. Assume that y possesses, approximately, a Poisson probability distribution. Calculate the probability that the patrolman will miss a given location during a half-hour period. What is the probability that he will visit it once? twice? at least once? (See Exercise 5.45 for the formula for the Poisson probability distribution.)

* 5.47 The average accident rate on a given stretch of highway is known to average three per week. What is the probability that no accidents will occur during a given week? (*Hint:* See Exercise 5.45.)

* 5.48 The number y of people entering the intensive care unit at a particular hospital on any one day possesses a Poisson probability distribution with mean equal to 5 persons per day.
(a) What is the probability that the number of people entering the intensive care unit on a particular day is equal to 2? less than or equal to 2?
(b) Is it likely that y will exceed 10? Explain. (*Note:* Use the information about the Poisson probability distribution given in Exercise 5.45.)

* 5.49 If a drop of water is placed on a slide and examined under a microscope, the number y of a particular type of bacteria present has been found to possess a Poisson probability distribution. Suppose that the maximum permissible count per water specimen for this type of bacteria is 5. If the mean count for your water supply is 2 and you test a single specimen, is it likely that the count will exceed the maximum permissible count? Explain. (Use the information about the Poisson probability distribution given in Exercise 5.45. *Note:* This exercise requires very little computation.)

* 5.50 The probability distribution for the number of well drillings y until the first success (Exercise 5.9) possesses a geometric probability distribution that is given by the formula

$$p(y) = p(1 - p)^{y-1},$$

where p is the probability of success for a single drilling. Further, it can be shown (proof omitted) that the mean and variance of p are equal to

$$E(y) = \frac{1}{p} \quad \text{and} \quad \sigma^2 = \frac{1 - p}{p^2}.$$

Suppose the company, Exercise 5.9, commenced drilling for oil. Would it be unlikely that it would take as many as 20 drillings before the company encountered its first success? Explain.

5.51 If you own common stock in a corporation, you can sometimes increase the return on your investment by selling an option. For a stated price, the purchaser of the option gains the right to buy your stock at any time up to a specified expiration date. Suppose you purchased 200 shares of a stock at $25 per share and you sell an option for the option purchaser to buy the stock at any time within the next six months for $30 per share. If the stock reaches $30 per share within the next six months, your stock will be sold and you will gain $5 per share plus $2 per share (from the dividends and the sale of the option) less $109 commission to the broker for selling your stock. If you do not sell, you will gain $2 per share. If the probability of the stock reaching $30 per share within the next six months is .7, what is the expected return (in dollars) from your 200 shares of stock? What is the expected annual rate of return, in percentage, on your investment? (*Note:* We have ignored a $15 commission on the sale of the option.)

* 5.52 Use the summation theorems, Theorems 2.1, 2.2, and 2.3, to prove the following theorems:

(a) Theorem 5.1: Let c be a constant and let y be a discrete random variable with probability distribution $p(y)$. Then the expected value of c is c. [That is, $E(c) = c$.]

(b) Theorem 5.2: Let c be a constant and let y be a discrete random variable with probability distribution $p(y)$. Then,

$$E(cy) = cE(y).$$

(c) Theorem 5.3: Let y be a discrete random variable with probability distribution $p(y)$ and let $g_1(y)$ and $g_2(y)$ be functions of the random variable y. Then,

$$E[g_1(y) + g_2(y)] = E[g_1(y)] + E[g_2(y)].$$

Note that this theorem implies that the expected value of a sum of functions of a random variable is equal to the sum of their respective expected values.

* 5.53 Use Theorems 5.1, 5.2, and 5.3 to show that

$$\sigma^2 = E[(y - \mu)^2] = E(y^2) - \mu^2.$$

(*Hint:* $E[(y - \mu)^2] = E(y^2 - 2\mu y + \mu^2)$.)

*5.54 Use the formula given in Exercise 5.53 to find σ^2 for Example 5.7.

*5.55 Use the formula given in Exercise 5.53 to find σ^2 for Exercise 5.20.

*5.56 Refer to Exercise 5.20 and find:
(a) $P[y = 2 | y > 0]$.
(b) $P[y \geq 3 | y > 0]$.
(*Hint:* Refer to the definition of a conditional probability, Section 4.5.)

EXPERIENCES WITH REAL DATA

We noted in Exercise 5.45 that a Poisson probability distribution (named for the famous French mathematician S. D. Poisson) provides a good model for the population relative frequency distribution for count data, where y, the count, represents the number of "rare events" observed in a given unit of time, distance, area, or volume. This probability distribution is given by the formula

$$p(y) = \frac{\mu^y e^{-\mu}}{y!}$$

where $y = 0, 1, 2, 3, \ldots$; $e = 2.718\ldots$; and μ is the population mean.

To verify that the Poisson probability distribution provides a good model for y, the number of "rare events" occurring in time, distance, area, or volume, let us collect some data on "rare events" and compare the relative frequency histogram for the data with the theoretical frequencies provided by the Poisson probability distribution.

Random variables that might possess a Poisson probability distribution and be suitable for our investigation are:

1. The number of children per family.
2. The number of automobile accidents which result in a fatality that occur in a single day and in a limited specified area (consult the records of your local police department).
3. The number of typing errors per page of typing.
4. The number of VW automobiles that pass a specific city or town intersection in a given minute. Your instructor may be able to suggest other random variables that would be good candidates for your experiment.

Randomly select $n = 200$ values of y (this could be a class project) for one of the preceding variables. Construct a table of the relative frequencies for $y = 0, 1, 2, \ldots$. Calculate $\bar{y}$ and use this as an estimate of μ. Then substitute this value into the formula for the Poisson probability distribution and calculate $p(0)$, $p(1)$, $p(2)$, ... Arrange these probabilities in the table opposite the corresponding observed relative frequencies. Construct the relative frequency histogram for the data and superimpose over it the probability histogram for y. Does it appear that the Poisson probability distribution is a good model for the population frequency distribution for your random variable?

CHAPTER OBJECTIVES

GENERAL OBJECTIVES A method for finding the probabilities associated with specific values of random variables was presented in Chapter 5. Specifically, we learned that discrete and continuous random variables required different probability distributions and that these distributions were subject to different interpretations. Now we turn to some specific applications of these notions and present a useful discrete random variable and its probability distribution. In concluding the chapter, we will show you how this probability distribution can be used in making inferences.

SPECIFIC OBJECTIVES

1. To describe the characteristics of a binomial experiment, to indicate types of data which represent measurements on binomial random variables, to give the formula for the binomial probability distribution and to show how it is used to calculate probabilities for a binomial random variable. *Sections 6.1, 6.2*
2. To present the formulas for the expected value and variance of a binomial random variable. These quantities will be used to describe a binomial probability distribution and will be used in Chapter 7 for a simple procedure for approximating binomial probabilities. *Section 6.3*
3. To show you how the binomial probability distribution is used to make inferences about a binomial population based on the information contained in a sample. *Sections 6.4, 6.5*
4. To introduce the concepts involved in a statistical test of an hypothesis. *Sections 6.5, 6.6, 6.7*

6 THE BINOMIAL PROBABILITY DISTRIBUTION

6.1

THE BINOMIAL EXPERIMENT

One of the most elementary, useful, and interesting discrete random variables is associated with the coin-tossing experiment described in Examples 4.3, 4.5, and 5.1. In an abstract sense, numerous coin-tossing experiments of practical importance are conducted daily in the social sciences, physical sciences, and industry.

To illustrate, we might consider a sample survey conducted to predict voter preferences in a political election. Interviewing a single voter bears a similarity, in many respects, to tossing a single coin because the voter's response may be in favor of our candidate—a "head"—or it may be opposition (or indicate indecision)—a "tail." In most cases, the fraction of voters favoring a particular candidate will not equal one-half, but even this similarity to the coin-tossing experiment is satisfied in national presidential elections. History demonstrates that the fraction of the total vote favoring the winning presidential candidate in most national elections is very near one-half.

Similar polls are conducted in the social sciences, in industry, in education. The sociologist is interested in the fraction of rural homes that have electricity; the cigarette manufacturer desires knowledge concerning the fraction of smokers who prefer his brand; the teacher is interested in the fraction of students who pass his course. Each person sampled is analogous to the toss of an unbalanced (since the probability of a "head" is usually not 1/2) coin.

Firing a projectile at a target is similar to a coin-tossing experiment if a "hit the target" and a "miss the target" are regarded as a head and a tail, respectively. A single missile will result in either a successful or unsuccessful launching. A new drug will prove either effective or noneffective when administered to a single patient, and a manufactured item selected from production will be either defective or nondefective. Although dissimilar in some respects, the experiments described above will often exhibit, to a reasonable degree of approximation, the characteristics of a binomial experiment.

DEFINITION

A binomial experiment is one that possesses the following properties:

1. **The experiment consists of n identical trials.**
2. **Each trial results in one of two outcomes. For lack of a better nomenclature, we will call the one outcome a success, S, and the other a failure, F.**
3. **The probability of success on a single trial is equal to p and remains the same from trial to trial. The probability of a failure is equal to $(1 - p) = q$.**
4. **The trials are independent.**
5. **We are interested in y, the number of successes observed during the n trials.**

EXAMPLE 6.1

Suppose that there are approximately 1,000,000 adults in a county and that an unknown proportion, p, favor and $(1 - p)$ oppose the Equal Rights Amendment (ERA). A random sample of 1,000 adults will be selected from among the 1,000,000 in the county and each will be asked whether he or she favors the Equal Rights Amendment. (The ultimate objective of this survey is to estimate the unknown proportion, p, a problem that we will learn how to solve in Chapter 8.) Is this a binomial experiment?

Solution To decide whether this is a binomial experiment, we must see if the sampling satisfies the five characteristics described in the definition.

1. The sampling consists of $n = 1{,}000$ identical trials. One trial represents the selection of a single adult from the 1,000,000 adults in the county.
2. Each trial will result in one of two outcomes. A person will either favor the amendment or will not. These two outcomes could be associated with the "success" and "failure" of a binomial experiment.
3. The probability of a success will equal the proportion of adults favoring the ERA. For example, if 500,000 of the 1,000,000 adults in the county favor the ERA, then the probability of selecting an adult favoring the ERA out of the 1,000,000 in the county is $p = .5$. For all practical purposes, this probability will remain the same from trial to trial even though adults selected in the earlier trials are not replaced as the sampling continues.
4. For all practical purposes, the probability of a success on any one trial will be unaffected by the outcome on any of the others (it will remain very close to p).
5. We are interested in the number y of adults in the sample of 1,000 who favor the Equal Rights Amendment.

Because the survey satisfies the five characteristics reasonably well, for all practical purposes it can be viewed as a binomial experiment.

EXAMPLE 6.2

A purchaser, who has received a boxcar containing 20 large electronic computers, wishes to sample 3 of the computers to see if they are in working order before he unloads the shipment. The 3 nearest the door of the boxcar are removed for testing and, afterward, are declared either defective or nondefective. Unknown to the purchaser, 2 of the computers are defective. Is this a binomial experiment?

Solution As for Example 6.1, we check the sampling procedure against the characteristics of a binomial experiment.

1. The experiment consists of $n = 3$ identical trials. Each trial represents the selection and testing of one computer from the total of 20.
2. Each trial results in one of two outcomes. Either a computer is defective (call this a "success") or it is not (a "failure").

3. Suppose that the computers were randomly loaded into the boxcar so that any one of the 20 computers could have been placed near the boxcar door. Then, the unconditional probability of drawing a defective computer on a given trial will be 3/20.
4. The condition of independence between trials *is not* satisfied because the probability of drawing a defective computer on the second and third trials will be dependent on the outcome of the first trial. For example, if the first trial results in a defective computer, then there are only 2 defectives left of the remaining 19 in the boxcar. Therefore, the conditional probability of success on trial 2, given a success on trial 1, is 2/19. This differs from the unconditional probability of a success on the second trial (which is 3/20). Therefore, the trials are dependent and the sampling does not represent a binomial experiment.

Example 6.1 illustrates an important point. Very few real-life situations will perfectly satisfy the requirements stated above, but this is of little consequence as long as the lack of agreement is moderate and does not affect the end result. For instance, the probability of drawing a voter favoring a particular candidate in a political poll remains approximately constant from trial to trial as long as the population of voters is relatively large in comparison with the sample. If 50 percent of a population of 1,000 voters prefer candidate A, then the probability of drawing an A on the first interview will be 1/2. The probability of an A on the second draw will equal 499/999 or 500/999, depending upon whether the first draw was favorable or unfavorable to A. Both numbers are near 1/2, for all practical purposes, and would continue to be for the third, fourth, and nth trial as long as n is not too large. Hence $P(A)$ remains approximately 1/2 from trial to trial and, for all practical purposes, we could regard the trials as independent. On the other hand, if the number in the population is 10, and 5 favor candidate A, then the probability of A on the first trial is 1/2; the probability of A on the second trial is 4/9 or 5/9, depending upon whether A was, or was not, drawn on the first trial. Thus, for small populations, the trial outcomes will represent dependent events and the resulting experiment will not be a binomial experiment.

6.2

THE BINOMIAL PROBABILITY DISTRIBUTION

Having defined the binomial experiment and suggested several practical applications, we now turn to a derivation of the probability distribution for the random variable, y, the number of successes observed in n trials. Rather than attempt a direct derivation, we will obtain $p(y)$ for experiments containing $n = 1$, 2, and 3 trials and leave the general formula to your intuition.

For $n = 1$ trial, we have two sample points, E_1 representing a success, S, and E_2 representing a failure, F, with probabilities p and $q = (1 - p)$, respectively. Since y is the number of successes for the one ($n = 1$) trial and since E_1 implies a success, we assign $y = 1$ to E_1. Similarly, since E_2 represents a failure

TABLE 6.1
$p(y)$ for a binomial experiment, $n = 1$

Sample Points		$P(E_i)$	y
E_1	S	p	1
E_2	F	q	0

y	$p(y)$
0	q
1	p

$$\sum_{y=0}^{1} p(y) = q + p = 1$$

for the single trial, we assign $y = 0$ to this sample point. The resulting probability distribution for y is given in Table 6.1. (*Note:* S represents a success on a single trial, F denotes a failure.)

The probability distribution for an experiment consisting of $n = 2$ trials is derived in a similar manner and is presented in Table 6.2. The four sample points associated with the experiments are presented in the first column with the notation SF implying a success on the first trial and a failure on the second.

The probabilities of the sample points are easily calculated because each point is an intersection of two independent events, namely, the outcomes of the first and second trials. Thus $P(E_i)$ can be obtained by applying the Multiplicative Law of Probability:

$$P(E_1) = P(SS) = P(S)P(S) = p^2,$$
$$P(E_2) = P(SF) = P(S)P(F) = pq,$$
$$P(E_3) = P(FS) = P(F)P(S) = qp,$$
$$P(E_4) = P(FF) = P(F)P(F) = q^2.$$

The value of y assigned to each sample point is given in the third column. You will note that the numerical event $y = 0$ contains sample point E_4, the event $y = 1$ contains sample points E_2 and E_3, and $y = 2$ contains sample point E_1. The probability distribution $p(y)$, presented to the right of Table 6.2, reveals a most

TABLE 6.2
$p(y)$ for a binomial experiment, $n = 2$

Sample Points		$P(E_i)$	y
E_1	SS	p^2	2
E_2	SF	pq	1
E_3	FS	qp	1
E_4	FF	q^2	0

y	$p(y)$
0	q^2
1	$2pq$
2	p^2

$$\sum_{y=0}^{2} p(y) = q^2 + 2pq + p^2$$
$$= (q + p)^2$$
$$= (1)^2 = 1$$

TABLE 6.3
$p(y)$ for a binomial experiment, $n = 3$

Sample Points		$P(E_i)$	y
E_1	SSS	p^3	3
E_2	SSF	p^2q	2
E_3	SFS	p^2q	2
E_4	SFF	pq^2	1
E_5	FSS	p^2q	2
E_6	FSF	pq^2	1
E_7	FFS	pq^2	1
E_8	FFF	q^3	0

y	$p(y)$
0	q^3
1	$3pq^2$
2	$3p^2q$
3	p^3

$$\sum_{y=0}^{3} p(y) = q^3 + 3pq^2 + 3p^2q + p^3$$
$$= (q + p)^3 = (1)^3$$
$$= 1$$

interesting consequence; the probabilities $p(y)$ are terms of the expansion of $(q + p)^2$.

Summing, we obtain

$$\sum_{y=0}^{2} p(y) = q^2 + 2pq + p^2$$
$$= (q + p)^2 = 1.$$

The point that we wish to make is now quite clear; the probability distribution for the binomial experiment consisting of n trials is obtained by expanding $(q + p)^n$. The proof of this statement is omitted but may be obtained by using combinatorial mathematics (optional Section 4.9) to count the appropriate sample points or can be proved by mathematical induction, a task which we leave to the student of mathematics. Those more easily convinced may acquire further evidence of the truth of our statement by observing the derivation of the probability distribution for a binomial experiment consisting of $n = 3$ trials presented in Table 6.3.

Since the probability associated with a particular value of y is simply the term involving p to the power y in the expansion of $(q + p)^n$, we may write the probability distribution for the binomial experiment as follows:

Binomial Probability Distribution

$$p(y) = C_y^n p^y q^{n-y},$$

where y may take values 0, 1, 2, 3, 4, . . . , n and C_y^n is a symbol used to represent

$$\frac{n!}{y!(n - y)!}.$$

You may recall that the factorial notation $n!$ (defined in Section 5.7) is used to represent the product, $n(n - 1)(n - 2) \cdots (3)(2)(1)$. For example, $5! = (5)(4)(3)(2)(1) = 120$, and $0!$ is defined to be equal to 1. The notation, C_y^n, is shorthand for $n!/y!(n - y)!$, an expression that appears in the formula for the binomial probability distribution.

Graphs of three binomial probability distributions are shown in Figure 6.1, the first for $n = 10$, $p = .1$, the second for $n = 10$, $p = .5$, and the third for $n = 20$, $p = .5$.

Let us now consider examples that illustrate applications of the binomial distribution.

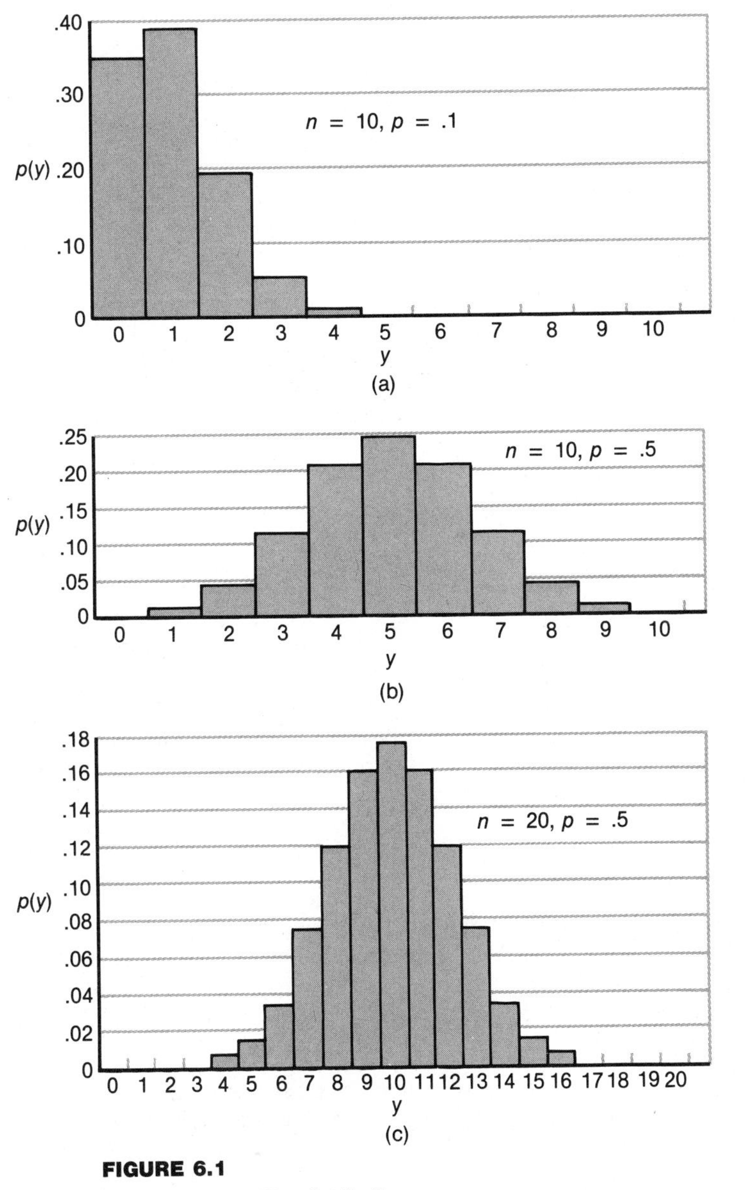

FIGURE 6.1

Binomial probability distributions

EXAMPLE 6.3

Over a long period of time it has been observed that a given rifleman can hit a target on a single trial with probability equal to .8. If he fires four shots at the target:

(a) What is the probability that he will hit the target exactly two times?

Solution Assuming that the trials are independent and that p remains constant from trial to trial, $n = 4$, $p = .8$, and

$$\begin{aligned} p(y) &= C_y^4(.8)^y(.2)^{4-y}, \\ p(2) &= C_2^4(.8)^2(.2)^{4-2} \\ &= \frac{4!}{2!2!}(.64)(.04) = \frac{4(3)(2)(1)}{2(1)(2)(1)}(.64)(.04) \\ &= .1536. \end{aligned}$$

(b) What is the probability that he will hit the target at least two times?

Solution

$$\begin{aligned} P(\text{at least two}) &= p(2) + p(3) + p(4) \\ &= 1 - p(0) - p(1) \\ &= 1 - C_0^4(.8)^0(.2)^4 - C_1^4(.8)(.2)^3 \\ &= 1 - .0016 - .0256 \\ &= .9728. \end{aligned}$$

(c) What is the probability that he will hit the target exactly four times?

Solution

$$\begin{aligned} p(4) &= C_4^4(.8)^4(.2)^0 \\ &= \frac{4!}{4!0!}(.8)^4(1) \\ &= .4096. \end{aligned}$$

Note that these probabilities would be incorrect if the rifleman could observe the location of each hit on the target; in that case, the trials would be dependent and p would likely increase from trial to trial.

EXAMPLE 6.4

Large lots of incoming product at a manufacturing plant are inspected for defectives by means of a sampling scheme. Ten items are to be examined and the lot rejected if two or more defectives are observed. If a lot contains exactly 5 percent defectives, what is the probability that the lot will be accepted? rejected?

Solution If y equals the number of defectives observed, then $n = 10$ and the probability of observing a defective on a single trial will be $p = .05$. Then,

$$p(y) = C_y^{10}(.05)^y(.95)^{10-y}$$

and

$$\begin{aligned} P(\text{accept}) &= p(0) + p(1) = C_0^{10}(.05)^0(.95)^{10} + C_1^{10}(.05)^1(.95)^9 \\ &= .914, \\ P(\text{reject}) &= 1 - P(\text{accept}) \\ &= 1 - .914 \\ &= .086. \end{aligned}$$

Although in a practical situation we would not know the exact value of p, we would want to know the probability of accepting bad lots (lots for which p is large) and good lots (lots for which p is small). This example shows how we can calculate this probability of acceptance for various values of p.

EXAMPLE 6.5

A new serum was tested to determine its effectiveness in preventing the common cold. Ten people were injected with the serum and observed for a period of one year. Eight survived the winter without a cold. Suppose that it is known that when serum is not used, the probability of surviving a winter without a cold is .5. What is the probability of observing eight or more survivors, given that the serum is ineffective in increasing the bodily resistance to colds?

Solution Assuming that the vaccine is ineffective, the probability of surviving the winter without a cold is $p = .5$. The probability distribution for y, the number of survivors, is

$$\begin{aligned} p(y) &= C_y^{10}(.5)^y(.5)^{10-y} = C_y^{10}(.5)^{10}, \\ P(8 \text{ or more}) &= p(8) + p(9) + p(10) \\ &= C_8^{10}(.5)^{10} + C_9^{10}(.5)^{10} + C_{10}^{10}(.5)^{10} \\ &= .055. \end{aligned}$$

Examples 6.3, 6.4, and 6.5 illustrate the use of the binomial probability distribution in calculating the probability associated with values of y, the number of successes in n trials defined for the binomial experiment. Thus we note that the probability distribution, $p(y) = C_y^n p^y q^{n-y}$, provides a simple formula for calculating the probabilities of numerical events, y, applicable to a broad class of experiments that occur in everyday life. This statement must be accompanied by a word of caution. The important point, of course, is that each physical application must be carefully checked against the defining characteristics of the binomial experiment, Section 6.1, to determine whether the binomial experiment is a valid model for the application of interest.

You will note that Examples 6.3, 6.4, and 6.5 were problems in probability rather than statistics. The composition of the binomial population, characterized by p, the probability of a success on a single trial, was assumed known and we were interested in calculating the probability of certain numerical events. As we proceed in this chapter, we will reverse the procedure; that is, we will assume that we possess a sample from the population and wish to make inferences concerning p. The physical settings for Examples 6.4 and 6.5 supply excellent practical situations in which the ultimate objective was statistical inference. We shall consider these two problems in greater detail in the succeeding sections.

EXERCISES

6.1 Records show that 30 percent of all patients admitted to a medical clinic fail to pay their bills and eventually the bills are forgiven. Suppose that $n = 4$ new patients represent a random selection from the large set of prospective patients served by the clinic. Find the probability that:
(a) All the patients' bills will eventually have to be forgiven.
(b) That one will have to be forgiven.
(c) That none will have to be forgiven.

6.2 Refer to Exercise 6.1 and let y equal the number of patients in the sample of $n = 4$ whose bills will have to be forgiven. Construct a probability histogram for $p(y)$.

6.3 Approximately 10 percent of all thunderstorms tracked over the past five years by the National Oceanic and Atmospheric Administrative Severe Storms Laboratory were "mesocyclones," storms that breed or turn into tornadoes (*Washington Post,* December 7, 1976). Suppose a particular community was subjected to 7 thunderstorms and 2 are judged to be "mesocyclonic." What is the probability of observing 2 mesocyclones out of a total of 7 thunderstorms? two or more? Would you regard 2 or more as a "rare event," an event that would occur with small probability if in fact the probability of observing a mesocyclonic thunderstorm is only 10 percent?

6.4 Calculate $p(y)$ for $y = 0, 1, 2, \ldots, 5, 6$ for $n = 6$ and $p = .1$. Graph $p(y)$. Repeat for binomial probability distributions for $p = .5$ and $p = .9$. Compare the graphs. How does the value of p affect the shape of $p(y)$? (*Note:* This exercise does not require extensive calculations. For example, you need calculate the coefficients C_y^n only once because they will be the same for all three probability distributions.)

6.5 Consider a metabolic defect that occurs in approximately 1 of every 100 births. If four infants are born in a particular hospital on a given day, what is the probability that:
(a) None has the defect?
(b) No more than one has the defect?

6.6 Early in the United States missile development program, government and industry defense officials were proclaiming that our missiles were highly reliable, and probabilities of successful firings in the neighborhood of .999 . . . were quoted. Such statements were made even though many missile firings resulted in failure (such as the Navy *Vanguard* missile of the 1950s). If the reliability (probability of a successful launch) is even as high as .9, what is the probability of observing as many as 3 failures in a total of 4 firings? As many as 1 failure? If you observed 2 or more failures out of 4, what would you think about the high claims of reliability of the missiles produced in the 1950s?

6.7 A new surgical procedure is said to be successful 80 percent of the time. If the operation is performed five times and if we can assume that the results are independent of one another:
(a) What is the probability that all five operations are successful?
(b) Exactly four?
(c) Less than two?

6.8 Refer to Exercise 6.7. If less than two operations were successful, how would you feel about the performance of the surgical team? (We shall have more to say about this type of reasoning in Section 6.5.)

6.9 Many utility companies have begun to promote energy conservation by offering discounted rates to consumers who keep their energy usage below certain established subsidy standards. A recent EPA report notes that 70 percent of the island residents of Puerto Rico have reduced their electricity usage sufficiently to qualify for discounted rates. If five residential subscribers are randomly selected from San Juan, Puerto Rico, find the probability that:
(a) All five qualify for the favorable rates.
(b) At least four qualify for the favorable rates.

6.10 A survey in a certain state indicated that nine out of ten automobiles carry automobile liability insurance. If four autos in that state are involved in accidents, what is the probability that:
(a) No more than two of the four drivers have liability insurance?
(b) Exactly two have liability insurance?

6.3
THE MEAN AND VARIANCE FOR THE BINOMIAL RANDOM VARIABLE

Because calculation of $p(y)$ becomes very tedious for large values of n, it is convenient to describe the binomial probability distribution using its mean and standard deviation. This will enable us to identify values of y that are highly improbable simply by using our knowledge of Tchebysheff's Theorem and the Empirical Rule. A more precise method for approximating binomial probabilities will be presented in Chapter 7, and this method will rely on knowledge of the mean and standard deviation of y, namely μ and σ. Consequently, we need to find the expected value and variance of the binomial random variable, y.

Our approach to the acquisition of formulas for μ and σ^2 will be similar to that employed in Section 6.2 for the determination of the probability distribution, $p(y)$. Since $\mu = E(y)$ and $\sigma^2 = E[(y - \mu)^2]$, we shall use the methods of Chapter 5 to obtain these expectations; that is, $E(y)$ and $E(y - \mu)^2$, for the simple cases $n = 1$, 2, and 3, and then give the formulas for the general case without proof. The interested reader can derive the formulas for the general case involving n trials by using the summation theorems of Chapter 2, some algebraic manipulation, and a bit of ingenuity (see Mendenhall and Scheaffer, Chapter 3).

The expected value of y for $n = 1$ and 2 can be derived using the probabilities given in Tables 6.1 and 6.2 along with the definition of an expectation presented in Section 5.6. Thus for $n = 1$ we obtain

$$\mu = E(y) = \sum_{y=0}^{1} yp(y) = (0)(q) + (1)(p) = p.$$

For $n = 2$,

$$\mu = E(y) = \sum_{y=0}^{2} yp(y) = (0)(q^2) + (1)(2pq) + (2)(p^2)$$
$$= 2p(q + p) = 2p.$$

Using Table 6.3, you can quickly show that $E(y)$ for $n = 3$ trials is equal to $3p$ and would surmise that this pattern would hold in general. Indeed, it can be shown that the expected value of y for a binomial experiment consisting of n trials is

Mean for a Binomial Random Variable

$$\mu = E(y) = np.$$

Similarly, we may obtain the variance of y for $n = 1$ and 2 trials as follows. For $n = 1$,

$$\sigma^2 = E(y - \mu)^2 = \sum_{y=0}^{1} (y - \mu)^2 p(y) = (0 - p)^2(q) + (1 - p)^2(p)$$
$$= p^2 q + q^2 p = pq(q + p).$$

Or, since $q + p = 1$,

$$\sigma^2 = pq.$$

For $n = 2$,

$$\sigma^2 = E(y - \mu)^2 = \sum_{y=0}^{2} (y - \mu)^2 p(y)$$
$$= (0 - 2p)^2(q^2) + (1 - 2p)^2(2pq) + (2 - 2p)^2(p^2)$$
$$= 4p^2q^2 + (1 - 4p + 4p^2)(2pq) + 4(1 - p)^2 p^2.$$

Using the substitution $q = 1 - p$ and a bit of algebraic manipulation, we find that the above reduces to

$$\sigma^2 = 2pq.$$

You may verify that the variance of y for $n = 3$ trials is $3pq$. In general, for n trials it can be shown that the variance of y is

Variance and Standard Deviation for a Binomial Random Variable

$$\sigma^2 = npq.$$
$$\sigma = \sqrt{npq}.$$

As we have previously mentioned, the formulas for μ and σ^2 can now be employed to compute the mean and variance for the binomial random variable, y, for a practical binomial experiment. Thus we can say something concerning the center of the distribution as well as the variability of y.

EXAMPLE 6.6

How do you evaluate scores on a multiple-choice test? A score of 0 on an objective test (questions requiring complete recall of the material) indicates that the person was unable to recall the test material at the time the test was given. In contrast, a person with little or no recall knowledge of the test material can score higher on a multiple-choice test because the person only needs to recognize (in contrast to recall) the correct answer and because some questions will be answered correctly, just by chance, even if the person does not know the correct answers. Consequently, the no-knowledge score for a multiple-choice test may be well above 0. If a multiple-choice test contains 100 questions, each with six possible answers, what is the expected score for a person who possesses no knowledge of the test material? Within what limits would you expect a no-knowledge score to be?

Solution Let p equal the probability of a correct choice on a single question and let y equal the number of correct responses out of the $n = 100$ questions. We will assume that no-knowledge means that a student will randomly select one of the six possible answers for each question and hence that $p = 1/6$. Then for $n = 100$ questions, a no-knowledge student's expected score would be $E(y)$, where,

$$E(y) = np = 100(1/6) = 16.7 \text{ correct questions.}$$

To evaluate the variation of no-knowledge scores, we need to know σ, where

$$\sigma = \sqrt{npq} = \sqrt{(100)(1/6)(5/6)} = 3.7.$$

Then from our knowledge of Tchebysheff's Theorem and the Empirical Rule,* we would expect most no-knowledge scores to lie in the interval, $\mu \pm 2\sigma$, or $16.7 \pm (2)(3.7)$, or from 9.3 to 24.1. This compares with a score of 0 for a no-knowledge student taking an objective recall test.

EXERCISES

6.11 Let us consider the medical payment problem (Exercise 6.1) in a more realistic setting. We were given the fact that 30 percent of all patients admitted to a medical clinic fail to pay their bills and that the bills are eventually forgiven. If over a year period of time the clinic treats 2,000 different patients, what is the mean (expected) number of bills that would have to be forgiven? If y is the number of forgiven bills in the group of 2,000 patients, find the variance and standard deviation of y. What can you say about the probability that y will exceed 700? (*Hint:* Use the values of μ and σ, along with Tchebysheff's Theorem, to answer this question.)

6.12 A national poll of 1,502 adults by pollster Louis Harris (*Environmental News,* EPA, September 1977) indicated that 71 percent "would rather live in an environment that is clean rather than in an area with a lot of jobs." If people really had no preference for environment over jobs, or vice versa, comment on the probability of observing this survey result (that is, observing 71 percent or more of a sample of 1,502 in favor of environment over jobs). What assumption must be made about the sampling proce-

*A histogram of $p(y)$ for $n = 100$, $p = 1/6$ will be mound-shaped. Hence we would expect the Empirical Rule to work very well. The reason for this will be explained in Chapter 7.

dure in order to calculate this probability? (*Hint:* Recall Tchebysheff's Theorem and the Empirical Rule.)

6.13 The administrator of the North Florida Regional Hospital (a private hospital) applied to the health planning council in 1975 for permission to make a 100-bed addition to the hospital, an application that was subsequently denied because of the uncertain impact that the addition would have on the existing public county hospital (*Gainesville Sun,* February 12, 1977). To support their application, and, presumably, those applications to be resubmitted in the future, the hospital administrators conducted a telephone survey of 180 people in the Gainesville area. In response to the question, "Should North Florida be allowed to expand . . . even if (there are) empty beds at AGH (the county hospital)," 62 percent answered yes. If the public is really split 50–50 on this question, is it likely that the telephone poll would result in as many as 62 percent in favor of the North Florida expansion? In order for the preceding probability calculation to be correct, what must you know about the sampling procedure? Do you think the poll supports North Florida's request? Explain. (*Hint:* Recall Tchebysheff's Theorem and the Empirical Rule.)

6.14 The Energy Policy Center of the EPA reports that 75 percent of the homes in New England are heated by oil-burning furnaces. If a certain New England community is known to have 2,500 homes, find the expected number of homes in the community that are heated by oil furnaces. If y is the number of homes in the community that are heated by oil, find the variance and standard deviation of y. Use Tchebysheff's Theorem to describe limits within which one could expect y to fall.

6.15 Suppose that 80 percent of a breed of hogs is infected with trichinosis. If a random sample of 1,000 hogs is examined by a state inspector:

(a) What is the expected value of y, the number of infected hogs in the sample of 1,000?

(b) What is the standard deviation of y?

(c) If the inspector observed $y = 900$ infected pigs in the sample, does it appear that the percentage of infected pigs in the population is really 80 percent? Explain.

6.16 If a political candidate possesses exactly 50 percent of the popular vote and the early election returns show a total of 10,000 votes:

(a) What is the expected value of y, the number of votes in favor of the candidate, if the 10,000 votes can be regarded as a random sample from the total population of voters?

(b) What is the standard deviation of y?

(c) Suppose that $y = 4{,}700$. Is this a value of y that might be expected with reasonable probability? How might you explain this observed result?

6.17 It is known that 10 percent of a brand of television tubes will burn out before their guarantee has expired. If 1,000 tubes are sold, find the expected value and variance of y, the number of original tubes that must be replaced. Within what limits would y be expected to fall? (*Hint:* Use Tchebysheff's Theorem.)

6.4

MAKING DECISIONS: LOT ACCEPTANCE SAMPLING FOR DEFECTIVES

A manufacturing plant may be regarded as an operation that transforms raw materials into a finished product, the raw materials entering the rear door of the plant and the product moving out the front. In order to operate efficiently, a manufacturer desires to minimize the amount of defective raw material received and, in the interest of quality, minimize the number of defective products shipped to customers. To accomplish this objective, the manufacturer will erect a "screen" at both doors in an attempt to prevent defectives from passing through.

To simplify our discussion, consider only the screening of incoming raw materials consisting of large lots (boxes) of items such as bolts, nails, or bearings. Diagrammatically, the screen would function in the manner indicated in Figure 6.2. The lots would proceed to the rear door of the plant, would be accepted if the fraction defective were small, and would be returned to the supplier if the fraction defective were large.

A screen could be constructed in a number of ways, the most obvious and seemingly perfect solution being a complete and careful inspection of each single item received. Unfortunately, the cost of total inspection is often enormous and hence economically unfeasible. A second disadvantage, not readily apparent, is that, even with complete inspection, defective items seem to slip through the screen. Humans become bored and lose perception when subjected to long hours of inspecting, particularly when the operation is conducted at high speed. Thus nondefective items are often rejected and defective items accepted.

A final disadvantage of total inspection is that some tests are by their very nature destructive. Testing a photoflash bulb to determine the quantity of light produced destroys the bulb. If all bulbs were tested in this manner, the manufacturer would have none left to sell.

A second type of screen, relatively inexpensive and lacking the tedium of total inspection, involves the use of a statistical sampling plan similar to the plan described in Example 6.4. A sample of n items is chosen at random from the lot of items and inspected for defectives. If the number of defectives, y, is less than or equal to a predetermined number, a, called the acceptance number, the lot is accepted. Otherwise the lot is rejected and returned to the supplier. The acceptance number for the plan described in Example 6.4 was $a = 1$.

You will note that this sampling plan operates in a completely objective manner and results in an inference concerning the population of items contained in the lot. If the lot is rejected, we infer that the fraction defective, p, is too large; if the lot is accepted, we infer that p is small and acceptable for use in the manufacturing process. Lot acceptance sampling plans provide an example of a statistical decision process which is, indeed, a method of statistical inference.

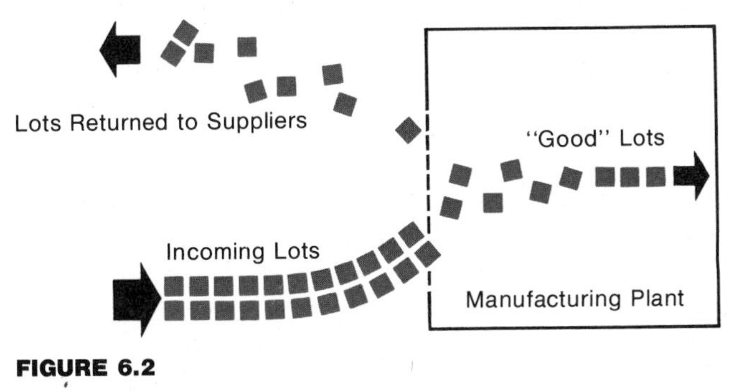

FIGURE 6.2
Screening for defectives

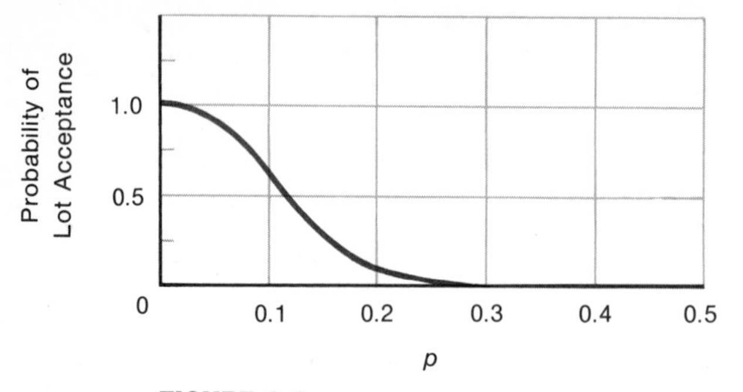

FIGURE 6.3

Typical operating characteristic curve for a sampling plan

Our discussion of lot acceptance sampling plans would be incomplete if we were to neglect some comment concerning the "goodness" of our inference-making procedure. Although the sampling plan described above is a decision-making process, it is not unique. We could change the number in the sample, n, change a, the acceptance number, or, for that matter, we could use some nonstatistical decision-making process, a procedure that is not uncommon in practice. In actuality, each individual is a decision maker relying on individual whims, preferences, and tastes. How can we compare these decision-making processes? The answer is immediately at hand; we choose the decision-making process which makes the correct decision most frequently, or, alternatively, makes the incorrect decision the smallest fraction of the time.

Quality control engineers characterize the goodness of a sampling plan by calculating the probability of lot acceptance for various lot fractions defective. The result is presented in a graphic form and is called the operating characteristic curve for the sampling plan. A typical operating characteristic curve is shown in Figure 6.3. In order for the screen to operate satisfactorily, we would like the probability of accepting lots with a low fraction defective to be high and the probability of accepting lots with a high fraction defective to be low. You will note that the probability of acceptance always drops as the fraction defective increases, a result that is in agreement with our intuition.

For instance, suppose that the supplier guarantees that lots will contain less than 1 percent defective and that the manufacturer can operate satisfactorily with lots containing less than 5 percent defective. Then the probability of accepting lots with less than 1 percent defective should be high. Otherwise the supplier will raise his price to cover the cost of "good lots" (less than 1 percent defective) which have been returned or will charge the manufacturer a fee for reinspection. On the other hand, the probability of accepting lots with 5 percent or more defective should be very low.

EXAMPLE 6.7

Calculate the probability of lot acceptance for a sampling plan with sample size $n = 5$ and acceptance number $a = 0$ for lot fraction defective $p = .1$, .3, and .5. Sketch the operating characteristic curve for the plan.

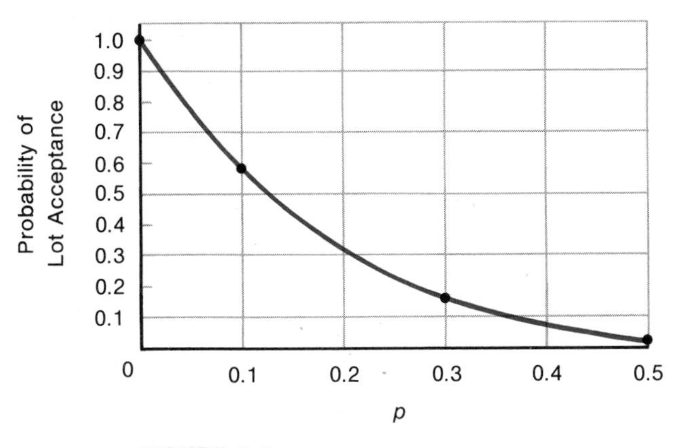

FIGURE 6.4

Operating characteristic curve, $n = 5$, $a = 0$

Solution

$$P(\text{accept}) = p(0) = C_0^5 p^0 q^5 = q^5,$$
$$P(\text{accept} \mid p = .1) = (.9)^5 = .590,$$
$$P(\text{accept} \mid p = .3) = (.7)^5 = .168,$$
$$P(\text{accept} \mid p = .5) = (.5)^5 = .031.$$

A sketch of the operating characteristic curve can be obtained by plotting the three points obtained from the above calculation. In addition, we know that the probability of acceptance must equal 1 when $p = 0$ and must equal zero when $p = 1$. The operating characteristic curve is given in Figure 6.4.

Calculating the binomial probabilities is a tedious task when n is large. To simplify our calculations, the sum of the binomial probabilities from $y = 0$ to $y = a$ is presented in Appendix II, Table 1, for sample sizes $n = 5$, 10, 15, 20, and 25. Why do we give the sum $\sum_{y=0}^{a} p(y)$ instead of the individual values of $p(y)$? The answer is that we will most often need sums of values of $p(y)$ to solve practical problems. If we gave individual values of $p(y)$ rather than the sums, you would be required to do more calculating when solving exercises.

To see how you use Table 1, suppose that you wish to find the sum of the binomial probabilities from $y = 0$ to $y = 3$ for $n = 5$ trials and $p = .6$. That is, you wish to find

$$\sum_{y=0}^{3} p(y) = p(0) + p(1) + p(2) + p(3)$$

where

$$p(y) = C_y^5 (.6)^y (.4)^{5-y}$$

Turn to Table 1(a), Appendix II, for $n = 5$. Since the tabulated values in the table give

$$\sum_{y=0}^{a} p(y),$$

you seek the tabulated value in the row corresponding to $a = 3$ and the column for $p = .6$. The tabulated value, .663, is shown below as it appears in Table 1(a):

$n = 5$

	p													
a	0.01	0.05	0.10	0.20	0.30	0.40	0.50	0.60	0.70	0.80	0.90	0.95	0.99	a
0	—	—	—	—	—	—	—	—	—	—	—	—	—	0
1	—	—	—	—	—	—	—	—	—	—	—	—	—	1
2	—	—	—	—	—	—	—	—	—	—	—	—	—	2
3	—	—	—	—	—	—	—	.663	—	—	—	—	—	3
4	—	—	—	—	—	—	—	—	—	—	—	—	—	4

Thus, $\sum_{y=0}^{3} p(y)$, the sum of the binomial probabilities from $y = 0$ to $a = 3$ (for $n = 5$, $p = .6$) is .663.

Table 1, Appendix II, can also be used to find an individual binomial probability, say $p(3)$ for $n = 5$, $p = .6$. We would calculate

$$\begin{aligned} p(3) &= [p(0) + p(1) + p(2) + p(3)] - [p(0) + p(1) + p(2)] \\ &= \sum_{y=0}^{3} p(y) - \sum_{y=0}^{2} p(y) \\ &= .663 - .317 \\ &= .346. \end{aligned}$$

Thus an individual value of $p(y)$ is equal to the difference between two sums which are adjacent entries given in a column of the table.

We will use Table 1 in the following example.

EXAMPLE 6.8

Construct the operating characteristic curve for a sampling plan with $n = 15$, $a = 1$.

Solution The probability of lot acceptance will be calculated for $p = .1, .2, .3, .5$.

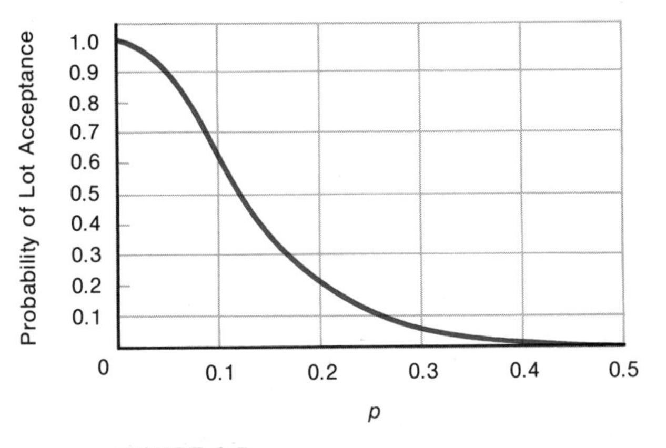

FIGURE 6.5
Operating characteristic curve, $n = 15$, $a = 1$

$$P(\text{accept}) = p(0) + p(1) = \sum_{y=0}^{a=1} p(y),$$

$$P(\text{accept} \mid p = .1) = .549,$$
$$P(\text{accept} \mid p = .2) = .167,$$
$$P(\text{accept} \mid p = .3) = .035,$$
$$P(\text{accept} \mid p = .5) = .000.$$

The operating characteristic curve for the sampling plan is given in Figure 6.5.

Sampling inspection plans are widely used in industry. Each sampling plan possesses its own unique operating characteristic curve which characterizes the plan and, in a sense, describes the size of the holes in the screen. The quality-control engineer will choose the plan that satisfies the requirements of his situation. Increasing the acceptance number increases the probability of acceptance and hence increases the size of the holes in the screen. Increasing the sample size provides more information upon which to base the decision and hence improves the discriminatory power of the decision procedure. Thus, when n is large, the operating characteristic curve will drop rapidly as p increases. You may verify these remarks by working the exercises at the end of the chapter.

Note that acceptance sampling is an example of statistical inference because the procedure implies a decision concerning the lot fraction defective, p. If you accept a lot, you infer that the true lot fraction defective, p, is some relatively small, acceptable value. If you reject a lot, it is clear that you think p is too large. Consequently, lot acceptance sampling is a procedure that implies inference making concerning the lot fraction defective. The operating characteristic curve for the sampling plan provides a measure of the goodness of this inferential procedure.

EXERCISES

6.18 A buyer and seller agree to use a sampling plan with sample size $n = 5$ and acceptance number $a = 0$. What is the probability that the buyer would accept a lot having the following fractions defective?
(a) $p = .1$ (b) $p = .3$ (c) $p = .5$
(d) $p = 0$ (e) $p = 1$
Construct the operating characteristic curve for this plan.

6.19 Repeat Exercise 6.18 for $n = 5$, $a = 1$.

6.20 Repeat Exercise 6.18 for $n = 10$, $a = 0$.

6.21 Repeat Exercise 6.18 for $n = 10$, $a = 1$.

6.22 Graph the operating characteristic curves for the four plans given in Exercises 6.18, 6.19, 6.20, and 6.21 on the same sheet of graph paper. What is the effect of increasing the acceptance number, a, when n is held constant? What is the effect of increasing the sample size, n, when a is held constant?

6.23 A radio and television manufacturer who buys large lots of transistors from an electronics supplier selects $n = 25$ transistors from each lot shipped by the supplier and notes the number of defectives.
(a) On the same sheet of graph paper, construct the operating characteristic curves for the sampling plans $n = 25$ with $a = 1$, 2, and 3.
(b) Which sampling plan best protects the supplier from having acceptable lots rejected and returned by the manufacturer?
(c) Which sampling plan best protects the manufacturer from accepting lots for which the fraction of defectives is exceedingly large?
(d) How might the sampling inspector arrive at an acceptance level that compromises between the risk to the producer and the risk to the consumer?

6.24 Refer to Exercise 6.23 and assume that the manufacturer wishes the probability to be at least .90 of his accepting lots containing 1 percent defective and the probability to be about .90 of rejecting any lot with 10 percent or more defective. If the manufacturer's sampling inspector samples $n = 25$ items from the supplier's incoming shipments, what is the acceptance number (a) that more nearly meets these requirements?

6.25 A buyer and a seller agree to use sampling plan ($n = 15$, $a = 0$) or sampling plan ($n = 25$, $a = 1$). Sketch the operating characteristic curves for the two sampling plans. If you were a buyer, which of the two sampling plans would you prefer? Why?

6.5

MAKING DECISIONS: A TEST OF AN HYPOTHESIS

The cold vaccine problem, Example 6.5, is illustrative of a statistical test of an hypothesis. The practical question to be answered concerns the effectiveness of the vaccine. Do the data contained in the sample present sufficient evidence to indicate that the vaccine is effective?

The reasoning employed in testing an hypothesis bears a striking resemblance to the procedure used in a court trial. In trying a man for theft, the court assumes the accused innocent until proved guilty. The prosecution collects and presents all available evidence in an attempt to contradict the "not guilty" hypothesis and hence to obtain a conviction. The statistical problem portrays the vaccine as the accused. The hypothesis to be tested, called the null hypothesis, is that the vaccine is ineffective. The evidence in the case is contained in the sample drawn from the population of potential vaccine customers. The experimenter, playing the role of the prosecutor, believes that his vaccine is really effective and hence attempts to use the evidence contained in a sample to reject the null hypothesis and thereby to support his contention that the vaccine is, in fact, a very successful cold vaccine. You will recognize this procedure as an essential feature of the scientific method, where all theories proposed must be compared with reality.

Intuitively, we would select the number of survivors, y, as a measure of the quantity of evidence in the sample. If y were very large, we would be inclined to reject the null hypothesis and conclude that the vaccine is effective. On the other hand, a small value of y would provide little evidence to support the rejection of the null hypothesis. As a matter of fact, if the null hypothesis were true and the vaccine were ineffective, the probability of surviving a winter without a cold would be $p = 1/2$ and the average value of y would be

$$E(y) = np = 10(1/2) = 5.$$

Most individuals utilizing their own built-in decision makers would have little difficulty arriving at a decision for the case $y = 10$ or $y = 5, 4, 3, 2,$ or 1, which, on the surface, appear to provide substantial evidence to support rejection or nonrejection, respectively. But what can be said concerning less obvious results, say $y = 7, 8,$ or 9? Clearly, whether we employ a subjective or an objective decision-making procedure, we would choose the procedure that gave the smallest probability of making an incorrect decision.

The statistician would test the null hypothesis in an objective manner similar to our intuitive procedure. A decision maker, commonly called a test statistic, is calculated from information contained in the sample. In our example, the number of survivors, y, would suffice for a test statistic. We would then consider all possible values which the test statistic may assume, for example, $y = 0, 1, 2, \ldots, 9, 10$. These values would be divided into two groups, as shown in Figure 6.6, one called the rejection region and the other the acceptance region. An experiment is then conducted and the decision maker, y, observed. If y takes a value in the rejection region, the hypothesis is rejected. Otherwise, it is accepted. (*Caution:* As you will subsequently learn, you will reject or accept the null hypothesis only if the risks of a wrong decision are known and are small for these two actions.) For example, we might choose $y = 8, 9,$ or 10 as the rejection region and assign the remaining values of y to the acceptance region. Since we observed $y = 8$ survivors, we would reject the null hypothesis that the vaccine is

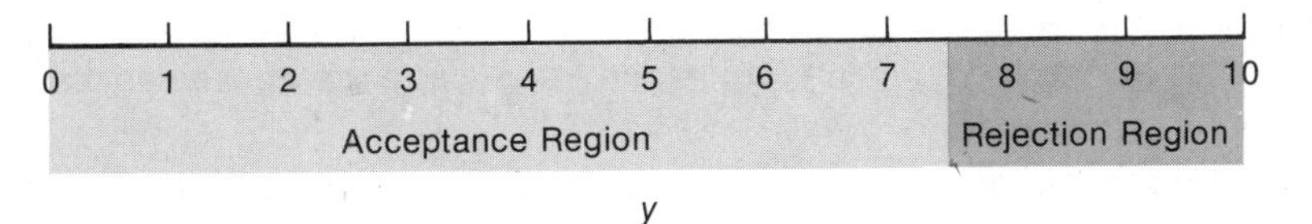

FIGURE 6.6

Possible values for the test statistic, y

ineffective and conclude that the probability of surviving the winter without a cold is greater than $p = 1/2$ when the vaccine is used. What is the probability that we will reject the null hypothesis when, in fact, it is true? The probability of falsely rejecting the null hypothesis is the probability that y will equal 8, 9, or 10, given that $p = 1/2$, and this is indeed the probability computed in Example 6.5 and found to equal .055. Since we have decided to reject the null hypothesis and note that this probability is small, we are reasonably confident that we have made the correct decision.

Upon reflection, you will observe that the cold vaccine manufacturer is faced with two possible types of error. On the one hand, he might reject the null hypothesis and falsely conclude that the vaccine was effective. Proceeding with a more thorough and expensive testing program or a pilot plant production of the vaccine would result in a financial loss. On the other hand, he might decide not to reject the null hypothesis and falsely conclude that the vaccine was ineffective. This error would result in the loss of potential profits which could be derived through the sale of a successful vaccine.

DEFINITION

Rejecting the null hypothesis when it is true is called a type I error for a statistical test. The probability of making a type I error is denoted by the symbol α.

The probability, α, will increase or decrease as we increase or decrease the size of the rejection region. Then why not decrease the size of the rejection region and make α as small as possible? For example, why not choose $y = 10$ as the rejection region? Unfortunately, decreasing α increases the probability of not rejecting when the null hypothesis is false and some alternative hypothesis is true. This second type of error is called the type II error for the statistical test and its probability is denoted by the symbol β.

DEFINITION

Accepting the null hypothesis when it is false is called a type II error for a statistical test. The probability of making a type II error when some specific alternative is true is denoted by the symbol β.

For a fixed sample size, n, α and β are inversely related; as one increases the other decreases. Increasing the sample size provides more information upon which to base the decision and hence reduces both α and β. In an experimental situation, the probabilities of the type I and type II errors for a test measure the risk of making an incorrect decision. The experimenter selects values for these probabilities, and the rejection region and sample size are chosen accordingly.

EXAMPLE 6.9

Refer to the cold vaccine study and the statistical test based on the rejection region shown in Figure 6.6 (i.e., $y = 8, 9, 10$).

(a) State the null hypothesis and the alternative hypothesis for the test.
(b) Find α for the test.
(c) Find β, the probability of accepting the null hypothesis when the probability of survival for a vaccinated person is $p = .9$.

Solution

(a) The null hypothesis is that $p = .5$ or, equivalently, that the vaccine is ineffective. The alternative hypothesis is that $p > .5$, that is, that the vaccine is effective.
(b) The probability of rejecting the null hypothesis when it is true ($p = .5$) is

$$\alpha = P(y = 8, 9, 10 \text{ given that } p = .5) = \sum_{y=8}^{10} p(y)$$

where $p(y)$ is a binomial probability distribution with $p = .5$. Then,

$$\alpha = \sum_{y=8}^{10} C_y^{10}(.5)^y(.5)^{10-y}$$

is found using Table 1, Appendix II, to equal .055. The probability distribution for $n = 10$, $p = .5$, is shown in Figure 6.7; α is represented by the shaded portion of the probability distribution.

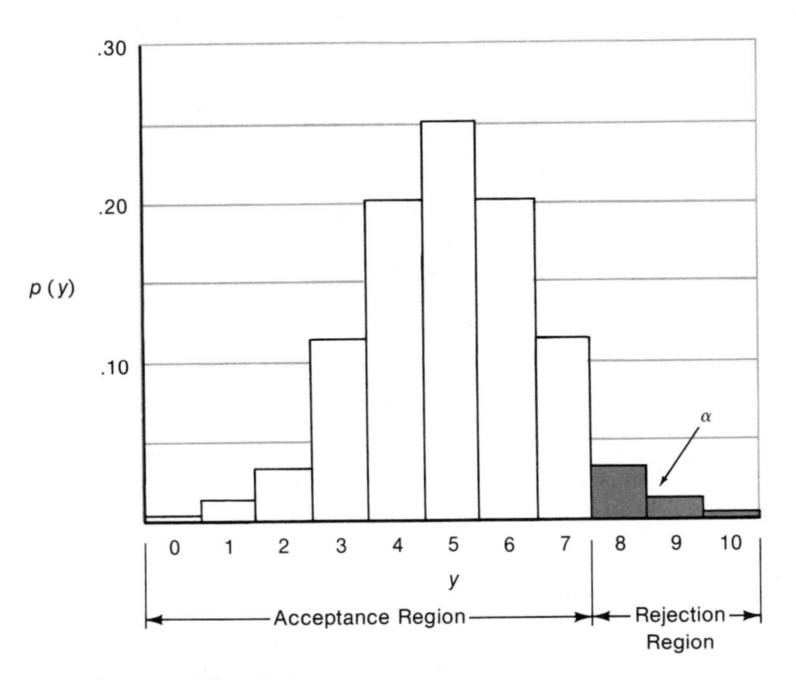

FIGURE 6.7

Binomial probability distribution, $n = 10$, $p = .5$

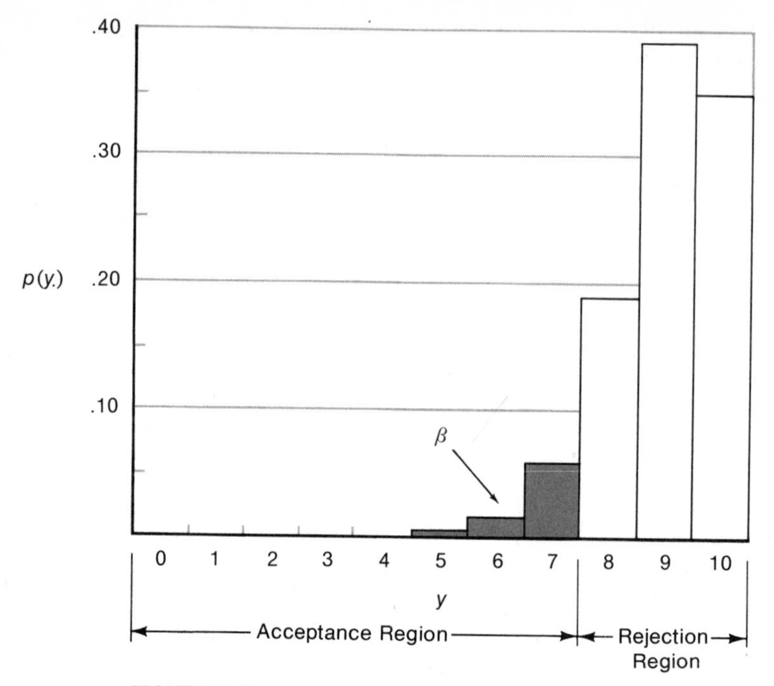

FIGURE 6.8

Binomial probability distribution, when the alternative hypothesis is true: $n = 10$, $p = .9$

(c) We wish to find the value of β when the null hypothesis is false (p is not equal to .5) and, in fact, $p = .9$. When $p = .9$, the probability distribution for y, the number of survivors, is as shown in Figure 6.8. Since

$$\begin{aligned}\beta &= P\text{ (accepting the null hypothesis when } p = .9)\\ &= P(y = 0, 1, 2, \ldots, 6, 7, \text{ given that } p = .9),\end{aligned}$$

β will equal the shaded portion of $p(y)$, Figure 6.8, the portion falling in the acceptance region. Thus,

$$\beta = \sum_{y=0}^{7} p(y), \qquad \text{given } p = .9,$$

where $p(y)$ is a binomial probability distribution with $p = .9$. This quantity can be obtained directly from Table 1, Appendix II, as

$$\beta = \sum_{y=0}^{7} C_y^{10}(.9)^y(.1)^{10-y} = .07.$$

To summarize the implications of (a), (b) and (c), Example 6.9, $\alpha = .055$ and $\beta = .07$ give measures of the risks of making the two (and only two) types of errors for this statistical test. The probability that the test statistic will, by chance, fall in the rejection region when the null hypothesis is true is only .055. That is, the probability of concluding that the vaccine was somewhat effective, when in fact it is worthless, is only .055. But suppose that the vaccine is really effective and that the probability of surviving a winter without a cold, when the vaccine has been used, is .9. What is the probability of accepting the null hypothesis that "the vaccine is ineffective"? We have shown that the risk of making this type II error is only $\beta = .07$.

6.6

CHOOSING THE NULL HYPOTHESIS

The reasoning employed in a statistical test of an hypothesis runs counter to our everyday method of thinking. That is, it is similar to the mathematical method of proof by contradiction. The hypothesis that the scientist wishes to "prove" (that is, support) is the alternative hypothesis (often called the "research hypothesis" by social scientists). To do this, he tests the converse (opposite) of the research hypothesis, the null hypothesis. He hopes that the data will support its rejection because this implies support for the alternative or "research hypothesis," which was his research objective. You can see that this is exactly what we did with the cold vaccine experiment. We showed support for the effectiveness of the cold vaccine by rejecting the null hypothesis that the vaccine was not effective.

Why employ this reverse type of thinking, gaining support for a theory by showing that there is little evidence to support its converse? Why not test the alternative or research hypothesis? The answer lies in the problem of evaluating the probabilities of incorrect decisions.

If the research hypothesis is true, the sample data will tend to support rejection of the null hypothesis (the converse of the research hypothesis). Then the probability of making an incorrect decision is readily at hand. It is α, a probability that was specified in setting up the rejection region. Thus if we reject the null hypothesis (which is what the researcher hopes will occur), we immediately know the probability of making an incorrect decision. This gives us a measure of confidence in our conclusion.

Suppose that we had taken the opposite tack, testing the alternative (research) hypothesis that the vaccine is effective. If the research hypothesis is true, the test statistic will most probably fall in the acceptance region, $y = 8, 9, 10$ (instead of the rejection region, $y = 0, 1, \ldots, 7$). Now, to find the probability of an incorrect decision, we must evaluate β, the probability of accepting the null hypothesis when it is false. Although this is not an insurmountable task for the cold vaccine problem, it is extra work. And for many statistical tests, it is very difficult to calculate β.

So, to summarize, it is a lot easier to follow the route of "proof by contradiction." Thus, the statistician will select the converse of the research hypothesis as the null hypothesis and hope the test leads to its rejection. If it does, the statistician knows α and immediately has a measure of the confidence that he or she can place in this conclusion.

6.7

A GENERAL COMMENT

A discussion of the theory of tests of hypotheses may seem a bit premature at this point, but it provides an introduction to a line of reasoning that is sometimes difficult to grasp and which is best presented when it is allowed to incubate in the mind of the student over a period of time. Thus, some of the exercises at the end of Chapter 6 involve the use of the binomial probability distribution and, at the same time, lead the student to utilize the reasoning involved in statistical tests of hypotheses. We shall take occasion to expand upon these ideas through examples and exercises in Chapter 7 and will discuss in detail the topic of statistical tests of hypotheses in Chapter 8 and succeeding chapters.

In closing, we direct your attention to the similarity of the lot acceptance sampling problem and the statistical test of an hypothesis. Theoretically, they are equivalent because each involves an inference, formulated as a decision, concerning the value of p, the unknown parameter of a binomial population.

EXERCISES

6.26 An experiment is conducted to test the hypothesis that a coin is balanced against the alternative hypothesis that the probability of observing a head, p, is not equal to .5. To test the hypothesis, a coin will be tossed four times and the number of heads observed. The hypothesis will be rejected if zero or four heads appear.
(a) What is the probability of the type I error for this test?
(b) If the coin really is biased and the probability of observing a head on a single trial is .7, what is the probability of the type II error for the test?

6.27 A pair of beetles is expected to produce blackeyed offspring 30 percent of the time. To test this theory, three are observed and it is found that all three possess blue eyes. Does this result present sufficient evidence to contradict the theory? Justify your answer statistically.

6.28 A manufacturer of floor wax has developed two new brands, A and B, which he wishes to subject to consumer evaluation to determine which of the two is superior. Both waxes, A and B, are applied to floor surfaces in each of 15 homes.
(a) If there is actually no difference in the quality of the brands, what is the probability that 10 or more consumers would state a preference for brand A?
(b) For either brand A or brand B?

6.29 Continuing Exercise 6.28, let p equal the probability that a consumer will choose brand B in preference to A and suppose that we wish to test the hypothesis that there is no observable difference between the brands—in other words that $p = 1/2$. Let y, the number of times that B is preferred to A, be the test statistic.
(a) Calculate the value of α for the test if the rejection region is chosen to include $y = 0, 1, 2, 3, 12, 13, 14,$ and 15.
(b) If p is really equal to .8, what is the value of β for the test defined in (a)? (Note that this is the probability that $y = 4, 5, \ldots, 10, 11$ given that $p = .8$.)

6.30 Continuing Exercise 6.29, suppose that the rejection region is enlarged to include $y = 0, 1, 2, 3, 4, 11, 12, 13, 14,$ and 15.
(a) What is the value of α for the test? Should this probability be larger or smaller than the answer given in Exercise 6.29?

(b) If p is really equal to .8, what is the value of β for the test? Compare with your answer to part (b), Exercise 6.29.

6.31 The number of defective fuses proceeding from each of two production lines, A and B, was recorded daily for a period of 10 days with the following results:

Day	Line A	Line B
1	172	201
2	165	179
3	206	159
4	184	192
5	174	177
6	142	170
7	190	182
8	169	179
9	161	169
10	200	210

Assume that both production lines produced the same daily output. Compare the number of defectives produced by A and B each day and let y equal the number of days when B exceeded A. Do the data present sufficient evidence to indicate that production line B produces more defectives, on the average, than A? State the null and alternative hypotheses. Use y as a test statistic.

6.32 In 1975, the General Accounting Office (GAO) examined 15 school districts in 14 states and concluded that it was debatable whether the multibillion dollar Title I program, aimed primarily at improving the reading ability of poor children, was effective. In particular, the GAO noted that the gap between the reading abilities of educationally deprived and of average children increased while the students were in the program (*Orlando Sentinel Star,* December 28, 1975). If the Title I program was completely ineffective, the change in the level of reading ability scores (before and after Title I) would depend on the random variation of individual student scores so that it would be reasonable to assume that P(increase in level) $= P$(decrease in level) $= p = .5$. Test the hypothesis that $p = .5$ using y, the number of school districts showing an increase in the reading ability gap, as a test statistic.
(a) State the null and the alternative hypotheses.
(b) Locate a rejection region for a value of α near .05 (assume that $n = 15$).
(c) Based on the GAO report, it appears that $y = 15$. What do you conclude concerning the effectiveness of the Title I program?

6.33 A number of psychological experiments are conducted as follows: A rat is attracted to the end of a ramp that divides, leading to one of two doors. The objective of the experiment, essentially, is to determine whether the rat possesses or acquires a preference for one of the two paths. For a given experiment consisting of 6 runs, the following results were observed:

Run	Door Chosen
1	2
2	1
3	2
4	2
5	2
6	2

(a) State the null hypothesis to be tested. State the alternative hypothesis.
(b) Let y equal the number of times the rat chose the second door. What is the value of α for the test if the rejection region includes $y = 0$ and $y = 6$?
(c) What is the value of β for the alternative, $p = .8$?

6.34 Establishing the value of an art object is subjective in nature and difficult at best. To determine whether two art appraisers tend to give different levels of appraisals, two appraisers, A and B, were asked to appraise each of 7 art objects. If appraiser A tends to give smaller appraisals (or give larger appraisals) than appraiser B, the probability p that the appraisal of A will exceed the appraisal for B will be less than 1/2 (or greater than 1/2). Under the assumption that there is no difference in the appraisal techniques of the two appraisers, $p = 1/2$. Let y, the number of art objects for which A's appraisal exceeds B's appraisal, be a test statistic.
(a) Find an appropriate rejection region to test the null hypothesis, $p = 1/2$, for $\alpha \approx .10$.
(b) If A tends to give conservative evaluations, so that p actually is equal to .9, calculate β for the test.
(c) If A appraises 5 of the 7 art objects to be worth more than B appraises them for, what do you conclude?

6.35 Most weather forecasters seem to protect themselves very well by attaching probabilities to their forecasts ("the probability of rain today is 40 percent"). Then if a particular forecast is incorrect, you are expected to attribute the error to the random behavior of the weather rather than the inaccuracy of the forecaster. To check the accuracy of a particular forecaster, records were checked only for days in which the forecaster predicted rain "with 30 percent probability." A check of 25 of these days indicated that it rained on 10 of the 25. Do these data disagree with the forecast of a "30 percent probability of rain"? To answer this question, you will want to detect a value of p that differs from .3, i.e., either $p > .3$ or $p < .3$. Consequently, you will reject the null hypothesis, $p = .3$, if you observe either very large or very small values of y. Suppose that you select $y \leq 3$ or $y \geq 12$ as the rejection region for the test.
(a) Find α.
(b) Complete the test and state your conclusions.

6.36 Refer to Exercise 6.35. If the probability of rain on a given day is really .6 when the weather forecaster forecasts .3, what is the probability that you will reject the null hypothesis, $p = .3$, based on a random sample of 25 such days? What is the value of β for the test?

6.37 A packaging experiment was conducted by placing two different package designs for a breakfast food side by side on a supermarket shelf. The objective of the experiment was to see if buyers indicated a preference for one of the two package designs. In a given day, five customers purchased a package from the supermarket, with one choosing package design 1 and four choosing design 2.
(a) State the hypothesis to be tested. (*Hint:* The null hypothesis should imply equal preference for the two designs.)
(b) Let y equal the number of buyers who choose the second package design. What is the value of α for the test if the rejection region includes $y = 0$ and $y = 5$?
(c) What is the value of β for the alternative $p = .9$ (that is, 90 percent of the buyers actually favor the second package design)?
(d) In the context of our problem, give a practical interpretation of the type I error and the type II error.

6.8

SUMMARY

A binomial experiment is typical of a large class of useful experiments encountered in real life which satisfy, to a reasonable degree of approximation the five defining characteristics stated in Section 6.1. The number of successes, y, observed in n trials is a discrete random variable with probability distribution

$$p(y) = C_y^n p^y q^{n-y},$$

where $q = 1 - p$ and $y = 0, 1, 2, 3, \ldots, n$. Statistically speaking, we are interested in making inferences concerning p, the parameter of a binomial population, as exemplified by the lot acceptance sampling, Section 6.4, and the test of the effectiveness of the cold vaccine, Section 6.5.

REFERENCES

Chapman, D. G., and R. A. Schaufele, *Elementary Probability Models and Statistical Inference*. New York: John Wiley & Sons, Inc., 1970.

Feller, W., *An Introduction to Probability Theory and Its Applications,* Vol. 1, 3rd ed. New York: John Wiley & Sons, Inc., 1968.

Handbook of Tables for Probability and Statistics, 2nd ed. Cleveland, Ohio: The Chemical Rubber Co., 1968.

Mendenhall, W., and R. L. Scheaffer, *Mathematical Statistics with Applications*. North Scituate, Mass.: Duxbury Press, 1973.

Mosteller, F., R. E. K. Rourke, and G. B. Thomas, Jr., *Probability with Statistical Applications,* 2nd ed. Reading, Mass.: Addison-Wesley Publishing Company, Inc., 1970.

National Bureau of Standards, *Tables of the Binomial Probability Distribution*. Washington, D.C.: Government Printing Office, 1949.

SUPPLEMENTARY EXERCISES

Starred (*) exercises are optional.

6.38 List the five identifying characteristics of the binomial experiment.

6.39 A balanced coin is tossed three times. Let y equal the number of heads observed.
- (a) Use the formula for the binomial probability distribution to calculate the probabilities associated with $y = 0, 1, 2,$ and 3.
- (b) Construct a probability distribution similar to Figure 5.1.
- (c) Find the expected value and standard deviation of y, using the formulas

$$E(y) = np,$$
$$\sigma = \sqrt{npq}.$$

- (d) Using the probability distribution, (b), find the fraction of the population measurements lying within one standard deviation of the mean. Repeat for two standard deviations. How do your results agree with Tchebysheff's Theorem and the Empirical Rule?

6.40 Refer to Exercise 6.39. Suppose that the coin was definitely unbalanced and that the probability of a head was equal to $p = .1$. Follow instructions (a), (b), (c), and (d). Note that the probability distribution loses its symmetry and becomes skewed when p is not equal to $1/2$.

6.41 The probability that a single radar set will detect an enemy plane is .9. If we have five radar sets, what is the probability that exactly four sets will detect the plane? (Assume that the sets operate independently of each other.) At least one set?

6.42 Suppose that the four engines of a commercial aircraft were arranged to operate independently and that the probability of in-flight failure of a single engine is .01. What is the probability that, on a given flight:
(a) No failures are observed?
(b) No more than one failure is observed?

6.43 A recent study of commuter trains shows that they run more than 35 minutes late with probability equal to .5. If we are using a commuter train that exhibits these characteristics and we randomly select five days within the past year, what is the probability that the train is always late? What is the probability the train is late more than three times out of five?

6.44 Suppose that 90 percent of the students taking English pass the course. What is the probability that at least 2 students in a class of 20 will fail the course? (Use Table I, Appendix II.)

6.45 Referring to Exercise 6.44, what is the expected value and standard deviation of y, the number of failures in a class of 20? Within what limits would y be expected to fall?

6.46 A city commissioner claims that 80 percent of all people in the city favor garbage collection by contract to a private concern (in contrast to collection by city employees). To check the theory that the proportion of the people in the city favoring private collection is .8, you randomly sample 25 people and find that y, the number of people who support the commissioner's claim, is 22.
(a) What is the probability of observing at least 22 who support the commissioner's claim if, in fact, $p = .80$?
(b) What is the probability that y is exactly equal to 22?

6.47 A handwriting analyst claims that he judges a person's sex on the basis of his handwriting. To test this claim he is given 15 pairs of handwriting samples, one of each pair being that of a male and the other that of a female. If he is not really able to judge sex on the basis of handwriting and is merely guessing, what is the probability that he correctly identifies the male handwriting in 10 or more pairs?

6.48 It is known that 90 percent of those who purchase a color television will have no claims covered by the guarantee during the duration of the guarantee. Suppose that 20 customers each buy a color television set from a certain appliance dealer. What is the probability that at least 2 of these 20 customers will have claims against the guarantee?

6.49 A multiple-choice test contains twenty questions and each question indicates five choices for the correct answer. If a student has absolutely no knowledge about the material and randomly selects the answer for each question:
(a) Give the probability distribution for y, the number of questions the student will answer correctly.
(b) Find the probability that the student will score at least 60 percent on the test.

6.50 Suppose that it is known that 1 out of 10 undergraduate college textbooks is an outstanding financial success. A publisher has selected 10 new textbooks for publication. What is the probability that
(a) Exactly one will be an outstanding financial success?
(b) At least one?
(c) At least two?

6.51 The proportion of residential households in Burlington, Vermont, that are heated by natural gas is approximately .2. A randomly selected city block within the Burlington city limits has 20 residential households. Assume that the properties of a binomial experiment are satisfied and find the probability that

(a) None of the households are heated by natural gas.

(b) No more than 4 of the 20 are heated by natural gas.

(c) Why might the binomial experiment not provide a good model for this sampling situation?

6.52 If a person is given the choice of a number from 0 to 9, is it more likely that the person will choose a number near the middle of the sequence? To answer this question, 20 persons are asked to select a number from 0 to 9 and 8 choose a 4, 5, or 6. If the choice of any one number is as likely as any other, what is the probability of this event? What is the probability of observing 8 or more choices of the interior numbers, 4, 5, or 6?

6.53 Refer to Exercise 6.31 and the General Accounting Office's report on the effectiveness of the Title I program designed to improve the reading ability of poor children. The report mentions that an analysis of student records showed the gap in reading ability between poor and average children increased for 60 percent of the poor children in the program, the gap was maintained for 6 percent, and the gap decreased for 34 percent. If the GAO examined 6,000 records, what is the expected value of y, the number showing an increase in the gap, if the program was really ineffective (i.e., $p = .5$)? If p is really equal to .5, is it likely that 60 percent of 6,000 poor students would show an increase in the gap? Since the purpose of the Title I program is to reduce the gap, you would expect to observe a relatively small value of y. Can you find an explanation for the peculiar results ($y = 3{,}600$) observed by the GAO?

6.54 A recent survey suggests that Americans anticipate a reduction in living standards and that a steadily increasing consumption no longer may be as important as it was in the past. Suppose that a poll of 2,000 people indicated 1,373 in favor of forcing a reduction in the size of American automobiles by legislative means. Would you expect to observe as many as 1,373 in favor of this proposition if, in fact, the general public was split 50–50 on the issue? Why?

6.55 A quality-control engineer wishes to study the alternative sampling plans $n = 5, a = 1$ and $n = 25, a = 5$. On the same sheet of graph paper, construct the operating characteristic curves for both plans, making use of acceptance probabilities at $p = .05$, $p = .10$, $p = .20$, $p = .30$, and $p = .40$ in each case.

(a) If you were a seller producing lots with fraction defective ranging from $p = 0$ to $p = .10$, which of the two sampling plans would you prefer?

(b) If you were a buyer wishing to be protected against accepting lots with fraction defective exceeding $p = .30$, which of the two sampling plans would you prefer?

6.56 Consider a lot acceptance plan with $n = 20$, $a = 1$. Calculate the probability of accepting lots having fraction defective

(a) $p = .01$, (b) $p = .05$, (c) $p = .10$, (d) $p = .20$.

Sketch the operating characteristic curve for the plan.

6.57 A chemist wishes to determine whether a new type of alloy is less susceptible to corrosion than alloys of two standard types. He makes up seven specimen strips of each type of alloy and forms seven groups of three dissimilar strips each. The strips in each group are subjected to attack by an acid and the number (y) of times that the new alloy is the least corroded is recorded. Let p be the probability that in a given group the strip of the new alloy suffers less corrosion than the other two strips.

(a) State the appropriate null and alternative hypotheses in the form of statements about p.

(b) Find α if the rejection region is $y = 5, 6, 7$.

(c) Find β if $p = .5$.
(d) If $p = .7$, would β be more than or less than the answer to part (c)?

6.58 A random sample of 400 people are interviewed relative to a given proposition and y, the number supporting the proposition, is recorded. Suppose that only 20 percent of the population supports the proposition.
(a) Find μ.
(b) Find σ^2.
(c) Within what limits can we expect the random variable, y, to lie with probability at least .75? (Use Tchebysheff's Theorem.)

6.59 Suppose that a given type battery will operate for at least 20 hours with probability .7. A piece of equipment uses three such batteries. Let y be the number of batteries (out of three) lasting longer than 20 hours.
(a) Write down the probability function for y.
(b) Find $p(3)$.
(c) Find $p(y \leq 2)$.

6.60 A shipment of 200 portable television sets is received by a retailer. To protect himself against a "bad" shipment, he will inspect 5 sets and accept the entire lot if he observes 0 or 1 defectives. Suppose there are actually 20 defective sets in the shipment.
(a) What is the probability that he accepts the entire shipment?
(b) Given that the retailer accepts the entire lot, what is the probability that he observed exactly 1 defective set?

6.61 A psychiatrist believes that 80 percent of all people who visit doctors have problems of a psychosomatic nature. He decides to select 25 patients at random to test his theory.
(a) Assuming that the psychiatrist's theory is true, what is the expected value of y, the number of the 25 patients who have psychosomatic problems?
(b) What is the variance of y, assuming that the theory is true.
(c) Find $p(y \leq 14)$. (Use tables and assume that the theory is true.)
(d) Based on the probability in part (c), if only 14 of the 25 sampled had psychosomatic problems, what conclusions would you make about the psychiatrist's theory? Explain.

6.62 A particular type of radar set has a probability of .2 of detecting and tracking an aircraft within a radius of 200 miles. If n sets are available for operation, how many sets would have to be operating simultaneously in order that the probability of detecting an aircraft would equal .9? (Assume that the sets operate independently and that each has a probability of detection equal to .2.)

6.63 In a random sample of 20 businessmen, 15 favor copy machine A over copy machine B. If the machines are equally desirable, the probability that a person will select A over B is .5. Test the null hypothesis $p = .5$, using $y = 0, 1, 2, 3, 4, 5, 15, 16, 17, 18, 19, 20$ as the rejection region. Calculate α. Find β if $p = .8$.

6.64 A student government states that 80 percent of all students favor an increase in student fees to subsidize a new recreational area. A random sample of $n = 25$ students produced 15 in favor of increased fees. What is the probability that 15 or fewer in the sample would favor the issue if student government is correct? Do the data support the student government's assertion, or does it appear that the percentage favoring an increase in fees is less than 80 percent?

6.65 From tape recordings of actual therapy sessions, 20 trained observers rate two therapists, A and B, on a 10-point "empathy" scale and the number of observers rating B higher than A is recorded. Let this number, y, be a test statistic to test a null hypothesis of "no difference in empathy" for the two therapists.

(a) Let p be the probability that B will be rated higher than A by a given observer. Express the null hypothesis in the form of a statement about p.

(b) If $y = 0, 1, 2, 3, 4, 5, 6, 14, 15, 16, 17, 18, 19$, and 20 is selected as the rejection region, find α.

(c) If the probability that B will be rated higher than A by a given observer is $p = .7$, find β.

6.66 Suppose that 1/4 of the mice of a particular hybrid strain are highly susceptible to audiogenic seizures. What is the minimum number of mice that would have to be tested in order that the probability of finding at least one highly susceptible mouse be greater than .9?

*6.67 Derive the formula for the binomial probability distribution by using mathematical induction.

*6.68 Derive the formula for the binomial probability distribution using the sample point approach and combinatorial mathematics.

*6.69 Use the formula for the binomial probability distribution along with the expectation theorems given in Exercises 5.52 and 5.53 to prove

$$E(y) = np,$$
$$\text{variance of } y = npq.$$

*6.70 Prove that the binomial probabilities always sum to 1; that is,

$$\sum_{y=0}^{n} p(y) = 1.$$

6.71 The manager of a large motor pool wished to compare the wearing qualities of two different types (type A and type B) of automobile tires. On each of 400 cars he replaced one rear tire with a new tire of type A and the other rear tire with a new tire of type B. When a given car had been driven 10,000 miles, he determined which of the two rear tires experienced the greater wear. Let y be the number of cars out of the 400 cars on which tire A showed the greater wear. Let p be the probability that on a given car the tire of type A will experience the greater wear. Using Tchebysheff's Theorem, find an upper bound to α for a test of the hypothesis that $p = 1/2$ if the rejection region includes the outcomes $y = 0, 1, 2, \ldots, 150, 250, 251, \ldots, 400$.

*6.72 The "power" of a test is defined to be $1 - \beta$. Find a lower bound to the power of the test given in Exercise 6.71 if $p = .8$.

EXPERIENCES WITH REAL DATA

It is a fact that early election returns often yield very accurate information concerning the prospective winner of an election. The news media employ sophisticated statistical techniques in their early evening forecasting, but such basic knowledge of the binomial experiment, as covered in Chapter 6, can provide a crude statistical tool for looking into the future and picking a winner. For example, the author noted that the early returns for a particular state in a recent presidential election showed (only 20 minutes after the polls had closed) 27,000 votes for candidate A and 33,000 for candidate B. Viewing such a result, why might we conclude that the state would go to candidate B? [*Hint:* Let y denote the number of voters in the sample of $n = 60,000$ that favor candidate B and assume that y has a binomial probability distribution with $p = .5$. Find μ and σ and use the parameters to characterize $p(y)$, assuming that B is *not* a winner.]

If you have managed to convince your instructor that the evidence supports the contention that B will win, note that your argument rests on the assumption that the selection of the 60,000 early returns represents a binomial experiment with a value of p that coincides with the proportion of voters in the state who favor candidate B. This assumption could be invalid because early returns could represent the votes solely from urban areas which might not represent the proportion of voters in the state who favor candidate B. Since the proportion of urban votes for the state was high, the author concluded that a binomial experiment might provide a crude model of reality. And, indeed, you have already noted that $y = 33{,}000$ lies a long way (many standard deviations) from its expected value, 30,000 (assuming that B is not a winner but is, at best, a tie).

CHAPTER OBJECTIVES

GENERAL OBJECTIVES An important discrete random variable, the binomial, and its probability distribution were presented in Chapter 6. This chapter presents the normal random variable, one of the most important and most common continuous random variables. We explain why the normal random variable occurs so frequently in practice, give its probability distribution, and show how the probability distribution can be used. We will take advantage of the normal distribution to reinforce your understanding of the basic concepts involved in a statistical test of an hypothesis.

SPECIFIC OBJECTIVES

1. To explain why normally distributed random variables occur so frequently in nature. The Central Limit Theorem is one of the reasons presented for this situation. *Sections 7.1, 7.3*
2. To present the normal probability distribution and to explain how to find the probability that a random variable will fall in a particular interval. *Section 7.2*
3. To present the Central Limit Theorem as a reason for studying the normal probability distribution and to explain why it plays an important role in statistics. *Section 7.3*
4. To demonstrate the applicability of the Central Limit Theorem by showing how the normal probability distribution can be used to approximate binomial probabilities when the number n of trials is large. *Section 7.4*
5. To use examples to reinforce the inferential ideas introduced in Chapter 6. *Section 7.4*

7 THE NORMAL PROBABILITY DISTRIBUTION

7.1

INTRODUCTION

Continuous random variables, as noted in Section 5.5, are associated with sample spaces representing the infinitely large number of sample points contained on a line interval. The heights and weights of humans, laboratory experimental measurement errors, and the length of life of light bulbs are typical examples of continuous random variables. Reviewing Section 5.5, we note that the probabilistic model for the frequency distribution of a continuous random variable involves the selection of a curve, usually smooth, called the probability distribution or probability density function. Although these distributions may assume a variety of shapes, it is interesting to note that a very large number of random variables observed in nature possess a frequency distribution which is approximately bell-shaped or, as the statistician would say, is approximately a normal probability distribution.

Mathematically speaking, the normal probability density function,

$$f(y) = \frac{e^{-\frac{(y-\mu)^2}{2\sigma^2}}}{\sigma\sqrt{2\pi}} \qquad (-\infty < y < \infty),$$

is the equation of the bell-shaped curve shown in Figure 7.1.

The symbols e and π represent irrational numbers whose values are approximately 2.7183 and 3.1416, respectively; μ and σ are the population mean and standard deviation. The equation for the density function is constructed such that the area under the curve will represent probability. Hence the total area is equal to 1.

In practice, we seldom encounter variables that range in value from "minus infinity" to "plus infinity," whatever meaning we may wish to attach to these phrases. Certainly the height of humans, the weight of a species of beetle, or the length of life of a light bulb do not satisfy this requirement. Nevertheless, a relative frequency histogram plotted for many types of measurements will generate a bell-shaped figure that may be approximated by the function shown in Figure 7.1. Why this particular phenomenon exists is a matter for conjecture. However, one explanation is provided by the Central Limit Theorem, a theorem that may be regarded as the most important in statistics. This theorem is discussed in Section 7.3.

7.2

TABULATED AREAS OF THE NORMAL PROBABILITY DISTRIBUTION

You will recall (Section 5.5) that the probability that a continuous random variable assumes a value in the interval, a to b, is the area under the probability density function between the points, a and b (see Figure 7.2).

The probability model for a continuous random variable differs greatly from the model for a discrete random variable when we talk about the probability that y

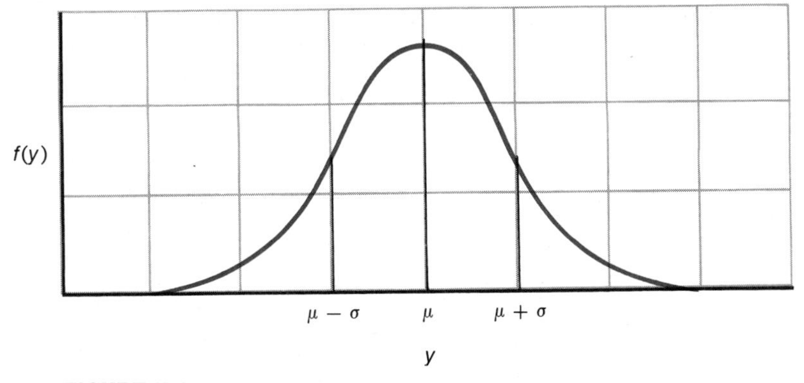

FIGURE 7.1

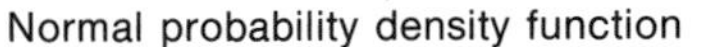
Normal probability density function

equals some particular value, say a. Since the area lying over any particular point, say $y = a$, is 0, it follows from our probability model that the probability that $y = a$ is 0. This means that the expression, $P(y \leq a)$, is the same as $P(y < a)$ because $P(y = a) = 0$. Similarly, $P(y \geq a) = P(y > a)$. This is, of course, not true for a discrete random variable because $P(y = a)$ may not equal 0.

To find areas under the normal curve, first note that the equation for the normal probability distribution, Section 7.1, is dependent upon the numerical values of μ and σ and that by supplying various values for these parameters, we could generate an infinitely large number of bell-shaped normal distributions. A separate table of areas for each of these curves is obviously impractical; rather we would like one table of areas applicable to all. The easiest way to do this is to work with areas lying within a specified number of standard deviations of the mean as was done in the case of the Empirical Rule. For instance, we know that approximately .68 of the area will lie within one standard deviation of the mean, .95 within two, and almost all within three. What fraction of the total area will lie within .7

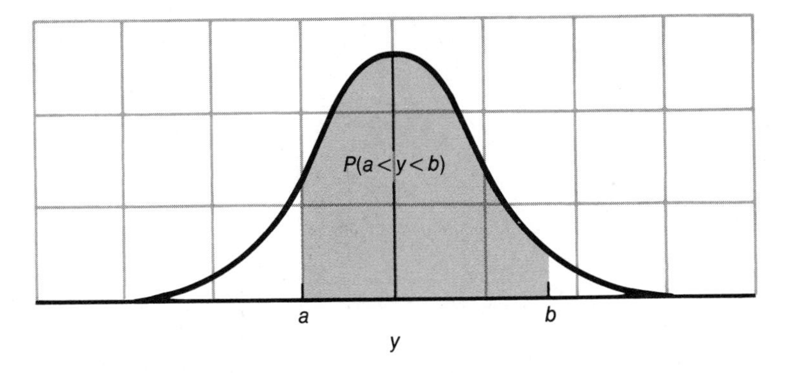

FIGURE 7.2

The probability $P(a < y < b)$ for a continuous random variable

standard deviations, for instance? This question, as well as others, will be answered by Table 3, Appendix II.

Inasmuch as the normal curve is symmetrical about the mean, half of the area under the curve will lie to the left of the mean and half to the right (see Figure 7.3). Also, because of the symmetry, we can simplify our table of areas by listing the areas between the mean and a specified number z of standard deviations to the right of μ. Areas to the left of the mean can be calculated by using the corresponding and equal area to the right of the mean. The distance from the mean to a given value of y is $(y - \mu)$. Expressing this distance in units of standard deviation σ, we obtain

$$z = \frac{y - \mu}{\sigma}.$$

Note that there is a one-to-one correspondence between z and y and particularly that $z = 0$ when $y = \mu$. z will be positive when y lies above the mean and negative when y lies below the mean. The probability distribution for z is often called the standardized normal distribution, because its mean is equal to zero and its standard deviation is equal to 1. It is shown in Figure 7.3. The area under the normal curve between the mean, $z = 0$, and a specified value of $z > 0$, say z_0, is the probability $P(0 \leq z \leq z_0)$. This area is recorded in Table 3, Appendix II, and is shown as the shaded area in Figure 7.3.

An abbreviated version of Table 3, Appendix II, is shown in Table 7.1. Note that z, correct to the nearest tenth, is recorded in the left-hand column. The second decimal place for z, corresponding to hundredths, is given across the top row. Thus the area between the mean and a point $z = .7$ standard deviation to the right, located in the second column of the table opposite $z = .7$, is found to equal .2580. Similarly, the area between the mean and $z = 1.0$ is .3413. The area lying within one standard deviation on either side of the mean would be two times the quantity .3413, or .6826. The area lying within two standard deviations of the mean, correct to four decimal places, is 2(.4772) = .9544. These numbers provide the approximate values, 68 percent and 95 percent, used in the Empirical Rule, Chapter 3. To find the area $z = .57$ standard deviation to the right of the mean,

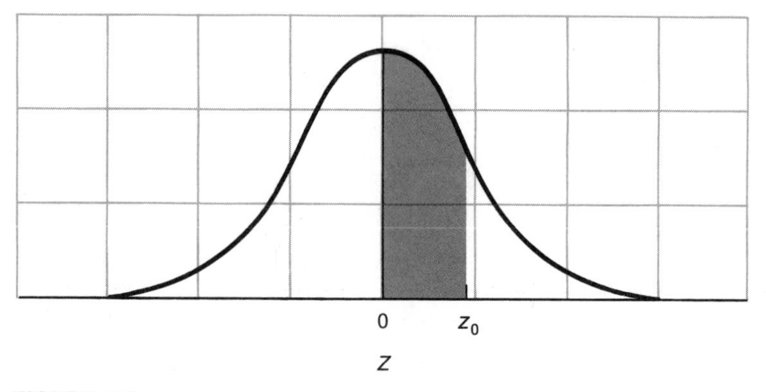

FIGURE 7.3
Standardized normal distribution

TABLE 7.1
Abbreviated version of Table 3, Appendix II

z	.00	.01	.02	.03	.04	.05	.06	.07	.08	.09
0.0	.0000	.0040	.0080	.0120	.0160	.0199	.0239	.0279	.0319	.0359
0.1	.0398	.0438	.0478	.0517	.0557	.0596	.0636	.0675	.0714	.0753
0.2	.0793	.0832	.0871	.0910	.0948	.0987	.1026	.1064	.1103	.1141
0.3	.1179	.1217	.1255	.1293	.1331	.1368	.1406	.1443	.1480	.1517
0.4	.1554	.1591	.1628	.1664	.1700	.1736	.1772	.1808	.1844	.1879
0.5	.1915	.1950	.1985	.2019	.2054	.2088	.2123	.2157	.2190	.2224
0.6	.2257	⋮	⋮	⋮	⋮	⋮	⋮	⋮	⋮	⋮
0.7	.2580									
⋮	⋮									
1.0	.3413									
⋮	⋮									
2.0	.4772									

proceed down the left-hand column to the 0.5 row. Then move across the top row of the table to the .07 column. The intersection of this row-column combination gives the approximate area, .2157. We conclude this section with some examples.

EXAMPLE 7.1

Find $P(0 \leq z \leq 1.63)$. This probability corresponds to the area between the mean ($z = 0$) and a point $z = 1.63$ standard deviations to the right of the mean (see Figure 7.4).

Solution The area is shaded and indicated by the symbol A in Figure 7.4. Since Table 3, Appendix II, gives areas under the normal curve to the right of the mean, we need only find the tabulated value corresponding to $z = 1.63$. Go down the left-hand column of the table to the row corresponding to $z = 1.6$ and across the top of the table to the column marked

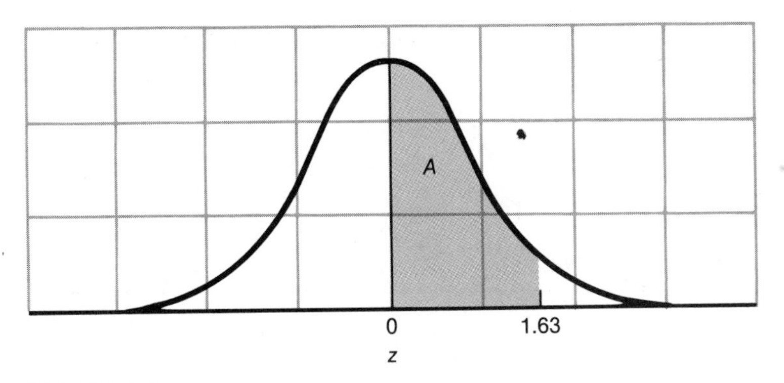

FIGURE 7.4
Probability required for Example 7.1

.03. The intersection of this row and column combination gives the area, $A = .4484$. Therefore, $P(0 < z < 1.63) = .4484$.

EXAMPLE 7.2

Find $P(-.5 \le z \le 1.0)$. This probability corresponds to the area between $z = -.5$ and $z = 1.0$, as shown in Figure 7.5.

Solution The area required is equal to the sum of A_1 and A_2 shown in Figure 7.5. From Table 3, Appendix II, we read $A_2 = .3413$. The area A_1 would equal the corresponding area between $z = 0$ and $z = .5$, or $A_1 = .1915$. Thus the total area is

$$\begin{aligned} A &= A_1 + A_2 \\ &= .1915 + .3413 \\ &= .5328. \end{aligned}$$

EXAMPLE 7.3

Find the value of z, say z_0, such that exactly (to four decimal places) .95 of the area is within $\pm z_0$ standard deviations of the mean.

Solution Half of the area, .95, will lie to the left of the mean and half to the right because the normal distribution is symmetrical. Thus we seek the value, z_0, corresponding to an area equal to .475. The area .475 falls in the row corresponding to $z = 1.9$ and the .06 column. Hence $z_0 = 1.96$. Note that this is very close to the approximate value, $z = 2$, used in the Empirical Rule.

EXAMPLE 7.4

Let y be a normally distributed random variable with mean equal to 10 and standard deviation equal to 2. Find the probability that y will lie between 11 and 13.6.

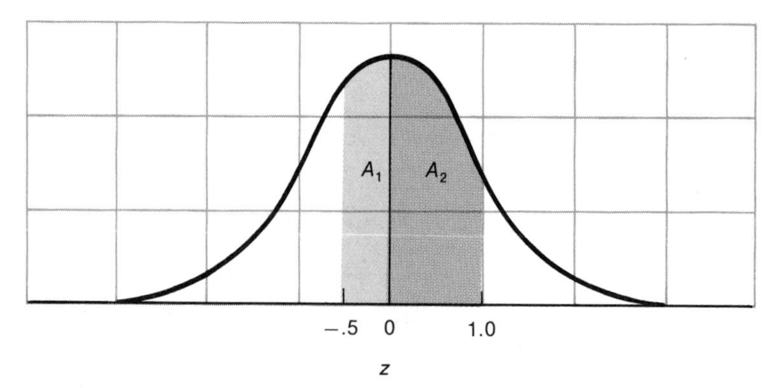

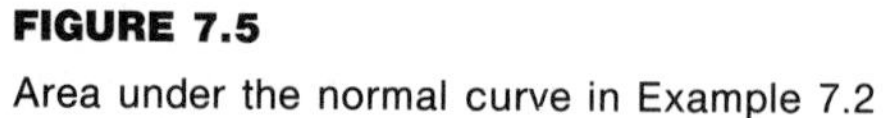

FIGURE 7.5

Area under the normal curve in Example 7.2

Solution As a first step, we must calculate the values of z corresponding to $y_1 = 11$ and $y_2 = 13.6$. Thus,

$$z_1 = \frac{y_1 - \mu}{\sigma} = \frac{11 - 10}{2} = .5,$$

$$z_2 = \frac{y_2 - \mu}{\sigma} = \frac{13.6 - 10}{2} = 1.80.$$

The probability desired, P, is therefore the area lying between z_1 and z_2, as shown in Figure 7.6. The areas between $z = 0$ and z_1, $A_1 = .1915$, and $z = 0$ and z_2, $A_2 = .4641$, are easily obtained from Table 3, Appendix II. The probability, P, is equal to the difference between A_1 and A_2; that is,

$$\begin{aligned} P &= A_2 - A_1 \\ &= .4641 - .1915 = .2726. \end{aligned}$$

EXAMPLE 7.5

Studies show that gasoline usage for compact cars sold in the United States is normally distributed with a mean usage of 25.5 miles per gallon (mpg) and a standard deviation of 4.5 mpg. What percentage of compacts obtain 30 or more miles per gallon?

Solution The proportion P of compacts obtaining 30 or more miles per gallon is given by the shaded area in Figure 7.7.

First we must find the z value corresponding to $y = 30$ mpg. Substituting into the formula for z, we obtain

$$z = \frac{y - \mu}{\sigma} = \frac{30 - 25.5}{4.5} = 1.0.$$

The area A to the right of the mean corresponding to $z = 1.0$ is .3413 (Table 3, Appendix II). Then the proportion of compacts having an mpg rating equal to or greater than 30 is equal to the entire area to the right of the mean, .5, less the area A. Thus,

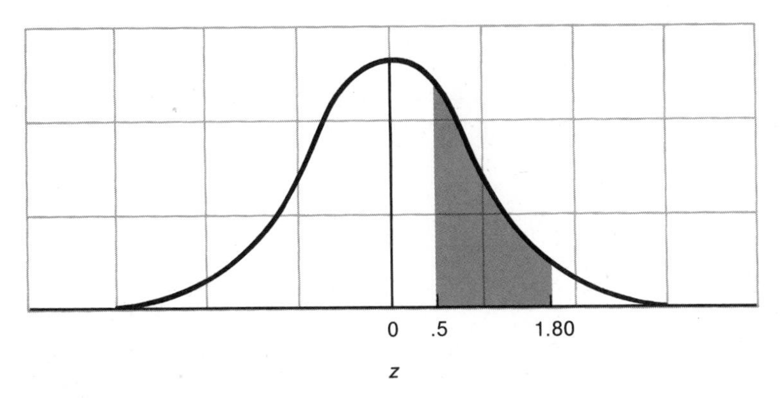

FIGURE 7.6

Area under the normal curve in Example 7.4

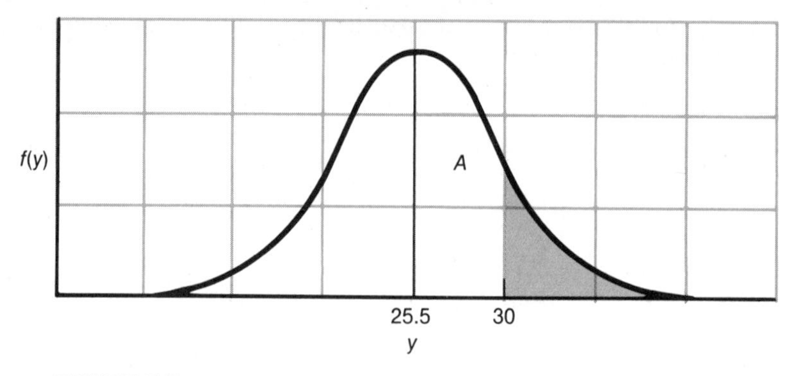

FIGURE 7.7
Area under the normal curve for Example 7.5

$$P = .5 - A = .5 - .3413 = .1587,$$

and the percentage exceeding 30 mpg is

$$100(.1587) = 15.87 \text{ percent.}$$

EXAMPLE 7.6

Refer to Example 7.5. If a manufacturer wishes to develop a compact car which outperforms 95 percent of the current compacts in fuel economy, what must be the gasoline usage rate for the new car?

Solution Let y be a normally distributed random variable with mean equal to 25.5 and standard deviation equal to 4.5. We want to find the value y_0 such that

$$P(y < y_0) = .95.$$

As a first step, we find

$$z_0 = \frac{y_0 - \mu}{\sigma} = \frac{y_0 - 25.5}{4.5},$$

and note that our required probability is the same as the area to the left of z_0 for the standardized normal distribution. Therefore,

$$P(z \leq z_0) = .95,$$

and from Table 3, we find

$$z_0 = \frac{y_0 - 25.5}{4.5} = 1.645,$$

and $y_0 = 32.9$. The manufacturer's new compact car must, therefore, obtain fuel economy of 32.9 mpg to outperform 95 percent of the compact cars currently available on the U.S. market.

EXERCISES

7.1 Using Table 3, Appendix II, calculate the area under the normal curve between
(a) $z = 0$ and $z = 1.4$ (b) $z = 0$ and $z = 1.75$

7.2 Repeat Exercise 7.1 for
(a) $z = 0$ and $z = .75$ (b) $z = 0$ and $z = -.75$

7.3 Repeat Exercise 7.1 for
(a) $z = -1.2$ and $z = 1.6$ (b) $z = .7$ and $z = 1.3$

7.4 Repeat Exercise 7.1 for
(a) $z = -1.0$ and $z = 1.0$ (b) $z = -2.0$ and $z = 2.0$
(c) $z = -3.0$ and $z = 3.0$

7.5 Find a z_0 such that $P(z > z_0) = .025$.

7.6 Find a z_0 such that $P(z < z_0) = .8051$.

7.7 Find a z_0 such that $P(z < z_0) = .1314$.

7.8 Find a z_0 such that $P(-z_0 < z < z_0) = .6046$.

7.9 Find a z_0 such that $P(z < z_0) = .05$.

7.10 Find a z_0 such that $P(-z_0 < z < z_0) = .90$.

7.11 Find a z_0 such that $P(-z_0 < z < z_0) = .99$.

7.12 The scores on a national achievement test were approximately normally distributed with a mean of 540 and a standard deviation of 110.
(a) If you achieved a score of 680, how far, in standard deviations, did your score depart from the mean?
(b) What percentage of those who took the examination scored higher than you?

7.13 Suppose that you must establish regulations concerning the maximum number of people who can occupy an elevator. A study of elevator occupancies indicates that if 8 people occupy the elevator, the probability distribution of the total weight of the 8 people possesses a mean equal to 1,200 pounds and a variance equal to 9,800 pounds. What is the probability that the total weight of 8 people exceeds 1,300 pounds? 1,500 pounds? (Assume that the probability distribution is approximately normal.)

7.14 The discharge of suspended solids from a phosphate mine is normally distributed, with a mean daily discharge of 27 milligrams per liter (mg/l) and a standard deviation of 14 mg/l. What proportion of days will the daily discharge exceed 50 mg/l?

7.15 The number of times y an adult human breathes per minute when at rest depends on the age of the human and varies greatly from person to person. Suppose that the probability distribution for y is approximately normal with mean equal to 16 and standard deviation equal to 4. If a person is selected at random and the number y of breaths per minute while at rest is recorded, what is the probability that y will exceed 22?

7.16 The daily sales (excepting Saturday) at a small restaurant has a probability distribution that is approximately normal with mean μ equal to \$530 per day and standard deviation σ equal to \$120.
(a) What is the probability that the sales will exceed \$700 on a given day?
(b) The restaurant must have at least \$300 sales per day in order to break even. What is the probability that on a given day the restaurant will not break even?

7.17 The length of life of a type of automatic washer is approximately normally distributed with mean and standard deviation equal to 3.1 and 1.2 years, respectively. If this type of washer is guaranteed for one year, what fraction of original sales will require replacement?

7.18 Suppose that the counts on the number of a particular type of bacteria in 1 milliliter of drinking water tend to be approximately normally distributed with a mean of 85 and a standard deviation of 9. What is the probability that a given 1-ml sample will contain more than 100 bacteria?

7.3

THE CENTRAL LIMIT THEOREM

In addition to being an important probability distribution in its own right, the normal probability distribution can often be used to approximate the probability distributions of other random variables because of the Central Limit Theorem. The Central Limit Theorem states that under rather general conditions, sums and means of samples of random measurements drawn from a population tend to possess, approximately, a bell-shaped distribution in repeated sampling. The significance of this statement is perhaps best illustrated by an example.

Consider a population of die throws generated by tossing a die an infinitely large number of times with resulting probability distribution given by Figure 7.8. Draw a sample of $n = 5$ measurements from the population by tossing a die five times and record each of the five observations as indicated in Table 7.2. Note that the numbers observed in the first sample were $y = 3, 5, 1, 3, 2$. Calculate the sum of the five measurements as well as the sample mean, $\bar{y}$. For experimental purposes, repeat the sampling procedure 100 times or preferably an even larger number of times. The results for 100 samples are given in Table 7.2 along with the corresponding values of $\sum_{i=1}^{5} y_i$ and $\bar{y}$. Construct a frequency histogram for $\bar{y}$ $\left(\text{or } \sum_{i=1}^{5} y_i\right)$ for the 100 samples and observe the resulting distribution in Figure 7.9. You will observe an interesting result, namely that although the values of y in the population ($y = 1, 2, 3, 4, 5, 6$) are equiprobable (Figure 7.8) and hence

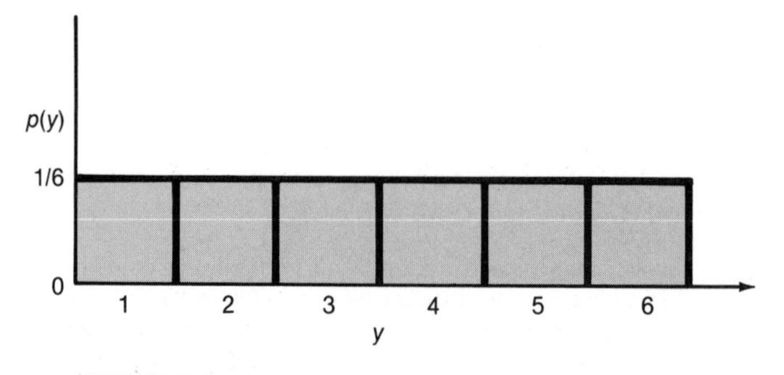

FIGURE 7.8

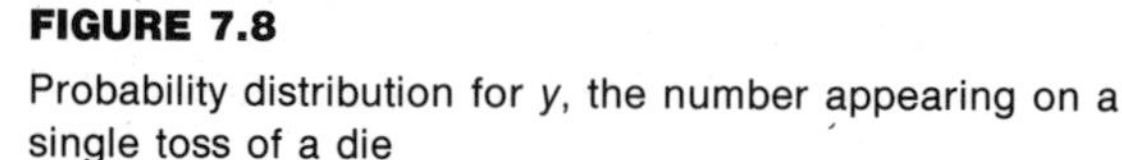

Probability distribution for y, the number appearing on a single toss of a die

TABLE 7.2
Sampling from the population of die throws

Sample Number	Sample Measurements	Σy_i	$\bar{y}$	Sample Number	Sample Measurements	Σy_i	$\bar{y}$
1	3, 5, 1, 3, 2	14	2.8	51	2, 3, 5, 3, 2	15	3.0
2	3, 1, 1, 4, 6	15	3.0	52	1, 1, 1, 2, 4	9	1.8
3	1, 3, 1, 6, 1	12	2.4	53	2, 6, 3, 4, 5	20	4.0
4	4, 5, 3, 3, 2	17	3.4	54	1, 2, 2, 1, 1	7	1.4
5	3, 1, 3, 5, 2	14	2.8	55	2, 4, 4, 6, 2	18	3.6
6	2, 4, 4, 2, 4	16	3.2	56	3, 2, 5, 4, 5	19	3.8
7	4, 2, 5, 5, 3	19	3.8	57	2, 4, 2, 4, 5	17	3.4
8	3, 5, 5, 5, 5	23	4.6	58	5, 5, 4, 3, 2	19	3.8
9	6, 5, 5, 1, 6	23	4.6	59	5, 4, 4, 6, 3	22	4.4
10	5, 1, 6, 1, 6	19	3.8	60	3, 2, 5, 3, 1	14	2.8
11	1, 1, 1, 5, 3	11	2.2	61	2, 1, 4, 1, 3	11	2.2
12	3, 4, 2, 4, 4	17	3.4	62	4, 1, 1, 5, 2	13	2.6
13	2, 6, 1, 5, 4	18	3.6	63	2, 3, 1, 2, 3	11	2.2
14	6, 3, 4, 2, 5	20	4.0	64	2, 3, 3, 2, 6	16	3.2
15	2, 6, 2, 1, 5	16	3.2	65	4, 3, 5, 2, 6	20	4.0
16	1, 5, 1, 2, 5	14	2.8	66	3, 1, 3, 3, 4	14	2.8
17	3, 5, 1, 1, 2	12	2.4	67	4, 6, 1, 3, 6	20	4.0
18	3, 2, 4, 3, 5	17	3.4	68	2, 4, 6, 6, 3	21	4.2
19	5, 1, 6, 3, 1	16	3.2	69	4, 1, 6, 5, 5	21	4.2
20	1, 6, 4, 4, 1	16	3.2	70	6, 6, 6, 4, 5	27	5.4
21	6, 4, 2, 3, 5	20	4.0	71	2, 2, 5, 6, 3	18	3.6
22	1, 3, 5, 4, 1	14	2.8	72	6, 6, 6, 1, 6	25	5.0
23	2, 6, 5, 2, 6	21	4.2	73	4, 4, 4, 3, 1	16	3.2
24	3, 5, 1, 3, 5	17	3.4	74	4, 4, 5, 4, 2	19	3.8
25	5, 2, 4, 4, 3	18	3.6	75	4, 5, 4, 1, 4	18	3.6
26	6, 1, 1, 1, 6	15	3.0	76	5, 3, 2, 3, 4	17	3.4
27	1, 4, 1, 2, 6	14	2.8	77	1, 3, 3, 1, 5	13	2.6
28	3, 1, 2, 1, 5	12	2.4	78	4, 1, 5, 5, 3	18	3.6
29	1, 5, 5, 4, 5	20	4.0	79	4, 5, 6, 5, 4	24	4.8
30	4, 5, 3, 5, 2	19	3.8	80	1, 5, 3, 4, 2	15	3.0
31	4, 1, 6, 1, 1	13	2.6	81	4, 3, 4, 6, 3	20	4.0
32	3, 6, 4, 1, 2	16	3.2	82	5, 4, 2, 1, 6	18	3.6
33	3, 5, 5, 2, 2	17	3.4	83	1, 3, 2, 2, 5	13	2.6
34	1, 1, 5, 6, 3	16	3.2	84	5, 4, 1, 4, 6	20	4.0
35	2, 6, 1, 6, 2	17	3.4	85	2, 4, 2, 5, 5	18	3.6
36	2, 4, 3, 1, 3	13	2.6	86	1, 6, 3, 1, 6	17	3.4
37	1, 5, 1, 5, 2	14	2.8	87	2, 2, 4, 3, 2	13	2.6
38	6, 6, 5, 3, 3	23	4.6	88	4, 4, 5, 4, 4	21	4.2
39	3, 3, 5, 2, 1	14	2.8	89	2, 5, 4, 3, 4	18	3.6
40	2, 6, 6, 6, 5	25	5.0	90	5, 1, 6, 4, 3	19	3.8
41	5, 5, 2, 3, 4	19	3.8	91	5, 2, 5, 6, 3	21	4.2
42	6, 4, 1, 6, 2	19	3.8	92	6, 4, 1, 2, 1	14	2.8
43	2, 5, 3, 1, 4	15	3.0	93	6, 3, 1, 5, 2	17	3.4
44	4, 2, 3, 2, 1	12	2.4	94	1, 3, 6, 4, 2	16	3.2
45	4, 4, 5, 4, 4	21	4.2	95	6, 1, 4, 2, 2	15	3.0
46	5, 4, 5, 5, 4	23	4.6	96	1, 1, 2, 3, 1	8	1.6
47	6, 6, 6, 2, 1	21	4.2	97	6, 2, 5, 1, 6	20	4.0
48	2, 1, 5, 5, 4	17	3.4	98	3, 1, 1, 4, 1	10	2.0
49	6, 4, 3, 1, 5	19	3.8	99	5, 2, 1, 6, 1	15	3.0
50	4, 4, 4, 4, 4	20	4.0	100	2, 4, 3, 4, 6	19	3.8

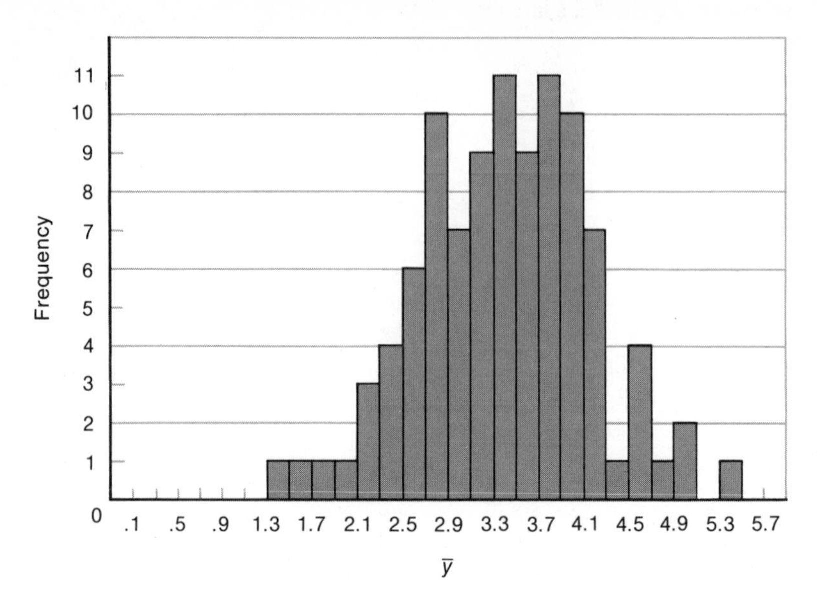

FIGURE 7.9

Histogram of sample means for the die-tossing experiments in Section 7.3

possess a probability distribution that is perfectly flat in shape, the distribution of the sample means (or sums) chosen from the population possesses a mound-shaped distribution (Figure 7.9). We shall add one further comment without proof. If we should repeat the experiment outlined above for a larger sample size, say $n = 10$, we would find that the distribution of the sample means tends to become more nearly bell-shaped.

Note that a proper evaluation of the form of the probability distribution of the sample means would require an infinitely large number of samples, or, at the very least, far more than the 100 samples contained in our experiment. Nevertheless, the 100 samples illustrate the basic idea involved in the Central Limit Theorem, which may be stated as follows.

The Central Limit Theorem: If random samples of n observations are drawn from a population with finite mean, μ, and standard deviation, σ, then, when n is large, the sample mean, $\bar{y}$, will be approximately normally distributed with mean equal to μ and standard deviation $\sigma/\sqrt{n}$. The approximation will become more and more accurate as n becomes large.

Note that the Central Limit Theorem could be restated to apply to the sum of the sample measurements, $\sum_{i=1}^{n} y_i$, which would also tend to possess a normal distribution, in repeated sampling, with mean equal to $n\mu$ and standard deviation $\sigma\sqrt{n}$, as n becomes large.

The Central Limit Theorem tells us that the mean and standard deviation of the distribution of sample means are definitely related to the mean and standard deviation of the sampled population as well as to the sample size n. The two distributions have the same mean μ and the standard deviation of the distribution of sample means is equal to the population standard deviation σ divided by $\sqrt{n}$ (it can be shown that this relationship is true regardless of the sample size n). Consequently, the spread of the distribution of sample means is considerably less ($1/\sqrt{n}$ as large) as the spread of the population distribution. But most important, the Central Limit Theorem tells us that the probability distribution of the sample mean is approximately normal for samples of moderate to large values of the sample size n.*

The significance of the Central Limit Theorem is twofold. First, it explains why some measurements tend to possess, approximately, a normal distribution. We might imagine the height of a human as being composed of a number of elements, each random, associated with such things as the height of the mother, the height of the father, the activity of a particular gland, the environment, and diet. If each of these effects tends to add to the others to yield the measurement of height, then height is the sum of a number of random variables and the Central Limit Theorem may become effective and yield a distribution of heights which is approximately normal. All of this is conjecture, of course, because we really do not know the true situation which exists. Nevertheless, the Central Limit Theorem, along with other theorems dealing with normally distributed random variables, provides an explanation of the rather common occurrence of normally distributed random variables in nature.

The second and most important contribution of the Central Limit Theorem is in statistical inference. Many estimators and decision makers that are used to make inferences about population parameters are sums or averages of the sample measurements. When this is true and when the sample size, n, is sufficiently large, we would expect the estimator or decision maker to possess a normal probability distribution in repeated sampling according to the Central Limit Theorem. We can then use the Empirical Rule discussed in Chapter 3 to describe the behavior of the inference maker. This aspect of the Central Limit Theorem will be utilized in Section 7.4 as well as in later chapters dealing with statistical inference.

One disturbing feature of the Central Limit Theorem, and of most approximation procedures, is that we must have some idea as to how large the sample size, n, must be in order for the approximation to give useful results. Unfortunately, there is no clear-cut answer to this question, as the appropriate value for n will depend upon the population probability distribution as well as the use we will make of the approximation. Although the preceding comment sidesteps the difficulty and suggests that we must rely solely upon experience, we may take comfort in the results of the die-tossing experiment discussed previously in this section. Note that the distribution of $\bar{y}$, in repeated sampling, based upon a sample of only $n = 5$ measurements, tends to be approximately bell-shaped. Generally speaking, the Central Limit Theorem functions very well, even for small

*The distribution of the sample means will be normally distributed, regardless of the sample size, for the special case when the population possesses a normal distribution.

samples, but this is not always true. We will observe an exception to this rule in Section 7.4. The appropriate sample size, n, will be given for specific applications of the Central Limit Theorem as they are encountered in Section 7.4 and later in the text.

EXAMPLE 7.7

To avoid difficulties with the Federal Trade Commission or state and local consumer protection agencies, a beverage bottler must make reasonably certain that 12-ounce bottles actually contain 12 ounces of beverage. To infer whether a bottling machine is working satisfactorily, one bottler randomly samples 10 bottles per hour and measures the amount of beverage in each bottle. The mean $\bar{y}$ of the 10 fill measurements is used to decide whether to readjust the amount of beverage delivered per bottle by the filling machine. If records show that the amount of fill per bottle possesses a standard deviation of .2 ounce and if the bottling machine is set to produce a mean fill per bottle of 12.1 ounces, what is the probability that the sample mean $\bar{y}$ of the ten test bottles is less than 12 ounces?

Solution The mean of the probability distribution of the sample mean $\bar{y}$ is identical to the mean of the population of bottle fills, namely, $\mu = 12.1$ ounces, and the standard deviation of the distribution, denoted by the symbol, $\sigma_{\bar{y}}$, is

$$\sigma_{\bar{y}} = \frac{\sigma}{\sqrt{n}} = \frac{.2}{\sqrt{10}} = .063.$$

(*Note:* σ is the standard deviation of the population of bottle fills and n is the number of bottles in the sample.) Even though n is as small as 10, it is likely, for this type of data, that the probability distribution of $\bar{y}$ will be approximately normal because of the Central Limit Theorem. Then the probability distribution of $\bar{y}$ will appear as shown in Figure 7.10.

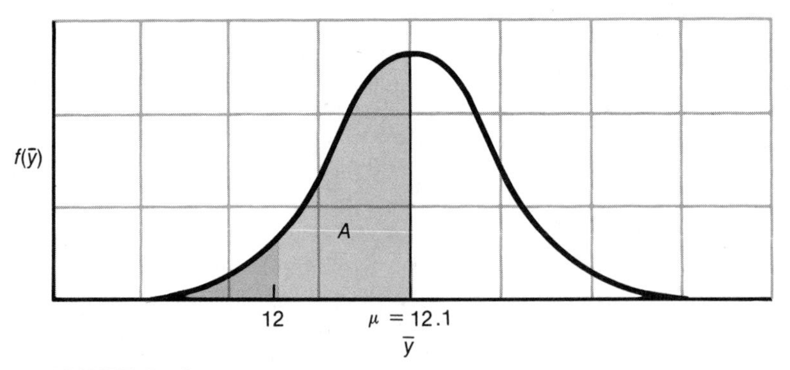

FIGURE 7.10

Probability distribution of $\bar{y}$, the mean of the $n = 10$ bottle fills

The probability that $\bar{y}$ will be less than 12 ounces is approximately equal to the shaded area under the normal curve, Figure 7.10. This area will equal $(.5 - A)$, where A is the area between 12 and the mean, $\mu = 12.1$. Expressing this distance in terms of z,

$$z = \frac{\bar{y} - \mu}{\sigma_{\bar{y}}} = \frac{12 - 12.1}{.063} = \frac{-.1}{.063} = -1.59.$$

[Note that we must use $\sigma_{\bar{y}}$ (not σ) in the formula for z because we are finding an area under the probability distribution for $\bar{y}$, *not* under the probability distribution for y.] Then, the area A over the interval, $12 \leq \bar{y} \leq 12.1$, is shown in Table 3, Appendix II, as .4441 and the probability that $\bar{y}$ will be less than 12 ounces is

$$\begin{aligned} P(\bar{y} < 12) &= .5 - A = .5 - .4441 \\ &= .0559 \\ &\approx .056. \end{aligned}$$

Or, if the fill machine is set to produce a mean load of 12.1 ounces, the mean fill $\bar{y}$ of a sample of 10 bottles will be less than 12 ounces with probability approximately equal to .056. When this danger signal occurs ($\bar{y}$ is less than 12), the bottler takes a larger sample to recheck the setting of the filling machine. Note that the Central Limit Theorem plays a role in the solution of this problem because it justifies the approximate normality of the probability distribution of the sample mean.

EXERCISES

7.19 Looking at the histogram of Figure 7.9, guess the value of its mean and standard deviation. (*Hint:* The Empirical Rule states that approximately 95 percent of the measurements associated with a mound-shaped distribution will lie within two standard deviations of the mean.)

7.20 Let y equal the number of dots observed when a single die is tossed. The mean value of y, Example 5.4, and standard deviation (Exercise 5.19) were found to equal $\mu = 3.5$ and $\sigma = 1.71$, respectively. Suppose that the sampling experiment, Section 7.3, were repeated over and over again for an infinitely large number of times, each sample consisting of $n = 5$ measurements. Find the mean and standard deviation for this distribution of sample means. (*Hint:* See the Central Limit Theorem.) Compare this solution with the solution to Exercise 7.19.

7.21 Suppose that you were to experiment by drawing thousands of samples where each sample involved tossing a die $n = 10$ times. If a histogram were constructed for the sample means, what would be the approximate value for the mean of the distribution? the standard deviation?

7.22 A lobster fisher's daily catch, y, is the total, in pounds, of lobster landed from a fixed number of lobster traps. What kind of probability distribution would you expect the daily catch to possess and why? If the mean catch per trap per day is 30 pounds with $\sigma = 5$ pounds, and the lobster fisher has 50 traps, give the mean and standard deviation of the probability distribution of the total daily catch y.

7.23 To obtain information on the volume of freight shipped by truck over a particular interstate highway, a state highway department monitored the highway for 25 one-hour periods randomly selected throughout a one-month period. The number of truck trailers was counted for each one-hour period and $\bar{y}$ was calculated for the sample of 25 individual one-hour periods. If the number of heavy-duty trailers per hour is approximately normally distributed, with $\mu = 50$ and $\sigma = 7$:

(a) What is the probability that the sample mean $\bar{y}$ for $n = 25$ one-hour periods is larger than 55?

(b) Suppose you were to count the truck trailers for each of $n = 4$ randomly selected one-hour periods. What is the probability that $\bar{y}$ would be larger than 55?*

(c) What is the probability that the total number of trucks for a four-hour period would exceed 180?*

7.24 A manufacturer of paper used for packaging requires a minimum strength of 20 pounds per square inch. To check on the quality of the paper, a random sample of 10 pieces of paper is selected each hour from the previous hour's production and a strength measurement is recorded for each. The standard deviation σ of the strength measurements, computed by pooling the sum of squares of deviations of many samples, is known to equal 2 pounds per square inch.

(a) What is the approximate probability distribution of the sample mean of $n = 10$ test pieces of paper?

(b) If the mean of the population of strength samples is 21 pounds per square inch, what is the probability that, for a random sample of $n = 10$ test pieces of paper, $\bar{y} < 20$?

(c) What value would you desire for the mean paper strength, μ, in order that $P(\bar{y} < 20)$ be equal to .001?

7.4
THE NORMAL APPROXIMATION OF THE BINOMIAL DISTRIBUTION

In Chapter 6 we considered several applications of the binomial probability distribution, all of which required that we calculate the probability that y, the number of successes in n trials, would fall in a given region. For the most part we restricted our attention to examples where n was small because of the tedious calculations necessary in the computations of $p(y)$. Let us now consider the problem of calculating $p(y)$, or the probability that y will fall in a given region, when n is large, say $n = 1{,}000$. A direct calculation of $p(y)$ for large values of n is not an impossibility, but it does provide a formidable task which we would prefer to avoid. Fortunately, the Central Limit Theorem provides a solution to this dilemma since we may view y, the number of successes in n trials, as a sum that satisfies the conditions of the Central Limit Theorem. Each trial results in either 0 or 1 success with probability q and p, respectively. Therefore, each of the n trials may be regarded as an independent observation drawn from a simpler binomial experiment consisting of one trial, and y, the total number of successes in n trials,

*The distribution of the sample means will be normally distributed, regardless of the sample size, for the special case when the population possesses a normal distribution.

is the sum of these n independent observations. Then, if n is sufficiently large, the binomial variable, y, will be approximately normally distributed with mean and variance (obtained in Chapter 6) np and npq, respectively. We may then use areas under a fitted normal curve to approximate the binomial probabilities.

For instance, consider a binomial probability distribution for y when $n = 10$ and $p = 1/2$. Then

$$\mu = np = 10(1/2) = 5 \qquad \text{and} \qquad \sigma = \sqrt{npq} = \sqrt{2.5} = 1.58.$$

Figure 7.11 shows the corresponding binomial probability distribution and the approximating normal curve on the same graph. A visual comparison of the figures would suggest that the approximation is reasonably good, even though a small sample, $n = 10$, was necessary for this graphic illustration.

The probability that $y = 2$, 3, or 4 is exactly equal to the area of the three rectangles lying over $y = 2$, 3, and 4. We may approximate this probability with the area under the normal curve from $y = 1.5$ to $y = 4.5$, which is shaded in Figure 7.11. Note that the area under the normal curve between $y = 2$ and $y = 4$ *would not* be a good approximation to the probability that $y = 2$, 3, or 4 because it would exclude one-half of the probability rectangles corresponding to $y = 2$ and $y = 4$. To get a good approximation you must remember to approximate the entire areas of the probability rectangles corresponding to $y = 2$ and $y = 4$ by including the area under the normal curve from $y = 1.5$ to $y = 4.5$.

Although the normal probability distribution provides a reasonably good approximation to the binomial probability distribution, Figure 7.11, this will not always be the case. When n is small and p is near 0 or 1, the binomial probability distribution will be nonsymmetrical; that is, its mean will be located near 0 or n. For example, when p is near zero, most values of y will be small, producing a distribution which is concentrated near $y = 0$ and which tails gradually toward n (see Figure 7.12). Certainly, when this is true, the normal distribution, symmetrical and bell-shaped, will provide a poor approximation to the binomial probability distribution. How, then, can we tell whether n and p are such that the binomial distribution will be symmetrical?

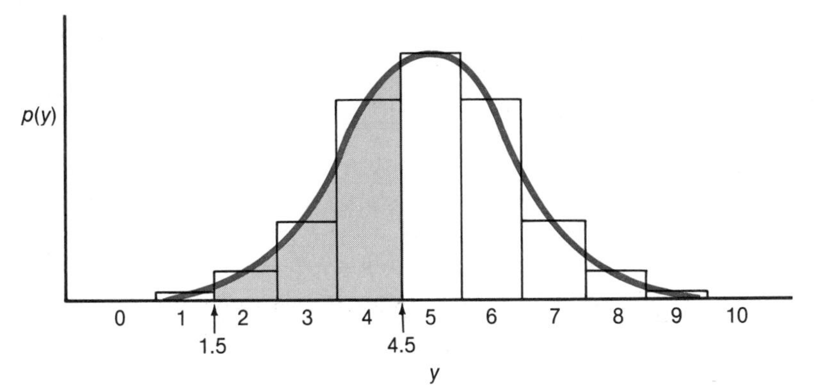

FIGURE 7.11

Comparison of a binomial probability distribution and the approximating normal distribution, $n = 10$, $p = 1/2$ ($\mu = np = 5$; $\sigma = \sqrt{npq} = 1.58$)

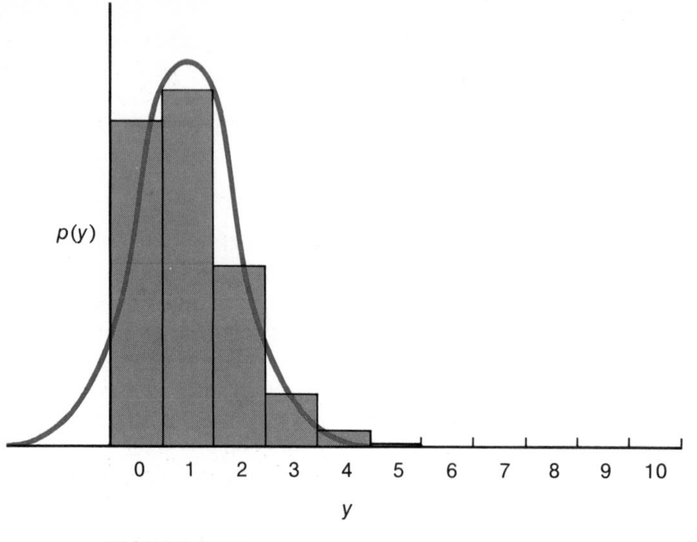

FIGURE 7.12

Comparison of a binomial probability distribution (shaded) and the approximating normal distribution, $n = 10$, $p = .1$ ($\mu = np = 1$; $\sigma = \sqrt{npq} = .95$)

Recalling the Empirical Rule, Chapter 3, approximately 95 percent of the measurements associated with a normal distribution will lie within two standard deviations of the mean and almost all will lie within three. We would suspect that the binomial probability distribution would be nearly symmetrical if the distribution were able to spread out a distance equal to two standard deviations on either side of the mean and this is, in fact, the case. Hence, to determine when the normal approximation will be adequate, calculate $\mu = np$ and $\sigma = \sqrt{npq}$. If the interval $\mu \pm 2\sigma$ lies within the binomial bounds, 0 and n, the approximation will be reasonably good. Note that this criterion is satisfied for the example, Figure 7.11, but is not satisfied for Figure 7.12.

EXAMPLE 7.8

Refer to the binomial experiment illustrated in Figure 7.11 where $n = 10$, $p = .5$. Calculate the probability that $y = 2$, 3, or 4 correct to three decimal places using Table 1, Appendix II. Then calculate the corresponding normal approximation to this probability.

Solution The exact probability, P_1, can be calculated using Table 1(b), Appendix II. Thus

$$P_1 = \sum_{y=2}^{4} p(y) = \sum_{y=0}^{4} p(y) - \sum_{y=0}^{1} p(y)$$
$$= .377 - .011$$
$$= .366.$$

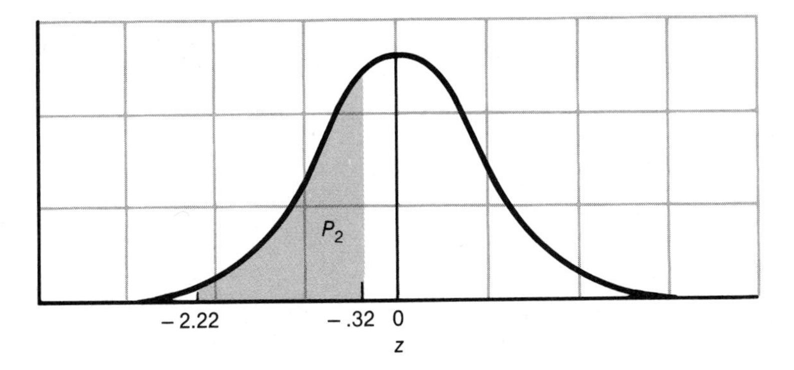

FIGURE 7.13
Area under the normal curve for Example 7.8

The normal approximation would require the area lying between $y_1 = 1.5$ and $y_2 = 4.5$ (see Figure 7.11), where $\mu = 5$ and $\sigma = 1.58$. The corresponding values of z are

$$z_1 = \frac{y_1 - \mu}{\sigma} = \frac{1.5 - 5}{1.58} = -2.22,$$

$$z_2 = \frac{y_2 - \mu}{\sigma} = \frac{4.5 - 5}{1.58} = -.32.$$

The probability, P_2, is shown in Figure 7.13. The area between $z = 0$ and $z = 2.22$ is $A_1 = .4868$. Likewise, the area between $z = 0$ and $z = .32$ is $A_2 = .1255$. It can be seen from Figure 7.13 that

$$\begin{aligned} P_2 &= A_1 - A_2 \\ &= .4868 - .1255 = .3613. \end{aligned}$$

Note that the normal approximation is quite close to the binomial probability obtained from Table 1.

You must be careful not to exclude half of the two extreme probability rectangles when using the normal approximation to the binomial probability distribution. This means that the y values used to calculate z values will always have a 5 in the tenths decimal place. To be certain that you include all the probability rectangles in your approximation, always draw a sketch similar to Figure 7.11.

EXAMPLE 7.9

The reliability of an electrical fuse is the probability that a fuse, chosen at random from production, will function under the conditions for which it has been designed. A random sample of 1,000 fuses was tested and $y = 27$ defectives were observed. Calculate the probability of observing 27 or more defectives, assuming that the fuse reliability is .98.

Solution The probability of observing a defective when a single fuse is tested is $p = .02$, given that the fuse reliability is .98. Then,

$$\mu = np = 1{,}000(.02) = 20,$$
$$\sigma = \sqrt{npq} = \sqrt{1{,}000(.02)(.98)} = 4.43.$$

The probability of 27 or more defective fuses, given $n = 1{,}000$, is

$$P = P(y \geq 27),$$
$$P = P(27) + P(28) + P(29) + \cdots + P(999) + P(1{,}000).$$

The normal approximation to P would be the area under the normal curve to the right of $y = 26.5$. (Note that we must use $y = 26.5$ rather than $y = 27$ so as to include the entire probability rectangle associated with $y = 27$.) The z value corresponding to $y = 26.5$ is

$$z = \frac{y - \mu}{\sigma} = \frac{26.5 - 20}{4.43} = \frac{6.5}{4.43} = 1.47$$

and the area between $z = 0$ and $z = 1.47$ is equal to .4292, as shown in Figure 7.14. Since the total area to the right of the mean is equal to .5,

$$P = .5 - .4292$$
$$= .0708.$$

EXAMPLE 7.10

A new serum was tested to determine its effectiveness in preventing the common cold. One hundred people were injected with the serum and observed for a period of one year. Sixty-eight survived the winter without a cold. Suppose that according to prior information it is known that the probability of surviving the winter without a cold is equal to .5 when the serum is not used. On the basis of the results of this experiment, what conclusions would you make regarding the effectiveness of the serum?

Solution Translating the question into an hypothesis concerning the parameter of the binomial population, we wish to test the null hypothesis

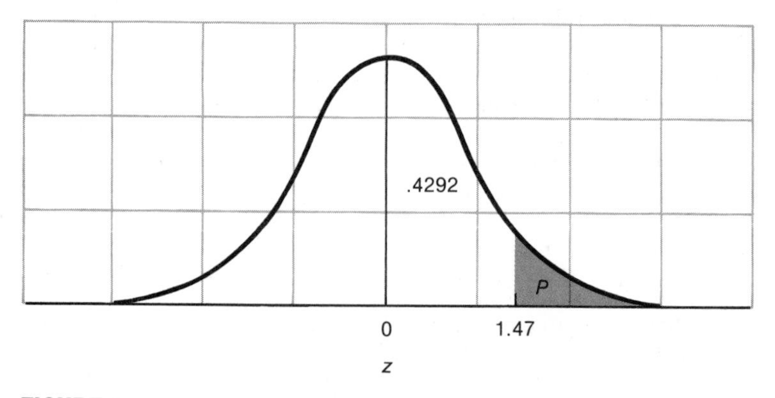

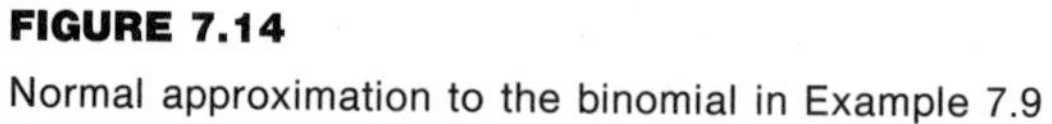

FIGURE 7.14

Normal approximation to the binomial in Example 7.9

that p, the probability of survival on a single trial, is equal to .5. Assume that the content of the serum is such that it could not increase the susceptibility to colds. Then the alternative to the null hypothesis would reject the null hypothesis when y, the number of survivors, is large.

Since the normal approximation to the binomial will be adequate for this example, we would interpret a large and improbable value of y to be one that lies several standard deviations away from the hypothesized mean, $\mu = np = 100(.5) = 50$.

Noting that

$$\sigma = \sqrt{npq} = \sqrt{100(.5)(.5)} = 5,$$

we may arrive at a conclusion without bothering to locate a specific rejection region. The observed value of y, 68, lies more than 3σ away from the hypothesized mean, $\mu = 50$. Specifically, y lies

$$z = \frac{y - \mu}{\sigma} = \frac{68 - 50}{5} = 3.6$$

standard deviations away from the hypothesized mean. This result is so improbable, assuming the serum ineffective, that we would reject the null hypothesis and conclude that the probability of surviving a winter without a cold is greater than $p = .5$ when the serum is used. (Observe that the area to the right of $z = 3.6$ is so small that it is not included in Table 3, Appendix II.)

Rejecting the null hypothesis raises additional questions. How effective is the serum and is it sufficiently effective, from an economic point of view, to warrant commercial production? The former question leads to an estimation problem, a topic discussed in Chapter 8, while the latter, involving a business decision, would utilize the results of our experiment as well as a study of consumer demand, sales and production costs, etc., to achieve an answer useful to the drug company.

EXAMPLE 7.11

The probability of a type I error, α, and location of the rejection region for a statistical test of an hypothesis are usually specified before the data are collected. Suppose that we wish to test the null hypothesis, $p = .5$, in a situation identical to the cold-serum problem in Example 7.10. Find the appropriate rejection region for the test if we wish α to be approximately equal to .05 (see Figure 7.15).

Solution We have previously stated in Example 7.10 that y, the number of survivors, would be used as a test statistic and that the rejection region would be located in the upper tail of the probability distribution for y. Desiring α approximately equal to .05, we seek a value of y, say y_α, such that

$$P(y \geq y_\alpha) \approx .05.$$

(*Note:* The symbol $\approx$ means "approximately equal to.") This can be determined by first finding the corresponding z_α, which gives the number of

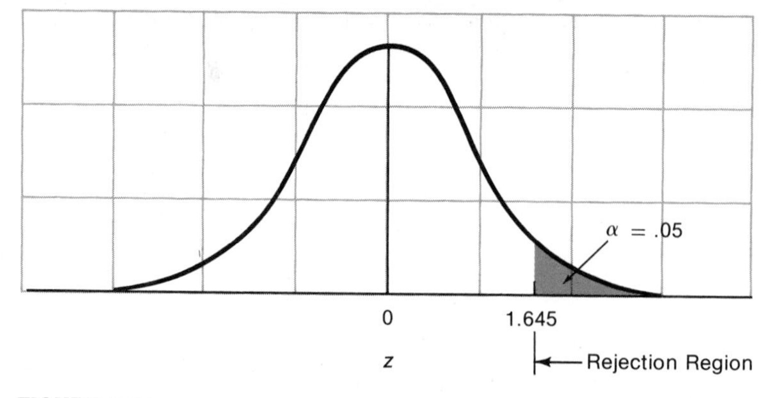

FIGURE 7.15
Location of the rejection region in Example 7.11

standard deviations between the mean, $\mu = 50$, and y_α. Since the total area to the right of $z = 0$ is .5, the area between $z = 0$ and z_α will equal .45. Checking Table 3, we find that $z = 1.64$ corresponds to an area equal to .4495 and $z = 1.65$ to an area of .4505. A linear interpolation between these values would give

$$z_\alpha = 1.645.$$

Recalling the relation between z and y,

$$z_\alpha = \frac{y_\alpha - \mu}{\sigma}$$

or

$$1.645 = \frac{y_\alpha - 50}{5}.$$

Solving for y_α, we obtain

$$y_\alpha = 58.225.$$

Obviously, we cannot observe $y = 58.225$ survivors and hence must choose 58 or 59 as the point where the rejection region commences. (Remember, the binomial is a discrete probability distribution; the normal is continuous.)

Suppose that we decide to reject when y is greater than or equal to 59. Then the actual probability of the type I error, α, for the test is

$$P(y \geq 59) = \alpha,$$

which can be approximated by using the area under the normal curve above $y = 58.5$, a problem similar to that encountered in Example 7.9. The z value corresponding to $y = 58.5$ is

$$z = \frac{y - \mu}{\sigma} = \frac{58.5 - 50}{5} = 1.7,$$

and the tabulated area between $z = 0$ to $z = 1.7$ is .4554:

$$\alpha = .5 - .4554$$
$$= .0446.$$

Although this method provides a more accurate value for α, there is very little practical difference between an α of .0446 and one equal to .05. When n is large, time and effort may be saved by using z as a test statistic rather than y. This method was employed in Example 7.11. We would then reject the null hypothesis when z is greater than or equal to 1.645.

EXAMPLE 7.12

A cigarette manufacturer believed that approximately 10 percent of all smokers favored his product, brand A. To test this belief, 2,500 smokers were selected at random from the population of cigarette smokers and questioned concerning their cigarette brand preference. A total of $y = 218$ expressed a preference for brand A. Do these data provide sufficient evidence to contradict the hypothesis that 10 percent of all smokers favor brand A? Conduct a statistical test using an α equal to .05.

Solution We wish to test the null hypothesis that p, the probability that a single smoker prefers brand A, is equal to .1 against the alternative that p is greater than or less than .1. The rejection region corresponding to an $\alpha = .05$ would be located as shown in Figure 7.16. We would reject the null hypothesis when $z > 1.96$ or $z < -1.96$. In other words, we would reject when y lies more than approximately two standard deviations away from its hypothesized mean. Note that half of α is placed in one tail of the distribution and half in the other because we wish to reject the null hypothesis when p is either larger or smaller than $p = .1$. This is called a two-tailed statistical test, in contrast to the one-tailed test discussed in Examples 7.10 and 7.11 when the alternative to the null hypothesis was only that p was larger than the hypothesized value.

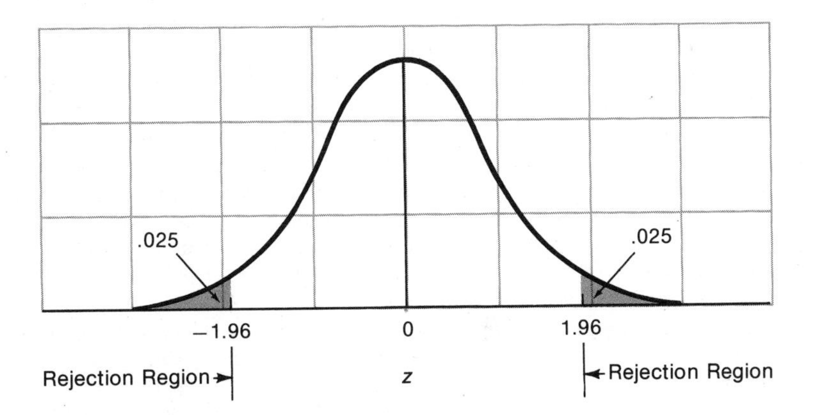

FIGURE 7.16
Location of the rejection region in Example 7.12

Assuming the null hypothesis to be true, the mean and standard deviation for y are

$$\mu = np = 2{,}500(.1) = 250,$$
$$\sigma = \sqrt{npq} = \sqrt{2{,}500(.1)(.9)} = 15.$$

The z value corresponding to the observed $y = 218$ is

$$z = \frac{y - \mu}{\sigma} = \frac{218 - 250}{15} = \frac{-32}{15} = -2.13.$$

Noting that z falls in the rejection region, we would reject the null hypothesis that $p = .1$. In fact, it appears that less than 10 percent of all smokers prefer brand A.

What is the probability that we have made an incorrect decision? The answer, of course, is either 1 or 0, depending upon whether our decision was correct or incorrect in this specific case. However, we know that if this statistical test were employed over and over again, the probability of rejecting the null hypothesis when it is true is only $\alpha = .05$. Hence we are reasonably certain that we have made the correct decision.

EXERCISES

7.25 To show how the Central Limit Theorem works, compare the shape of the binomial probability distributions for $n = 10$, $p = .2$ and $n = 25$, $p = .2$. Use Table 1, Appendix II, to construct the probability histograms. Note the lack of symmetry for $n = 10$, $p = .2$. In contrast, the probability distribution for $n = 25$, $p = .2$ is more symmetric and tending to a bell shape. The larger the value of n, the more closely the binomial probability distribution will approach the normal distribution.

7.26 Let y be a binomial random variable for $n = 25$, $p = .2$ (see Exercise 7.25).
(a) Use Table 1, Appendix II, to calculate $P(4 \leq y \leq 6)$.
(b) Find μ and σ for the binomial probability distribution and use the normal distribution to approximate the probability, $P(4 \leq y \leq 6)$. Note that this value is a good approximation to the exact value of $P(4 \leq y \leq 6)$.

7.27 Consider a binomial experiment with $n = 20$, $p = .4$. Calculate $P(y \geq 10)$ by use of:
(a) Table 1, Appendix II.
(b) The normal approximation to the binomial probability distribution.

7.28 Briggs and King developed the technique of nuclear transplantation in which a nucleus of a cell from one of the later stages of the development of an embryo is transplanted into a zygote (a single cell, fertilized egg) to see if the nucleus can support normal development. If the probability that a single transplant from the early gastrula stage will be successful is .65, what is the probability that more than 70 transplants out of 100 will be successful?

7.29 In a study of primary liver tumors Vana and colleagues* comment on the possible link between liver tumors and the use of oral contraceptives and report on a survey of 477 hospitals that revealed 378 cases of primary liver tumors in females. Of these 187 were known to have used oral contraceptives. Suppose that the proportion of

J. Amer. Med. Assoc., *238* (1977), pp. 2154–2158.

females in the United States who have used oral contraceptives is p and, for the sake of argument, suppose $p = .4$. If the 378 female patients with liver tumors can be viewed as a random sample from among all adult females, what is the probability that the number of patients who have used oral contraceptives is as large as 187? If p really is equal to .4, is the observation of 187 in the "random sample" of 378 a rare event? Do the 378 tumor patients represent a random sample selected from among all adult females in the United States?

7.30 According to a recent article (*Gainesville Sun,* November 25, 1977), television may be dangerous to your diet. Psychologists believe that excessive eating may be associated with emotional states (being upset, bored, etc.) and environmental cues (TV, reading, etc.). To test this theory, suppose that we were to randomly select 60 overweight persons and match them by weight and sex in pairs. For a period of two weeks, one of each pair is required to spend evenings reading novels of interest to them. The other member of each pair spends each evening watching television. The calorie count for all snack and drink intake for the evenings is recorded for each person and we record $y = 19$, the number of pairs for which the TV watchers calorie intake exceeded the intake of the readers. If there is no difference in the effect of TV and reading on calorie intake, the probability, p, that the calorie intake of one member of a pair exceeds that of the other member is .5. (If there is a difference, $p \neq .5$.) Do these data provide sufficient evidence to indicate a difference between the effects of TV watching and reading on calorie intake?

7.31 Airlines and hotels often grant reservations in excess of capacity to minimize losses due to no-shows. Suppose that the records of a motel show that, on the average, 10 percent of their prospective guests will not claim their reservation. If the motel accepts 215 reservations and there are only 200 rooms in the motel, what is the probability that all guests who arrive to claim a room will receive one?

7.32 In Exercise 6.13, Chapter 6, we commented on a telephone survey conducted by the administration of the North Florida Regional Hospital, Gainesville, Florida. The purpose of the survey was to determine public attitude toward the hospital's desire to expand even though the expansion might have a negative effect on the occupancy of beds in the local county hospital. Of the 180 persons polled, 112 stated that they favored the expansion. What is the probability that y, the number of people favoring the expansion, is as large or larger than 112, given that the proportion of adults in the community who favor the expansion is only .5?

7.33 Compilation of large masses of data on lung cancer show that approximately 1 of every 40 adults acquire the disease. Workers in a certain occupation are known to work in an air-polluted environment that may cause an increased rate of lung cancer. A random sample of $n = 400$ workers shows 19 with identifiable cases of lung cancer. Do the data provide sufficient evidence to indicate a higher rate of lung cancer for these workers than for the national average?

7.5

SUMMARY

Many continuous random variables observed in nature possess a probability distribution which is bell-shaped and which may be approximated by the normal probability distribution discussed in Section 7.1. The common occurrence of normally distributed random variables may be partly explained by the Central Limit Theorem, which states that, under rather general conditions, the sum or the mean of a random sample of n measurements drawn from a population will be approximately normally distributed in repeated sampling when n is large.

As a case in point, the number of successes, y, associated with a binomial experiment may be regarded as a sum of n sample measurements that will possess, approximately, a normal probability distribution when n, the total number of trials, is large. This application of the Central Limit Theorem provides a method for calculating, with reasonable accuracy, the probabilities of the binomial probability distribution by using corresponding areas under the normal probability distribution. While other applications of the Central Limit Theorem and the normal distribution will be encountered in succeeding chapters, we particularly note that the Central Limit Theorem provides justification for the use of the Empirical Rule, Chapter 3. Furthermore, we observe that the contents of this chapter provide an extension and refinement of the thought embodied in the Empirical Rule.

Tips on Problem Solving

1. Always sketch a normal curve and locate the probability areas pertinent to the exercise. If you are approximating a binomial probability distribution, sketch in the probability rectangles as well as the normal curve.
2. Read each exercise carefully to see whether the data come from a binomial experiment or whether they possess a distribution that, by its very nature, is approximately normal. If you are approximating a binomial probability distribution, do not forget to make a half-unit correction so that you will include the half rectangles at the ends of the interval. If the distribution is not binomial, *do not* make the half-unit corrections. If you make a sketch (as suggested in step 1), you will see why the half-unit correction is or is not needed.

REFERENCES

Chapman, D. G., and R. A. Schaufele, *Elementary Probability Models and Statistical Inference.* New York: John Wiley & Sons, Inc., 1970.

Hoel, P. G., *Elementary Statistics,* 4th ed. New York: John Wiley & Sons, Inc., 1976.

Huntsberger, D. V., and P. Billingsley, *Elements of Statistical Inference,* 4th ed. Boston: Allyn and Bacon, Inc., 1977.

SUPPLEMENTARY EXERCISES

7.34 Using Table 3, Appendix II, calculate the area under the normal curve between (a) $z = 0$ and $z = 1.2$, (b) $z = 0$ and $z = -.9$.

7.35 Repeat Exercise 7.34:
(a) $z = 0$ and $z = 1.6$, (b) $z = 0$ and $z = .75$.

7.36 Repeat Exercise 7.34:
(a) $z = 0$ and $z = 1.46$, (b) $z = 0$ and $z = -.42$.

7.37 Repeat Exercise 7.34:
(a) $z = 0$ and $z = -1.44$, (b) $z = 0$ and $z = 2.01$.

7.38 Repeat Exercise 7.34:
(a) $z = .3$ and $z = 1.56$, (b) $z = .2$ and $z = -.2$.

7.39 Repeat Exercise 7.34:
(a) $z = .88$ and $z = 1.85$, (b) $z = -.31$ and $z = 1.63$.

7.40 Repeat Exercise 7.34:
(a) $z = 1.21$ and $z = 1.75$, (b) $z = -1.3$ and $z = 1.74$.

7.41 Find the probability that z is greater than $-.75$.

7.42 Find the probability that z is less than 1.35.

7.43 Find a z_0 such that $P(z > z_0) = .5$.

7.44 Find a z_0 such that $P(z < z_0) = .8643$.

7.45 Find the probability that z lies between $z = .7$ and $z = 1.63$.

7.46 Let y be a normally distributed random variable with mean equal to 7 and standard deviation equal to 1.5. If a value of y is chosen at random from the population, find the probability that y falls between $y = 8$ and $y = 9$.

7.47 Find the probability that z lies between $z = -.2$ and $z = 1.83$.

7.48 Find the probability that z lies between $z = -1.48$ and $z = 1.48$.

7.49 Find a z_0 such that $P(-z_0 < z < z_0) = .5$.

7.50 The influx of new ideas into a college or university, introduced primarily by hiring new young faculty, is becoming a matter of concern because of the increasing ages of faculty members. That is, the distribution of faculty ages is shifting upward, due most likely to a shortage of vacant positions and an oversupply of PhD's. Thus faculty members are more reluctant to move and give up a secure position. If the retirement age at most universities is 65, would you expect the distribution of faculty ages to be normal?

7.51 The scores on a national achievement test are normally distributed with mean and standard deviation equal to 500 and 100, respectively. If Jones scored 650, what fraction of students exceeded Jones' score?

7.52 The grade-point averages of a large population of college students are approximately normally distributed with mean equal to 2.4 and standard deviation equal to .8. What fraction of the students will possess a grade-point average in excess of 3.0?

7.53 Refer to Exercise 7.52. If students possessing a grade-point average equal to or less than 1.9 are dropped from college, what percentage of the students will be dropped?

7.54 A machine operation produces bearings whose diameters are normally distributed with mean and standard deviation equal to .498 and .002, respectively. If specifications require that the bearing diameter equal .500 inch plus or minus .004 inch, what fraction of the production will be unacceptable?

7.55 Consider a binomial experiment with $n = 25$, $p = .4$. Calculate $P(8 \leq y \leq 11)$ by use of:
(a) The binomial probabilities, Table 1, Appendix II.
(b) The normal approximation to the binomial.

7.56 Consider a binomial experiment with $n = 25$, $p = .2$. Calculate $P(y \leq 4)$ by use of:
(a) Table 1, Appendix II.
(b) The normal approximation to the binomial.

7.57 The average length of time required for a college achievement test was found to equal 70 minutes, with a standard deviation of 12 minutes. When should the test be terminated if we wish to allow sufficient time for 90 percent of the students to complete the test? (Assume that the time required to complete the test is normally distributed.)

7.58 An advertising agency has stated that 20 percent of all television viewers watch a particular program. In a random sample of 1,000 viewers, $y = 184$ viewers were watching the program. Do these data present sufficient evidence to contradict the advertiser's claim?

7.59 A salesman has found that, on the average, the probability of a sale on a single contact is equal to .3. If the salesman contacts 50 customers, what is the probability that at least 10 will buy? (Assume that y, the number of sales, follows a binomial probability distribution.)

7.60 One thousand flash bulbs were selected from a large production lot and tested. Sixty-three were found to be defective. Does the sample present sufficient evidence to indicate that more than 5 percent of the bulbs in the lot are defective?

7.61 Data collected over a long period of time show that a particular genetic defect occurs in 1 of every 1,000 children. The records of a medical clinic show $y = 60$ children with the defect in a total of 50,000 examined. If the 50,000 children were a random sample from the population of children represented by past records, what is the probability of observing a value of y equal to 60 or more? Would you say that the observation of $y = 60$ children with genetic defects represents a rare event?

7.62 A soft drink machine can be regulated so that it discharges an average of μ ounces per cup. If the ounces of fill are normally distributed with standard deviation equal to .3 ounce, give the setting for μ so that 8-ounce cups will overflow only 1 percent of the time.

7.63 Voters in a certain city were sampled concerning their voting preference in a primary election. Suppose that candidate *A* could win if he could pull 40 percent of the vote. If 920 of a sample of 2,500 voters favored *A*, does this contradict the hypothesis that *A* will win?

7.64 A statistical test is to be conducted to test the hypothesis that p, the parameter of a binomial population, is equal to .1. If the sample size is $n = 400$ and we wish α to be approximately equal to .05 (two-tailed test), locate the rejection region if the test statistic is (a) z, (b) y.

7.65 Calculate β for the test in Exercise 7.64 if p really is equal to .15.

7.66 A newly designed portable radio was styled on the assumption that 50 percent of all purchasers are female. If a random sample of 400 purchasers is selected, what is the probability that the number of female purchasers in the sample will be greater than 175?

7.67 Refer to Exercise 7.66. Suppose we wish to test the hypothesis that 50 percent of all purchasers are female against the alternative that this percentage is less than 50. To do this, we shall select a random sample of 400 purchasers and reject the hypothesis if the number of female purchasers in the sample is less than or equal to 175.
(a) Find α for this test.
(b) Find β if the true fraction of female purchasers is .4.

7.68 A manufacturing plant utilizes 3,000 electric light bulbs that have a length of life that is normally distributed with mean and standard deviation equal to 500 and 50 hours, respectively. In order to minimize the number of bulbs that burn out during operating hours, all the bulbs are replaced after a given period of operation. How often should the bulbs be replaced if we wish no more than 1 percent of the bulbs to burn out between replacement periods?

7.69 The admissions office of a small college is asked to accept deposits from a number of qualified prospective freshmen so that with probability about .95 the size of the freshman class will be less than or equal to 120. Consider that the applicants comprise a random sample from a population of applicants 80 percent of whom would actually enter the freshman class if accepted.

(a) How many deposits should the admissions counselor accept?

(b) If applicants in the number determined in part (a) are accepted, what is the probability that the freshman class size will be less than 105?

7.70 The Central Limit Theorem implies that a sample mean, $\bar{y}$, is approximately normally distributed for large values of n. Consider that a sample of size $n = 100$ is drawn from a population with mean $\mu = 40$ and $\sigma = 4$.

(a) What is $E(\bar{y})$?

(b) What is the standard deviation of $\bar{y}$?

(c) What is $P(\bar{y} > 41)$?

7.71 An airline finds that 5 percent of the persons making reservations on a certain flight will not show up for the flight. If the airline sells 160 tickets for a flight with only 155 seats, what is the probability that a seat will be available for every person holding a reservation and planning to fly?

7.72 Suppose that a population of measurements has a distribution that is normal, with $\mu = 10$ and $\sigma = 2$.

(a) What is the probability that a measurement randomly drawn from this population will be greater than 14?

(b) If two measurements are *independently* and randomly drawn from this population, what is the probability that both measurements will be greater than 14?

7.73 The length of life of a certain fuse is known to be normally distributed with mean 1,000 hours and standard deviation 50 hours. Find the probability that one of the fuses chosen at random will last between 1,020 and 1,110 hours.

7.74 A certain kind of automobile battery is known to have a length of life that is normally distributed with mean 1,200 days and standard deviation 100 days. How long should the guarantee time be if the manufacturer wants to replace only 10 percent of the batteries sold?

7.75 It is known that 30 percent of all calls coming into a telephone exchange are long-distance calls. If 200 calls come into the exchange, what is the probability that at least 50 will be long-distance calls?

7.76 A sample of $n = 64$ measurements is drawn from a population with mean $\mu = 70$ and standard deviation $\sigma = 32$. Find $P(\bar{y} \leq 77)$, where $\bar{y}$ is the mean of this sample.

7.77 Suppose that the random variable y has a binomial distribution corresponding to $n = 20$ and $p = .30$. Use Table 1, Appendix II, to find

(a) $P(y = 5)$.

(b) $P(y \geq 7)$.

7.78 Refer to Exercise 7.77. Use the normal approximation to calculate $P(y = 5)$ and $P(y \geq 7)$. Compare with the exact values obtained from Table 1.

7.79 The probability that a certain kind of component will fail in 1,000 hours or less is .20. Let y be the random number of components that fail in a sample of size 100.

(a) What is the expected value of y?

(b) What is the standard deviation of y?

(c) Use the normal approximation to find $P(y < 30)$.

7.80 The safety requirements for hard hats worn by construction workers and others, established by the American National Standards Institute (ANSI), specifies that each of three hats pass the following test.* A hat is mounted on an aluminum head form. An 8-pound steel ball is dropped on the hat from a height of 5 feet and the resulting

*Source: "Job-Safety Equipment Comes Under Fire," *Wall Street Journal,* November 18, 1977.

force is measured at the bottom of the head form. The force exerted on the head form by each of the three hats must be less than 1,000 pounds and the average of the three must be less than 850 pounds. (The relationship between this test and actual human head damage is unknown.) Suppose that the exerted force is normally distributed and hence that a sample mean of three force measurements is normally distributed. If a random sample of three hats is selected from a shipment with a mean equal to 900 and $\sigma = 100$, what is the probability that the sample mean will satisfy the ANSI standard?

7.81 A purchaser of electric relays is supplied by two suppliers, *A* and *B*. It is known that 2 of every 3 relays used by the company come from supplier *A*. If 75 relays are selected at random from those in use by the company, find the probability that at most 48 of these relays come from supplier *A*. Assume that the company uses a large number of relays.

7.82 A national poll claims that 60 percent of all American voters favor a popular vote for presidential nominees. Investigation indicates that only 100 voters were sampled and, of those, $y = 54$ favored the popular vote. Suppose that the fraction of voters, p, favoring a popular vote is actually .6. Find the probability that $y \leq 54$ given that $p = .60$.

7.83 The average personal yearly income in a given state is \$6,200, with a standard deviation of \$400.

(a) If a sample of 64 people is randomly chosen from this state, find the probability that the mean income for the sample exceeds \$6,300.

(b) If a second independent sample of 64 people is randomly chosen from this state, find the probability that both sample means exceed \$6,300.

7.84 The maximum load (with a generous safety factor) for the elevator in an office building is 2,000 pounds. The relative frequency distribution of the weights of all of the men and women using the elevator is mound-shaped (slightly skewed to the heavy weights) with mean μ equal to 150 pounds and standard deviation σ equal to 35 pounds. What is the largest number of people you can allow on the elevator if you want their total weight to exceed the maximum weight with a small probability (say near .01)? (*Hint:* If $y_1, y_2, \ldots, y_n$ are independent observations made on a random variable y and if y possesses mean μ and variance σ^2, then the mean and variance of $\sum_{i=1}^{n} y_i$ are $n\mu$ and $n\sigma^2$, respectively. This result was given in Section 7.3.)

EXPERIENCES WITH REAL DATA

Measurements, whether in the biological, physical, or social sciences, can be viewed as the sum of a number of random components, one due to the random variability attributed to the object of measurement and the others due to various sources of measurement error (caused by environmental and human factors). If each source of measurement error possessed a normal probability distribution (and the resulting measurement really was equal to the *sum* of the individual components), the measurement would possess a normal probability distribution (proof omitted). This result does not imply that all measurements will possess a normal probability distribution but it, along with the Central Limit Theorem, provides odds in favor of the occurrence of normally distributed data.

Most students have had some exposure to a laboratory science or have access to data originating in their field of interest. Let each student collect $n = 100$ observations on a continuous random variable in a scientific area of interest. Construct a relative frequency histogram for the data. Compare the histograms for the class and note the number that (if constructed properly) show a mound-shaped, and even bell-shaped, distribution. This tends to confirm our argument that distributions of real data very often can be modeled by a normal probability distribution. It also helps to explain why the Empirical Rule frequently provides a good description of data.

CHAPTER OBJECTIVES

GENERAL OBJECTIVE To present the basic concepts of statistical inference using four very practical inference-making situations to illustrate the ideas involved; to show how the Central Limit Theorem, Chapter 7, justifies each of these inference-making procedures.

SPECIFIC OBJECTIVES

1. To relate the preceding chapters to a discussion of statistical inference, the topic of Chapter 8. *Sections 8.1, 8.2*
2. To describe two types of estimators and to explain how we can measure their reliability. *Sections 8.3, 8.4, 8.5*
3. To present a general formula for calculating a large-sample interval estimate of a population mean, a population proportion, or the difference between two population means or proportions. *Sections 8.6, 8.7, 8.8, 8.9*
4. To explain how to select the sample size to obtain an estimator with a predetermined measure of reliability. *Section 8.10*
5. To review the basic concepts of a statistical test of an hypothesis. *Sections 8.11, 8.13*
6. To show how the Central Limit Theorem, Chapter 7, provides justification for a large-sample test of an hypothesis concerning a population mean, a population proportion, or the difference between two population means or proportions. *Section 8.12*
7. To show you how statistical inference can be used to solve applied problems in business and in the biological, physical, and social sciences. *Sections 8.4, 8.5, 8.7, 8.8, 8.9, 8.10, 8.12*

8 LARGE-SAMPLE STATISTICAL INFERENCE

8.1

A BRIEF SUMMARY

The preceding seven chapters set the stage for the objective of this text, developing an understanding of statistical inference and how it can be applied to the solution of practical problems. In Chapter 1 we stated that statisticians are concerned with making inferences about populations of measurements based on information contained in samples. We showed you how to phrase an inference—that is, how you describe a set of measurements—in Chapter 3. We discussed probability, the mechanism for making inferences, in Chapter 4, and we followed that with three chapters about probability distributions—a general presentation in Chapter 5, the binomial probability distribution in Chapter 6, and the normal distribution in Chapter 7.

To get you thinking about statistical inference, we introduced to you, in Chapter 6, the useful application of the binomial probability distribution to lot acceptance sampling and the test of an hypothesis concerning the effectiveness of a cold vaccine. We touched lightly on these topics again in the examples and exercises of Chapter 7. Now we are ready to utilize the foundation we have laid—to study the basic concepts involved in statistical inference.

Perhaps the most important contribution to our preparation for a study of statistical inference is the Central Limit Theorem of Chapter 7. This theorem, which justifies the approximate normality of the probability distribution of sample means for large samples, was used to justify the normal approximation to the binomial probability distribution in Chapter 7. But more important, it will be used to justify the approximate normality of the probability distributions of estimators and decision makers encountered in this chapter.

8.2

INFERENCE: THE OBJECTIVE OF STATISTICS

Inference, specifically decision making and prediction, is centuries old and plays a very important role in our individual lives. Each of us is faced with daily personal decisions and situations that require predictions concerning the future. The government is concerned with predicting the flow of gold to Europe. The broker wishes knowledge concerning the behavior of the stock market. The metallurgist seeks to use the results of an experiment to infer whether or not a new type of steel is more resistant to temperature changes than another. The housewife wishes to know whether detergent *A* is more effective than detergent *B* in her washing machine. Hopefully, these inferences are based upon relevant bits of available factual information that we would call observations or data.

In many practical situations the relevant information is abundant, seemingly inconsistent, and, in many respects, overwhelming. As a result, our carefully considered decision or prediction is often little better than an outright guess. The reader need only refer to the "Market Views" section of the *Wall Street Journal* to observe the diversity of expert opinion concerning future stock market behavior. Similarly, a visual analysis of data by scientists and engineers will often yield conflicting opinions regarding conclusions to be drawn from an experiment. Although many individuals tend to feel that their own built-in inference-making equipment is quite good, experience would suggest that most people are incapable of utilizing large amounts of data, mentally weighing each bit of relevant information, and arriving at a good inference. (You may test your individual inference-making equipment using the exercises in Chapters 8 and 9. Scan the data and make an inference before using the appropriate statistical procedure. Compare the results.) Certainly, a study of inference-making systems is desirable, and this is the objective of the mathematical statistician. Although we have purposely touched upon some of the notions involved in statistical inference in preceding chapters, it will be beneficial to collect our ideas at this point as we attempt an elementary presentation of some of the basic ideas involved in statistical inference.

The objective of statistics is to make inferences about a population based upon information contained in a sample. Inasmuch as populations are characterized by numerical descriptive measures called parameters, statistical inference is concerned with making inferences about population parameters. Typical population parameters are the mean, the standard deviation, the area under the probability distribution above or below some value of the random variable, or the area between two values of the variable. Indeed, the practical problems mentioned in the first paragraph of this section can be restated in the framework of a population with a specified parameter of interest.

Methods for making inferences about parameters fall into one of two categories. We may make decisions concerning the value of the parameter, as exemplified by the lot acceptance sampling and test of an hypothesis described in Chapter 6. Or, we may estimate or predict the value of the parameter. While some statisticians view estimation as a decision-making problem, it will be convenient for us to retain the two categories and, particularly, to concentrate on estimation and tests of hypotheses.

A statement of the objective and types of statistical inference would be incomplete without reference to a measure of goodness of inferential procedures. We may define numerous objective methods for making inferences in addition to our own individual procedures based upon intuition. Certainly a measure of goodness must be defined so that one procedure may be compared with another. More than that, we would like to state the goodness of a particular inference in a given physical situation. Thus, to say we predict that the price of a stock will be \$80 next Monday would be insufficient and would stimulate few of us to take action to buy or sell. Indeed, we ask whether the estimate is correct to within $\pm$\$1, \$2, or \$10. Statistical inference in a practical situation contains two elements: (1) the inference and (2) a measure of its goodness.

Before concluding this introductory discussion of inference, it would be well to dispose of a question that frequently disturbs the beginner. Which method of inference should be used; that is, should the parameter be estimated or should we test an hypothesis concerning its value? The answer to this question is dictated by the practical question that has been posed and very often is determined by personal preference. Some people like to test theories concerning parameters; others prefer to express their inference as an estimate. We will find that there are actually two methods of estimation, the choice of which, once again, is a matter of personal preference. Inasmuch as both estimation and tests of hypotheses are frequently used in scientific literature, we would be remiss in excluding one or the other from our discussion.

8.3

TYPES OF ESTIMATORS

Estimation procedures may be divided into two types, point estimation and interval estimation. Suppose that we wish to estimate the grade-point average of a particular student at Bucknell University. The estimate might be given as a single number, for instance 2.9, or we might estimate that the grade-point average would fall in an interval, for instance 2.7 to 3.2. The first type of estimate is called a point estimate because the single number, representing the estimate, may be associated with a point on a line. The second type, involving two points and defining an interval on a line, is called an interval estimate. We shall consider each of these methods of estimation.

A point estimation procedure utilizes information in a sample to arrive at a single number or point that estimates the parameter of interest. The actual estimation is accomplished by an estimator.

DEFINITION

An estimator is a rule that tells us how to calculate the estimate based upon information in the sample and is generally expressed as a formula.

For example, the sample mean,

$$\bar{y} = \frac{\sum_{i=1}^{n} y_i}{n},$$

is an estimator of the population mean, μ, and explains exactly how the actual numerical value of the estimate may be obtained once the sample values, $y_1, y_2, \ldots, y_n$, are known. On the other hand, an interval estimator uses the data in the sample to calculate *two* points which are intended to enclose the true value of the parameter estimated.

An investigation of the reasoning used in calculating the goodness of a point estimator is facilitated by considering an analogy. Point estimation is similar, in many respects, to firing a revolver at a target. The estimator, generating estimates, is analogous to the revolver, a particular estimate to the bullet, and the parameter of interest to the bull's-eye. Drawing a sample from the population and estimating the value of the parameter is equivalent to firing a single shot at the target.

Suppose that a man fires a single shot at a target and that the shot pierces the bull's-eye. Do we conclude that he is an excellent shot? Obviously, the answer is no because not one of us would consent to hold the target while a second shot was fired. On the other hand, if 1,000,000 shots in succession hit the bull's-eye, we might acquire sufficient confidence in the marksman to hold the target for the next shot, if the compensation were adequate. The point we wish to make is certainly clear. We cannot evaluate the goodness of an estimation procedure on the basis of a single estimate; rather, we must observe the results when the estimation procedure is used over and over again, many, many times—we then observe how closely the shots are distributed about the bull's-eye. In fact, since the estimates are numbers, we would evaluate the goodness of the estimator by constructing a frequency distribution of the estimates obtained in repeated sampling and note how closely the distribution centers about the parameter of interest.

This point is aptly illustrated by considering the results of the die-tossing experiment, Section 7.3, where 100 samples of $n = 5$ measurements were drawn from the die-tossing population, which possessed a mean and standard deviation equal to $\mu = 3.5$ and $\sigma = 1.71$, respectively. The distribution of the 100 sample means, each representing an estimate, is given in Figure 7.9. A glance at the distribution tells us that the estimates tend to pile up about the mean, $\mu = 3.5$, and also gives an indication as to the error of estimation that might be expected. Although the distribution, Figure 7.9, is informative, we would like to have the distribution of estimates based upon an infinitely large number of samples, thus generating the probability distribution for the estimator. Fortunately, this task is not too difficult. Mathematical methods are available for deriving the probability distribution of estimators, but these techniques are beyond the scope of this course. A second and very powerful method utilizes a high-speed electronic computer to draw the extremely large number of samples required, calculate the corresponding estimates, and record the results in the form of a frequency distribution.

Suppose, then, that we wish to estimate some population parameter, which, for convenience, we will call θ. The estimator of θ will be indicated by the symbol $\hat{\theta}$, where the "hat" indicates that we are estimating the parameter immediately beneath. Now, with the revolver-firing example in mind, we see that the desirable properties of a good estimator are quite obvious. We would like the distribution of the estimates to center about the parameter estimated as shown in Figure 8.1 and,

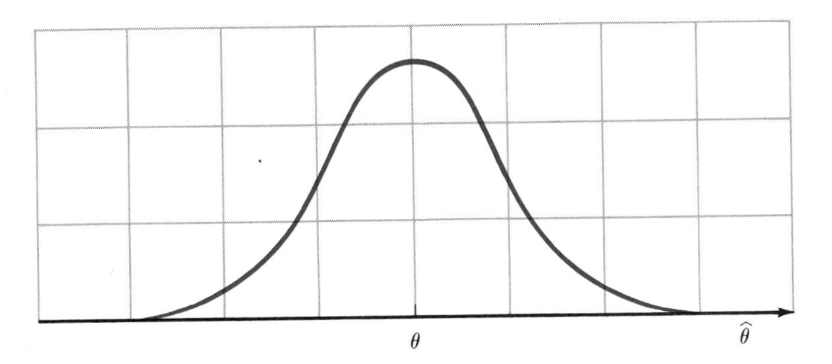

FIGURE 8.1
Distribution of estimates

in addition, we would like the spread of the distribution to be as small as possible. In other words, we would like the mean or expected value of the distribution of estimates to equal the parameter estimated.

DEFINITION

If $\hat{\theta}$ is an estimator of a parameter θ and if the mean of the distribution of $\hat{\theta}$ is θ, that is,

$$E(\hat{\theta}) = \theta$$

then $\hat{\theta}$ is said to be unbiased. Otherwise, $\hat{\theta}$ is said to be biased.

The frequency distributions for an unbiased estimator and a biased estimator are shown in Figure 8.2(a) and (b).

Also we desire the variance or standard deviation for the estimator—that is, the distribution of estimates—to be a minimum. Thus the distribution of estimates in Figure 8.3(a) is preferable to that shown in Figure 8.3(b).

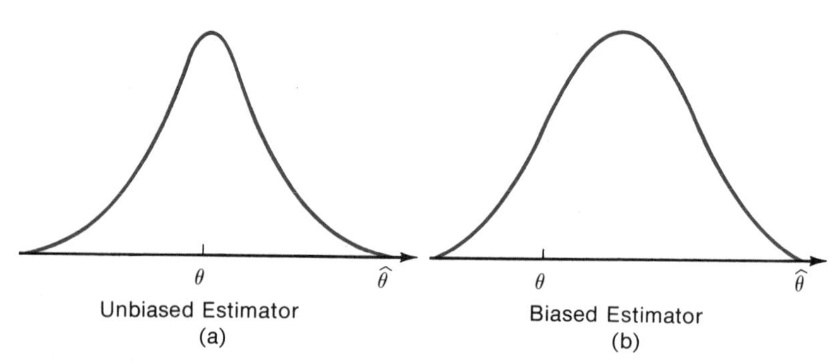

FIGURE 8.2
Distributions for unbiased and biased estimators

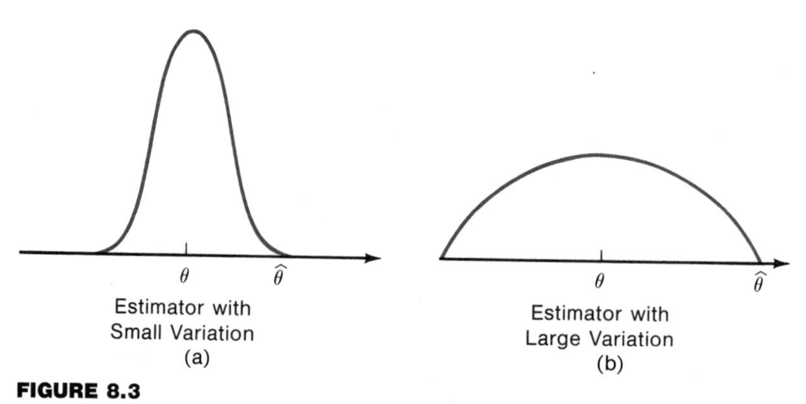

FIGURE 8.3

Comparison of estimator variability

The goodness of an interval estimator is analyzed in much the same manner as is a point estimator. Samples of the same size are repeatedly drawn from the population and the interval estimate is calculated on each occasion. This process will generate a large number of intervals rather than points. A good interval estimate would successfully enclose the true value of the parameter a large fraction of the time. This fraction is called the confidence coefficient for the estimator; the estimator itself is often called a confidence interval.

The selection of a "best" estimator—the proper formula to use in calculating the estimates—involves the comparison of various methods of estimation. This is the task of the theoretical statistician and is beyond the scope of this text. Throughout the remainder of this chapter and succeeding chapters, populations and parameters of interest will be defined and the appropriate estimator indicated along with its expected value and standard deviation.

8.4 POINT ESTIMATION OF A POPULATION MEAN

Practical problems very often lead to the estimation of a population mean, μ. We are concerned with the average achievement of college students in a particular university, in the average strength of a new type of steel, in the average number of deaths per capita in a given social class, and in the average demands for a new product. Conveniently, the estimation of μ serves as a very practical application of statistical inference as well as an excellent illustration of the principles of estimation discussed in Section 8.3. Many estimators are available for estimating the population mean, μ, including the sample median, the average between the largest and smallest measurements in the sample, and the sample mean, $\bar{y}$. Each would generate a probability distribution in repeated sampling and, depending upon the population and practical problem involved, would possess certain advantages and disadvantages. Although the sample median and the average of the sample extremes are easier to calculate, the sample mean, $\bar{y}$, is usually superior in that, for some populations, its variance is a minimum and, furthermore, regardless of the population, it is always unbiased.

Three facts emerge from a study of the probability distribution of $\bar{y}$ in repeated random sampling of n measurements from a population with mean equal to μ and variance equal to σ^2. Regardless of the probability distribution of the population, the following facts hold true (proof omitted):

Properties of the Sample Mean

1. The expected value of $\bar{y}$ is equal to μ, the population mean.
2. The standard deviation of $\bar{y}$ is equal to

$$\sigma_{\bar{y}} = \frac{\sigma}{\sqrt{n}}\sqrt{\frac{N-n}{N-1}},$$

where N is equal to the number of measurements in the population. In the following discussion we shall assume that N is large relative to the sample size, n, and hence that $\sqrt{\frac{N-n}{N-1}}$ is approximately equal to 1. Then,

$$\sigma_{\bar{y}} = \frac{\sigma}{\sqrt{n}}.$$

3. When n is large, $\bar{y}$ will be approximately normally distributed according to the Central Limit Theorem (assuming that μ and σ are finite numbers).

Thus $\bar{y}$ is an unbiased estimator of μ with a standard deviation that is proportional to the population standard deviation, σ, and inversely proportional to the square root of the sample size, n. Although we give no proof of these results, we suggest that they are intuitively reasonable.* Certainly, the more variable the population data, measured by σ, the more variable will be $\bar{y}$. On the other hand, more information will be available for estimating μ as n becomes large. Hence the estimates should fall closer to μ, and $\sigma_{\bar{y}}$ should decrease.

In addition to knowledge of the mean and standard deviation of the probability distribution for $\bar{y}$, the Central Limit Theorem provides information on its form. That is, when the sample size, n, is large, the distribution of $\bar{y}$ will be approximately normal. The probability distributions for $\bar{y}$ based on random samples of $n = 5$, $n = 20$, and $n = 80$ from a normal distribution are shown in Figure 8.4. Notice how these distributions center about μ and how the spread of the distributions decreases as n increases.

With the results above in mind, suppose that we draw a single sample of n measurements from a population and calculate the sample mean, $\bar{y}$. How good

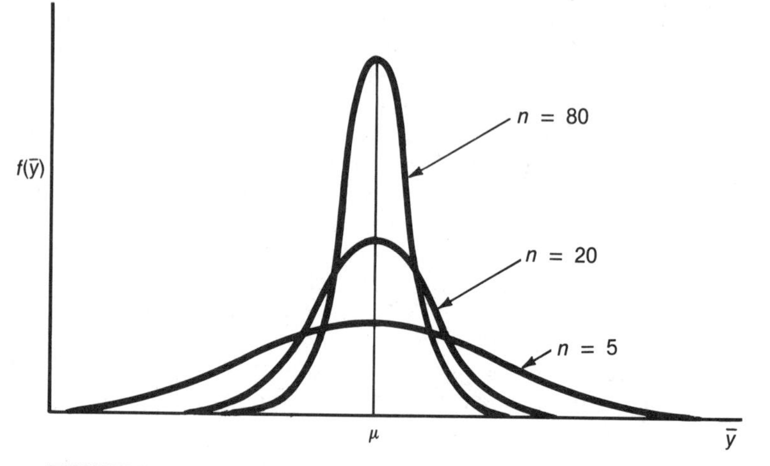

FIGURE 8.4

Probability distributions for $\bar{y}$ based on random sampling from a normal distribution, $n = 5$, 20, and 80

*Proofs of these results are based on mathematics beyond the scope of this text. They can be found in most of the standard texts on mathematical statistics. See the references at the end of this chapter.

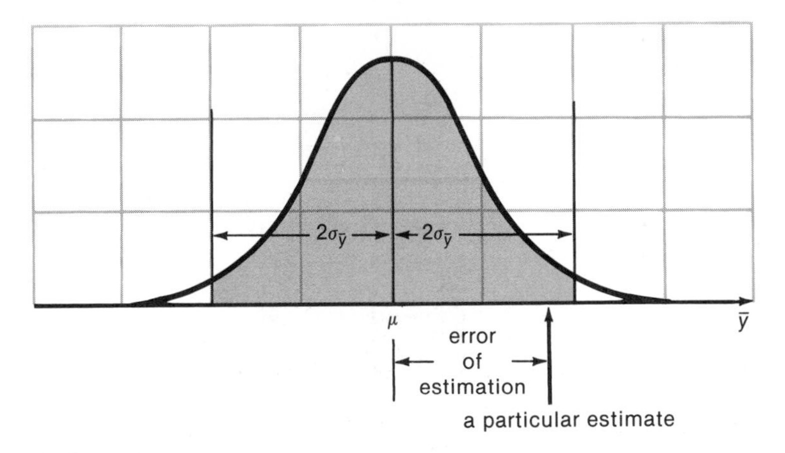

FIGURE 8.5

Distribution of $\bar{y}$ for large n

will this estimate of μ be; that is, how far will it deviate from the mean, μ? Although we cannot state that $\bar{y}$ will *definitely* lie within a specified distance of μ, Tchebysheff's Theorem states that if we were to draw many samples of the same size n from the population and compute $\bar{y}$ each time, at least three-fourths of the values of $\bar{y}$ would lie within $2\sigma_{\bar{y}}$ of the mean of the distribution of $\bar{y}$'s, that is, μ. See Figure 8.5.

The relative frequency distribution of the 100 values of $\bar{y}$, each calculated from a sample $n = 5$ die tosses, illustrates this concept. As noted, Section 8.3, the mean and standard deviation of the population of die tosses is $\mu = 3.5$ and $\sigma = 1.71$. Therefore,

$$2\sigma_{\bar{y}} = \frac{2\sigma}{\sqrt{n}} = \frac{2(1.71)}{\sqrt{5}} = \frac{2(1.71)}{2.24} = 1.53.$$

You can see that most of the sample means fall within 1.53 of the mean, $\mu = 3.5$, in Figure 8.6.

The difference between a particular estimate and the parameter it estimates is called the error of estimation. Consequently, we could regard $2\sigma_{\bar{y}}$ as an approximate bound on this error. By this we mean that at least three-fourths of the estimates, and most likely 95 percent, will deviate from the mean by less than $2\sigma_{\bar{y}}$ (see Figure 8.5). Although the use of two standard deviations rather than three is not sacred, two would seem to provide a reasonable measure of the reliability of an estimate for most practical problems. (Why do we not use 1.96 in place of 2 standard deviations of the estimator to acquire the bound on the error of estimation? The answer is that other factors involved in our computations produce errors that overshadow any error that might occur by replacing 1.96 by 2.* For example, most estimators do not possess a probability distribution that is exactly normal. Further, we create another error by substituting s for σ.)

*Most practical applications of estimation procedures in sample surveys employ two-standard-deviation bounds on the error of estimation. [See *Elementary Survey Sampling* (Mendenhall, Ott, and Scheaffer, 1971).]

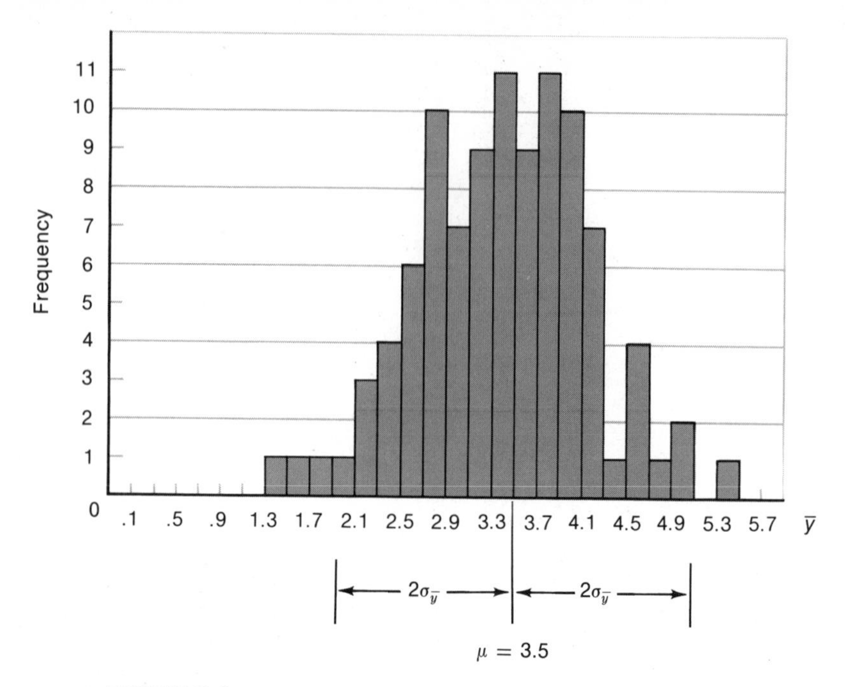

FIGURE 8.6

Histogram of sample means for the die-tossing experiment, Section 7.3

Point Estimator of a Population Mean

Estimator: $\bar{y}$.

Bound on error: $2\sigma_{\bar{y}} = \dfrac{2\sigma}{\sqrt{n}}$.

Now consider the following example of point estimation.

EXAMPLE 8.1

Suppose that we wish to estimate the average daily yield of a chemical manufactured in a chemical plant. The daily yield, recorded for $n = 50$ days, produced a mean and standard deviation equal to

$$\bar{y} = 871 \text{ tons},$$
$$s = 21 \text{ tons}.$$

Estimate the average daily yield, μ.

Solution The estimate of the daily yield is $\bar{y} = 871$ tons. The bound on the error of estimation is

$$2\sigma_{\bar{y}} = \frac{2\sigma}{\sqrt{n}} = \frac{2\sigma}{\sqrt{50}}.$$

Although σ is unknown, we may approximate its value by using s, the estimator of σ. Thus the bound on the error of estimation is approximately

$$\frac{2s}{\sqrt{n}} = \frac{2(21)}{\sqrt{50}} = \frac{42}{7.07} = 5.94.$$

We would feel fairly confident that our estimate of 871 tons is within 5.94 tons of the true average yield.

Example 8.1 deserves further comment in regard to two points. The erroneous use of 2σ as a bound on the error of estimation rather than $2\sigma_{\bar{y}}$ is common to beginners. Certainly, if we wish to discuss the distribution of $\bar{y}$, we must use its standard deviation, $\sigma_{\bar{y}}$, to describe its variability. Care must be taken not to confuse the descriptive measures of one distribution with another.

A second point of interest concerns the use of s to approximate σ. This approximation will be reasonably good when n is large, say, 30 or greater. If the sample size is small, two techniques are available. Sometimes experience or data obtained from previous experiments will provide a good estimate of σ. When this is not available, we may resort to a small-sample procedure described in Chapter 9. The choice of $n = 30$ as the division between "large" and "small" samples is arbitrary. The reasoning for its selection will become apparent in Chapter 9.

EXERCISES

8.1 The Environmental Protection Agency (EPA) has released initial findings (*Environment News,* October 1976) on the occurrence of polychlorinated biphenyls (PCBs), a hazardous substance, in the milk of nursing mothers. An analysis of milk samples taken from 67 women in 10 states showed the presence of PCBs in the milk of 65. The sample mean for the 65 was reported to be 1.7 parts per million (ppm) and the largest reading was 10.6. To evaluate this estimate of the mean level of contamination in the sampled population, you need to know σ, the measure of variation of the contamination readings.

(a) Although the article does not give the value of σ, it is unlikely that σ would be larger than 2.7 and it probably is nearer in value to 2. Why?

(b) Assume that $\sigma = 2.5$. Use this value to assess the accuracy of the estimate, $\bar{y} = 1.7$ ppm, of the population mean, μ. Interpret your result.

8.2 Geologists are interested in shifts and movements of the earth's surface indicated by fractures (cracks) in the earth's crust. One of the most famous large fractures is the San Andreas fault (moving fracture) in California. A geologist attempting to study the movement of the relative shifts in the earth's crust at a particular location found many fractures in the local rock structure. In an attempt to determine the mean angle of the breaks, he sampled $n = 50$ fractures and found the sample mean and standard deviation to be 39.8° and 17.2°, respectively. Estimate the mean angular direction of the fractures and place a bound on the error of estimation.

8.3 An increase in the rate of consumer savings is frequently tied to a lack of confidence in the economy and is said to be an indicator of a recessional tendency in the economy. A random sampling of $n = 200$ savings accounts in a local community showed a mean increase in savings account values of 7.2 percent over the past 12 months with a standard deviation of 5.6 percent. Estimate the mean percent increase in savings account values over the past 12 months for depositors in the community. Place a bound on your error of estimation.

8.4 The mean and standard deviation for the life of a random sample of 100 light bulbs were calculated to be 1,280 and 142 hours, respectively. Estimate the mean life of the population of light bulbs from which the sample was drawn and place bounds on the error of estimation.

8.5 Suppose that the population mean, Exercise 8.4, were really 1,285 hours, with $\sigma = 150$ hours. What is the probability that the mean of a random sample of $n = 100$ measurements would exceed 1,300 hours?

8.6 In Section 8.3 we stated that for the die-toss population, $\mu = 3.5$ and $\sigma = 1.71$.

(a) If you were to toss a die $n = 5$ times, what is the approximate probability that the mean $\bar{y}$ of the sample would fall in the interval, $2.5 \leq \bar{y} \leq 4.5$?

(b) If you were to toss a die $n = 10$ times, what is the approximate probability that $\bar{y}$ will fall in the interval, $2.5 \leq \bar{y} \leq 4.5$?

(c) You will note [part (b)] that $P(2.5 \leq \bar{y} \leq 4.5)$ is large. If you have access to a die, toss it $n = 10$ times and calculate $\bar{y}$. Does $\bar{y}$ fall in the interval, $2.5 \leq \bar{y} \leq 4.5$? Suppose that the mean of the population, $\mu = 3.5$, were unknown and that you were using your mean to estimate μ. What is your error of estimation?

8.5 INTERVAL ESTIMATION OF A POPULATION MEAN

Constructing an interval estimate is like attempting to rope an immobile steer. In this case, the parameter that you wish to estimate corresponds to the steer and the interval to the loop formed by the cowboy's lariat. Each time you draw a sample, you construct a confidence interval for a parameter and you hope to "rope it," that is, include it in the interval. You will not be successful for every sample. The probability that an interval will enclose the estimated parameter is the confidence coefficient.

DEFINITION

The probability that a confidence interval will enclose the estimated parameter is called the confidence coefficient.

To consider a practical example, suppose that you wish to estimate the mean number of bacteria per cubic centimeter in a polluted stream. If we were to draw 10 samples, each containing $n = 20$ observations, and construct a confidence interval for the population mean, μ, for each sample, the intervals might appear as shown in Figure 8.7. The horizontal line segments represent the 10 intervals and the vertical line represents the location of the true mean number of bacteria per cubic centimeter. Note that the parameter is fixed and that the interval location and width vary from sample to sample. Thus we speak of "the probability that the interval encloses μ" not "the probability that μ falls in the

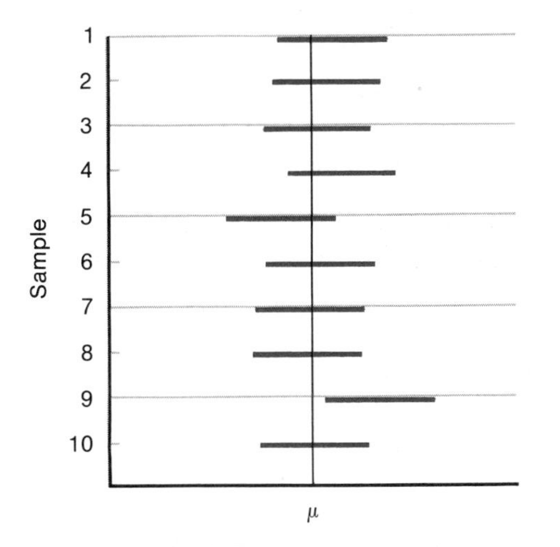

FIGURE 8.7

Ten confidence intervals for the mean number of bacteria per cubic centimeter (each based on a sample of $n = 20$ observations)

interval'' because μ is fixed. The interval is random. Having grasped the concept of a confidence interval, let us now consider how to find the confidence interval for a population mean, μ, based on a random sample of n observations.

An interval estimator, or confidence interval, for a population mean can be obtained from the result of Section 8.4. It is possible that $\bar{y}$ might lie either above or below the population mean, although we would not expect it to deviate more than approximately $2\sigma_{\bar{y}}$ from μ. Hence, if we choose $(\bar{y} - 2\sigma_{\bar{y}})$ as the lower point of the interval, called the lower confidence limit, LCL, and $(\bar{y} + 2\sigma_{\bar{y}})$ as the upper point or upper confidence limit, UCL, the interval most probably will enclose the true population mean, μ. (See Figure 8.8.) In fact, if n is large and the distribution of $\bar{y}$ is approximately normal, we would expect approximately 95 percent of the intervals obtained in repeated sampling to enclose the population mean, μ.

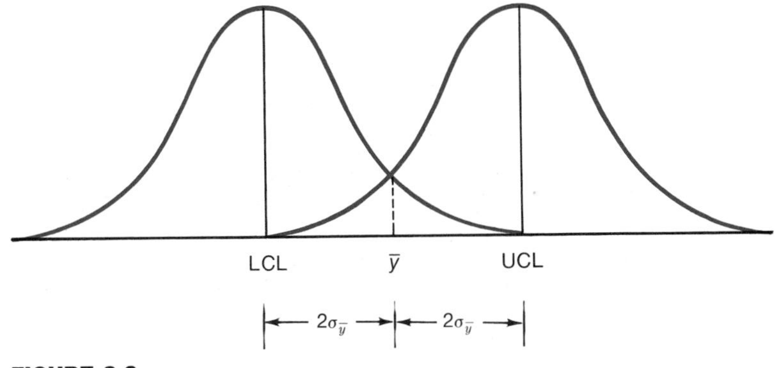

FIGURE 8.8

Confidence limits for μ

The confidence interval described is called a large-sample confidence interval (or confidence limits) because n must be large enough for the Central Limit Theorem to be effective and hence for the distribution of $\bar{y}$ to be approximately normal. Inasmuch as σ is usually unknown, the sample standard deviation must be used to estimate σ. As a rule of thumb, this confidence interval would be appropriate when $n = 30$ or more.

The confidence coefficient, .95, corresponds to $\pm 2\sigma_{\bar{y}}$, or, more exactly, $\pm 1.96\sigma_{\bar{y}}$. Recalling that .90 of the measurements in a normal distribution will fall within $z = 1.645$ standard deviations of the mean (Table 3, Appendix II), we could construct 90 percent confidence intervals by using

$$\text{LCL} = \bar{y} - 1.645\sigma_{\bar{y}} = \bar{y} - \frac{1.645\sigma}{\sqrt{n}},$$

and

$$\text{UCL} = \bar{y} + 1.645\sigma_{\bar{y}} = \bar{y} + \frac{1.645\sigma}{\sqrt{n}}$$

In general, we can construct confidence intervals for μ corresponding to any desired confidence coefficient, say $(1 - \alpha)$, by use of the following:

Large-Sample Confidence Interval for a Population Mean, μ

$$\bar{y} \pm \frac{z_{\alpha/2}\sigma}{\sqrt{n}}.$$

We shall define the quantity $z_{\alpha/2}$ to be the value in the z-table such that the area to the right of $z_{\alpha/2}$ is equal to $\alpha/2$ (see Figure 8.9); that is, $P(z > z_{\alpha/2}) = \alpha/2$. Thus, a confidence coefficient equal to .95 would imply that $\alpha = .05$ and $z_{.025} = 1.96$. The value of z employed for a 90 percent confidence interval would be $z_{.05} = 1.645$.

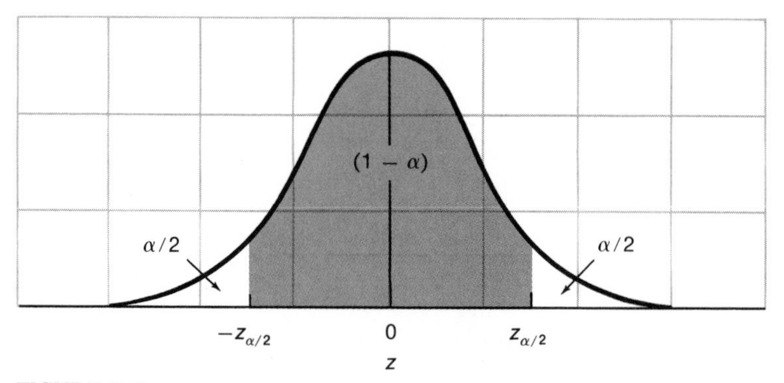

FIGURE 8.9
Location of $z_{\alpha/2}$

EXAMPLE 8.2

Find a 90 percent confidence interval for the population mean of Example 8.1. Recall that $\bar{y} = 871$ tons and $s = 21$ tons.

Solution The 90 percent confidence limits would be

$$\bar{y} \pm \frac{1.645\sigma}{\sqrt{n}}.$$

Using s to estimate σ, we obtain

$$871 \pm (1.645)\frac{(21)}{\sqrt{50}}$$

or

$$871 \pm 4.89.$$

Therefore, we estimate that the average daily yield, μ, lies in the interval 866.11 to 875.89 tons. The confidence coefficient, .90, implies that in repeated sampling, 90 percent of the confidence intervals would enclose μ.

Note that the width of the confidence interval increases as the confidence coefficient increases, a result that is in agreement with our intuition. Certainly if we wish to be more confident that the interval will enclose μ, we would increase the width of the interval. Confidence limits corresponding to some of the commonly used confidence coefficients are tabulated in Table 8.1.

TABLE 8.1
Confidence limits for μ

Confidence Coefficient	α	$z_{\alpha/2}$	LCL	UCL
.90	.10	1.645	$\bar{y} - 1.645\sigma/\sqrt{n}$	$\bar{y} + 1.645\sigma/\sqrt{n}$
.95	.05	1.96	$\bar{y} - 1.96\sigma/\sqrt{n}$	$\bar{y} + 1.96\sigma/\sqrt{n}$
.99	.01	2.58	$\bar{y} - 2.58\sigma/\sqrt{n}$	$\bar{y} + 2.58\sigma/\sqrt{n}$

The choice of the confidence coefficient to be used in a given situation is made by the experimenter and will depend upon the degree of confidence that he wishes to place in his estimate. As we have pointed out, the larger the confidence coefficient, the wider the interval. As a result of this freedom of choice, it has become the custom of many experimenters to use a .95 confidence coefficient, although there is no logical foundation for its popularity.

The frequent use of the .95 confidence coefficient introduces a question asked by many beginners. Should one use $z = 1.96$ or $z = 2$ in the confidence interval? The answer is that it does not really make much difference which value is used. The value $z = 1.96$ is more exact for a .95 confidence coefficient, but the error introduced by using $z = 2$ will be very small. The use of $z = 2$ simplifies the calculations, particularly when the computing is done manually. We will agree to use two standard deviations when placing bounds on the error of a point estimator but will use $z = 1.96$ when constructing a confidence interval, simply to remind you that this is the z value obtained from the table of areas under the normal curve.

At this point you may see little distinction between point estimators and interval estimators. When we place bounds on the error of a point estimate, for all practical purposes, we construct an interval estimate. Furthermore, the point estimate falls in the middle of the interval estimate when a population mean is being estimated. While this close relationship will exist for most of the parameters estimated in this text, it is not generally true. It is not obvious that the best point estimator will fall in the middle of the best interval estimator—in many cases it does not. Furthermore, it is not a foregone conclusion that the best interval estimator will even be a function of the best point estimator. Although these problems are of a theoretical nature, they are important and worth mentioning. From a practical point of view, point and interval estimation are closely related and the choice between the point and the interval estimator in an actual problem depends upon the preference of the experimenter.

EXERCISES

8.7 Some researchers think that vitamin C may be useful in reducing the buildup of cholesterol deposits on the inner walls of arteries, thereby reducing the possibility of heart attacks (*Gainesville Sun,* June 14, 1976). The cholesterol level of each of 50 persons (with higher than normal cholesterol levels) was recorded before and then after a one-month daily regime of 500 milligrams (mg) of vitamin C per day. The data collected for this sample showed the mean and standard deviation of the drop in cholesterol level to be $\bar{y} = 64.3$ mg per 100 milliliters (ml) and $s = 18.9$ mg per 100 ml. Estimate the mean drop per person in cholesterol level by using a 95 percent confidence interval.

8.8 Owing to a variation in laboratory techniques, impurities in materials, and other unknown factors, the results of an experiment in a chemistry laboratory will not always yield the same numerical answer. In an electrolysis experiment, a class measured the amount of copper precipitated from a saturated solution of copper sulfate over a 30-minute period. The $n = 30$ students acquired a sample mean and standard deviation equal to .145 and .0051 mole, respectively. Find a 90 percent confidence interval for the mean amount of copper precipitated from the solution over the period of time.

8.9 According to *Environment News* (September 1975), acid rain, caused by the reaction of certain air pollutants with rainwater, appears to be a growing problem in the northeastern United States. (Acid rain affects the soil and causes corrosion on exposed metal surfaces.) Pure rain falling through clean air registers a pH value (pH is a measure of acidity; 0 is acid, 14 is alkaline) of 5.7. Suppose that water samples from 40 rainfalls are analyzed for pH and that $\bar{y}$ and s are equal to 3.7 and .5, respectively. Find a 99 percent confidence interval for the mean pH in rainfalls and interpret the interval. What assumption must be made in order that the confidence interval be valid?

8.10 In a psychology study of susceptibility to perceptual illusions, 50 male subjects judged the length of an illusory figure. When each judgment was scored in terms of magnitude of the deviation from the actual length, the resultant scores had a mean and standard deviation equal to

$\bar{y} = 81$ millimeters
$s = 12$ millimeters

Find a 95 percent confidence interval for the mean magnitude of deviation, μ.

8.6
ESTIMATION FROM LARGE SAMPLES

Estimation of a population mean, Sections 8.4 and 8.5, sets the stage for the other estimation problems to be discussed in this chapter. A thread of unity runs through all, which, once it is observed, will simplify the learning process for the beginner. The following conditions will be satisfied for all estimation problems discussed in this chapter. Each point estimator of a parameter, say θ, will be unbiased. That is, the mean of the distribution of estimates obtained in repeated sampling will equal the parameter estimated. The standard deviation of the estimator will be given so that we may place a two-standard deviation, $2\sigma_{\hat{\theta}}$, bound on the error of estimation. In each case, the point estimator will be approximately normally distributed by the Central Limit Theorem, when n is large, and the probability that the error will be less than the bound, $2\sigma_{\hat{\theta}}$, will be approximately .95.

The corresponding interval estimators will assume that the sample is large enough for the Central Limit Theorem to imply normality in the distribution of the point estimator of θ as well as to provide a good estimate of any other unknown (for example, σ). Then the confidence intervals for any confidence coefficient, $(1 - \alpha)$, will equal

Large-Sample Confidence Interval for θ

$$\hat{\theta} \pm z_{\alpha/2}\sigma_{\hat{\theta}}.$$

8.7
ESTIMATING THE DIFFERENCE BETWEEN TWO MEANS

A problem of equal importance to the estimation of population means is the comparison of two population means. For instance, we might wish to compare the effectiveness of two teaching methods. Students would be randomly divided into two groups, the first subjected to method 1 and the second to method 2. We would then make inferences concerning the difference in average student achievement as measured by some testing procedure.

Or, we might wish to compare the average yield in a chemical plant using raw materials furnished by two suppliers, A and B. Samples of daily yield, one for each of the two raw materials, would be recorded and used to make inferences concerning the difference in mean yield.

For each of these examples we postulate two populations, the first with mean and variance μ_1 and σ_1^2, and the second with mean and variance μ_2 and σ_2^2. A random sample of n_1 measurements is drawn from population I and n_2 from population II, where the samples are assumed to have been drawn independently of one another. Finally the estimates of the population parameters, $\bar{y}_1, s_1^2, \bar{y}_2$, and s_2^2, are calculated from the sample data.

The point estimator of the difference between the population means, $(\mu_1 - \mu_2)$, is $(\bar{y}_1 - \bar{y}_2)$, the difference between the sample means. If repeated pairs of samples of n_1 and n_2 measurements are drawn from the two populations and the estimate, $(\bar{y}_1 - \bar{y}_2)$, calculated for each pair, a distribution of estimates will result. The probability distribution of the estimator, $(\bar{y}_1 - \bar{y}_2)$, will be approximately normally distributed* for large samples (say n_1 and n_2 both equal to 30 or more), with the mean and standard deviation

Mean and Standard Deviation of $(\bar{y}_1 - \bar{y}_2)$

$$E(\bar{y}_1 - \bar{y}_2) = \mu_1 - \mu_2,$$

$$\sigma_{(\bar{y}_1 - \bar{y}_2)} = \sqrt{\frac{\sigma_1^2}{n_1} + \frac{\sigma_2^2}{n_2}}.$$

Although the formula for the standard deviation of $(\bar{y}_1 - \bar{y}_2)$ may appear to be complicated, a result derived in mathematical statistics will assist in its memorization. Certainly the variability of a difference between two independent random variables would seem, intuitively, to be greater than the variability of either of the two variables, since one may be extremely large at the same time that the other is extremely small. Hence each contributes a portion of its variability to the variability of the difference. This intuitive explanation is supported by a theorem in mathematical statistics which states that the variance of either the sum or the difference of two independent random variables is equal to the sum of their respective variances. That is,

$$\sigma^2_{(y_1 + y_2)} = \sigma^2_{y_1} + \sigma^2_{y_2}$$

and

$$\sigma^2_{(y_1 - y_2)} = \sigma^2_{y_1} + \sigma^2_{y_2}.$$

Therefore,

$$\sigma^2_{(\bar{y}_1 - \bar{y}_2)} = \sigma^2_{\bar{y}_1} + \sigma^2_{\bar{y}_2} = \frac{\sigma_1^2}{n_1} + \frac{\sigma_2^2}{n_2},$$

and the standard deviation is

$$\sigma_{(\bar{y}_1 - \bar{y}_2)} = \sqrt{\frac{\sigma_1^2}{n_1} + \frac{\sigma_2^2}{n_2}}.$$

*When n_1 and n_2 are large, the probability distributions of $\bar{y}_1$ and $\bar{y}_2$ are approximately normally distributed because of the Central Limit Theorem. It can be shown (proof omitted) that the difference between two normally distributed random variables possesses a normal probability distribution. This justifies the approximate normality of the probability distribution of $(\bar{y}_1 - \bar{y}_2)$ when n_1 and n_2 are large.

We shall have occasion to use this result again in Section 8.9.

The bound on the error of the point estimate is

$$2\sqrt{\frac{\sigma_1^2}{n_1}+\frac{\sigma_2^2}{n_2}}$$

The sample variances, s_1^2 and s_2^2, may be used to estimate σ_1^2 and σ_2^2 when these parameters are unknown. This approximation will be reasonably good when n_1 and n_2 are equal to 30 or more.

EXAMPLE 8.3

A comparison of the wearing quality of two types of automobile tires was obtained by road testing samples of $n_1 = n_2 = 100$ tires for each type. The number of miles until wear-out was recorded, where wear-out was defined as a specific amount of tire wear. The test results were as follows:

$$\bar{y}_1 = 26{,}400 \text{ miles}, \quad \bar{y}_2 = 25{,}100 \text{ miles};$$
$$s_1^2 = 1{,}440{,}000, \quad s_2^2 = 1{,}960{,}000.$$

Estimate the difference in mean time to wear-out and place bounds on the error of estimation.

Solution The point estimate of $(\mu_1 - \mu_2)$ is

$$(\bar{y}_1 - \bar{y}_2) = 26{,}400 - 25{,}100 = 1{,}300 \text{ miles}$$

and

$$\begin{aligned}\sigma_{(\bar{y}_1-\bar{y}_2)} &= \sqrt{\frac{\sigma_1^2}{n_1}+\frac{\sigma_2^2}{n_2}} \\ &\approx \sqrt{\frac{s_1^2}{n_1}+\frac{s_2^2}{n_2}} = \sqrt{\frac{1{,}440{,}000}{100}+\frac{1{,}960{,}000}{100}} \\ &= \sqrt{34{,}000} = 184 \text{ miles.}\end{aligned}$$

We would expect the error of estimation to be less than $2\sigma_{(\bar{y}_1-\bar{y}_2)}$, or 368 miles. Therefore, it would appear that tire type 1 is superior to type 2 in wearing quality when subjected to the road test.

A confidence interval for $(\mu_1 - \mu_2)$ with confidence coefficient $(1 - \alpha)$ can be obtained by using

Large-Sample Confidence Interval for $(\mu_1 - \mu_2)$

$$(\bar{y}_1 - \bar{y}_2) \pm z_{\alpha/2}\sqrt{\frac{\sigma_1^2}{n_1}+\frac{\sigma_2^2}{n_2}}.$$

As a rule of thumb, we will require both n_1 and n_2 to be equal to 30 or more in order that s_1^2 and s_2^2 provide good estimates of their respective population variances.

EXAMPLE 8.4

Place a confidence interval on the difference in mean time to wear-out for the problem described in Example 8.3. Use a confidence coefficient of .99.

Solution The confidence interval will be

$$(\bar{y}_1 - \bar{y}_2) \pm 2.58 \sqrt{\frac{\sigma_1^2}{n_1} + \frac{\sigma_2^2}{n_2}}.$$

When we use the results of Example 8.3, we find that the confidence interval is

$$1{,}300 \pm 2.58(184).$$

Therefore, LCL = 825, UCL = 1,775, and the difference in mean time to wear-out is estimated to lie between these two points. Note that the confidence interval is wider than the $\pm 2\sigma_{(\bar{y}_1 - \bar{y}_2)}$ used in Example 8.3 because we have chosen a larger confidence coefficient.

EXERCISES

8.11 An experiment was conducted to compare two diets, A and B, designed for weight reduction. Two groups of 30 overweight dieters each were randomly selected. One group was placed on diet A and the other on diet B and their weight losses were recorded over a 30-day period. The means and standard deviations of the weight loss measurements for the two groups are shown in the accompanying table. Find a 95 percent confidence interval for the difference in mean weight loss for the two diets. Interpret your confidence interval.

Diet A	Diet B
$\bar{y}_A = 21.3$	$\bar{y}_B = 13.4$
$s_A = 2.6$	$s_B = 1.9$

8.12 One method for solving the electric power shortage employs the construction of floating nuclear power plants located a few miles offshore in the ocean. Because there is concern about the possibility of a ship collision with the floating (but anchored plant), an estimate of the density of ship traffic in the area is needed. The number of ships passing within 10 miles of the proposed power-plant location per day, recorded for $n = 60$ days during July and August, possessed sample mean and variance equal to

$$\bar{y} = 7.2$$
$$s^2 = 8.8$$

(a) Find a 95 percent confidence interval for the mean number of ships passing within 10 miles of the proposed power-plant location during a one-day time period.

(b) The density of ship traffic was expected to decrease during the winter months. A sample of $n = 90$ daily recordings of ship sightings for December, January, and February gave the following mean and variance:

$$\bar{y} = 4.7$$
$$s^2 = 4.9$$

Find a 90 percent confidence interval for the difference in mean density of ship traffic between the summer and winter months.

(c) What is the population associated with your estimate, part (b)? What could be wrong with the sampling procedure, parts (a) and (b)?

8.13 The in-city gasoline mileage per gallon was computed for two economy automobiles, each for 40 tanks of gasoline. The mean and standard deviation of the mile-per-gallon readings were as given in the table. Carefully define the populations associated with the mean mile-per-gallon ratings, μ_1 and μ_2, associated with the two automobiles. Find a 90 percent confidence interval for the difference in mean miles per gallon for the two automobiles. Interpret the confidence interval.

Auto 1	Auto 2
$\bar{y}_1 = 22.3$	$\bar{y}_2 = 25.1$
$s_1 = 1.1$	$s_2 = 1.3$

8.8

ESTIMATING THE PARAMETER OF A BINOMIAL POPULATION

The best point estimator of the binomial parameter, p, is also the estimator that would be chosen intuitively. That is, the estimator, $\hat{p}$, would equal

$$\hat{p} = \frac{y}{n},$$

the total number of successes divided by the total number of trials. By "best" we mean that $\hat{p}$ is unbiased and possesses a minimum variance compared with other possible estimators.

We recall that, according to the Central Limit Theorem, y is approximately normally distributed when n is large. Inasmuch as n is a constant, we would suspect that $\hat{p}$ is also normally distributed when n is large, and this is indeed true. Furthermore, the expected value and standard deviation of $\hat{p}$ can be shown to equal

Mean and Standard Deviation of $\hat{p}$

$$E(\hat{p}) = p,$$

$$\sigma_{\hat{p}} = \sqrt{\frac{pq}{n}}.$$

Bounds on the error of a point estimate will be

$$2\sqrt{\frac{pq}{n}},$$

and the $100(1 - \alpha)$ percent confidence interval, appropriate for large n, is

Large-Sample Confidence Interval for p

$$\hat{p} \pm z_{\alpha/2}\sqrt{\frac{\hat{p}\hat{q}}{n}}.$$

The sample size will be considered large when we can assume that $\hat{p}$ is approximately normally distributed. These conditions were discussed in Section 7.4.

The only difficulty encountered in our procedure will be in calculating $\sigma_{\hat{p}}$, which involves p (and $q = 1 - p$), which is unknown. Note that we have substituted $\hat{p}$ for the parameter p in the standard deviation, $\sqrt{pq/n}$. When n is large, little error will be introduced by this substitution. As a matter of fact, the standard deviation changes only slightly as p changes. This can be observed in Table 8.2, where $\sqrt{pq}$ is recorded for several values of p. Note that $\sqrt{pq}$ changes very little as p changes, especially when p is near .5.

TABLE 8.2
Some calculated values of $\sqrt{pq}$

p	$\sqrt{pq}$
.5	.50
.4	.49
.3	.46
.2	.40
.1	.30

EXAMPLE 8.5

A random sample of $n = 100$ voters in a community produced $y = 59$ voters in favor of candidate A. Estimate the fraction of the voting population favoring A and place a bound on the error of estimation.

Solution The point estimate is

$$\hat{p} = \frac{y}{n} = \frac{59}{100} = .59,$$

and the bound on the error of estimation is

$$2\sigma_{\hat{p}} = 2\sqrt{\frac{pq}{n}} \approx 2\sqrt{\frac{(.59)(.41)}{100}} = .098.$$

A 95 percent confidence interval for p would be

$$\hat{p} \pm 1.96\sqrt{\frac{\hat{p}\hat{q}}{n}}$$

or

$$.59 \pm 1.96(.049).$$

Thus we would estimate that p lies in the interval .494 to .686 with confidence coefficient .95.

EXERCISES

8.14 A published report on a Gallup Youth Survey (*Gainesville Sun,* May 18, 1977) states that 1,069 teenagers were questioned concerning their opinion on what they consider to be the key problems facing youth today. The key problems were topped by drug use and abuse (27%), getting along with and communicating with parents (20%), alcohol use and abuse (7%), and finding employment (6%). If the 1,069 teenagers can be regarded as a random sample of all teenagers in the United States, estimate the fraction that regard drug use and abuse as the number one problem. Use a 99 percent confidence interval.

8.15 Osteomyelitis (bone infection) is an ailment that often fails to respond to treatment. Doctors at the Naval Regional Medical Center in Long Beach, California, have been testing the effect of a pressurized oxygen environment on patients (*Gainesville Sun,* October 21, 1977). Of 70 patients treated with the pressurized oxygen environment, all improved and 63 percent have remained free of the disease. If the treated patients represent a random sample from the population of all patients with osteomyelitis, estimate the proportion of treated patients that will remain free of the disease. Use a 95 percent confidence interval.

8.16 The *Orlando Sentinel Star* (December 1, 1977) reported on a survey they conducted to investigate the attitudes of television viewers concerning a certain national sports commentator. They state that of 380 letters received, 325 wanted the sports commentator "off the air." The remaining 55 were in favor of retaining him.

(a) The validity of a confidence interval calculated for the proportion p of the newspaper's readers who favor retention of the commentator is dependent on how the survey was conducted. Suppose that the viewer responses were obtained by requesting that readers send in their opinions on the question of retention or firing. Would this method of sampling yield a valid confidence interval for p?

(b) Suppose that the 380 respondents represented a random sample of the newspaper's readers. Find a 95 percent confidence interval for p and interpret it.

8.17 A sample of 400 human subjects produced $y = 280$ students who were classified as right-eye-dominant on the basis of a sighting task. Estimate the fraction of the entire population who are right-eye-dominant; use a 95 percent confidence interval.

8.18 A random sample of $n = 1{,}500$ consumers showed 20 percent planning to buy new cars during the coming year. Estimate the fraction of consumers planning to buy new cars; use a 90 percent confidence interval.

8.19 A new type of photoflash bulb was tested to estimate the probability, p, that the new bulb would produce the required light output at the appropriate time. A sample of 1,000 bulbs was tested and 920 were observed to function according to specifications. Estimate p and place bounds on the error of estimation.

8.9

ESTIMATING THE DIFFERENCE BETWEEN TWO BINOMIAL PARAMETERS

The fourth and final estimation problem considered in this chapter is the estimation of the difference between the parameters of two binomial populations. Assume that the two populations I and II possess parameters p_1 and p_2, respectively. Independent random samples consisting of n_1 and n_2 trials are drawn from their respective populations and the estimates $\hat{p}_1$ and $\hat{p}_2$ are calculated.

The point estimator of $(p_1 - p_2)$, $(\hat{p}_1 - \hat{p}_2)$, is an unbiased estimator with mean and standard deviation as follows:

Mean and Standard Deviation of $(\hat{p}_1 - \hat{p}_2)$

$$E(\hat{p}_1 - \hat{p}_2) = (p_1 - p_2),$$

with standard deviation*

$$\sigma_{(\hat{p}_1 - \hat{p}_2)} = \sqrt{\frac{p_1 q_1}{n_1} + \frac{p_2 q_2}{n_2}}.$$

Therefore, the bound on the error of estimation is

$$2\sqrt{\frac{p_1 q_1}{n_1} + \frac{p_2 q_2}{n_2}},$$

where the estimates, $\hat{p}_1$ and $\hat{p}_2$, may be substituted for p_1 and p_2.

The $100(1 - \alpha)$ percent confidence interval, appropriate when n_1 and n_2 are large, is

Large-Sample Confidence Interval for $(p_1 - p_2)$

$$(\hat{p}_1 - \hat{p}_2) \pm z_{\alpha/2}\sqrt{\frac{\hat{p}_1 \hat{q}_1}{n_1} + \frac{\hat{p}_2 \hat{q}_2}{n_2}}.$$

EXAMPLE 8.6

A manufacturer of fly sprays wished to compare two new concoctions, I and II. Two rooms of equal size, each containing 1,000 flies, were employed in the experiment, one treated with fly spray I and the other treated with an equal amount of fly spray II. A total of 825 and 760 flies succumbed to sprays I and II, respectively. Estimate the difference in the rate of kill for the two sprays when used in the test environment.

Solution The point estimate of $(p_1 - p_2)$ is

$$(\hat{p}_1 - \hat{p}_2) = .825 - .760 = .065.$$

The bound on the error of estimation is

$$2\sqrt{\frac{p_1 q_1}{n_1} + \frac{p_2 q_2}{n_2}} \approx 2\sqrt{\frac{(.825)(.175)}{1{,}000} + \frac{(.76)(.24)}{1{,}000}}$$
$$= .036.$$

*The formula for the standard deviation of $(\hat{p}_1 - \hat{p}_2)$ follows from our discussion in Section 8.7. Since the variance of the difference of two independent random variables is equal to the sum of their respective variances, the variance of $(\hat{p}_1 - \hat{p}_2)$ is equal to $\sigma^2_{\hat{p}_1} + \sigma^2_{\hat{p}_2} = (p_1 q_1/n_1) + (p_2 q_2/n_2)$. The standard deviation of $(\hat{p}_1 - \hat{p}_2)$ is equal to the square root of this quantity.

The corresponding confidence interval, using confidence coefficient .95, is

$$(\hat{p}_1 - \hat{p}_2) \pm 1.96\sqrt{\frac{\hat{p}_1\hat{q}_1}{n_1} + \frac{\hat{p}_2\hat{q}_2}{n_2}}.$$

The resulting confidence interval is $.065 \pm .035$.

Hence we estimate that the difference between the rates of kill, $(p_1 - p_2)$, will fall in the interval .030 to .100. We are fairly confident of this estimate because we know that if our sampling procedure were repeated over and over again, each time generating an interval estimate, approximately 95 percent of the estimates would enclose the quantity $(p_1 - p_2)$.

EXERCISES

8.20 According to *Environment News* (April 1975), "The continuing analysis of lead levels in the drinking water of several Boston communities has verified elevated lead levels in the water supplies of Somerville, Brighton and Beacon Hill. . . ." Preliminary results of a study carried out in 1974 found that "20 percent of 248 households tested in those communities showed levels exceeding the U.S. Public Health Service standard of 50 parts per million." In contrast, in Cambridge, which adds anticorrosives to its water, "only 5 percent of the 110 households tested showed lead levels exceeding the standard." Find a 95 percent confidence interval for the difference in the proportions of households which have lead levels exceeding the standard between the communities of Somerville, Brighton, and Beacon Hill and the community of Cambridge.

8.21 A sampling of political candidates, 200 randomly chosen from the West and 200 from the East, was classified according to whether the candidate received backing by a national labor union and whether the candidate won. A summary of the data is shown below:

	West	East
Winners backed by union	120	142

Find a 95 percent confidence interval for the difference in the proportions of union-backed winners between the West and the East.

8.22 The percentage of D's and F's awarded to students by two college history professors was duly noted by the dean. Professor I achieved a rate equal to 32 percent as opposed to 21 percent for professor II, based upon 200 and 180 students, respectively. Estimate the difference in the percentage of D's and F's awarded by the professors. Place bounds on the error of estimation.

8.23 In a study of the relationship between birth order and college success, an investigator found that 126 in a sample of 180 college graduates were first-born or only children; in a sample of 100 nongraduates of comparable age and socioeconomic background the number of first-born or only children was 54. Estimate the difference in proportion of first-born or only children for the two populations from which these samples were drawn. Use a 90 percent confidence interval.

8.24 Two different vaccines developed to prevent a disease in poultry showed 80 percent and 88 percent effectiveness in preventing the disease in samples each containing $n = 400$ vaccinated chickens exposed to the disease. Construct a 95 percent confidence interval for the difference in proportions of chickens surviving the disease for the two types of vaccine.

8.10

CHOOSING THE SAMPLE SIZE

The design of an experiment is essentially a plan for purchasing a quantity of information which, like any other commodity, may be acquired at varying prices depending upon the manner in which the data are obtained. Some measurements contain a large amount of information concerning the parameter of interest; others may contain little or none. Since the sole product of research is information, we want to make its purchase at minimum cost.

The sampling procedure, or experimental design as it is usually called, affects the quantity of information per measurement. This, along with the sample size, n, controls the total amount of relevant information in a sample. With few exceptions we shall be concerned with the simplest sampling situation, random sampling from a relatively large population, and will devote our attention to the selection of the sample size, n.

The researcher makes little progress in planning an experiment before encountering the problem of selecting the sample size. Indeed, perhaps one of the most frequent questions asked of the statistician is: How many measurements should be included in the sample? Unfortunately, the statistician cannot answer this question without knowing how much information the experimenter wishes to buy. Certainly, the total amount of information in the sample will affect the measure of goodness of the method of inference and must be specified by the experimenter. Referring specifically to estimation, we would like to know how accurate the experimenter wishes the estimate to be. This may be stated by specifying a bound on the error of estimation.

For instance, suppose that we wish to estimate the average daily yield of a chemical, μ (Example 8.1), and we wish the error of estimation to be less than 4 tons with a probability of .95. Since approximately 95 percent of the sample means will lie within $2\sigma_{\bar{y}}$ of μ in repeated sampling,* we are asking that $2\sigma_{\bar{y}}$ equal 4 tons (see Figure 8.10). Then,

$$2\sigma_{\bar{y}} = 4$$

or

$$\frac{2\sigma}{\sqrt{n}} = 4.$$

Solving for n, we obtain

$$n = \frac{\sigma^2}{4}.$$

You will quickly note that we cannot obtain a numerical value for n unless the population standard deviation, σ, is known. And certainly, this is exactly what we would expect because the variability of $\bar{y}$ depends upon the variability of the population from which the sample was drawn.

Lacking an exact value for σ, we would use the best approximation available, such as an estimate, s, obtained from a previous sample or knowledge of the

*The probability distribution of $\bar{y}$ will be approximately normal because of the Central Limit Theorem.

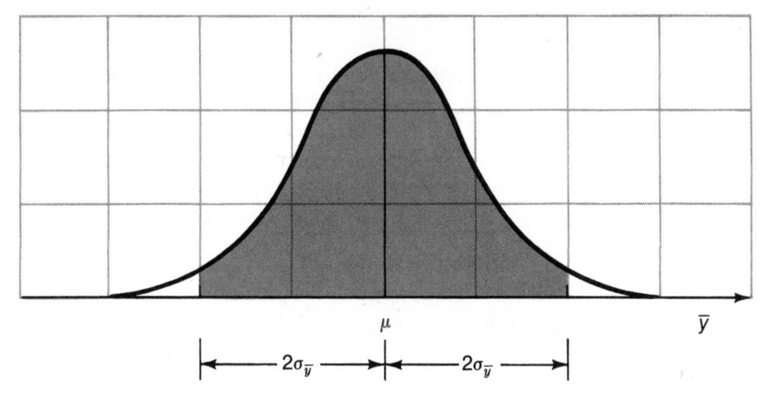

FIGURE 8.10

Approximate distribution of $\bar{y}$ for large samples

range in which the measurements will fall. Since the range is approximately equal to 4σ (the Empirical Rule), one-fourth of the range will provide an approximate value for σ. For our example we would use the results of Example 8.1, which provided a reasonably accurate estimate of σ equal to $s = 21$. Then,

$$n = \frac{\sigma^2}{4} \approx \frac{(21)^2}{4} = 110.25$$

or

$$n = 111.$$

Using a sample size $n = 111$, we would be reasonably certain (with probability approximately equal to .95) that our estimate will lie within $2\sigma_{\bar{y}} = 4$ tons of the true average daily yield.

Actually we would expect the error of estimation to be much less than 4 tons. According to the Empirical Rule, the probability is approximately equal to .68 that the error of estimation would be less than $\sigma_{\bar{y}} = 2$ tons. The reader will note that the probabilities .95 and .68 used in these statements will be inexact, owing to the fact that s was substituted for σ. While this method of choosing the sample size is approximate for a specified desired accuracy of estimation, it is the best available and is certainly better than selecting the sample size on the basis of our intuition.

The method of choosing the sample size for all the large-sample estimation procedures discussed in preceding sections is identical to that described above. The experimenter must specify a desired bound on the error of estimation and an associated confidence level, $(1 - \alpha)$. For example, if the parameter is θ and the desired bound is B, we would equate

$$z_{\alpha/2}\sigma_{\hat{\theta}} = B,$$

where $z_{\alpha/2}$ is the z value defined in Section 8.5; that is,

$$P(z > z_{\alpha/2}) = \frac{\alpha}{2}.$$

We shall illustrate with examples.

EXAMPLE 8.7

The reaction of an individual to a stimulus in a psychological experiment may take one of two forms, *A* or *B*. If an experimenter wishes to estimate the probability, p, that a person will react in favor of *A*, how many people must be included in the experiment? Assume that he will be satisfied if the error of estimation is less than .04 with probability equal to .90. Assume also that he expects p to lie somewhere in the neighborhood of .6.

Solution Since the confidence coefficient is $1 - \alpha = .90$, α must equal .10 and $\alpha/2 = .05$. The z value corresponding to an area equal to .05 in the upper tail of the z distribution is $z_{\alpha/2} = 1.645$.* We then require

$$1.645\sigma_{\hat{p}} = .04$$

or

$$1.645\sqrt{\frac{pq}{n}} = .04.$$

Since the variability of $\hat{p}$ is dependent upon p, which is unknown, we must use the guessed value of $p = .6$ provided by the experimenter as an approximation. Then,

$$1.645\sqrt{\frac{(.6)(.4)}{n}} = .04$$

or

$$\sqrt{n} = \frac{1.645}{.04}\sqrt{(.6)(.4)}$$

and

$$n = 406.$$

Remember, we used an approximate value of p to obtain this value of n. Consequently, the value, $n = 406$, is not exact. It is only an approximation to the required sample size. Therefore, it is reasonable to say that a sample of approximately $n = 406$ people will allow you to estimate p with an error of estimation less than .04 with probability equal to .90.

EXAMPLE 8.8

An experimenter wishes to compare the effectiveness of two methods of training industrial employees to perform a certain assembly operation. A number of employees is to be divided into two equal groups, the first receiving training method 1 and the second training method 2. Each will perform the assembly operation, and the length of assembly time will be recorded. It is expected that the measurements for both groups will have a range of approximately 8 minutes. If the estimate of the difference in mean

*Note: It is not necessary to carry three decimal places on $z_{\alpha/2}$ for these calculations. The calculated value of n will be very approximate because we are using an approximate value for p in the calculations.

time to assemble is desired correct to within 1 minute with probability equal to .95, how many workers must be included in each training group?

Solution Equating $2\sigma_{(\bar{y}_1 - \bar{y}_2)}$ to 1 minute, we obtain

$$2\sqrt{\frac{\sigma_1^2}{n_1} + \frac{\sigma_2^2}{n_2}} = 1.$$

Or, since we desire n_1 to equal n_2, we may let $n_1 = n_2 = n$ and obtain the equation

$$2\sqrt{\frac{\sigma_1^2}{n} + \frac{\sigma_2^2}{n}} = 1.$$

As noted above, the variability of each method of assembly is approximately the same and hence $\sigma_1^2 = \sigma_2^2 = \sigma^2$. Since the range, equal to 8 minutes, is approximately equal to 4σ, then

$$4\sigma = 8$$

and

$$\sigma = 2.$$

Substituting this value for σ_1 and σ_2 in the above equation, we obtain

$$2\sqrt{\frac{(2)^2}{n} + \frac{(2)^2}{n}} = 1;$$

or

$$2\sqrt{8/n} = 1$$

and

$$\sqrt{n} = 2\sqrt{8}.$$

Solving, we have $n = 32$. Thus each group should contain approximately $n = 32$ members.

EXERCISES

8.25 How many voters must be included in a sample collected to estimate the fraction of the popular vote favorable to a presidential candidate in a national election if the estimate is desired correct to within .005 with probability approximately equal to .95? Assume that the true fraction will lie somewhere in the neighborhood of .5.

8.26 If a wildlife service wishes to estimate the mean number of days hunting per hunter for all hunters licensed in the state during a given season, with a bound on the error of estimation equal to 2 hunting days, how many hunters must be included in the survey? Assume that data collected in earlier surveys has shown σ to be approximately equal to 10.

8.27 The Federal Trade Commission samples and tests cigarettes to determine whether the nicotine and tar contents of the cigarettes agrees with those claimed by the manufacturer. An *FTC News Summary* (October 29, 1976) states that for a test of True king size, filter, soft-pack cigarettes, the tar and nicotine contents were 5 and .4 milligrams (mg) per cigarette. The report does not state how many cigarettes were analyzed to obtain these figures nor does it give a measure of the cigarette-to-cigarette variation in the measurements. Suppose that the standard deviation of the tar content measurements is approximately equal to 1 mg per cigarette. If the FTC wishes to estimate the mean tar content per cigarette correct to within .1 mg, how many cigarettes would the FTC have to analyze? (Assume that the FTC wishes the error of estimation to be less than .1 mg with probability equal to .99.)

8.28 It is desired to estimate the difference in grade-point average between two groups of college students accurate to within .2 grade point with probability approximately equal to .95. If the standard deviation of the grade-point measurements is approximately equal to .6, how many students must be included in each group? (Assume that the groups will be of equal size.)

8.29 An experimenter has prepared a drug-dose level designed to induce sleep in 60 percent of all cases treated. How large a sample should be treated if he wishes to estimate the true fraction within .02, with probability .95?

8.30 Refer to Exercise 8.9 and the measurement of acidity in rainwater and suppose that you wish to estimate the mean pH of rainfalls in an area that suffers heavy pollution due to the discharge of smoke from a power plant. Assume you know that σ is in the neighborhood of .5 pH and that you wish your estimate to lie within .1 of μ with probability near .95. Approximately how many rainfalls must be included in your sample (one pH reading per rainfall)? Would it be valid to select all of your water specimens from a single rainfall? Explain.

8.31 Refer to Exercise 8.30. Suppose that you wish to estimate the difference between the mean acidity for rainfalls at two different locations, one in a relatively unpolluted area along the ocean and the other in an area subject to heavy air pollution. If you wish your estimate to be correct to the nearest .1 pH with probability near .90, approximately how many rainfalls (pH values) would have to be included in each sample? (Assume that the variance of the pH measurements is approximately .25 at both locations and that the samples will be of equal size.)

8.32 Refer to the Exercise 8.7 and the experiment conducted to investigate the link between vitamin C and cholesterol level. Approximately how many persons with moderately high cholesterol levels should be included in the experiment if you wish to estimate the mean drop in cholesterol level correct to within 2 mg per 100 ml with probability .95?

8.11

A STATISTICAL TEST OF AN HYPOTHESIS

The basic reasoning employed in a statistical test of an hypothesis was outlined in Section 6.5 in connection with the test of the effectiveness of a cold vaccine. In this section we shall attempt a condensation of the basic points involved and refer you to Section 6.5 for an intuitive presentation of the subject.

The objective of a statistical test is to test an hypothesis concerning the values of one or more population parameters. We will generally have a theory, a research hypothesis, about the parameter(s) that we wish to support. For example, we might wish to show that the mean life μ_1 of a new type of automobile tire is greater than the mean life μ_2 of a competitor's tire. Support for this research hypothesis, called the alternative hypothesis by statisticians, is obtained by showing (using the sample data as evidence) that the converse of the research hypothesis, the null hypothesis, is false. Thus support for one theory is obtained by showing lack of support for its converse, in a sense a "proof" by contradiction. For the tire example, the converse of the research (alternative) hypothesis $\mu_1 > \mu_2$ is $\mu_1 < \mu_2$. If we can show that the sample data support rejection of the null hypothesis, $\mu_1 = \mu_2$ (the means are equal), in favor of the alternative hypothesis, $\mu_1 > \mu_2$, we have achieved our research objective.* Although it is common to speak of testing a null hypothesis, keep in mind that the research objective is always to show support for the alternative hypothesis, if support is warranted.

A statistical test involves four elements:

1. Null hypothesis.
2. Alternative hypothesis.
3. Test statistic.
4. Rejection region.

Note that the specification of these four elements defines a particular test and that changing one or more creates a new test.

The null hypothesis, indicated symbolically as H_0, states the hypothesis to be tested. Thus H_0 will specify hypothesized values for one or more population parameters. For example, we might wish to test the hypothesis that a population mean is equal to 50, or that two population means, say μ_1 and μ_2, are equal.

The decision to reject or accept the null hypothesis is based upon information contained in a sample drawn from the population of interest. The sample values are used to compute a single number, corresponding to a point on a line, which operates as a decision maker and which is called the test statistic. The entire set of values that the test statistic may assume is divided into two sets or regions, one corresponding to the rejection region and the other to the acceptance region (see Figure 8.11). If the test statistic computed from a particular sample assumes a value in the rejection region, the null hypothesis is rejected and you decide in favor of the alternative hypothesis. If the test statistic falls in the acceptance region, the null hypothesis is accepted, that is, you decide in favor of the null hypothesis.

The decision procedure just described is subject to two types of errors which are prevalent in any two-choice decision problem. We may reject the null hypothesis when, in fact, it is true, or we may accept H_0 when it is false and some *alternative hypothesis* is true. These errors are called the type I and type II errors, respectively, for the statistical test. The two states for the null hypothesis, that is, true or false, along with the two decisions which the experimenter may make are indicated in the two-way table, Table 8.3. The occurrences of the type I and type II errors are indicated in the appropriate cells.

The goodness of a statistical test of an hypothesis is measured by the

*Note that if the test rejects the null hypothesis, $\mu_1 = \mu_2$, in favor of the alternative, $\mu_1 > \mu_2$, then it should (our test will) be even more likely to reject a null hypothesis for values of μ_1 and μ_2 such that $\mu_1 < \mu_2$ (since this is even more contradictory to the proposition, $\mu_1 > \mu_2$, than the proposition that $\mu_1 = \mu_2$). For this reason, we usually say that we are testing the null hypothesis, $\mu_1 = \mu_2$.

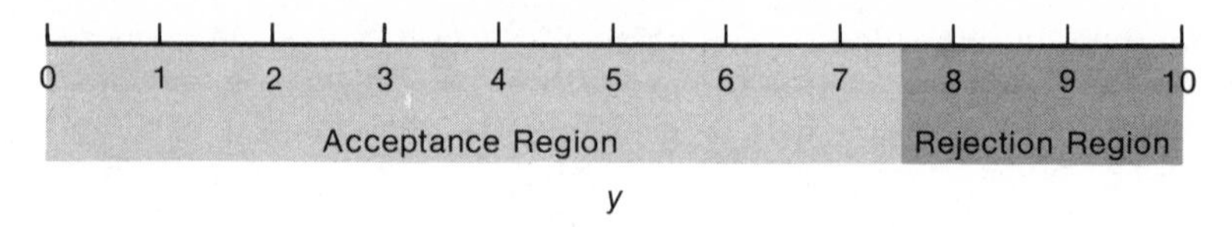

FIGURE 8.11
Possible values for the test statistic, y

probabilities of making a type I or a type II error, denoted by the symbols α and β, respectively. These probabilities, calculated for the elementary statistical tests presented in the exercises in Chapter 6, illustrate the basic relationship among α, β, and the sample size, n. Since α is the probability that the test statistic will fall in the rejection region, assuming H_0 to be true, an increase in the size of the rejection region will increase α and, at the same time, decrease β for a fixed sample size. Reducing the size of the rejection region will decrease α and increase β. If the sample size, n, is increased, more information will be available upon which to base the decision and both α and β will decrease.

The probability of making a type II error, β, varies depending upon the true value of the population parameter. For instance, suppose that we wish to test the null hypothesis that the binomial parameter, p, is equal to $p_0 = .4$. (We shall use a subscript 0 to indicate the parameter value specified in the null hypothesis, H_0.) Furthermore, suppose that H_0 is false and that p is really equal to an alternative value, say p_a. What will be more easily detected, a $p_a = .4001$ or a $p_a = 1.0$? Certainly, if p is really equal to 1.0, every single trial will result in a success and the sample results will produce strong evidence to support a rejection of H_0: $p_0 = .4$. On the other hand, $p_a = .4001$ lies so close to $p_0 = .4$ that it would be extremely difficult to detect without a very large sample. In other words, the probability β of accepting H_0 will vary depending upon the difference between the true value of p and the hypothesized value, p_0. A graph of the probability of a type II error, β, as a function of the true value of the parameter is called the operating characteristic curve for the statistical test. Note that the operating characteristic curves for the lot acceptance sampling plans, Chapter 6, were really graphs expressing β as a function of p.

Since the rejection region is specified and remains constant for a given test, α will also remain constant and, as in lot acceptance sampling, the operating characteristic curve will describe the characteristics of the statistical test. An increase in the sample size, n, will decrease β and reduce its value for all alternative values of the parameter tested. Thus we will possess an operating

TABLE 8.3
Decision table

	Null Hypothesis	
Decision	True	False
Reject	Type I error	Correct decision
Accept	Correct decision	Type II error

characteristic curve corresponding to each sample size. This property of the operating characteristic curve was illustrated in the exercises in Chapter 6.

Ideally, experimenters will have in mind some values, α and β, that measure the risks of the respective errors they are willing to tolerate. They will also have in mind some deviation from the hypothesized value of the parameter, which they consider of *practical* importance and which they wish to detect. The rejection region for the test will then be located in accordance with the specified value of α. Finally, they will choose the sample size necessary to achieve an acceptable value of β for the specified deviation that they wish to detect. This could be done by consulting the operating characteristic curves, corresponding to various sample sizes, for the chosen test.

In conclusion, keep in mind that "accepting" a particular hypothesis means deciding in its favor. Regardless of the outcome of a test, you are never *certain* that the hypothesis you "accept" is true. There is always a risk of being wrong (measured by α and β). Consequently, you never "accept" H_0 if β is unknown or its value is unacceptable to you. When this situation occurs, you should withhold judgment and collect more data.

8.12 A LARGE-SAMPLE STATISTICAL TEST

Large-sample tests of hypotheses concerning the population parameters discussed in Sections 8.4 to 8.9 are based upon a normally distributed test statistic and, for that reason, may be regarded as one and the same test. We shall present the reasoning in a very general manner, referring to the parameter of interest as θ. Thus we could imagine θ as representing μ, $(\mu_1 - \mu_2)$, p, or $(p_1 - p_2)$. The specific tests for each will be illustrated by examples.

Suppose that we wish to test an hypothesis concerning a parameter θ and that an unbiased point estimator, $\hat{\theta}$, is available and known to be normally distributed with standard deviation, $\sigma_{\hat{\theta}}$. If the null hypothesis, H_0: $\theta = \theta_0$, is true, then $\hat{\theta}$ will be normally distributed about θ_0 as shown in Figure 8.12.

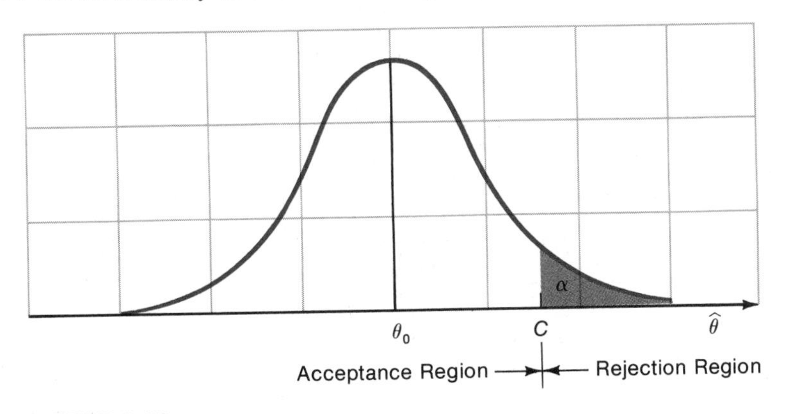

FIGURE 8.12
Distribution of $\hat{\theta}$ when H_0 is true

Suppose that, from a *practical* point of view, we are primarily concerned with the rejection of H_0 when θ is greater than θ_0. Then the alternative hypothesis would be H_a: $\theta > \theta_0$ and we would reject the null hypothesis when $\hat{\theta}$ is too large. "Too large," of course, means too many standard deviations, $\sigma_{\hat{\theta}}$, away from θ_0. The rejection region for the test is shown in Figure 8.12.

DEFINITION

The value of $\hat{\theta}$, *C*, which separates the rejection and acceptance regions is called the critical value of the test statistic.

The probability of rejecting, assuming the null hypothesis to be true, would equal the area under the normal curve lying above the rejection region. Thus if we desire $\alpha = .05$, we would reject when $\hat{\theta}$ is more than $1.645\sigma_{\hat{\theta}}$ to the right of θ_0. A test rejecting in one tail of the distribution of the test statistic is called a one-tailed statistical test.

If we wish to detect departures *either* greater than or less than θ_0, the alternative hypothesis would be

$$H_a: \quad \theta \neq \theta_0;$$

that is,

$$\theta > \theta_0$$

or

$$\theta < \theta_0.$$

The probability of a type I error, α, would be equally divided between the two tails of the normal distribution and we would reject H_0 for values of $\hat{\theta}$ greater than some critical value, *C*, or less than $-C$. This is called a two-tailed statistical test.

The calculation of β for the one-tailed statistical test described above can be facilitated by considering Figure 8.13. When H_0 is false and $\theta = \theta_a$, the test statistic, $\hat{\theta}$, will be normally distributed about a mean θ_a, rather than θ_0. The distribution of $\hat{\theta}$, assuming $\theta = \theta_a$, is shown by the black curve. The hypothesized

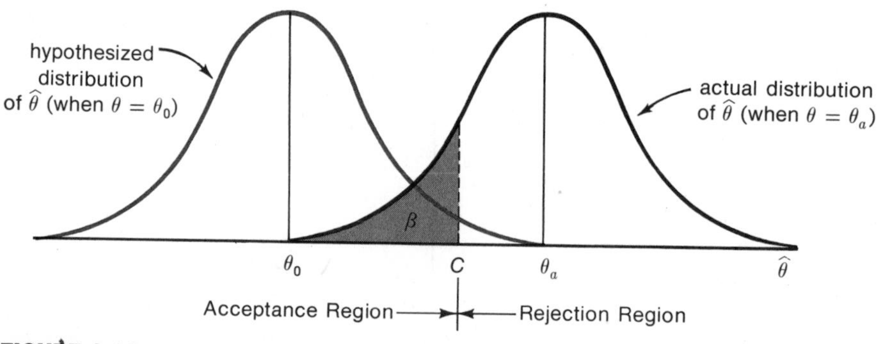

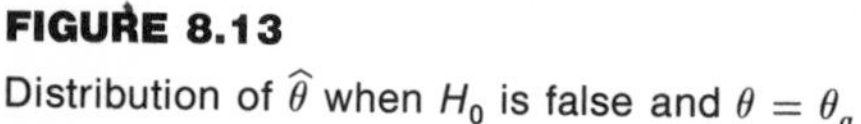

FIGURE 8.13

Distribution of $\hat{\theta}$ when H_0 is false and $\theta = \theta_a$

distribution of $\widehat{\theta}$, shown by the colored curve, locates the rejection region and the critical value of $\widehat{\theta}$, C. Since β is the probability of accepting H_0, given $\theta = \theta_a$, β would equal the area under the black curve located above the acceptance region. This area, which is shaded, could be easily calculated using the methods described in Chapter 7.

You will quickly note that all the point estimators discussed in the preceding section satisfy the requirements of the test described above when the sample size, n, is large. That is, the sample size must be large enough so that the point estimator will be approximately normally distributed, by the Central Limit Theorem, and also must permit a reasonably good estimate of its standard deviation. We may therefore test hypotheses concerning μ, p, $(\mu_1 - \mu_2)$, and $(p_1 - p_2)$.

The mechanics of testing are simplified by using z as a test statistic, as noted in Example 7.10

Large-Sample Test Statistic

$$z = \frac{\widehat{\theta} - \theta_0}{\sigma_{\widehat{\theta}}}.$$

Note that z is simply the deviation of a normally distributed random variable, $\widehat{\theta}$, from θ_0, expressed in units of $\sigma_{\widehat{\theta}}$. Thus for a two-tailed test with $\alpha = .05$ we would reject H_0 when $z > 1.96$ or $z < -1.96$.

As we have previously stated, the method of inference used in a given situation will often depend upon the preference of the experimenter. Some people wish to express an inference as an estimate; others prefer to test an hypothesis concerning the parameter of interest. The following example will demonstrate the use of the z test in testing an hypothesis concerning a population mean and, at the same time, will illustrate the close relationship between the statistical test and the large-sample confidence intervals discussed in the preceding sections.

EXAMPLE 8.9

Refer to Example 8.1, Section 8.4. Test the hypothesis that the average daily yield of the chemical is $\mu = 880$ tons per day against the alternative that μ is either greater or less than 880 tons per day. The sample (Example 8.1), based upon $n = 50$ measurements, yielded $\bar{y} = 871$ and $s = 21$ tons.

Solution The point estimate for μ is $\bar{y}$. Therefore, the test statistic is

$$z = \frac{\bar{y} - \mu_0}{\sigma_{\bar{y}}} = \frac{\bar{y} - \mu_0}{\sigma/\sqrt{n}}.$$

Using s to approximate σ, we obtain

$$z = \frac{871 - 880}{21/\sqrt{50}} = -3.03.$$

For $\alpha = .05$, the rejection region is $z > 1.96$ or $z < -1.96$. See Figure 8.14. Since the calculated value of z falls in the rejection region, we reject

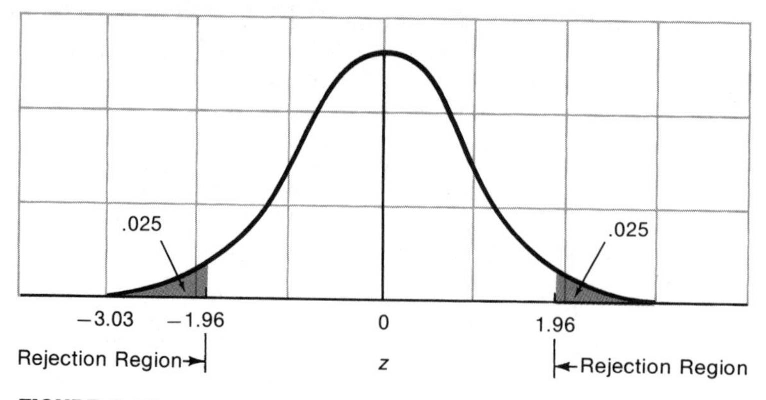

FIGURE 8.14
Location of the rejection region in Example 8.9

the hypothesis that $\mu = 880$ tons. (In fact, it appears that the mean yield is less than 880 tons per day.) The probability of rejecting, assuming H_0 to be true, is only $\alpha = .05$. Hence we are reasonably confident that our decision is correct.

The statistical test based upon a normally distributed test statistic, with given α, and the $(1 - \alpha)$ confidence interval, Section 8.5, are clearly related. The interval $\bar{y} \pm 1.96\sigma/\sqrt{n}$, or approximately 871 ± 5.82, is constructed such that, in repeated sampling, $(1 - \alpha)$ of the intervals will enclose μ. Noting that $\mu = 880$ does not fall in the interval, we would be inclined to reject $\mu = 880$ as a likely value and conclude that the mean daily yield was, indeed, less.

The following example will demonstrate the calculation of β for the statistical test, Example 8.9.

EXAMPLE 8.10

Referring to Example 8.9, calculate the probability, β, of accepting H_0 when μ is actually equal to 870 tons.

Solution The acceptance region for the test, Example 8.9, is located in the interval $\mu_0 \pm 1.96\sigma_{\bar{y}}$. Substituting numerical values, we obtain

$$880 \pm 1.96(21/\sqrt{50})$$

or

$$874.18 \text{ to } 885.82.$$

The probability of accepting H_0, given $\mu = 870$, is equal to the area under the frequency distribution for the test statistic, $\bar{y}$, above the interval 874.18 to 885.82. Since $\bar{y}$ will be normally distributed with mean equal to 870 and $\sigma_{\bar{y}} = 21/\sqrt{50} = 2.97$, β is equal to the area under the normal curve (Figure

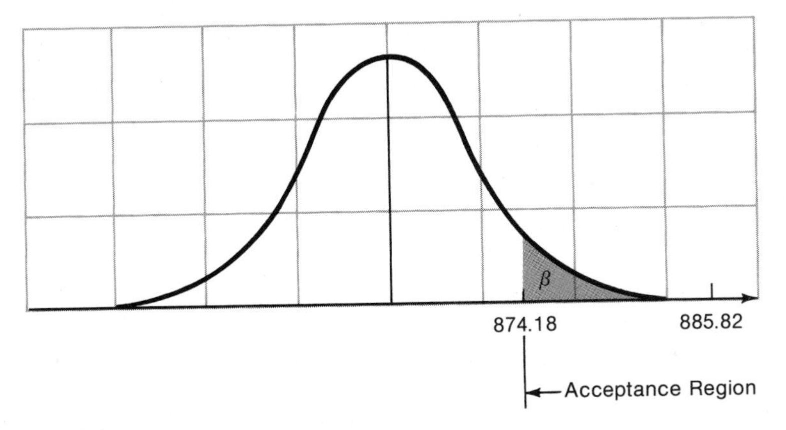

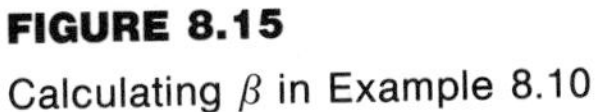

FIGURE 8.15
Calculating β in Example 8.10

8.15) located to the right of 874.18 (because the area to the right of 885.82 is negligible). Calculating the z value corresponding to 874.18, we obtain

$$z = \frac{\bar{y} - \mu}{\sigma / \sqrt{n}} = \frac{874.18 - 870}{21 / \sqrt{50}} = 1.41.$$

We see from Table 3, Appendix II, that the area between $z = 0$ and $z = 1.41$ is .4207. Therefore,

$$\beta = .5 - .4207 = .0793.$$

Thus the probability of accepting H_0, given that μ is really equal to 870, is .0793 or, approximately, 8 chances in 100.

EXAMPLE 8.11

It is known that approximately 1 in 10 smokers favor cigarette brand *A*. After a promotional campaign in a given sales region, a sample of 200 cigarette smokers were interviewed to determine the effectiveness of the campaign. The result of the survey showed that a total of 26 people expressed a preference for brand *A*. Do these data present sufficient evidence to indicate an increase in the acceptance of brand *A* in the region? (Note that, for all practical purposes, this problem is identical to the cold-serum problem given in Example 7.10.)

Solution It is assumed that the sample satisfies the requirements of a binomial experiment. The question posed may be answered by testing the hypothesis

$$H_0\colon\ p = .10$$

against the alternative

$$H_a\colon\ p > .10.$$

A one-tailed statistical test would be utilized because we are primarily concerned with detecting a value of p greater than .10. For this situation it can be shown that the probability of a type II error, β, is minimized by placing the entire rejection region in the upper tail of the distribution of the test statistic.

The point estimator of p is $\hat{p} = y/n$ and the test statistic would be

$$z = \frac{\hat{p} - p_0}{\sigma_{\hat{p}}} = \frac{\hat{p} - p_0}{\sqrt{p_0 q_0 / n}}.$$

Or, multiplying numerator and denominator by n, we obtain

$$z = \frac{y - np_0}{\sqrt{np_0 q_0}},$$

which is the test statistic used in Example 7.12. Note that the two test statistics are equivalent.

Once again we require a value of p so that $\sigma_{\hat{p}} = \sqrt{pq/n}$, appearing in the denominator of z, may be calculated. Since we have hypothesized that $p = p_0$, it would seem reasonable to use p_0 as an approximation for p. Note that this differs from the estimation procedure, where, lacking knowledge of p, we chose $\hat{p}$ as the best approximation. This apparent inconsistency will have negligible effect on the inference, whether it is the result of a test or of estimation, when n is large.

Choosing $\alpha = .05$, we would reject H_0 when $z > 1.645$. See Figure 8.16. Substituting the numerical values into the test statistic, we obtain

$$z = \frac{\hat{p} - p_0}{\sqrt{p_0 q_0 / n}} = \frac{.13 - .10}{\sqrt{\frac{(.10)(.90)}{200}}} = 1.41.$$

The calculated value, $z = 1.41$, does not fall in the rejection region, and hence *we do not reject* H_0.

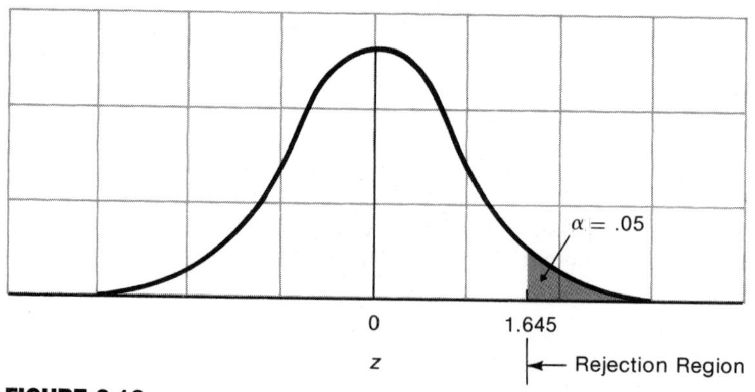

FIGURE 8.16

Location of the rejection region in Example 8.11

Do we accept H_0? No, not until we have stated an alternative value of p that is larger than $p_0 = .10$ and considered to be of *practical* significance. The probability of a type II error, β, should be calculated for this alternative. If β is sufficiently small, we would accept H_0 and would do so with the risk of an erroneous decision fully known.

Tips on Problem Solving: When testing an hypothesis concerning p, use p_0 (not $\hat{p}$) to calculate $\sigma_{\hat{p}}$ in the denominator of the z statistic. The reason for this is that the rejection region is determined by the distribution of $\hat{p}$ when the null hypothesis is true, namely, when $p = p_0$.

Examples 8.9 and 8.11 illustrate an important point. If the data present sufficient evidence to reject H_0, the probability of an erroneous conclusion, α, is known in advance because α is used in locating the rejection region. Since α is usually small, we are fairly certain that we have made a correct decision. On the other hand, if the data present insufficient evidence to reject H_0, the conclusions are not so obvious. Ideally, following the statistical test procedure outlined in Section 8.11, we would have specified a practically significant alternative, p_a, in advance and chosen n such that β would be small. Unfortunately, many experiments are not conducted in this ideal manner. Someone chooses a sample size and the experimenter or statistician is left to evaluate the evidence.

The calculation of β is not too difficult for the statistical test procedure outlined in this section but may be extremely difficult, if not beyond the capability of the beginner, in other test situations. A much simpler procedure is to *not reject* H_0, rather than to accept it; then estimate using a confidence interval. The interval will give you a range of possible values for p.

EXAMPLE 8.12

A university investigation, conducted to determine whether car ownership was detrimental to academic achievement, was based upon two random samples of 100 male students, each drawn from the student body. The grade-point average for the $n_1 = 100$ non–car owners possessed an average and variance equal to $\bar{y}_1 = 2.70$ and $s_1^2 = .36$ as opposed to a $\bar{y}_2 = 2.54$ and $s_2^2 = .40$ for the $n_2 = 100$ car owners. Do the data present sufficient evidence to indicate a difference in the mean achievement between car owners and non–car owners?

Solution We wish to test the null hypothesis that the difference between two population means, $(\mu_1 - \mu_2)$, equals some specified value, say D_0. For our example, we hypothesize that $D_0 = 0$. The alternative hypothesis is that the means differ, that is, $(\mu_1 - \mu_2) \neq 0$.

Recall that $(\bar{y}_1 - \bar{y}_2)$ is an unbiased point estimator of $(\mu_1 - \mu_2)$ that will be approximately normally distributed in repeated sampling when n_1 and n_2 are large. Furthermore, the standard deviation of $(\bar{y}_1 - \bar{y}_2)$ is

$$\sigma_{(\bar{y}_1 - \bar{y}_2)} = \sqrt{\frac{\sigma_1^2}{n_1} + \frac{\sigma_2^2}{n_2}}.$$

Then,

$$z = \frac{(\bar{y}_1 - \bar{y}_2) - D_0}{\sqrt{\dfrac{\sigma_1^2}{n_1} + \dfrac{\sigma_2^2}{n_2}}}$$

will serve as a test statistic when σ_1^2 and σ_2^2 are known or when s_1^2 and s_2^2 provide a good approximation for σ_1^2 and σ_2^2 (that is, when n_1 and n_2 are larger than 30). For our example,

$$H_0: \mu_1 - \mu_2 = D_0 = 0 \quad \text{and} \quad H_a: \mu_1 - \mu_2 \neq 0.$$

Substituting into the formula for the test statistic, we obtain

$$z = \frac{(\bar{y}_1 - \bar{y}_2) - D_0}{\sqrt{\dfrac{\sigma_1^2}{n_1} + \dfrac{\sigma_2^2}{n_2}}} = \frac{2.70 - 2.54}{\sqrt{\dfrac{.36}{100} + \dfrac{.40}{100}}} = 1.84.$$

Using a two-tailed test with $\alpha = .05$, we would reject when $z > 1.96$ or $z < -1.96$. Since z does not fall in the rejection region, we do not reject the null hypothesis. See Figure 8.17. Note, however, that if we choose $\alpha = .10$, the rejection region would be $z > 1.645$ or $z < -1.645$ and the null hypothesis would be rejected.

The decision to reject or accept would, of course, depend upon the risk that we would be willing to tolerate. If we choose $\alpha = .05$, the null hypothesis would not be rejected, but we could not accept H_0 (that is, $\mu_1 = \mu_2$) without investigating the probability of a type II error. If α were chosen equal to .10, the null hypothesis would be rejected. With no other information given, we would be inclined to reject the null hypothesis that there is no difference in the average academic achievement of car owners versus non-car owners. The chance of rejecting H_0, assuming H_0 true, is only $\alpha = .10$, and hence we would be inclined to think that we have made a reasonably good decision.

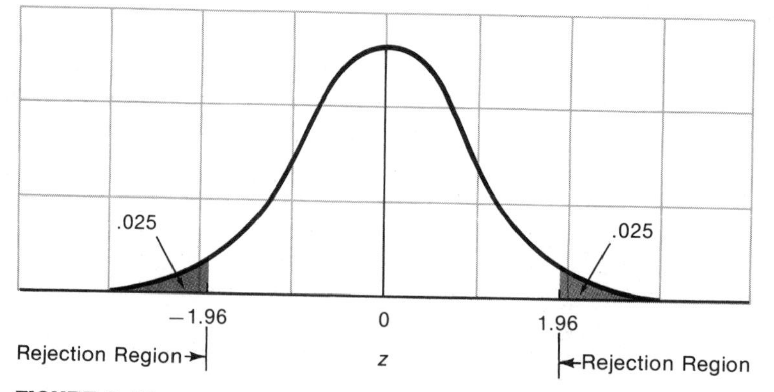

FIGURE 8.17

Location of the rejection region in Example 8.12

EXAMPLE 8.13

The records of a hospital show that 52 men in a sample of 1,000 men versus 23 women in a sample of 1,000 women were admitted because of heart disease. Do these data present sufficient evidence to indicate a higher rate of heart disease among men admitted to the hospital?

Solution We shall assume that the number of patients admitted for heart disease will follow approximately a binomial probability distribution for both men and women with parameters p_1 and p_2, respectively. Stated generally, we wish to decide whether a difference exists between p_1 and p_2, say that $(p_1 - p_2) = D_0$. (For our example, we wish to test the hypothesis that $D_0 = 0$.) Recall that for large samples, the point estimator of $(p_1 - p_2)$, $(\hat{p}_1 - \hat{p}_2)$, is approximately normally distributed in repeated sampling with mean equal to $(p_1 - p_2)$ and standard deviation

$$\sigma_{(\hat{p}_1-\hat{p}_2)} = \sqrt{\frac{p_1 q_1}{n_1} + \frac{p_2 q_2}{n_2}}.$$

Then,

$$z = \frac{(\hat{p}_1 - \hat{p}_2) - (p_1 - p_2)}{\sigma_{(\hat{p}_1-\hat{p}_2)}}$$

would possess a standardized normal distribution in repeated sampling. Hence z could be employed as a test statistic to test

$$H_0\colon (p_1 - p_2) = D_0$$

when suitable approximations are used for p_1 and p_2 which appear in $\sigma_{(\hat{p}_1-\hat{p}_2)}$. Approximations are available for two cases.

Case I: If we hypothesize that p_1 equals p_2, that is,

$$H_0\colon p_1 = p_2$$

or, equivalently, that

$$(p_1 - p_2) = 0,$$

then $p_1 = p_2 = p$ and the best estimate of p is obtained by pooling the data from both samples. Thus, if y_1 and y_2 are the numbers of successes obtained from the two samples, then

$$\hat{p} = \frac{y_1 + y_2}{n_1 + n_2}.$$

The test statistic would be

$$z = \frac{(\hat{p} - \hat{p}_2) - 0}{\sqrt{\frac{\hat{p}\hat{q}}{n_1} + \frac{\hat{p}\hat{q}}{n_2}}}$$

or

$$z = \frac{\hat{p}_1 - \hat{p}_2}{\sqrt{\hat{p}\hat{q}\left(\frac{1}{n_1} + \frac{1}{n_2}\right)}}$$

Case II: On the other hand, if we hypothesize that D_0 is *not equal* to zero, that is,

$$H_0\colon (p_1 - p_2) = D_0,$$

where $D_0 \neq 0$, then the best estimates of p_1 and p_2 are $\hat{p}_1$ and $\hat{p}_2$, respectively. The test statistic would be

$$z = \frac{(\hat{p}_1 - \hat{p}_2) - D_0}{\sqrt{\dfrac{\hat{p}_1\hat{q}_1}{n_1} + \dfrac{\hat{p}_2\hat{q}_2}{n_2}}}.$$

For most practical problems that involve the comparison of two binomial populations, the experimenter will wish to test the null hypothesis that $(p_1 - p_2) = D_0 = 0$. Thus, for our example, we test

$$H_0\colon (p_1 - p_2) = 0$$

against the alternative

$$H_a\colon (p_1 - p_2) > 0.$$

Note that a one-tailed statistical test will be employed because, if a difference exists, we wish to detect $p_1 > p_2$. Choosing $\alpha = .05$, we will reject H_0 when $z > 1.645$. See Figure 8.18.

The pooled estimate of p required for $\sigma_{(\hat{p}_1 - \hat{p}_2)}$ is

$$\hat{p} = \frac{y_1 + y_2}{n_1 + n_2} = \frac{52 + 23}{1{,}000 + 1{,}000} = .0375.$$

The test statistic is

$$z = \frac{\hat{p}_1 - \hat{p}_2}{\sqrt{\hat{p}\hat{q}\left(\dfrac{1}{n_1} + \dfrac{1}{n_2}\right)}} = \frac{.052 - .023}{\sqrt{(.0375)(.9625)\left(\dfrac{1}{1{,}000} + \dfrac{1}{1{,}000}\right)}}$$

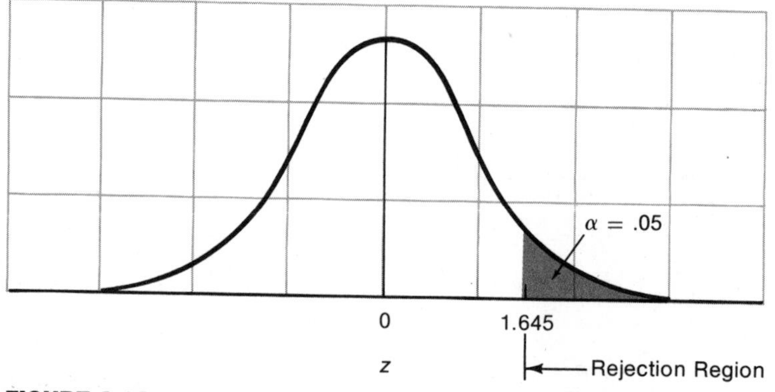

FIGURE 8.18
Location of the rejection region in Example 8.13

or

$$z = 3.41.$$

Since the computed value of z falls in the rejection region, we reject the hypothesis that $p_1 = p_2$ and conclude that the data present sufficient evidence to indicate that the percentage of men entering the hospital because of heart disease is higher than that of women. (*Note:* This does not imply that the *incidence* of heart disease is higher in men. Perhaps fewer women enter the hospital when afflicted with the disease!)

EXERCISES

8.33 A drug manufacturer claimed that the mean potency of one of its antibiotics was 80 percent. A random sample of $n = 100$ capsules was tested and produced a sample mean of $\bar{y} = 79.7$ percent, with a standard deviation $s = .8$ percent. Do the data present sufficient evidence to refute the manufacturer's claim? Let $\alpha = .05$.
(a) State the null hypothesis to be tested.
(b) State the alternative hypothesis.
(c) Conduct a statistical test of the null hypothesis and state your conclusion.

8.34 Refer to Exercise 8.9 and the collection of water samples to estimate the mean acidity (in pH) of rainfalls in the northeastern United States. As noted, the pH for pure rain falling through clean air is approximately 5.7. The sample of $n = 40$ rainfalls produced pH readings with $\bar{y} = 3.7$ and $s = .5$. Do the data provide sufficient evidence to indicate that the mean pH for rainfalls is more acidic (H_a: $p < 5.7$ pH) than pure rainwater? Test using $\alpha = .05$. Note that this inference is appropriate only for the area in which the rainwater specimens were collected.

8.35 An editorial appearing in the *Journal of the American Medical Association* (June 1977) comments on the findings of a study group that recently toured mainland China. The group reported that Chinese doctors claimed 90 percent overall effectiveness of acupuncture as an anesthesia for surgery. Of 48 patients the Americans personally observed, only 73 percent experienced satisfactory relief with acupuncture. Are the American's observations inconsistent with the claims of the Chinese physicians?

8.36 According to the *Washington Post* (December 7, 1976), federal scientists in Oklahoma "have been able to forecast tornadoes in Oklahoma with an accuracy that is better than 90 percent." Suppose that the scientists tracked 274 storms that did not result in tornadoes and 37 that did. Of these 311 storms, they correctly forecast the type of storm in 273 cases. Would this disagree with the claimed forecasting ability of the federal scientists?

8.37 A manufacturer claimed that at least 20 percent of the public preferred his product. A sample of 100 persons is taken to check his claim. With an $\alpha = .05$, how small would the sample percentage need to be before the claim could be rightfully refuted? (Note that this would require a one-tailed test of an hypothesis.)

8.38 To test the effects of a new fertilizer on wheat production, a tract of land was divided into 60 squares of equal areas, all portions having similar qualities in respect to soil, exposure to sunlight, etc. The new fertilizer was applied to 30 squares and the old fertilizer was applied to the remaining squares. The mean number of bushels of wheat harvested per square of land using the new fertilizer was 18 bushels, with a standard deviation of .6 bushel. The corresponding mean and standard deviation for the squares using the old fertilizer were 17 and .5 bushels, respectively. Using a significance level of .05, test the hypothesis that there is no difference between the fertilizers against the alternative that the new fertilizer is better than the old.

8.39 In a study to assess various effects of using a female model in automobile advertising, each of 100 male subjects was shown photographs of two automobiles matched for price, color, and size but of different makes. One of the automobiles was shown with a female model to 50 of the subjects (group *A*) and both automobiles were shown without the model to the other 50 subjects (group *B*). In group *A*, the automobile shown with the model was judged as more expensive by 37 subjects, while in group *B* the same automobile was judged as the more expensive by 23 subjects. Do these results indicate that using a female model influences the perceived expensiveness of an automobile? Use a one-tailed test with $\alpha = .05$.

8.40 A surprising event occurred during the 1976 presidential race. The Gallup and the Harris polls published conflicting results. The Gallup Poll of 1,078 registered voters showed Ford beating Humphrey by 51 to 39 percent. The Harris Poll (taken slightly later) of 950 people gave Humphrey a 52 to 41 percent margin over Ford. If you were to compare the percentages in the two samples favoring Ford and if the two samples were randomly selected from the same population, what is the probability that the sample percentages favoring Ford would differ by as much as 10 percent (51 versus 41 percent)? Do the data provide sufficient evidence to indicate that the two samples were selected from different populations? Or could the disparity in the conclusions of the pollsters be due to the methods of selecting the samples?

8.41 To curb the spread of the swine flu, President Ford ordered a massive vaccination program in 1976. Shortly thereafter, numerous cases of the paralyzing Guillain-Barre syndrome were reported. In December 1976, fearful that the vaccine was causing the syndrome, the vaccination program was suspended to allow for an analysis of the data. Of 220 million Americans, approximately 50 million received swine flu shots. Of 383 persons who contracted the Guillain-Barre syndrome, 202 received the swine flu shots. Do you think these data suggest a dependency between persons who received the swine flu shot and the contraction of the Guillain-Barre syndrome? Test the null hypothesis that the probability of contracting the Guillain-Barre syndrome is the same for the vaccinated portion of the general public as for the unvaccinated portion. Test using $\alpha = .05$.

8.13

SOME COMMENTS ON THE THEORY OF TESTS OF HYPOTHESES

The theory of a statistical test of an hypothesis outlined in Section 8.11 is indeed a very clear-cut procedure enabling the experimenter to either reject or accept the null hypothesis with measured risks, α and β. Unfortunately, the theoretical framework does not suffice for all practical situations.

The crux of the theory requires that we be able to specify a meaningful alternative hypothesis that permits the calculation of the probability of a type II error, β, for all alternative values of the parameter(s). This indeed can be done for many statistical tests, including the test discussed in Section 8.11, although the calculation of β for various alternatives and sample sizes may, in some cases, be a formidable task. On the other hand, it is extremely difficult, in some test situations, to clearly specify alternatives to H_0 that have *practical* significance. This may occur when we wish to test an hypothesis concerning the values of a *set* of parameters, a situation we shall encounter in Chapter 11 in analyzing enumerative data.

The obstacle that we mention does not invalidate the use of statistical tests. Rather, it urges caution in drawing conclusions when insufficient evidence is available to reject the null hypothesis. It, together with the difficulty encountered in the calculation and tabulation for β for other than the simplest statistical tests, justifies skirting this issue in an introductory text. Hence, we shall agree to adopt the procedure described in Example 8.10 when tabulated values of β (the operating characteristic curve) are unavailable for the test. When the test statistic falls in the acceptance region, we will "*not reject*" rather than "*accept*" the null hypothesis. Further conclusions may be made by calculating an interval estimate for the parameter or by consulting one of the several published statistical handbooks for tabulated values of β. We shall not be too surprised to learn that these tabulations are inaccessible, if not completely unavailable, for some of the more complicated statistical tests.

The probability of making a type I error, α, is often called the significance level of the statistical test, a term that originated in the following way. The probability of the observed value of the test statistic, or some value even more contradictory to the null hypothesis, measures, in a sense, the weight of evidence favoring rejection. Thus, some experimenters report test results as being *significant* (we would reject) at the 5 percent significance level but not at the 1 percent level. This means that we would reject H_0 if α were .05 but not if it were .01. This line of thought does not conflict with the procedure of choosing the test in advance of the data collection. Rather, it presents a convenient way of publishing the statistical results of a scientific investigation, permitting the reader to choose his own α and β as he pleases.

Finally, we might comment on the choice between a one- or two-tailed test for a given situation. We emphasize that this choice is dictated by the practical aspects of the problem and will depend upon the alternative value of the parameter, say θ, which the experimenter is trying to detect. Thus, if we were to sustain a large financial loss if θ were greater than θ_0 but not if it were less, we would concentrate our attention on the detection of values of θ greater than θ_0. Hence, we would reject in the upper tail of the distribution for the test statistics previously discussed. On the other hand, if we are equally interested in detecting values of θ that are either less than or greater than θ_0, we would employ a two-tailed test.

8.14

SUMMARY

The material presented in Chapter 8 was directed toward two objectives. First, we wanted to discuss the various methods of inference along with procedures for evaluating their goodness. Second, we wished to present a number of estimation procedures and statistical tests of hypotheses which, owing to the Central Limit Theorem, make use of the results of Chapter 7. The resulting techniques possess practical value and, at the same time, illustrate the principles involved in statistical inference.

Inferences concerning the parameter(s) of a population may be made by estimating or testing hypotheses concerning their value. A parameter may be estimated using either a point or an interval estimator with the confidence coefficient and width of the interval measuring the goodness of the procedure.

A statistical test of an hypothesis or theory concerning the population parameter(s), ideally, will result in its rejection or acceptance. Practically, we may be forced to view this decision in terms of rejection or nonrejection. The probabilities of making the two possible incorrect decisions, resulting in the type I and type II errors, measure the goodness of the decision procedure. While a test of an hypothesis may be best suited for some physical situations (for example, lot acceptance sampling), it would seem that estimation would be the eventual goal of many experimental investigations and hence would be desirable if one were permitted an option in his choice of a method of inference.

All the confidence intervals and statistical tests described in this chapter were based upon the Central Limit Theorem and hence apply to large samples. When n is large, each of the respective estimators and test statistics will possess, for all practical purposes, a normal distribution in repeated sampling. This result, along with the properties of the normal distribution studies in Chapter 7, permits the construction of the confidence intervals and the calculation of α and β for the statistical tests.

REFERENCES

Barr, A. J., J. H. Goodnight, J. P. Sall, and J. T. Helwig, *A User's Guide to SAS 76*. Raleigh, N.C.: SAS Institute, Inc., 1976.

Brown, M. B., ed., *Biomedical Computer Programs*. Berkeley, Calif.: University of California, 1977.

Dixon, W. J., and F. J. Massey, Jr., *Introduction to Statistical Analysis,* 3rd ed. New York: McGraw-Hill Book Company, 1969.

Freund, J. E., *Mathematical Statistics,* 2nd ed. Englewood Cliffs, N.J.: Prentice-Hall, Inc., 1971.

Hogg, R. V., and A. T. Craig, *Introduction to Mathematical Statistics,* 3rd ed. New York: Macmillan Publishing Co., Inc., 1970.

Mendenhall, W., L. Ott, and R. Scheaffer, *Elementary Survey Sampling,* 2nd ed. North Scituate, Mass.: Duxbury Press, 1979.

Mendenhall, W., and R. Scheaffer, *Mathematical Statistics with Applications*. North Scituate, Mass.: Duxbury Press, 1973.

Nie, N., C. H. Hull, J. G. Jenkins, K. Steinbrenner, and D. H. Bent, *Statistical Package for the Social Sciences,* 2nd ed. New York: McGraw-Hill Book Company, 1975.

Walpole, R. E., *Introduction to Statistics,* 2nd ed. New York: Macmillan Publishing Co., Inc., 1974.

Tips on Problem Solving: In solving the exercises in this chapter, you will be required to answer a practical question of interest to a businessperson, a professional person, a scientist, or a layperson. To find the answer to the question, you will need to make an inference about one or more population parameters. Consequently, the first step in solving a problem is in deciding on the objective of the exercise. What parameters do you wish to make an inference about? Answering the following two questions will help you reach a decision.

1. What *type of data* is involved? This will help you decide the type of parameters about which you will wish to make inferences, binomial proportions (p's) or population means (μ's): Check to see if the data are of the yes-no (two-possibility) variety. If they are, the data are probably binomial and you will be interested in proportions. If not, the data probably represent measurements on one or more quantitative random variables and you will be interested in means. To aid you, look for key words such as "proportions," "fractions," etc., which indicate binomial data. Binomial data often (but not exclusively) evolve from a "sample survey."

2. Do I wish to make an inference about a *single parameter,* p or μ, or about the *difference between two parameters,* $(p_1 - p_2)$ or $(\mu_1 - \mu_2)$? This is an easy question to answer. Check on the number of samples involved. One sample implies an inference about a single parameter; two samples imply a comparison of two parameters. The answers to questions 1 and 2 identify the parameter.

After identifying the parameter(s) involved in the exercise, you must identify the exercise objective. It will be one of these three:

1. Choosing the sample size required to estimate a parameter with a specified bound on the error of estimation.
2. Estimating a parameter (or difference between two parameters).
3. Making a decision about one or more parameters (a test of an hypothesis).

The objective will be very clear if it is 1 because the question will ask for or direct you to find the "sample size." Objective 2 will be clear because the exercise will specifically direct you to estimate a parameter (or the difference between two parameters). If you are required to make a decision (other than a choice of sample size), the objective is most likely a test of an hypothesis.

To summarize these tips, your thought process should follow the decision tree shown on page 266.

SUPPLEMENTARY EXERCISES

8.42 State the Central Limit Theorem. Of what value is the Central Limit Theorem in statistical inference?

8.43 An experiment was conducted to estimate the effect of smoking on the blood pressure of a group of 25 college-age cigarette smokers. The difference for each participant was obtained by taking the difference in the blood-pressure readings at the time of graduation and again 5 years later. The sample mean increase, measured in millimeters of mercury, was $\bar{y} = 9.7$. The sample standard deviation was $s = 5.8$.

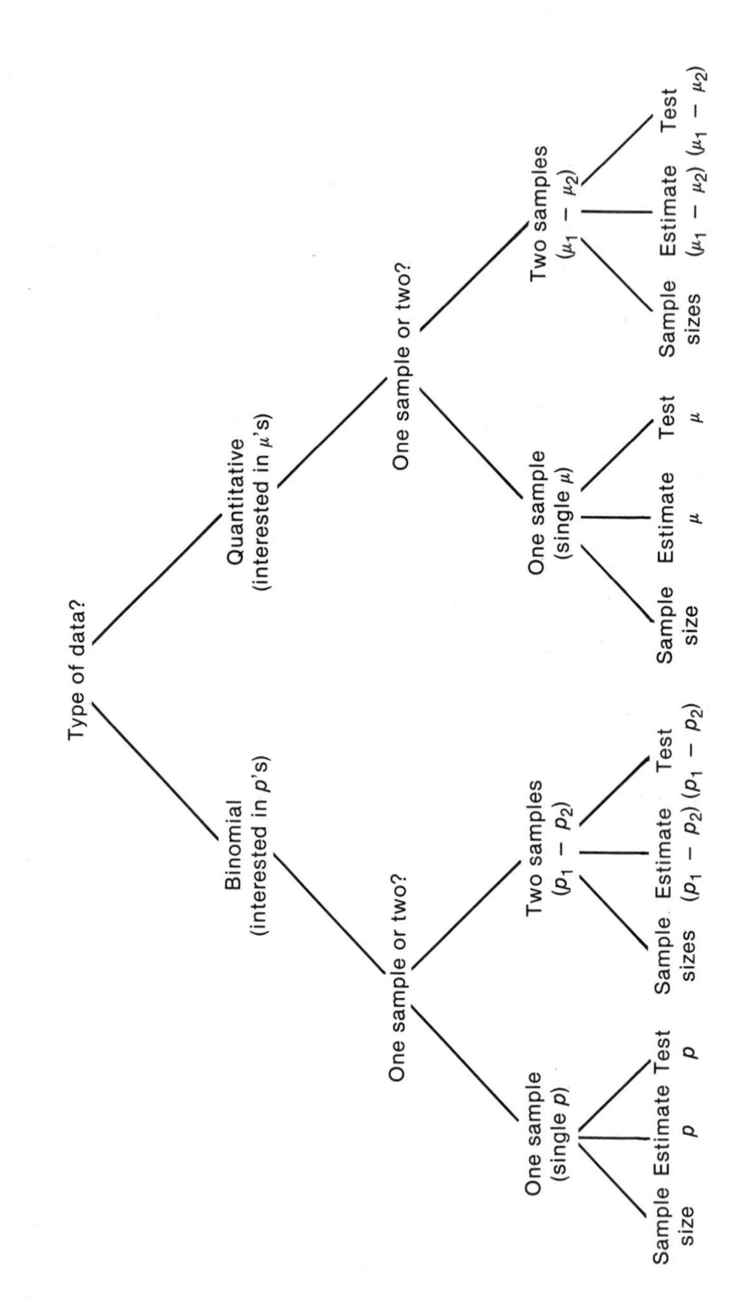

Type of data?
Binomial (interested in p's)
Quantitative (interested in μ's)
One sample or two?
One sample or two?
One sample (single p)
Two samples ($p_1 - p_2$)
One sample (single μ)
Two samples ($\mu_1 - \mu_2$)
Sample size
Estimate p
Test p
Sample sizes
Estimate ($p_1 - p_2$)
Test ($p_1 - p_2$)
Sample size
Estimate μ
Test μ
Sample sizes
Estimate ($\mu_1 - \mu_2$)
Test ($\mu_1 - \mu_2$)

Estimate the mean increase in blood pressure that one would expect for cigarette smokers over the time span indicated by the experiment. Place a bound on the error of estimation. Describe the population associated with the mean that you have estimated.

8.44 Using a confidence coefficient equal to .90, place a confidence interval on the mean increase in blood pressure for Exercise 8.43.

8.45 A random sample of 400 radio tubes was tested and 40 tubes were found to be defective. With confidence coefficient equal to .90, estimate the interval within which the true fraction defective lies.

8.46 An experiment was conducted to compare the depth of penetration for two different hydraulic mining nozzles. The rock structure and the length of drilling time were the same for both nozzles. With nozzle A the average penetration was 10.8 inches with a standard deviation of 1.2 inches for a sample of 50 holes. With nozzle B the average and standard deviation of the penetration measurements were 9.1 and 1.6 inches, respectively, for a sample of 80 holes. Estimate the difference in mean penetration and place a bound on the error of estimation.

8.47 Construct a confidence interval for the difference between the population means of Exercise 8.46 using a confidence coefficient equal to .90.

8.48 The mean of a sample of 43 measurements is 26.3 and the standard deviation is 1.9. Find the 98 percent confidence limits for the mean of the population.

8.49 A chemist has prepared a product designed to kill 60 percent of a particular type of insect. How large a sample should be used if he desires to be 95 percent confident that he is within .02 of the true fraction of insects killed?

8.50 Past experience shows that the standard deviation of the yearly income of textile workers in a certain state was \$500. How large a sample of textile workers would one need to take if one wished to estimate the population mean to within \$50.00, with a probability of .95 of being correct? Given that the mean of the sample in this problem is \$6,800, determine 95 percent confidence limits for the population mean.

8.51 In a poll taken among college students, 300 of 500 fraternity men favored a certain proposition, whereas 64 of 100 nonfraternity men favored it. Estimate the difference in the fractions favoring the proposition and place a bound upon the error of estimation.

8.52 Refer to Exercise 8.51. How many fraternity and nonfraternity students must be included in a poll if we wish to estimate the difference in the fractions correct to within .05? Assume that the groups will be of equal size and that $p = .6$ will suffice as an approximation to both fractions.

8.53 From each of two normal populations with identical means and with standard deviations of 6.40 and 7.20, independent random samples of 64 variants are drawn. Find the probability that the difference between the means of the samples exceeds .60 in absolute value.

8.54 If it is assumed that the heights of men are normally distributed with a standard deviation of 2.5 inches, how large a sample should be taken in order to be fairly sure (probability .95) that the sample mean does not differ from the true mean (population mean) by more than .50 in absolute value?

8.55 It is desired to use the sample mean, $\bar{y}$, to estimate the mean of a normally distributed population with an error of less than .5 with probability .9. If it is known that the variance of the population is equal to 4, how large should the sample be to achieve the accuracy stated above?

8.56 Define α and β for a statistical test of an hypothesis.

8.57 What is the level of significance of a statistical test of an hypothesis?

8.58 The daily wages in a particular industry are normally distributed with a mean of \$23.20

and a standard deviation of \$4.50. If a company in this industry employing 40 workers pays these workers on the average \$21.20, can this company be accused of paying inferior wages at the 1 percent level of significance?

8.59 Two sets of 50 elementary school children were taught to read by two different methods. At the conclusion of the instructional period, a reading test gave the results $\bar{y}_1 = 74, \bar{y}_2 = 71, s_1 = 9, s_2 = 10$. Test to see if there is evidence of a real difference between the two population means. (Use $\alpha = .10$.)

8.60 A manufacturer of automatic washers provides a particular model in one of three colors, *A, B,* or *C*. Of the first 1,000 washers sold, it is noticed that 400 of the washers were of color *A*. Would you conclude that more than 1/3 of all customers have a preference for color *A*? Justify your answer.

8.61 Refer to Exercise 8.37. Sixteen people in the sample of 100 consumers expressed a preference for the manufacturer's product. Does this present sufficient evidence to reject the manufacturer's claim? Test at the 10 percent level of significance.

8.62 What conditions must be met in order that the *z* test may be used to test an hypothesis concerning a population mean, μ?

8.63 A manufacturer claimed that at least 95 percent of the equipment that he supplied to a factory conformed to specification. An examination of a sample of 700 pieces of equipment revealed that 53 were faulty. Test his claim at a significance level of .05.

8.64 Refer to Exercise 8.43. Test the hypothesis that the mean increase in blood pressure is 5 mm of mercury against the alternative that the mean is larger. Use $\alpha = .05$.

8.65 Refer to Exercise 8.46. Do the data present sufficient evidence to indicate a difference in the mean depth of penetration obtained using the two nozzles?

8.66 Random samples of 200 bolts manufactured by machine *A* and 200 bolts manufactured by machine *B* showed 16 and 8 defective bolts, respectively. Do these data present sufficient evidence to suggest a difference in the performance of the machines? Use a .05 level of significance.

8.67 In measuring reaction time a psychologist estimates that the standard deviation is .05 second. How large a sample of measurements must he take in order to be 90 percent confident that the error of his estimate will not exceed .01 second?

8.68 The mean lifetime of a sample of 100 fluorescent bulbs produced by a company is computed to be 1,570 hours, with a standard deviation of 120 hours. If μ is the mean lifetime of all the bulbs produced by the company, test the hypothesis $\mu = 1{,}600$ hours against the alternative hypothesis $\mu < 1{,}600$. Use a level of significance of .05.

8.69 Suppose that the true fraction *p* in favor of the death penalty is the same for Democrats as it is for Republicans. Independent random samples are selected, one consisting of 800 Republicans and the other of 800 Democrats. Find the probability that the sample fraction of Republicans favoring the death penalty exceeds that of the Democrats by more than .03 if $p = .5$. What is this probability if $p = .1$?

8.70 A social scientist believes that the fraction p_1 of Republicans in favor of the death penalty is greater than the fraction p_2 of Democrats in favor of the death penalty. She acquired independent random samples of 200 Republicans and 200 Democrats, respectively, and found 46 Republicans and 34 Democrats favoring the death penalty. Does this evidence provide statistical support at the .05 level of significance for the social scientist's belief?

8.71 Refer to Exercise 8.70. Some thought should have been given to designing a test for which β is tolerably low when p_1 exceeds p_2 by an important amount. For example, find a common sample size *n* for a test with $\alpha = .05$ and $\beta \leq .20$ when in fact p_1 exceeds p_2 by .1. [*Hint:* The maximum value of $p(1 - p)$ is .25.]

8.72 A survey is conducted to determine what difference may exist between the fractions of married and single persons in the 20–30 age group who smoke. A sample of 200 persons from each group is polled and 64 married persons and 80 single persons are found to smoke. Do the data provide sufficient evidence to indicate a difference in the fractions of smokers for the two populations? (Use $\alpha = .10$.)

8.73 In an experiment 320 out of 400 seeds germinate. Find a confidence interval for the true fraction of germinating seeds; use a confidence coefficient of .98.

8.74 Presently 20 percent of potential customers buy a certain brand of soap, say brand A. To increase sales, an extensive advertising campaign will be conducted. At the end of the campaign, a sample of 400 potential customers will be interviewed to determine if the campaign was successful.

(a) State H_0 and H_a in terms of p, the probability that a customer prefers soap brand A.

(b) It is decided to conclude that the advertising campaign was a success if at least 92 of the 400 customers interviewed prefer brand A. Find α. (Use the normal approximation to the binomial distribution to evaluate the desired probability.)

8.75 In comparing the mean weight loss for two diets the following sample data were obtained:

	Diet I	Diet II
Sample size, n	40	40
Sample mean, $\bar{y}$	10 lb	8 lb
Sample variance, s^2	4.3	5.7

Do the data provide sufficient evidence to indicate that diet I produces a greater mean weight loss than diet II? Use $\alpha = .05$.

8.76 To estimate the proportion of unemployed workers in Panama, an economist selected at random 400 persons from the working class. Of these, 25 were unemployed.

(a) Estimate the true proportion of unemployed workers and place bounds on the error of estimation.

(b) How many persons must be sampled to reduce the bound on error to .02?

8.77 A random sample of 36 cigarettes of a certain brand were tested for nicotine content. The sample gave a mean of 22 and a standard deviation of 4 milligrams. Find a 98 percent confidence interval for μ, the true mean nicotine content of the brand.

8.78 How large a sample is necessary to estimate a binomial parameter p to within .01 of its true value with probability .95? Assume the value of p is approximately .5.

8.79 Sixty of 87 housewives prefer detergent A. If the 87 housewives represent a random sample from the population of all potential purchasers, estimate the fraction of total housewives favoring detergent A. Use a 90 percent confidence interval.

8.80 A dean of men wishes to estimate the average cost of the freshman year at a particular college correct to within \$200.00 with a probability of .95. If a random sample of freshmen is to be selected and requested to keep financial data, how many must be included in the sample? Assume that the dean knows only that the range of expenditure will vary from approximately \$2,200 to \$4,000.

8.81 In the past, a chemical plant has produced an average of 1,100 pounds of chemical per day. The records for the past year, based on 260 operating days, show the following:

$$\bar{y} = 1{,}060 \text{ lb/day},$$
$$s = 340 \text{ lb/day}.$$

It is desired to test whether or not the average daily production has dropped significantly over the past year.

(a) Give the appropriate null and alternative hypotheses.
(b) If z is used as a test statistic, determine the rejection region corresponding to a level of significance $\alpha = .05$.
(c) Do the data provide sufficient evidence to indicate a drop in average daily production?

8.82 It is required to estimate the fraction of automobiles on Florida highways that have defective brakes.
(a) A random sample of $n = 100$ automobiles is selected for inspection. Of these 20 are found to have defective brakes. Construct a 95 percent confidence interval for the fraction of automobiles on Florida highways that have defective brakes.
(b) Suppose that a more accurate estimate than the one found in (a) is wanted. How large a sample would be necessary to reduce the bound on error to .04?

8.83 An experimenter wishes to estimate the difference, $(p_1 - p_2)$, for two binomial populations to within .1 of the true difference. Both p_1 and p_2 are expected to assume values between .3 and .7. If samples of equal size are to be selected from the populations to estimate $(p_1 - p_2)$, how large should they be?

8.84 A test of the breaking strengths of two different types of cables was conducted using samples of $n_1 = n_2 = 100$ pieces of each type of cable.

Cable I	Cable II
$\bar{y}_1 = 1{,}925$	$\bar{y}_2 = 1{,}905$
$s_1 = 40$	$s_2 = 30$

Do the data provide sufficient evidence to indicate a difference between the mean breaking strengths of the two cables? Use $\alpha = .10$.

8.85 An experimenter fed different rations to two groups of 100 chicks each. Assume that all factors other than rations are the same for both groups. A record of mortality for each group is as follows:

Chicks	Ration *A*	Ration *B*
n	100	100
Number died	13	6

Construct a 98 percent confidence interval for the true difference in mortality rates for the two rations.

8.86 The braking ability was compared for two types of 1978 automobiles. Random samples of 64 automobiles were tested for each type. The recorded measurement was the distance (in feet) required to stop when the brakes were applied at 40 miles per hour. The computed sample means and variances were:

$$\bar{y}_1 = 118, \quad \bar{y}_2 = 109;$$
$$s_1^2 = 102, \quad s_2^2 = 87.$$

Do the data provide sufficient evidence to indicate a difference in the mean stopping distance for the two types of automobiles?

8.87 It is desired to estimate the mean hourly yield for a process manufacturing an antibiotic. The process is observed for 100 hourly periods chosen at random, with the following results:

$$\bar{y} = 34 \text{ oz/hr}; \quad s = 3.$$

Estimate the mean hourly yield for the process using a 95 percent confidence interval.

8.88 It is assumed that heights of men at the University of Florida have a standard deviation of 4 inches. If one wants to estimate the average height of all men at the university to within 1 inch with a .95 confidence coefficient, how many observations should he take?

8.89 A quality-control engineer wants to estimate the fraction of defectives in a large lot of light bulbs. From previous experience, he feels that the actual fraction of defectives should be somewhere around .2. How large a sample should he take if he wants to estimate the true fraction to within .01 using a 95 percent confidence interval?

8.90 A fruit grower wishes to test a new spray which a manufacturer claims will *reduce* the loss due to damage by a certain insect. To test the claim, the grower sprays 200 trees with the new spray and 200 trees with the standard spray. The following data were recorded:

	New Spray	Standard Spray
Mean yield per tree, $\bar{y}$ (lb)	240	227
Variance, s^2	980	820

(a) Do the data provide sufficient evidence to conclude that the new spray is better than the old? (Use $\alpha = .05$.)

(b) Set up a 95 percent confidence interval for the difference in mean yields for the two sprays.

8.91 It is thought that two-tenths of the individuals in a large population have a certain genetic defect. To test whether this fraction of "defectives" is larger than .2, a sample of 400 individuals is taken and y, the number of individuals having this defect, is recorded. Let p be the probability of observing the genetic defect.

(a) State H_0 and H_a in terms of p.

(b) α is the probability of a type I error. Define a type I error.

(c) It is decided to reject H_0 if 95 or more individuals in the sample exhibit the genetic defect. Use the normal approximation to binomial probabilities with the correction for continuity to find α, the probability of a type I error.

(d) If 101 individuals in the sample exhibited the defect, what would your conclusion be?

8.92 Samples of 400 radio tubes were selected from each of two production lines, *A* and *B*. The number of defectives in the sample were:

Line	Number of Defectives
A	40
B	80

Estimate the difference in the actual fractions of defectives for the two lines with a confidence coefficient of .90.

8.93 The mean daily yield was compared for two chemical processes. The yields of the two processes were each observed for 72 days with the following results:

$$\bar{y}_1 = 834, \qquad \bar{y}_2 = 808;$$
$$s_1^2 = 346, \qquad s_2^2 = 302.$$

Do the data provide sufficient evidence to indicate a difference in mean yield for the two processes?

8.94 A sample of $n = 200$ items was taken from each of two binomial populations where $p_1 = p_2 = .5$. What is the probability that $(\hat{p}_1 - \hat{p}_2)$ is greater than .1 in absolute value?

8.95 Refer to Exercise 8.37 and calculate the value of β for an alternative, $p_a = .15$.

8.96 Refer to Exercise 8.64. What is the probability of not rejecting the null hypothesis, $\mu_0 = 5$ mm of mercury, if the mean blood pressure is $\mu = 7$ mm? Assume that σ is approximately equal to $s = 5.8$.

8.97 Refer to Exercise 8.45. Suppose that 10 samples of $n = 400$ radio tubes were tested and a confidence interval constructed for p for each of the 10 samples. What is the probability that exactly one of the intervals will not enclose the true value of p? At least one?

8.98 Refer to Example 8.10. Use the procedure described in Example 8.10 to calculate β for several alternative values of μ. (For example, $\mu = 873$, 875, and 877.) Use the three computed values of β along with the value computed in Example 8.10 to construct an operating characteristic curve for the statistical test. The resulting graph will be similar to that shown in Figure 8.19.

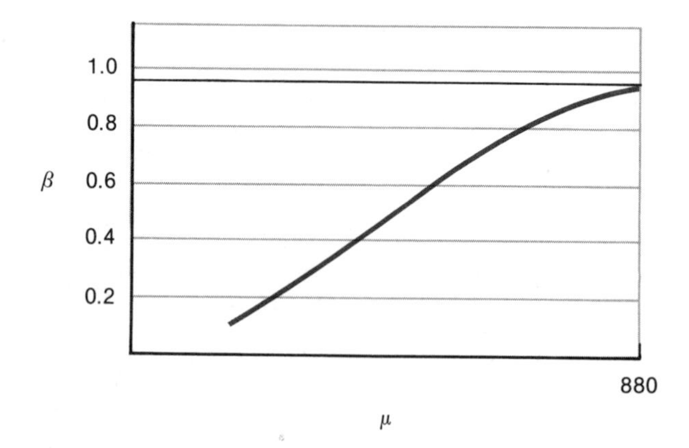

FIGURE 8.19

8.99 Repeat the procedure described in the preceding exercise for a sample size $n = 25$ (as opposed to $n = 50$ used in Exercise 8.98) and compare the two operating characteristic curves.

8.100 A biologist hypothesizes that high concentrations of actinomysin D inhibit RNA synthesis in cells and hence the production of proteins as well. An experiment conducted to test this theory compared the RNA synthesis in cells treated with two concentrations of actinomysin D, .6 and .7 microgram per milliliter, respectively. Cells treated with the lower concentration (.6) of actinomysin D showed that 55 of 70 developed normally, whereas only 23 of 70 appeared to develop normally for the higher concentration (.7). Do these data provide sufficient evidence to indicate a difference in the rate of normal RNA synthesis for cells exposed to the two different concentrations of actinomysin D?

8.101 During the 1976–1977 academic year, 253 University of Florida graduates applied for admission to colleges of medicine and 105 (42 percent) were admitted. This compares with a national admission rate of 38 percent. Do these data provide sufficient evidence to indicate that the admission rate of University of Florida students differs from the national average? Test using $\alpha = .05$. Interpret your results.

8.102 Refer to Exercise 8.101. Find a 95 percent confidence interval for the University of Florida acceptance rate. Interpret the confidence interval.

8.103 The results of a Gallup Poll (*Orlando Sentinel Star,* December 2, 1976) showed that among 8 physical afflictions, including cancer, heart disease, and blindness, 58 percent of those polled named cancer as the most feared affliction. If this estimate was based on a random sample of 2,000 adults, how accurate is the estimate? Find a 90 percent confidence interval for the proportion of American adults who rate cancer as the most feared affliction.

8.104 The 1976–1977 national statistics on admissions to medical colleges, classified according to undergraduate major programs, are shown in the accompanying table.

Major	Number of Admissions	Percentage of Total of Major Applications Admitted
biology	15,488	36
chemistry	4,813	45
zoology	3,025	35
premed	2,243	38
biochemistry	1,387	50
interdisciplinary	172	52

(a) Suppose that prior to viewing this data, you decided to compare the chances of admission between majors in biology and chemistry. Do the data provide sufficient evidence to indicate a difference in the probabilities of admission for these two majors? Explain.

(b) Find a 95 percent confidence interval for the difference in the probabilities of admission between biology majors and biochemistry majors. Interpret the confidence interval.

(*Note:* In Chapter 11 we shall address the general question of whether the data provide sufficient evidence to indicate a difference in the probabilities of admission among the six major program categories.)

8.105 A hospital wished to estimate the average number of days required for treatment of patients between the ages of 25 and 34. A random sample of 500 hospital patients between these ages produced a mean and standard deviation equal to 5.4 and 3.1 days, respectively. Estimate the mean length of stay for the population of patients from which the sample was drawn. Place bounds on the error of estimation.

8.106 Because of our infatuation with expensive automobiles, the Bureau of Labor Statistics claims that the American family has reached the point of spending proportionately more money on transportation than on food. In fact, it notes that the mean number of cars in 1972 increased to 1.3 over a value of 1.0 a decade earlier (*Orlando Sentinel Star,* May 15, 1977). In deciding where to place its emphasis in advertising, the market research department for a major automobile manufacturer wished to compare the mean number of automobiles per family in two regions of the United States. Suppose that a preliminary study of the number of cars per family for $n = 200$ families from each of the two regions gave the means and variances for the two samples as shown in the table.

	Area 1	Area 2
Sample size	200	200
Sample mean	1.30	1.37
Sample variance	.53	.64

(a) Note that a small increase in the mean number of automobiles per family can represent a very large number of automobiles for a region. Do the data provide sufficient evidence to indicate a difference in the mean number of automobiles per family for the two regions?

(b) Picture the data associated with either of the two populations. What values will y assume? Imagine the probability distributions for these two populations. Will their nature violate the conditions necessary in order that the test, part (a) be valid? Explain.

(c) Find a 95 percent confidence interval for the mean number of automobiles per family for region 2. Interpret the interval.

(d) Find a 90 percent confidence interval for the difference in the mean number of automobiles for the two regions. Interpret the interval.

Ψ 8.107 "Schizophrenia, a disease marked by withdrawal from reality, confusion, flat and often inappropriate emotions and, often, hallucinations," may be linked to a long-standing viral disease (*Orlando Sentinel Star,* October 21, 1977). Researchers have found that a higher proportion of schizophrenics are born during months in which viral infections are prevalent. Particularly, they found that of a total of 53,584 schizophrenics born between 1920 and 1955, 6.9 percent more were born in the peak months March and April than in the other months. Suppose that we interpret this statement to mean that during any of the other two-month periods, the observed percentage of schizophrenic births was 16.48 and that during the March–April period, the percentage was 6.9 percent larger, or 17.61 percent. Is the percentage of schizophrenic births observed by the researchers during the March–April period statistically significant? That is, do the data provide sufficient evidence to indicate that the proportion of schizophrenics born during March–April is higher than might be expected if schizophrenia and birth date are unrelated? Test using $\alpha = .05$. (*Hint:* If the proportion p of schizophrenics born during any two-month period is independent of the months included in the pair, then $p = 1/6$.)

EXPERIENCES WITH REAL DATA

Conduct a sample survey to determine the attitude of your student body on a major political, social, or campus issue. Use the method of Section 12.3 to select a large random sample of n students (say $n = 400$ or more) from the student directory. Since each of the n students in the sample must be contacted by telephone to determine his or her attitude concerning the question, this chore should be divided so that each student (or a small group of students) would be responsible for contacting a fixed number, say 25 of the total.

When the data have been collected, pool the n student responses and obtain an estimate of proportion of students in the student body who favor the issue. Place a bound on your error of estimation.

To observe sampling variation, let each student (or team) use the $n = 25$ students whom he or she was to contact to estimate the proportion in favor of the issue. Collect the estimates from each team and construct a histogram of the estimates. Locate the estimate based on all n students on the graph. Notice how the estimates based on the samples containing 25 students tend to vary and that the large-sample estimate (based on all n students) falls near the center of the histogram.

The calculations for this experiment can easily be accomplished on an electronic desk calculator. However, for other analyses considered in this chapter, you might wish to use packaged programs and an electronic computer to perform the calculations. Useful packaged programs are available in the Biomed (Biomedical Programs), the SAS (Statistical Analysis Systems), and the SPSS (Statistical Package for the Social Sciences) program libraries (see the references). Others are also available.

CHAPTER OBJECTIVES

GENERAL OBJECTIVE The basic concepts of statistical estimation were presented in Chapter 8 along with a summary of the concepts involved in a statistical test of an hypothesis. Large-sample estimation and test procedures for population means and proportions were used to illustrate concepts as well as to give you some useful tools for solving some practical problems. Because all these techniques rely on the Central Limit Theorem to justify the normality of the estimators and test statistics, they apply only when the sample sizes are large. Consequently, the objective of Chapter 9 is to supplement the results of Chapter 8 by presenting small-sample statistical test and estimation procedures for population means and variances. These techniques differ substantially from those of Chapter 8 because they require that the relative frequency distributions of the sampled populations be approximately normal.

SPECIFIC OBJECTIVES

1. To present the Student's t statistic and identify the relationship between it and the z statistic of Chapter 8. *Section 9.2*
2. To show the relationship between the normal distribution and a Student's t distribution and to explain how to use the table values for the Student's t statistic. *Section 9.2*
3. To show how the t statistic is used to construct small-sample confidence intervals and tests of hypotheses for population means and the difference between two population means. *Sections 9.3, 9.4*
4. To give our first example of an experiment designed to reduce the cost of experimentation, the paired-difference experiment. We shall show how the techniques of Section 9.2 can be used to analyze these data. *Section 9.5*
5. To show how to estimate and test hypotheses concerning population variances and to give examples and exercises that indicate the applications of these techniques to the solution of certain practical problems. *Sections 9.6, 9.7*
6. To emphasize the assumptions you must make in order for the estimation and test procedures described in this chapter to be valid. *Section 9.8*

9

INFERENCE FROM SMALL SAMPLES

9.1

INTRODUCTION

Large-sample methods for making inferences concerning population means and the difference between two means were discussed with examples in Chapter 8. Frequently cost, available time, and other factors limit the size of the sample that may be acquired. When this occurs, the large-sample procedures of Chapter 8 are inadequate and other tests and estimation procedures must be employed. In this chapter we shall study several small-sample inferential procedures which are closely related to the large-sample methods presented in Chapter 8. Specifically, we shall consider methods for estimating and testing hypotheses concerning population means, the difference between two means, a population variance, and a comparison of two population variances. Small-sample tests and confidence intervals for binomial parameters will be omitted from our discussion.*

9.2

STUDENT'S t DISTRIBUTION

We introduce our topic by considering the following problem. A very costly experiment has been conducted to evaluate a new process for producing synthetic diamonds. Six diamonds have been generated by the new process with recorded weights .46, .61, .52, .48, .57, and .54 karat.

A study of the process costs indicates that the average weight of the diamonds must be greater than .5 karat in order that the process be operated at a profitable level. Do the six diamond-weight measurements present sufficient evidence to indicate that the average weight of the diamonds produced by the process is in excess of .5 karat?

Recall that, according to the Central Limit Theorem,

$$z = \frac{\bar{y} - \mu}{\sigma/\sqrt{n}}$$

possesses approximately a normal distribution in repeated sampling when n is large. For $\alpha = .05$, we would employ a one-tailed statistical test and reject when $z > 1.645$. This, of course, assumes that σ is known or that a good estimate, s, is available and is based upon a reasonably large sample (we have suggested $n \geq 30$). Unfortunately, this latter requirement will not be satisfied for the $n = 6$ diamond-weight measurements. How, then, may we test the hypothesis that $\mu = .5$ against the alternative that $\mu > .5$ when we have a small sample?

The problem that we pose is not new; rather, it is one that received serious attention by statisticians and experimenters at the turn of the century. If a sample standard deviation, s, were substituted for σ in z, would the resulting quantity possess approximately a standardized normal distribution in repeated sampling?

*A small-sample test for the binomial parameter p was presented in Chapter 6.

More specifically, would the rejection region $z > 1.645$ be appropriate; that is, would approximately 5 percent of the values of the test statistic, computed in repeated sampling, exceed 1.645? The answer to these questions, not unlike many of the problems encountered in the sciences, may be resolved by experimentation. That is, we could draw a small sample, say $n = 6$ measurements, and compute the value of the test statistic. Then we would repeat this process over and over again a very large number of times and construct a frequency distribution for the computed values of the test statistic. The general shape of the distribution and the location of the rejection region would then be evident.

The distribution of the test statistic

$$t = \frac{\bar{y} - \mu}{s/\sqrt{n}}$$

for samples drawn from a normally distributed population was discovered by W. S. Gosset and published (1908) under the pen name of Student. He referred to the quantity under study as t and it has ever since been known as Student's t. We omit the complicated mathematical expression for the density function for t but describe some of its characteristics.

The distribution of the test statistic

$$t = \frac{\bar{y} - \mu}{s/\sqrt{n}}$$

in repeated sampling is, like z, mound-shaped and perfectly symmetrical about $t = 0$. Unlike z, it is much more variable, tailing rapidly out to the right and left, a phenomenon that may readily be explained. The variability of z in repeated sampling is due solely to $\bar{y}$; the other quantities appearing in z (n and σ) are nonrandom. On the other hand, the variability of t is contributed by *two* random quantities, $\bar{y}$ and s, which can be shown to be independent of one another. Thus when $\bar{y}$ is very large, s may be very small, and vice versa. As a result, t will be more variable than z in repeated sampling (see Figure 9.1). Finally, as we might surmise, the variability of t decreases as n increases because the estimate s of σ will be based upon more and more information. When n is infinitely large, the t

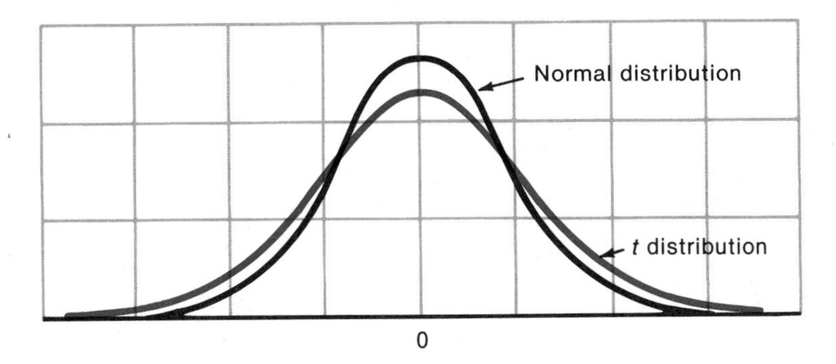

FIGURE 9.1

Standard normal, z, and the t distribution based on $n = 6$ measurements (5 d.f.)

and z distributions will be identical. Thus Gosset discovered that the distribution of t depended upon the sample size, n.

The divisor of the sum of squares of deviations, $(n - 1)$, that appears in the formula for s^2 is called the number of degrees of freedom associated with s^2. The origin of the term "degrees of freedom" is linked to the statistical theory underlying the probability distribution of s^2. We shall not pursue this point further except to note that one may say that the test statistic t is based upon a sample of n measurements or that it possesses $(n - 1)$ degrees of freedom.

The critical values of t which separate the rejection and acceptance regions for the statistical test are presented in Table 4, Appendix II. Table 4 is partially reproduced in Table 9.1. The tabulated value, t_α, records the value of t such that an area, α, lies to its right as shown in Figure 9.2. The degrees of freedom associated with s^2, d.f., is shown in the first and last columns of the table (see Table 9.1), and the t_α, corresponding to various values of α, appear in the top row. Thus, if we wish to find the value of t, such that 5 percent of the area lies to its right, we would use the column marked $t_{.05}$. The critical value of t for our example, found in that $t_{.05}$ column opposite d.f. $= (n - 1) = (6 - 1) = 5$, is $t = 2.015$ (shaded in Table 9.1). Thus we would reject H_0: $\mu = .5$ in favor of H_a: $\mu > .5$ when $t > 2.015$.

Note that the critical value of t will always be larger than the corresponding critical value of z for a specified α. For example, where $\alpha = .05$, the critical value of t for $n = 2$ (d.f. $= 1$) is $t = 6.314$, which is very large when compared with the corresponding $z_{.05} = 1.645$. Proceeding down the $t_{.05}$ column, we note that the critical value of t decreases, reflecting the effect of a larger sample size (larger degrees of freedom) on the estimation of σ. Finally, when n is infinitely large, the critical value of t will equal 1.645.

TABLE 9.1

Format of the Student's t table, Table 4, Appendix II

d.f.	$t_{.100}$	$t_{.050}$	$t_{.025}$	$t_{.010}$	$t_{.005}$	d.f.
1	3.078	6.314	12.706	31.821	63.657	1
2	1.886	2.920	4.303	6.965	9.925	2
3	1.638	2.353	3.182	4.541	5.841	3
4	1.533	2.132	2.776	3.747	4.604	4
5	1.476	2.015	2.571	3.365	4.032	5
6	1.440	1.943	2.447	3.143	3.707	6
7	1.415	1.895	2.365	2.998	3.499	7
8	1.397	1.860	2.306	2.896	3.355	8
9	1.383	1.833	2.262	2.821	3.250	9
⋮	⋮	⋮	⋮	⋮	⋮	⋮
26	1.315	1.706	2.056	2.479	2.779	26
27	1.314	1.703	2.052	2.473	2.771	27
28	1.313	1.701	2.048	2.467	2.763	28
29	1.311	1.699	2.045	2.462	2.756	29
inf.	1.282	1.645	1.960	2.326	2.576	inf.

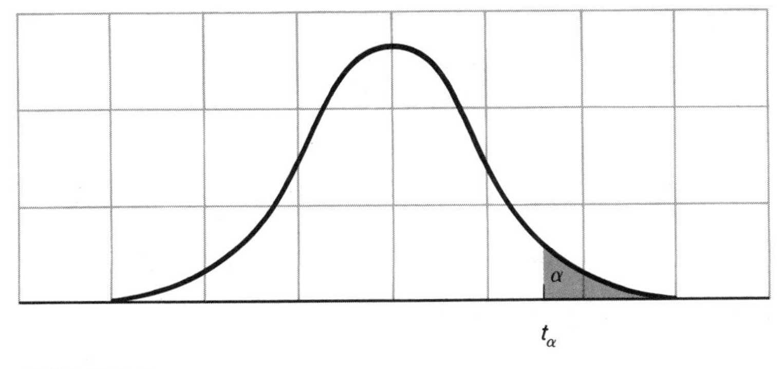

FIGURE 9.2
Tabulated values of Student's *t*

The reason for choosing $n = 30$ (an arbitrary choice) as dividing line between large and small samples is apparent. For $n = 30$ (d.f. $= 29$), the critical value of $t_{.05} = 1.699$ is numerically quite close to $z_{.05} = 1.645$. For a two-tailed test based upon $n = 30$ measurements and $\alpha = .05$, we would place .025 in each tail of the t distribution and reject H_0: $\mu = \mu_0$ when $t > 2.045$ or $t < -2.045$. Note that this is very close to the $z_{.025} = 1.96$ employed in the z test.

It is important to note that the Student's t and corresponding tabulated critical values are based upon the assumption that the sampled population possesses a normal probability distribution. This indeed is a very restrictive assumption because, in many sampling situations, the properties of the population will be completely unknown and may well be nonnormal. If this were to seriously affect the distribution of the t statistic, the application of the t test would be very limited. Fortunately, this point is of little consequence, as it can be shown that the distribution of the t statistic possesses nearly the same shape as the theoretical t distribution for populations that are nonnormal but possess a mound-shaped probability distribution. This property of the t statistic and the common occurrence of mound-shaped distributions of data in nature enhance the value of Student's t for use in statistical inference.

Second, we would note that $\bar{y}$ and s^2 must be independent (in a probabilistic sense) in order that the quantity

$$\frac{\bar{y} - \mu}{s/\sqrt{n}}$$

possess a t distribution in repeated sampling. As mentioned previously, this requirement will automatically be satisfied when the sample has been randomly drawn from a normal population.

Having discussed the origin of Student's t and the tabulated critical values, Table 4, Appendix II, we now return to the problem of making an inference about the mean diamond weight based upon our sample of $n = 6$ measurements. Prior to considering the solution, you may wish to test your built-in inference-making equipment by glancing at the six measurements and arriving at a conclusion concerning the significance of the data.

9.3

SMALL-SAMPLE INFERENCES CONCERNING A POPULATION MEAN

The statistical test of an hypothesis concerning a population mean may be stated as follows:

Test of an Hypothesis Concerning a Population Mean

Null Hypothesis, H_0: $\mu = \mu_0$.

Alternative Hypothesis, H_a: Specified by the experimenter depending upon the alternative values of μ which he wishes to detect.

Test Statistic: $t = \dfrac{\bar{y} - \mu_0}{s/\sqrt{n}}$.

Rejection Region: See the critical values of t, Table 4, Appendix II.

To apply this test to the diamond-weight data, we must first calculate the sample mean, $\bar{y}$, and standard deviation, s. This latter quantity is calculated using the formula of Section 3.8.

You can verify that the mean and standard deviation for the six diamond weights are .53 and .0559, respectively.

Recall that we wish to test the null hypothesis that the mean diamond weight is .5 against the alternative hypothesis that it is greater than .5. Then the elements of the test as defined above are:

H_0: $\mu = .5$.

H_a: $\mu > .5$.

Test Statistic: $t = \dfrac{\bar{y} - \mu_0}{s/\sqrt{n}} = \dfrac{.53 - .5}{.0559/\sqrt{6}} = 1.31.$

Rejection Region: The rejection region for $\alpha = .05$ and $(n - 1) = (6 - 1) = 5$ degrees of freedom is $t > 2.015$. Noting that the calculated value of the test statistic does not fall in the rejection region, we do not reject H_0. This implies that the data do not present sufficient evidence to indicate that the mean diamond weight exceeds .5 karat.

The calculation of the probability of a type II error, β, for the t test is very difficult and is beyond the scope of this text. Therefore, we shall avoid this problem and obtain an interval estimate for μ, as noted in Section 8.13.

We recall that the large-sample confidence interval for μ is

$$\bar{y} \pm \frac{z_{\alpha/2}\sigma}{\sqrt{n}},$$

where $z_{\alpha/2} = 1.96$ for a confidence coefficient equal to .95. This result assumes that σ is known and simply involves a measurement of $1.96\sigma_{\bar{y}}$ (or approximately $2\sigma_{\bar{y}}$) on either side of $\bar{y}$ in conformity with the Empirical Rule. When σ is unknown and must be estimated by a small-sample standard deviation, s, the large-sample confidence interval will not enclose μ 95 percent of the time in repeated sampling. Although we omit the derivation, it seems fairly clear that the corresponding small-sample confidence interval for μ, with confidence coefficient $(1 - \alpha)$, will be

Small-Sample Confidence Interval for μ

$$\bar{y} \pm \frac{t_{\alpha/2}s}{\sqrt{n}},$$

where $s/\sqrt{n}$ is the *estimated* standard deviation of $\bar{y}$.

For our example,

$$\bar{y} \pm \frac{t_{\alpha/2}\, s}{\sqrt{n}} = .53 \pm 2.571\,\frac{.0559}{\sqrt{6}} \qquad \text{or} \qquad .53 \pm .059.$$

The interval estimate for μ is therefore .471 to .589 with confidence coefficient equal to .95. If the experimenter wishes to detect a small increase in mean diamond weight in excess of .5 karat, the width of the interval must be reduced by obtaining more diamond-weight measurements. This will decrease both $1/\sqrt{n}$ and $t_{\alpha/2}$ and thereby decrease the width of the interval. Or, looking at it from the standpoint of a statistical test of an hypothesis, more information will be available upon which to base a decision and the probability of making a type II error will decrease.

EXAMPLE 9.1

A manufacturer of gunpowder has developed a new powder which is designed to produce a muzzle velocity equal to 3,000 feet per second. Eight shells are loaded with the charge and the muzzle velocities measured. The resulting velocities are shown in Table 9.2. Do the data present sufficient evidence to indicate that the average velocity differs from 3,000 feet per second?

Solution Testing the null hypothesis that $\mu = 3{,}000$ feet per second against the alternative that μ is either greater than or less than 3,000 feet per second will result in a two-tailed statistical test. Thus,

$$H_0\colon \mu = 3{,}000,$$
$$H_a\colon \mu \neq 3{,}000.$$

TABLE 9.2

Muzzle Velocity (ft/sec)	
3,005	2,995
2,925	3,005
2,935	2,935
2,965	2,905

Using $\alpha = .05$ and placing .025 in each tail of the t distribution, we find that the critical value of t for $n = 8$ measurements [or $(n - 1) = 7$ d.f.] is $t = 2.365$. Hence we will reject H_0 if $t > 2.365$ or $t < -2.365$. (Recall that the t distribution is symmetrical about $t = 0$.)

The sample mean and standard deviation for the recorded data are

$$\bar{y} = 2{,}958.75 \qquad \text{and} \qquad s = 39.26.$$

Then

$$t = \frac{\bar{y} - \mu_0}{s/\sqrt{n}} = \frac{2{,}958.75 - 3{,}000}{39.26/\sqrt{8}} = -2.97.$$

Since the observed value of t falls in the rejection region, we will reject H_0 and conclude that the average velocity differs from 3,000 feet per second. Furthermore, we will be reasonably confident that we have made the correct decision. When we use our procedure, we should erroneously reject H_0 only $\alpha = .05$ of the time in repeated applications of the statistical test.

A 95 percent confidence interval will provide additional information concerning μ. Calculating

$$\bar{y} \pm \frac{t_{\alpha/2}\, s}{\sqrt{n}},$$

we obtain

$$2{,}958.75 \pm (2.365)\frac{39.26}{\sqrt{8}}$$

or

$$2{,}958.75 \pm 32.83.$$

Thus, we estimate the average muzzle velocity to lie in the interval 2,926 to 2,992 feet per second. A more accurate interval estimate (a shorter interval) can be obtained by increasing the sample size.

EXERCISES

9.1 Noise limits, established by the EPA for heavy trucks, will be set at a level of 83 decibels in 1978 (*Environment News,* October 1976). If a random sample of six heavy trucks in 1977 produced the following noise levels (in decibels),*

85.4, 86.8, 86.1, 85.3, 84.8, 86.0,

how far from the 1978 limit is the mean noise level in 1977? Express your estimate as a 95 percent confidence interval. Interpret your findings.

9.2 The tremendous growth of the Florida lobster (called spiny lobster) industry over the past twenty years has made it the state's second most valuable fishery industry. A

*Decibels are measured on a logarithmic scale. An increase of 3 decibels represents a doubling of the actual noise energy.

recent declaration by the Bahamian government that prohibits U.S. lobstermen from fishing on the Bahamian portion of the continental shelf is expected to produce a dramatic reduction in the landings in pounds per lobster per trap. According to the records, the mean landings per trap is 30.31 pounds. A random sampling of 20 lobster traps since the Bahamian fishing restriction went into effect gave the following results (in pounds):

17.4	18.9	39.6	34.4	19.6
33.7	37.2	43.4	41.7	27.5
24.1	39.6	12.2	25.5	22.1
29.3	21.1	23.8	43.2	24.4

Do these landings provide sufficient evidence to support the contention that the mean landings per trap has decreased since imposition of the Bahamian restrictions? Test using $\alpha = .05$.

9.3 Refer to Exercise 9.2 and find a 90 percent confidence interval for the mean landing per lobster trap (in pounds) and interpret your result.

9.4 PCBs (polychlorinated biphenyls), used in the manufacture of large electrical power transformers and capacitors, are known to be hazardous substances which have serious health effects. *Environment News* (October 1976) reports that the federal and state governments are investigating PCB contamination in the Acushnet River in the New Bedford, Massachusetts, area. As part of their investigation, they found that analyses of three clam specimens showed PCB levels of 21, 23, and 53 parts per million (ppm) (the FDA tolerance level for PCBs is 5 ppm). Use these three clam PCB concentration levels to estimate the mean concentration in clams in the Acushnet River area. Use a 90 percent confidence interval. Interpret your results. What assumptions must be made in order that your inference be valid?

9.5 In an earlier exercise, we mentioned the Bureau of Labor Statistic's surveys of consumer spending habits, particularly focusing on the estimate of the mean number of automobiles per family in the United States. Suppose that you were to run a very small survey for your community. You randomly sample $n = 20$ households and for each you record the number of automobiles per household. Suppose that $\bar{y} = 1.75$ and $s = .55$. Is it valid to use the t statistic to form a 90 percent confidence interval for the mean number of automobiles per household for the community? Explain.

9.6 A cigarette manufacturer claimed that his cigarettes contained no more than 25 milligrams of nicotine. A sample of 16 cigarettes yielded a mean and standard deviation equal to 26.4 and 2, respectively. Do the data provide sufficient evidence to refute the manufacturer's claim? (Use $\alpha = .10$.)

9.7 Refer to Exercise 9.6. Estimate the mean nicotine content of this brand of cigarette using a 90 percent confidence interval.

9.8 Measurements of water intake, obtained from a sample of 17 rats that had been injected with a sodium chloride solution, produced a mean and standard deviation of 31.0 and 6.2 cubic centimeters, respectively. Given that the average water intake for noninjected rats observed over a comparable period of time is 22.0 cubic centimeters, do the data indicate that injected rats drink more water than noninjected rats? Test at the 5 percent level of significance.

9.9 Refer to Exercise 9.8. Find a 90 percent confidence interval for the mean water intake for injected rats.

9.10 Organic chemists often purify organic compounds by a method known as fractional crystallization. An experimenter desired to prepare and purify 4.85 grams of aniline. Ten 4.85-gram quantities of aniline were individually prepared and purified to acetanilide. The following dry yields were recorded:

3.85	3.80
3.88	3.85
3.90	3.36
3.62	4.01
3.72	3.83

Estimate the mean number of grams of acetanilide that could be recovered from an initial amount of 4.85 grams of aniline. Use a 95 percent confidence interval.

9.11 An experimenter, interested in determining the mean thickness of the cortex of the sea urchin egg, employed an experimental procedure developed by Sakai. The thickness of the cortex was measured for $n = 10$ sea urchin eggs. The following measurements were obtained:

4.5	5.2
6.1	2.6
3.2	3.7
3.9	4.6
4.7	4.1

Estimate the mean thickness of the cortex using a 95 percent confidence interval.

9.12 Owing to the variability of trade-in allowance, the profit per new car sold by an automobile dealer varies from car to car. The profit per sale, tabulated for the past week, was:

Profit per Sale (hundreds of dollars)	
2.1	6.2
3.0	4.5
1.2	5.1

Do these data present sufficient evidence to indicate that the average profit per sale is less than $480? Test at an $\alpha = .05$ level of significance.

9.13 Find a 90 percent confidence interval for the mean profit per sale in Exercise 9.12.

9.14 Industrial wastes and sewage dumped into our rivers and streams absorb oxygen and thereby reduce the amount of dissolved oxygen available for fish and other forms of aquatic life. One state agency requires a minimum of 5 parts per million of dissolved oxygen in order that the oxygen content be sufficient to support aquatic life. Six water specimens taken from a river at a specific location during the low-water season (July) gave readings of 4.9, 5.1, 4.9, 5.0, 5.0, and 4.7 parts per million of dissolved oxygen. Do the data provide sufficient evidence to indicate that the dissolved oxygen content is less than 5 parts per million? Test using $\alpha = .05$.

9.15 A manufacturer of television sets claimed that his product possessed an average defect-free life of 3 years. Three households in a community have purchased the sets and all three sets are observed to fail before 3 years, with failure times equal to 2.5, 1.9, and 2.9 years, respectively. Do these data present sufficient evidence to contradict the manufacturer's claim? Test at an $\alpha = .05$ level of significance.

9.16 Calculate a 90 percent confidence interval for the mean life of the television sets in Exercise 9.15.

9.17 Refer to Exercises 9.15 and 9.16. Approximately how many observations would be required to estimate the mean life of the television sets correct to within two-tenths of a year with probability equal to .90?

9.4

SMALL-SAMPLE INFERENCES CONCERNING THE DIFFERENCE BETWEEN TWO MEANS

The physical setting for the problem that we consider is identical to that discussed in Section 8.7. Independent random samples of n_1 and n_2 measurements, respectively, are drawn from two populations which possess means and variances μ_1, σ_1^2 and μ_2, σ_2^2. Our objective is to make inferences concerning the difference between the two population means, $(\mu_1 - \mu_2)$.

The following small-sample methods for testing hypotheses and placing a confidence interval on the difference between two means are, like the case for a single mean, founded upon assumptions regarding the probability distributions of the sampled populations. Specifically, we shall assume that both populations possess a normal probability distribution and, also, that the population variances, σ_1^2 and σ_2^2, are equal. In other words, we assume that the variability of the measurements in the two populations is the same and can be measured by a common variance which we will designate as σ^2, that is, $\sigma_1^2 = \sigma_2^2 = \sigma^2$.

The point estimator of $\mu_1 - \mu_2$, $(\bar{y}_1 - \bar{y}_2)$, the difference between the sample means, was discussed in Section 8.7, where it was observed to be unbiased and to possess a standard deviation,

$$\sigma_{(\bar{y}_1-\bar{y}_2)} = \sqrt{\frac{\sigma_1^2}{n_1} + \frac{\sigma_2^2}{n_2}},$$

in repeated sampling. This result was used in placing bounds on the error of estimation, the construction of a large-sample confidence interval, and the z test statistic,

$$z = \frac{(\bar{y}_1 - \bar{y}_2) - D_0}{\sqrt{\dfrac{\sigma_1^2}{n_1} + \dfrac{\sigma_2^2}{n_2}}},$$

for testing an hypothesis, H_0: $\mu_1 - \mu_2 = D_0$ (i.e., D_0 is the hypothesized difference between μ_1 and μ_2). Utilizing the assumption that $\sigma_1^2 = \sigma_2^2 = \sigma^2$, the z test statistic could be simplified as follows:

$$z = \frac{(\bar{y}_1 - \bar{y}_2) - D_0}{\sqrt{\dfrac{\sigma^2}{n_1} + \dfrac{\sigma^2}{n_2}}} = \frac{(\bar{y}_1 - \bar{y}_2) - D_0}{\sigma\sqrt{\dfrac{1}{n_1} + \dfrac{1}{n_2}}}.$$

For small-sample tests of the hypothesis H_0: $\mu_1 - \mu_2 = D_0$, it would seem reasonable to use the test statistic

$$t = \frac{(\bar{y}_1 - \bar{y}_2) - D_0}{s\sqrt{\dfrac{1}{n_1} + \dfrac{1}{n_2}}};$$

that is, we would substitute a sample standard deviation s for σ. Surprisingly enough, this test statistic will possess a Student's t distribution in repeated sampling when the stated assumptions are satisfied, a fact that can be proved mathematically or verified by experimental sampling from two normal populations.

The estimate, s, to be used in the t statistic could be either s_1 or s_2, the standard deviations for the two samples, although the use of either would be wasteful since both estimate σ. Since we wish to obtain the best estimate available, it would seem reasonable to use an estimator that would pool the information from both samples. This estimator, utilizing the sums of squares of the deviations about the mean for both samples, is

Pooled Estimator of σ^2

$$s^2 = \frac{\sum_{i=1}^{n_1}(y_i - \bar{y}_1)^2 + \sum_{i=1}^{n_2}(y_i - \bar{y}_2)^2}{n_1 + n_2 - 2}.$$

Note that the pooled estimator may also be written as

$$s^2 = \frac{(n_1 - 1)s_1^2 + (n_2 - 1)s_2^2}{n_1 + n_2 - 2},$$

because

$$s_1^2 = \frac{\sum_{i=1}^{n_1}(y_i - \bar{y}_1)^2}{n_1 - 1}$$

and

$$s_2^2 = \frac{\sum_{i=1}^{n_2}(y_i - \bar{y}_2)^2}{n_2 - 1}.$$

As in the case for the single sample, the denominator in the formula for s^2, $(n_1 + n_2 - 2)$, is called the "number of degrees of freedom" associated with s^2. It can be proved either mathematically or experimentally that the expected value of the pooled estimator, s^2, is equal to σ^2 and hence s^2 is an unbiased estimator of the common population variance. Finally, we note that the divisors of the sums of squares of deviations in s_1^2 and s_2^2, $(n_1 - 1)$ and $(n_2 - 1)$, respectively, are the numbers of degrees of freedom associated with these two independent estimators of σ^2. It is interesting to note that an estimator that uses the pooled information from both samples would possess $(n_1 - 1) + (n_2 - 1)$ or $(n_1 + n_2 - 2)$ degrees of freedom.

Summarizing, the small-sample statistical test for the difference between two means is as follows.

Test of an Hypothesis Concerning the Difference Between Two Means

Null Hypothesis, H_0: $(\mu_1 - \mu_2) = D_0$.

Alternative Hypothesis, H_a: $(\mu_1 - \mu_2) \neq D_0$, for a two-tailed statistical test.

Test Statistic: $$t = \frac{(\bar{y}_1 - \bar{y}_2) - D_0}{s\sqrt{\frac{1}{n_1} + \frac{1}{n_2}}}.$$

Rejection Region: For a two-tailed test, reject H_0 if $t < -t_{\alpha/2}$ or $t > t_{\alpha/2}$, where $t_{\alpha/2}$ is based on $n_1 + n_2 - 2$ degrees of freedom.

The critical value of t can be obtained from Table 4, Appendix II. Thus if $n_1 = 10$ and $n_2 = 12$, we would use the t value corresponding to $(n_1 + n_2 - 2) = 20$ degrees of freedom. The following example will serve as an illustration.

EXAMPLE 9.2

An assembly operation in a manufacturing plant requires approximately a one-month training period for a new employee to reach maximum efficiency. A new method of training was suggested and a test was conducted to compare the new method with the standard procedure. Two groups of nine new employees were trained for a period of three weeks, one group using the new method and the other following standard training procedure. The length of time in minutes required for each employee to assemble the device was recorded at the end of the three-week period. These measurements appear in Table 9.3. Do the data present sufficient evidence to indicate that the mean time to assemble at the end of a three-week training period is less for the new training procedure?

TABLE 9.3

Standard Procedure	New Procedure
32	35
37	31
35	29
28	25
41	34
44	40
35	27
31	32
34	31

Solution Let μ_1 and μ_2 equal the mean time to assemble for the standard and the new assembly procedures, respectively. Also, assume that the variability in mean time to assemble is essentially a function of individual differences and that the variability for the two populations of measurements will be approximately equal.

The sample means and sums of squares of deviations are

$$\bar{y}_1 = 35.22,$$

$$\sum_{i=1}^{9} (y_i - \bar{y}_1)^2 = 195.56;$$

$$\bar{y}_2 = 31.56,$$

$$\sum_{i=1}^{9} (y_i - \bar{y}_2)^2 = 160.22.$$

Then the pooled estimate of the common variance is

$$s^2 = \frac{\sum_{i=1}^{9} (y_i - \bar{y}_1)^2 + \sum_{i=1}^{9} (y_i - \bar{y}_2)^2}{n_1 + n_2 - 2}$$

$$= \frac{195.56 + 160.22}{9 + 9 - 2}$$

$$= 22.24,$$

and the standard deviation is $s = 4.72$.

The null hypothesis to be tested is

$$H_0\colon \mu_1 - \mu_2 = 0.$$

Suppose that we are concerned only with detecting whether the new method reduces the assembly time, and therefore that the alternative hypothesis is

$$H_a\colon \mu_1 - \mu_2 > 0.$$

This would imply that we should use a one-tailed statistical test and that the rejection region for the test will be located in the upper tail of the t distribution. Referring to Table 4, Appendix II, we note that the critical value of t for $\alpha = .05$ and $(n_1 + n_2 - 2) = 16$ degrees of freedom is 1.746. Therefore, we will reject when $t > 1.746$.

The calculated value of the test statistic is

$$t = \frac{\bar{y}_1 - \bar{y}_2}{s\sqrt{\frac{1}{n_1} + \frac{1}{n_2}}} = \frac{35.22 - 31.56}{4.72\sqrt{\frac{1}{9} + \frac{1}{9}}}$$

$$= 1.64.$$

Comparing this with the critical value, $t = 1.746$, we note that the calculated value does not fall in the rejection region. Therefore, we must conclude that there is insufficient evidence to indicate that the new method of training is superior at the .05 level of significance. As we shall show in the next example, the variability of the data is sufficient to mask a difference in μ_1 and μ_2, if in fact it exists. Consequently, the plant manager may wish to increase the number of employees in each sample and repeat the test.

The small-sample confidence interval for $(\mu_1 - \mu_2)$ is based upon the same assumption as was the statistical test procedure. This confidence interval, with confidence coefficient $(1 - \alpha)$, is given by the formula

Small-Sample Confidence Interval for $(\mu_1 - \mu_2)$

$$(\bar{y}_1 - \bar{y}_2) \pm t_{\alpha/2}\, s\sqrt{\frac{1}{n_1} + \frac{1}{n_2}}.$$

Note the similarity in the procedures for constructing the confidence intervals for a single mean, Section 9.2, and the difference between two means. In both cases, the interval is constructed by using the appropriate point estimator and then adding and subtracting an amount equal to $t_{\alpha/2}$ times the *estimated* standard deviation of the point estimator.

EXAMPLE 9.3

Find an interval estimate for $(\mu_1 - \mu_2)$, Example 9.2, using a confidence coefficient equal to .95.

Solution Substituting into the formula

$$(\bar{y}_1 - \bar{y}_2) \pm t_{\alpha/2}\, s\sqrt{\frac{1}{n_1} + \frac{1}{n_2}},$$

we find the interval estimate (or 95 percent confidence interval) to be

$$(35.22 - 31.56) \pm (2.120)(4.72)\sqrt{\frac{1}{9} + \frac{1}{9}}$$

or

$$3.66 \pm 4.72.$$

Thus, we estimate the difference in mean time to assemble, $(\mu_1 - \mu_2)$, to fall in the interval -1.06 to 8.38. Note that the interval width is considerable and that it would seem advisable to increase the size of the samples and reestimate.

Before concluding our discussion it is necessary to comment on the two assumptions upon which our inferential procedures are based. Moderate departures from the assumption that the populations possess a normal probability distribution do not seriously affect the distribution of the test statistic and the confidence coefficient for the corresponding confidence interval. On the other hand, the population variances should be nearly equal in order that the aforementioned procedures be valid. A procedure will be presented in Section 9.7 for testing an hypothesis concerning the equality of two population variances.

If there is reason to believe that the population variances are unequal or that the normality assumptions have been violated, you can test for a shift in location of two population distributions using the nonparametric Mann–Whitney

U test of Chapter 14. This test procedure, which requires no assumptions concerning the nature of the population probability distributions, is almost as sensitive in detecting a difference in population means when the conditions necessary for the t test are satisfied. It may be more sensitive when the assumptions are not satisfied.

EXERCISES

9.18 Two methods for teaching reading were applied to two randomly selected groups of elementary school children and compared on the basis of a reading comprehension test given at the end of the learning period. The sample means and variances computed from the test scores are as follows:

Method	Number of Children in Group	$\bar{y}$	s^2
1	11	64	52
2	14	69	71

Do the data present sufficient evidence to indicate a difference in the mean scores for the populations associated with the two teaching methods? Test at an $\alpha = .05$ level of significance.

9.19 As *Environment Midwest* (December 1976) notes, air pollution spread by air currents is becoming an increasing problem in rural as well as urban areas, particularly in the east. The data shown in the accompanying table presume to show the maximum amount of ozone recorded at each of eight Ohio air-pollution monitoring stations as well as the number of violations of the maximum allowable limit at each location recorded over the period, June 14 to August 31, 1974. Do the data provide sufficient evidence to indicate that there is a difference in the mean number of violations between urban and rural areas? What must you assume in order to make your inference?

City	Designation	Maximum Concentration (ppm)	Days Exceeding Standard (%)	Number of Violations (Total)
Canton	urban	.14	44	148
Cincinnati	urban	.18	44	54
Cleveland	urban	.14	26	51
Columbus	urban	.15	27	113
Dayton	urban	.13	35	114
McConnelsville	rural	.16	56	239
Wilmington	rural	.18	58	259
Wooster	rural	.17	55	262

9.20 An experiment is conducted to compare two new automobile designs. Twenty people are randomly selected and each person is asked to rate each design on a scale of 1 (poor) to 10 (excellent). The resulting ratings will be used to test the null hypothesis that the mean level of approval is the same for both designs against the alternative hypothesis that one of the automobile designs is preferred. Would these data satisfy the assumptions required for the Student's t test of Section 9.4? Explain.

9.21 The length of time to recovery was recorded for patients randomly assigned and subjected to two different surgical procedures. The data, recorded in days, are as follows:

Procedure I	Procedure II
$n_1 = 21$	$n_2 = 23$
$\bar{y}_1 = 7.3$	$\bar{y}_2 = 8.9$
$s_1^2 = 1.23$	$s_2^2 = 1.49$

Do the data present sufficient evidence to indicate a difference in mean recovery time for the two surgical procedures? Test using $\alpha = .05$.

9.22 Refer to Exercise 9.14 where we measured the dissolved oxygen content in river water to determine whether a stream possessed sufficient oxygen to support aquatic life. A pollution control inspector suspected that a river community was releasing amounts of semitreated sewage into a river. To check his theory, he drew five randomly selected specimens of river water at a location above the town and another five below. The dissolved oxygen readings, in parts per million, are as follows:

Above Town	4.8	5.2	5.0	4.9	5.1
Below Town	5.0	4.7	4.9	4.8	4.9

(a) Do the data provide sufficient evidence to indicate that the mean oxygen content between locations above and below the town is less than the mean oxygen content above? Test using $\alpha = .05$.

(b) Suppose that you prefer estimation as a method of inference. Estimate the difference in mean dissolved oxygen content between locations above and below the town. Use a 95 percent confidence interval.

9.23 An experiment was conducted to compare the mean lengths of time required for the bodily absorption of two drugs, A and B. Ten people were randomly selected and assigned to each drug treatment. Each person received an oral dosage of the assigned drug and the length of time (in minutes) for the drug to reach a specified level in the blood was recorded. The means and variances for the two samples are as follows:

Drug A	Drug B
$\bar{y}_1 = 27.2$	$\bar{y}_2 = 33.5$
$s_1^2 = 16.36$	$s_2^2 = 18.92$

Do the data provide sufficient evidence to indicate a difference in mean times to absorption for the two drugs? Test using $\alpha = .10$.

9.24 Refer to Exercise 9.23. Find a 95 percent confidence interval for the difference in mean times to absorption.

9.25 Refer to Exercise 9.24. Suppose that you wished to estimate the difference in mean times to absorption correct to within 1 minute with probability approximately equal to .95.

(a) Approximately, how large a sample would be required for each drug (assume that the sample sizes will be equal)?

(b) To conduct the experiment using the sample sizes of part (a) will require a large amount of time and money. Can anything be done to reduce the sample sizes and still achieve the 1-minute bound on the error of estimation?

9.26 In an article concerning the adverse effects of alcohol abuse on offspring during pregnancy, Ouellette* and colleagues comment on the mean age of the heavy drinkers (group A) versus the mean age of the moderate drinkers or abstinents (group B). The age data for the females included in the experiment are shown in the accompanying table. The authors utilize a t test to test the null hypothesis, $\mu_1 - \mu_2$, against the alternative that $\mu_1 \neq \mu_2$.

Group A Heavy Drinkers	Group B Abstinent or Moderate Drinkers
$n_A = 58$	$n_B = 575$
$\bar{y}_A = 25.7$ years	$\bar{y}_B = 22.8$ years
$s_A = 5.9$ years	$s_B = 5.5$ years

(a) Is it necessary to use the t test for the analysis of this data? Explain.

(b) The authors state that "($P < .001$ by t test)" or, equivalently, that the results are statistically significant at the .001 level. This means that if t_0 is the computed (observed) value of the t statistic, $P(t > t_0 \text{ or } t < -t_0) < .001$. Do you agree with this statement? Check their calculations.

9.5

A PAIRED-DIFFERENCE TEST

A manufacturer wished to compare the wearing qualities of two different types of automobile tires, A and B. To make the comparison, a tire of type A and one of type B were randomly assigned and mounted on the rear wheels of each of five automobiles. The automobiles were then operated for a specified number of miles and the amount of wear was recorded for each tire. These measurements appear in Table 9.4. Do the data present sufficient evidence to indicate a difference in the average wear for the two tire types?

Analyzing the data, we note that the difference between the two sample means is $(\bar{y}_1 - \bar{y}_2) = .48$, a rather small quantity, considering the variability of the data and the small number of measurements involved. At first glance it would seem that there is little evidence to indicate a difference between the population means, a conjecture that we may check by the method outlined in Section 9.4.

*Ouellette, E. M., H. L. Rosett, N. P. Rosman, and L. Weiner, "Adverse Effects on Offspring of Maternal Alcohol Abuse During Pregnancy," *New England J. of Med.*, *297* (September 8, 1977), pp. 528–530.

TABLE 9.4

Automobile	Tire A	Tire B
1	10.6	10.2
2	9.8	9.4
3	12.3	11.8
4	9.7	9.1
5	8.8	8.3
	$\bar{y}_1 = 10.24$	$\bar{y}_2 = 9.76$

The pooled estimate of the common variance, σ^2, is

$$s^2 = \frac{\sum_{i=1}^{n_1}(y_i - \bar{y}_1)^2 + \sum_{i=1}^{n_2}(y_i - \bar{y}_2)^2}{n_1 + n_2 - 2} = \frac{6.932 + 7.052}{5 + 5 - 2} = 1.748,$$

and

$$s = 1.32.$$

The calculated value of t used to test the hypothesis that $\mu_1 = \mu_2$ is

$$t = \frac{\bar{y}_1 - \bar{y}_2}{s\sqrt{\frac{1}{n_1} + \frac{1}{n_2}}} = \frac{10.24 - 9.76}{1.32\sqrt{\frac{1}{5} + \frac{1}{5}}} = .57,$$

a value that is not nearly large enough to reject the hypothesis that $\mu_1 = \mu_2$.

The corresponding 95 percent confidence interval is

$$(\bar{y}_1 - \bar{y}_2) \pm t_{\alpha/2}s\sqrt{\frac{1}{n_1} + \frac{1}{n_2}} = (10.24 - 9.76) \pm (2.306)(1.32)\sqrt{\frac{1}{5} + \frac{1}{5}}$$

or -1.45 to 2.41. Note that the interval is quite wide, considering the small difference between the sample means.

A second glance at the data reveals a marked inconsistency with this conclusion. We note that the wear measurement for the type A is larger than the corresponding value for type B for *each* of the five automobiles. These differences, recorded as $d = A - B$, are as follows:

Automobile	$d = A - B$
1	.4
2	.4
3	.5
4	.6
5	.5
	$\bar{d} = .48$

Suppose that we were to use y, the number of times that A is larger than B, as a test statistic, as was done in Exercise 6.34. Then the probability that A would be larger than B on a given automobile, assuming no difference between the wearing quality of the tires, would be $p = 1/2$, and y would be a binomial random variable. If the null hypothesis were true, the expected value of y would be $\mu = np = 5(1/2) = 2.5$.

If we choose the most extreme values of y, $y = 0$ and $y = 5$, as the rejection region for a two-tailed test, than $\alpha = P(0) + P(5) = 2(1/2)^5 = 1/16$. We would then reject H_0: $(\mu_1 = \mu_2)$ with a probability of type I error equal to $\alpha = 1/16$. Certainly this is evidence to indicate that a difference exists in the mean wear of the two tire types.

You will note that we have employed two different statistical tests to test the same hypothesis. Is it not peculiar that the t test, which utilizes more information (the actual sample measurements) than the binomial test, fails to supply sufficient evidence for rejection of the hypothesis $\mu_1 = \mu_2$?

The explanation of this seeming inconsistency is quite simple. The t test described in Section 9.4 is *not* the proper statistical test to be used for our example. The statistical test procedure, Section 9.4, required that the two samples be *independent* and random. Certainly, the independence requirement was violated by the manner in which the experiment was conducted. The (pair of) measurements, an A and B, for a particular automobile are definitely related. A glance at the data will show that the readings are of approximately the same magnitude for a particular automobile but vary from one automobile to another. This, of course, is exactly what we might expect. Tire wear, in a large part, is determined by driver habits, the balance of the wheels, and the road surface. Since each automobile had a different driver, we might expect a large amount of variability in the data from one automobile to another.

The familiarity that we have gained with interval estimation has shown that the width of the large- and small-sample confidence intervals will depend upon the magnitude of the standard deviation of the point estimator of the parameter. The smaller its value, the better the estimate and the more likely that the test statistic will reject the null hypothesis if it is, in fact, false. Knowledge of this phenomenon was utilized in *designing* the tire wear experiment.

The experimenter would realize that the wear measurements would vary greatly from auto to auto and that this variability could not be separated from the data if the tires were assigned to the 10 wheels in a *random* manner. (A random assignment of the tires would have implied that the data be analyzed according to the procedure of Section 9.4.) Instead, a comparison of the wear between the tire types A and B made on each automobile resulted in the five difference measurements. This design eliminates the effect of the car-to-car variability and yields more information on the mean difference in the wearing quality for the two tire types.

The proper analysis of the data would utilize the five difference measurements to test the hypothesis that the average difference μ_d is equal to zero, or, equivalently, to test the null hypothesis H_0: $\mu_d = \mu_1 - \mu_2 = 0$ against the alternative hypothesis, $\mu_d = \mu_1 - \mu_2 \neq 0$.

You may verify that the average and standard deviation of the five difference measurements are

$$\bar{d} = .48,$$
$$s_d = .0837.$$

Then,

$$H_0\colon \mu_d = 0$$

and

$$t = \frac{\bar{d} - 0}{s_d/\sqrt{n}} = \frac{.48}{.0837/\sqrt{5}} = 12.8.$$

The critical value of t for a two-tailed statistical test, $\alpha = .05$ and four degrees of freedom, is 2.776. Certainly, the observed value of $t = 12.8$ is extremely large and highly significant. Hence we would conclude that the average amount of wear for tire type B is less than that for type A.

A 95 percent confidence interval for the difference between the mean wear would be

$$\bar{d} \pm \frac{t_{\alpha/2}s_d}{\sqrt{n}} = .48 \pm (2.776)\frac{.0837}{\sqrt{5}}$$

or $.48 \pm .10$.

The statistical design of the tire experiment represents a simple example of a randomized block design and the resulting statistical test is often called a paired-difference test. Note that the pairing occurred when the experiment was planned and *not* after the data were collected. Comparisons of tire wear were made within relatively homogeneous blocks (automobiles) with the tire types randomly assigned to the two automobile wheels.

An indication of the gain in the amount of information obtained by blocking the tire experiment may be observed by comparing the calculated confidence interval for the unpaired (and incorrect) analysis with the interval obtained for the paired-difference analysis. The confidence interval for $(\mu_1 - \mu_2)$ that might have been calculated, had the tires been randomly assigned to the 10 wheels (unpaired), is unknown but likely would have been of the same magnitude as the interval -1.45 to 2.41 calculated by analyzing the observed data in an unpaired manner. Pairing the tire types on the automobiles (blocking) and the resulting analysis of the differences produced the interval estimate .38 to .58. Note the difference in the width of the intervals indicating the very sizable increase in information obtained by blocking in this experiment.

While blocking proved to be very beneficial in the tire experiment, this may not always be the case. We observe that the degrees of freedom available for estimating σ^2 are less for the paired than for the corresponding unpaired experiment. If there were actually no difference between the blocks, the reduction in the degrees of freedom would produce a moderate increase in the $t_{\alpha/2}$ employed in the confidence interval and hence increase the width of the interval. This, of course, did not occur in the tire experiment because the large reduction in the standard deviation of $\bar{d}$ more than compensated for the loss in degrees of freedom.

Before concluding, we want to reemphasize a point. Once you have used a paired design for an experiment, you no longer have the option of using the unpaired analysis of Section 9.4. The assumptions upon which that test are based have been violated. Your only alternative is to use the correct method of analysis, the paired-difference test (and associated confidence interval) of this section.

EXERCISES

9.27 The following data were collected on lost-time accidents (the figures given are mean man-hours lost per month over a period of 1 year) both before and after an industrial safety program was put into effect. Data were recorded for six industrial plants. Do the data provide sufficient evidence to indicate whether the safety program was effective in reducing lost-time accidents? (Use $\alpha = .10$.)

	Plant Number					
	1	2	3	4	5	6
Before program	38	64	42	70	58	30
After program	31	58	43	65	52	29

9.28 The earth's temperature (which affects seed germination, crop survival in bad weather, and many other aspects of agricultural production) can be measured using either ground-based sensors or infrared sensing devices mounted in aircraft or space satellites. Ground-based sensoring is tedious, requiring many replications to obtain an accurate estimate of ground temperature. On the other hand, airplane or satellite sensoring of infrared waves appears to introduce a bias in the temperature readings. To determine the bias, readings were obtained at six different locations using both ground- and air-based temperature sensors. The readings, in degrees Celsius, were as follows:

Location	Ground	Air
1	46.9	47.3
2	45.4	48.1
3	36.3	37.9
4	31.0	32.7
5	24.7	26.2

(a) Do the data present sufficient evidence to indicate a bias in the air-collected temperature readings? Explain. (Use $\alpha = .05$.)

(b) Estimate the difference in mean temperature between ground- and air-based sensors using a 95 percent confidence interval.

9.29 Why use paired observations to estimate the difference between two population means in preference to estimation based on independent random samples selected from the two populations? Is a paired experiment always preferable? Explain.

9.30 Refer to Exercise 9.28. How many paired observations would be required to estimate the difference in mean temperature between ground- and air-based sensors correct to within .2 degree Celsius with probability approximately equal to .95?

9.31 To compare the demand for two different entrees, the manager of a cafeteria recorded the number of purchases for each entree on seven consecutive days. The data are shown in the table. Do the data provide sufficient evidence to indicate a greater mean demand for one of the entrees?

Day	A	B
Mon.	420	391
Tues.	374	343
Wed.	434	469
Thurs.	395	412
Fri.	637	538
Sat.	594	521
Sun.	679	625

9.32 In response to a complaint that a particular tax assessor (A) was biased, an experiment was conducted to compare the assessor named in the complaint with another tax assessor (B) from the same office. Eight properties were selected and each was assessed by both assessors. The assessments (in thousands) are shown in the table.

Property	Assessor A	Assessor B
1	36.3	35.1
2	48.4	46.8
3	40.2	37.3
4	54.7	50.6
5	28.7	29.1
6	42.8	41.0
7	36.1	35.3
8	39.0	39.1

(a) Do the data provide sufficient evidence to indicate that assessor A tends to give higher assessments than assessor B? Test with $\alpha = .05$.

(b) Estimate the difference in mean assessments for the two assessors.

(c) What assumptions must you make in order that the inferences, (a) and (b), are valid?

(d) Suppose the assessor A had been compared with a more stable standard, say the average, $\bar{y}$, of the assessments given by four assessors selected from the tax office. Thus each property would be assessed by A and also by each of the four other assessors and $y_A - \bar{y}$ would be calculated. If the test, part (a), is valid, could you use the paired-difference t test to test the hypothesis that the bias, the mean difference between A's assessments and the mean of the assessments of the four assessors, is equal to 0? Explain.

9.33 Seven obese persons were placed on a diet for one month and the weights, at the beginning and at the end of the month were recorded. They are shown in the table.

	Weights	
Subjects	Initial	Final
1	310	263
2	295	251
3	287	249
4	305	259
5	270	233
6	323	267
7	277	242
8	299	265

Estimate the mean weight loss for obese persons when placed on the diet for a one-month period. Use a 95 percent confidence interval and interpret your results. What assumptions must you make in order that your inference be valid?

9.6

INFERENCES CONCERNING A POPULATION VARIANCE

We have seen in the preceding sections that an estimate of the population variance, σ^2, is fundamental to procedures for making inferences about population means. Moreover, there are many practical situations where σ^2 is the primary objective of an experimental investigation; thus it may assume a position of far greater importance than that of the population mean.

Scientific measuring instruments must provide unbiased readings with a very small error of measurement. An aircraft altimeter that measured the correct altitude on the *average* would be of little value if the standard deviation of the error of measurement were 5,000 feet. Indeed, bias in a measuring instrument can often be corrected, but the precision of the instrument, measured by the standard deviation of the error of measurement, is usually a function of the design of the instrument itself and cannot be controlled.

Machined parts in a manufacturing process must be produced with minimum variability in order to reduce out-of-size and hence defective products. And, in general, it is desirable to maintain a minimum variance in the measurements of the quality characteristics of an industrial product in order to achieve process control and therefore minimize the percentage of poor-quality product.

The sample variance,

$$s^2 = \frac{\sum_{i=1}^{n}(y_i - \bar{y})^2}{n-1},$$

is an unbiased estimator of the population variance, σ^2. Thus the distribution of sample variances generated by repeated sampling will have a probability distribution that commences at $s^2 = 0$ (since s^2 cannot be negative) with a mean equal to σ^2. Unlike the distribution of $\bar{y}$, the distribution of s^2 is nonsymmetrical, the exact form being dependent upon the probability distribution of the population.

For the methodology that follows we will assume that the sample is drawn from a normal population and that s^2 is based upon a random sample of n measurements. Or, using the terminology of Section 9.2, we would say that s^2 possesses $(n - 1)$ degrees of freedom.

The next and obvious step would be to consider the distribution of s^2 in repeated sampling from a specified normal distribution—one with a specific mean and variance—and to tabulate the critical values of s^2 for some of the commonly used tail areas. If this is done, we shall find that the distribution of s^2 is independent of the population mean, μ, but possesses a distribution for each sample size and each value of σ^2. This task would be quite laborious, but fortunately it may be simplified by *standardizing,* as was done by using z in the normal tables.

The quantity

$$\chi^2 = \frac{(n-1)s^2}{\sigma^2},$$

called a chi-square variable by statisticians, admirably suits our purposes. Its distribution in repeated sampling is called, as we might suspect, a chi-square probability distribution. The equation of the density function for the chi-square distribution is well known to statisticians who have tabulated critical values corresponding to various tail areas of the distribution. These values are presented in Table 5, Appendix II.

The shape of the chi-square distribution, like that of the t distribution, will vary with the sample size, or, equivalently, with the degrees of freedom associated with s^2. Thus Table 5, Appendix II, is constructed in exactly the same manner as the t table, with the degrees of freedom shown in the last column. A partial reproduction of Table 5, Appendix II, is shown in Table 9.5. The symbol χ^2_α indicates that the tabulated χ^2 value is such that an area, α, lies to its right. (See Figure 9.3.) Stated in probabilistic terms,

$$P(\chi^2 > \chi^2_\alpha) = \alpha.$$

Thus, 99 percent of the area under the χ^2 distribution would lie to the right of $\chi^2_{.99}$. We note that the extreme values of χ^2 must be tabulated for both the lower and upper tail of the distribution because it is nonsymmetrical.

You may check your ability to use the table by verifying the following statements. The probability that χ^2, based upon $n = 16$ measurements (d.f. = 15), will exceed 24.9958 is .05. For a sample of $n = 6$ measurements (d.f. = 5), 95 percent of the area under the χ^2 distribution will lie to the right of $\chi^2 = 1.145476$. These values of χ^2 are shaded in Table 9.5. The statistical test of a null hypothesis concerning a population variance,

$$H_0\colon \sigma^2 = \sigma_0^2,$$

TABLE 9.5

Format of the chi-square table, Table 5, Appendix II

d.f.	$\chi^2_{0.995}$	...	$\chi^2_{0.950}$	$\chi^2_{0.900}$	$\chi^2_{0.100}$	$\chi^2_{0.050}$	...	$\chi^2_{0.005}$	d.f.
1	0.0000393		0.0039321	0.0157908	2.70554	3.84146		7.87944	1
2	0.0100251		0.102587	0.210720	4.60517	5.99147		10.5966	2
3	0.0717212		0.351846	0.584375	6.25139	7.81473		12.8381	3
4	0.206990		0.710721	1.063623	7.77944	9.48773		14.8602	4
5	0.411740		1.145476	1.61031	9.23635	11.0705		16.7496	5
6	0.675727		1.63539	2.20413	10.6446	12.5916		18.5476	6
⋮	⋮		⋮	⋮	⋮	⋮		⋮	⋮
15	4.60094		7.26094	8.54675	22.3072	24.9958		32.8013	15
16	5.14224		7.96164	9.31223	23.5418	26.2962		34.2672	16
17	5.69724		8.67176	10.0852	24.7690	27.5871		35.7185	17
18	6.26481		9.39046	10.8649	25.9894	28.8693		37.1564	18
19	6.84398		10.1170	11.6509	27.2036	30.1435		38.5822	19
⋮	⋮		⋮	⋮	⋮	⋮		⋮	⋮

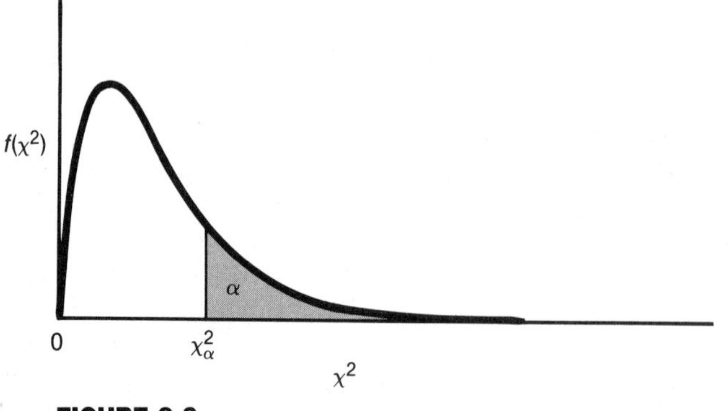

FIGURE 9.3

A chi-square distribution

will employ the test statistic

$$\chi^2 = \frac{(n-1)s^2}{\sigma_0^2}.$$

If σ^2 is really greater than the hypothesized value, σ_0^2, then the test statistic will tend to be large and will likely fall toward the upper tail of the distribution. If $\sigma^2 < \sigma_0^2$, the test statistic will tend to be small and will likely fall toward the lower tail of the χ^2 distribution. As in other statistical tests, we may use either a one- or two-tailed statistical test, depending upon the alternative hypothesis that we choose.

Test of an Hypothesis Concerning a Population Variance

Null Hypothesis, H_0: $\sigma^2 = \sigma_0^2$.

Alternative Hypothesis, H_a: $\sigma^2 \neq \sigma_0^2$, for a two-tailed statistical test.

Test Statistic: $$\chi^2 = \frac{(n-1)s^2}{\sigma_0^2}.$$

Rejection Region: For a two-tailed test, reject H_0 if $\chi^2 < \chi^2_{1-\alpha/2}$ or $\chi^2 > \chi^2_{\alpha/2}$, where $\chi^2_{1-\alpha/2}$ is that value of χ^2 such that $100(1 - \alpha/2)$ percent of the area under the χ^2 distribution, with $n - 1$ degrees of freedom, will lie to the right.

For alternatives, H_a: $\sigma^2 > \sigma_0^2$ or H_a: $\sigma^2 < \sigma_0^2$, use a one-tailed test and place all of α in the appropriate tail (upper or lower) of the χ^2 distribution.

EXAMPLE 9.4

A cement manufacturer claimed that concrete prepared from his product would possess a relatively stable compressive strength and that the strength measured in kilograms per square centimeter would lie within a range of 40 kilograms per square centimeter. A sample of $n = 10$ measurements produced a mean and variance equal to, respectively,

$$\bar{y} = 312,$$
$$s^2 = 195.$$

Do these data present sufficient evidence to reject the manufacturer's claim?

Solution As stated, the manufacturer claimed that the range of the strength measurements would equal 40 kilograms per square centimeter. We will suppose that he meant that the measurements would lie within this range 95 percent of the time and, therefore, that the range would equal approximately 4σ and that $\sigma = 10$. We would then wish to test the null hypothesis

$$H_0: \sigma^2 = (10)^2 = 100$$

against the alternative

$$H_a: \sigma^2 > 100.$$

The alternative hypothesis would require a one-tailed statistical test with the entire rejection region located in the upper tail of the χ^2 distribution. The critical value of χ^2 for $\alpha = .05$, $n = 10$, is $\chi^2 = 16.9190$, which implies that we will reject H_0 if the test statistic exceeds this value.

Calculating, we obtain

$$\chi^2 = \frac{(n-1)s^2}{\sigma_0^2} = \frac{1{,}755}{100} = 17.55.$$

Since the value of the test statistic falls in the rejection region, we conclude that the null hypothesis is false and that the range of concrete-strength measurements will exceed the manufacturer's claim.

A confidence interval for σ^2 with a $(1-\alpha)$ confidence coefficient can be shown to be

Confidence Interval for σ^2

$$\frac{(n-1)s^2}{\chi_U^2} < \sigma^2 < \frac{(n-1)s^2}{\chi_L^2},$$

where χ_L^2 and χ_U^2 are the lower and upper χ^2 values which would locate one-half of α in each tail of the chi-square distribution.

For example, a 90 percent confidence interval for σ^2, Example 9.4, would use

$$\chi_L^2 = \chi_{.95}^2 = 3.32511,$$
$$\chi_U^2 = \chi_{.05}^2 = 16.9190.$$

Then the interval estimate for σ^2 would be

$$\frac{9(195)}{16.9190} < \sigma^2 < \frac{9(195)}{3.32511}$$

or

$$103.73 < \sigma^2 < 527.80.$$

EXAMPLE 9.5

An experimenter was convinced that his measuring equipment possessed a variability measured by a standard deviation, $\sigma = 2$. During an experiment he recorded the measurements 4.1, 5.2, 10.2. Do these data disagree with his assumption? Test the hypothesis, H_0: $\sigma = 2$ or $\sigma^2 = 4$. Then place a 90 percent confidence interval on σ^2.

Solution The calculated sample variance is $s^2 = 10.57$. Since we wish to detect $\sigma^2 > 4$ as well as $\sigma^2 < 4$, we should employ a two-tailed test. When we use $\alpha = .10$ and place .05 in each tail, we will reject when $\chi^2 > 5.99147$ or $\chi^2 < .102587$.

The calculated value of the test statistic is

$$\chi^2 = \frac{(n-1)s^2}{\sigma_0^2} = \frac{2(10.57)}{4} = 5.29.$$

Since the test statistic does not fall in the rejection region, the data do not provide sufficient evidence to reject the null hypothesis, H_0: $\sigma^2 = 4$.

The corresponding 90 percent confidence interval is

$$\frac{(n-1)s^2}{\chi_U^2} < \sigma^2 < \frac{(n-1)s^2}{\chi_L^2}.$$

The values of χ_L^2 and χ_U^2 are

$$\chi_L^2 = \chi_{.95}^2 = .102587,$$
$$\chi_U^2 = \chi_{.05}^2 = 5.99147.$$

Substituting these values into the formula for the interval estimate, we obtain

$$\frac{2(10.57)}{5.99147} < \sigma^2 < \frac{2(10.57)}{.102587}$$

or

$$3.53 < \sigma^2 < 206.07.$$

Thus we estimate the population variance to fall in the interval, 3.53 to 206.07. This very wide confidence interval indicates how little information on the population variance is obtained in a sample of only three measurements. Consequently, it is not surprising that there was insufficient evidence to reject the null hypothesis, $\sigma^2 = 4$. To obtain more information on σ^2, the experimenter needs to increase the sample size.

EXERCISES

9.34 In Exercise 9.1 we mentioned that the EPA set a maximum noise level for heavy trucks in 1978 at 83 decibels. How this limit will be applied will greatly affect the industry and the public. One way to apply the limit would be to require all trucks to conform to the noise limit. A second but less satisfactory method would be to require the truck fleet mean noise level to be less than the limit. If the latter were the rule, variation in the noise level from truck to truck would be important because a large value of σ^2 would imply many trucks exceeding the limit, even if the mean fleet level was 83 decibels. The data for the six trucks, Exercise 9.1, in decibels, were

85.4, 86.8, 86.1, 85.3, 84.8, 86.0.

Use these data to construct a 90 percent confidence interval for σ^2, the variance of the truck noise emission readings. Interpret your results.

9.35 A precision instrument is guaranteed to read accurately to within 2 units. A sample of four instrument readings on the same object yielded the measurements 353, 351, 351, and 355. Test the null hypothesis that $\sigma = .7$ against the alternative $\sigma > .7$. Conduct the test at the $\alpha = .05$ level of significance.

9.36 Find a 90 percent confidence interval for the population variance in Exercise 9.35.

9.37 In order to properly treat patients, drugs prescribed by physicians must possess a potency that is accurately defined. Consequently, the distribution of potency values for shipments of a drug must not only possess a mean value as specified on the drug's container, but the variation in potency must be small. Otherwise, pharmacists would be distributing drug prescriptions that could be harmfully potent or possess a low potency that would be ineffective. A drug manufacturer claims that his drug is marketed with a potency of $5 \pm .1$ milligram per cubic centimeter. A random sample of four containers gave potency readings equal to 4.94, 5.09, 5.03, and 4.90 mg/cc.

(a) Do the data present sufficient evidence to indicate that the mean potency differs from 5 mg/cc?

(b) Do the data present sufficient evidence to indicate that the variation in potency differs from the error limits specified by the manufacturer? [*Hint:* It is sometimes difficult to determine exactly what is meant by limits on potency as specified by a manufacturer. Since he implies that the potency values will fall in the interval $5 \pm .1$ mg/cc with very high probability—the implication is "always"—let us assume that the range .2(.49 to .51) represents 6σ, as suggested by the Empirical Rule. Note that letting the range equal 6σ rather than 4σ places a stringent interpretation on the manufacturer's claim. We want the potency to fall in the interval $5.0 \pm .1$ with very high probability.]

9.38 Refer to Exercise 9.37. Testing of an additional 60 randomly selected containers of the drug gave a sample mean and variance equal to 5.04 and .0063 (for the total of $n = 64$ containers). Using a 95 percent confidence interval, estimate the variance of the manufacturer's potency measurements.

9.39 A manufacturer of hard safety hats for construction workers is concerned about the mean and the variation of the forces helmets transmit to wearers when subjected to a standard external force. The manufacturer desires the mean force transmitted by helmets to be 800 pounds (or less), well under the legal 1,000-pound limit, and σ to be less than 40. A random sample of $n = 40$ helmets was tested and the sample mean and variance were found to be equal to 825 pounds and 2,350 (pounds)^2, respectively.

(a) If $\mu = 800$ and $\sigma = 40$, is it likely that any helmet, subjected to the standard external force, will transmit a force to a wearer in excess of 1,000 pounds? Explain.

(b) Do the data provide sufficient evidence to indicate that when subjected to the standard external force, the mean force transmitted by the helmets exceeds 800 pounds?

(c) Do the data provide sufficient evidence to indicate that σ exceeds 40?

9.40 The EPA limit on the allowable discharge of suspended solids into rivers and streams is 60 milligrams per liter (mg per l) per day. A study of water samples selected from the discharge at a phosphate mine shows that over a long period of time, the mean daily discharge of suspended solids is 48 mg per l but the day-to-day discharge readings are very variable. State inspectors measured the discharge rates of suspended solids for $n = 20$ days and found $s^2 = 39$ (mg/l)^2. Find a 90 percent confidence interval for σ^2. Interpret your results.

COMPARING TWO POPULATION VARIANCES

The need for statistical methods to compare two population variances is readily apparent from the discussion in Section 9.6. We may frequently wish to compare the precision of one measuring device with that of another, the stability of one manufacturing process with that of another, or even the variability in the grading procedure of one college professor with that of another.

Intuitively, we might compare two population variances, σ_1^2 and σ_2^2, using the ratio of the sample variances s_1^2/s_2^2. If s_1^2/s_2^2 is nearly equal to 1, we would find little evidence to indicate that σ_1^2 and σ_2^2 are unequal. On the other hand, a very large or very small value for s_1^2/s_2^2 would provide evidence of a difference in the population variances.

How large or small must s_1^2/s_2^2 be in order that sufficient evidence exists to reject the null hypothesis,

$$H_0\colon \sigma_1^2 = \sigma_2^2?$$

The answer to this question may be acquired by studying the distribution of s_1^2/s_2^2 in repeated sampling.

When independent random samples are drawn from two normal populations with equal variances, that is, $\sigma_1^2 = \sigma_2^2$, then s_1^2/s_2^2 possesses a probability distribution in repeated sampling that is known to statisticians as an F distribution. We need not concern ourselves with the equation of the density function for F except to state that, as we might surmise, it is reasonably complex. For our purposes it will suffice to accept the fact that the distribution is well known and that critical values have been tabulated. These appear in Tables 6 and 7, Appendix II.

The shape of the F distribution is nonsymmetrical and will depend upon the number of degrees of freedom associated with s_1^2 and s_2^2. We will represent these quantities as ν_1 and ν_2, respectively. This fact complicates the tabulation of critical values for the F distribution and necessitates the construction of a table for each value that we may choose for a tail area, α. Thus, Tables 6 and 7 (Appendix II) present critical values corresponding to $\alpha = .05$ and .01, respectively. A typical F probability distribution ($\nu_1 = 10$ and $\nu_2 = 10$) is shown in Figure 9.4.

A partial reproduction of Table 6, Appendix II, is shown in Table 9.6. Table 6 records the value $F_{.05}$ such that the probability that F will exceed $F_{.05}$ is .05. Another way of saying this is that 5 percent of the area under the F distribution lies to the right of $F_{.05}$. The degrees of freedom for s_1^2, ν_1, are indicated across the top of the table, and the degrees of freedom for s_2^2, ν_2, appear in the first column on the left.

Referring to Table 9.6 we note that $F_{.05}$ for sample sizes $n_1 = 7$ and $n_2 = 10$ (that is, $\nu_1 = 6$, $\nu_2 = 9$) is 3.37. Likewise, the critical value, $F_{.05}$, for sample sizes

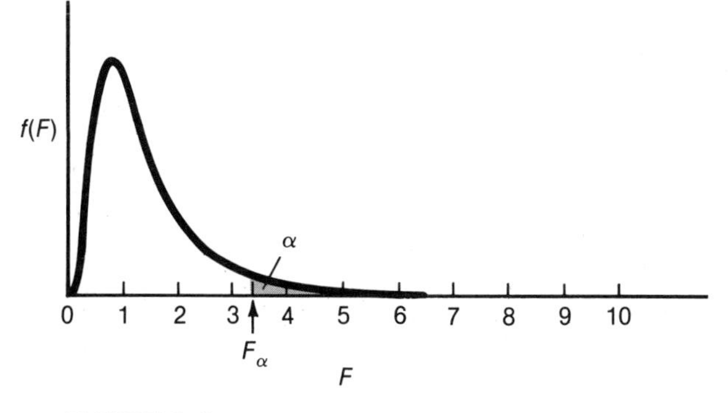

FIGURE 9.4

An F distribution with $\nu_1 = 10$ and $\nu_2 = 10$

$n_1 = 9$ and $n_2 = 16$ ($\nu_1 = 8$, $\nu_2 = 15$) is 2.64. These values of F are shaded in Table 9.6.

In a similar manner, the critical values for a tail area, $\alpha = .01$, are presented in Table 7, Appendix II. Thus

$$P(F > F_{.01}) = .01.$$

The statistical test of the null hypothesis,

$$H_0\colon \sigma_1^2 = \sigma_2^2,$$

utilizes the test statistic,

$$F = \frac{s_1^2}{s_2^2}.$$

TABLE 9.6

Format of the F table, Table 6, Appendix II, $\alpha = .05$

Numerator Degrees of Freedom; Denominator Degrees of Freedom (ν_2)

ν_2 \ ν_1	1	2	3	4	5	6	7	8	9	...	60	120	∞	ν_1 / ν_2
1	161.4	199.5	215.7	224.6	230.2	234.0	236.8	238.9	240.5		252.2	253.3	254.3	1
2	18.51	19.00	19.16	19.25	19.30	19.33	19.35	19.37	19.38		19.48	19.49	19.50	2
3	10.13	9.55	9.28	9.12	9.01	8.94	8.89	8.85	8.81		8.57	8.55	8.53	3
4	7.71	6.94	6.59	6.39	6.26	6.16	6.09	6.04	6.00		5.69	5.66	5.63	4
5	6.61	5.79	5.41	5.19	5.05	4.95	4.88	4.82	4.77		4.43	4.40	4.36	5
6	5.99	5.14	4.76	4.53	4.39	4.28	4.21	4.15	4.10		3.74	3.70	3.67	6
7	5.59	4.74	4.35	4.12	3.97	3.87	3.79	3.73	3.68		3.30	3.27	3.23	7
8	5.32	4.46	4.07	3.84	3.69	3.58	3.50	3.44	3.39		3.01	2.97	2.93	8
9	5.12	4.26	3.86	3.63	3.48	**3.37**	3.29	3.23	3.18		2.79	2.75	2.71	9
⋮	⋮	⋮	⋮	⋮	⋮	⋮	⋮	⋮	⋮		⋮	⋮	⋮	⋮
15	4.54	3.68	3.29	3.06	2.90	2.79	2.71	**2.64**	2.59		2.16	2.11	2.07	15
16	4.49	3.63	3.24	3.01	2.85	2.74	2.66	2.59	2.54		2.11	2.06	2.01	16
17	4.45	3.59	3.20	2.96	2.81	2.70	2.61	2.55	2.49		2.06	2.01	1.96	17
18	4.41	3.55	3.16	2.93	2.77	2.66	2.58	2.51	2.46		2.02	1.97	1.92	18
19	4.38	3.52	3.13	2.90	2.74	2.63	2.54	2.48	2.42		1.98	1.93	1.88	19
⋮	⋮	⋮	⋮	⋮	⋮	⋮	⋮	⋮	⋮		⋮	⋮	⋮	⋮

When the alternative hypothesis implies a one-tailed test, that is,

$$H_a: \sigma_1^2 > \sigma_2^2,$$

we may use the tables directly. However, when the alternative hypothesis requires a two-tailed test,

$$H_a: \sigma_1^2 \neq \sigma_2^2,$$

we note that the rejection region will be divided between the lower and upper tail of the F distribution and that tables of critical values for the lower tail are conspicuously missing. The reason for their absence is explained as follows: We are at liberty to identify either of the two populations as population I. If the population with the larger sample variance is designated as population II, then $s_2^2 > s_1^2$ and we will be concerned with rejection in the lower tail of the F distribution. Since the identification of the populations was arbitrary, we may avoid this difficulty by designating the population with the larger sample variance as population I. In other words, always place the larger sample variance in the numerator of

$$F = \frac{s_1^2}{s_2^2}$$

and designate that population as I. Then, since the area in the right-hand tail will represent only $\alpha/2$, we double this value to obtain the correct value for the probability of a type I error, α. Hence, if we use Table 6 for a two-tailed test, the probability of a type I error will be $\alpha = .10$.

Test of an Hypothesis Concerning the Equality of Two Population Variances

Null Hypothesis, $H_0: \sigma_1^2 = \sigma_2^2$.

Alternative Hypothesis, $H_a: \sigma_1^2 \neq \sigma_2^2$, for a two-tailed statistical test.

Test Statistic: $F = s_1^2/s_2^2$, where s_1^2 is the larger sample variance.

Rejection Region: For a two-tailed test, reject H_0 if $F > F_{\alpha/2}$, where $F_{\alpha/2}$ is based on $n_1 - 1$ and $n_2 - 1$ degrees of freedom and the value $\alpha/2$ is doubled to obtain the correct value for the probability of a type I error, α. (If we use Table 6 for a two-tailed test, the probability of a type I error will be $\alpha = .10$.)

We illustrate with examples.

EXAMPLE 9.6

Two samples consisting of 10 and 8 measurements each were observed to possess sample variances equal to $s_1^2 = 7.14$ and $s_2^2 = 3.21$, respectively. Do the sample variances present sufficient evidence to indicate that the population variances are unequal?

Solution Assume that the populations possess probability distributions that are reasonably mound-shaped and hence will satisfy, for all practical purposes, the assumption that the populations are normal.

We wish to test the null hypothesis,

$$H_0: \sigma_1^2 = \sigma_2^2,$$

against the alternative,

$$H_a: \sigma_1^2 \neq \sigma_2^2.$$

Using Table 6, Appendix II, and doubling the tail area, we will reject when $F > 3.68$ with $\alpha = .10$.

The calculated value of the test statistic is

$$F = \frac{s_1^2}{s_2^2} = \frac{7.14}{3.21} = 2.22.$$

Noting that the test statistic does not fall in the rejection region, we do not reject $H_0: \sigma_1^2 = \sigma_2^2$. Thus, there is insufficient evidence to indicate a difference in the population variances.

The confidence interval, with confidence coefficient $(1 - \alpha)$, for the ratio between two population variances can be shown to equal

A Confidence Interval for σ_1^2/σ_2^2

$$\frac{s_1^2}{s_2^2}\frac{1}{F_{v_1,v_2}} < \frac{\sigma_1^2}{\sigma_2^2} < \frac{s_1^2}{s_2^2}F_{v_2,v_1},$$

where F_{v_1,v_2} is the tabulated critical value of F corresponding to ν_1 and ν_2 degrees of freedom in the numerator and denominator of F, respectively. Similar to the two-tailed test, the α will be double the tabulated value. Thus F values extracted from Tables 6 and 7 will be appropriate for confidence coefficients equal to .90 and .98, respectively.

The 90 percent confidence interval for σ_1^2/σ_2^2, Example 9.6, is therefore

$$\frac{s_1^2}{s_2^2}\frac{1}{F_{v_1,v_2}} < \frac{\sigma_1^2}{\sigma_2^2} < \frac{s_1^2}{s_2^2}F_{v_2,v_1}.$$

Noting that

$$\nu_1 = (n_1 - 1) = 9 \quad \text{and} \quad \nu_2 = (n_2 - 1) = 7,$$

$$F_{v_1,v_2} = F_{9,7} = 3.68,$$

and

$$F_{v_2,v_1} = F_{7,9} = 3.29.$$

Substituting these values along with the sample variances into the formula for the confidence interval we obtain

$$\frac{7.14}{3.21}\frac{1}{3.68} < \frac{\sigma_1^2}{\sigma_2^2} < \frac{(7.14)(3.29)}{3.21}$$

or

$$.60 < \frac{\sigma_1^2}{\sigma_2^2} < 7.32.$$

The calculated interval estimate, .60 to 7.32, is observed to include 1.0, the value hypothesized in H_0. This indicates that it is quite possible that $\sigma_1^2 = \sigma_2^2$ and therefore agrees with our test conclusions. Do not reject H_0: $\sigma_1^2 = \sigma_2^2$.

EXAMPLE 9.7

The variability in the amount of impurities present in a batch of chemical used for a particular process depends upon the length of time the process is in operation. A manufacturer using two production lines, 1 and 2, has made a slight adjustment to process 2, hoping to reduce the variability as well as the average amount of impurities in the chemical. Samples of $n_1 = 25$ and $n_2 = 25$ measurements from the two batches yield means and variances as follows:

$$\bar{y}_1 = 3.2, \quad s_1^2 = 1.04;$$
$$\bar{y}_2 = 3.0, \quad s_2^2 = .51.$$

Do the data present sufficient evidence to indicate that the process variability is less for process 2? Test the null hypothesis, H_0: $\sigma_1^2 = \sigma_2^2$.

Solution The practical implications of Example 9.7 are illustrated in Figure 9.5. We believe that the mean levels of impurities in the two production lines are nearly equal (in fact, that they may be equal) but that there is a possibility that the variation in the level of impurities is substantially less for line 2. Then distributions of impurity measurements for the two production lines would have nearly the same mean level but they would differ in their variation. A large variance for the level of impurities increases the probability of producing shipments of chemical with an unacceptably high level of impurities. Consequently, we hope to show that the process change in line 2 has made σ_2^2 less than σ_1^2.

Testing the null hypothesis,

$$H_0: \sigma_1^2 = \sigma_2^2,$$

against the alternative,

$$H_a: \sigma_1^2 > \sigma_2^2,$$

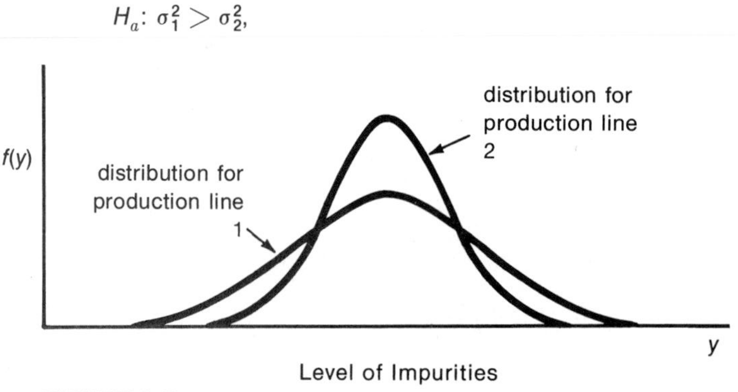

FIGURE 9.5

Distributions of impurity measurements for two production lines at an $\alpha = .05$ significance level, we will reject H_0 when F is greater than $F_{.05} = 1.98$; that is, we shall employ a one-tailed statistical test.

We readily observe that the calculated value of the test statistic,

$$F = \frac{s_1^2}{s_2^2} = \frac{1.04}{.51} = 2.04,$$

falls in the rejection region, and hence we conclude that the variability of process 2 is less than that for process 1.

The 90 percent confidence interval for the ratio σ_1^2/σ_2^2 is

$$\frac{s_1^2}{s_2^2}\frac{1}{F_{v_1,v_2}} < \frac{\sigma_1^2}{\sigma_2^2} < \frac{s_1^2}{s_2^2}F_{v_2,v_1}$$

$$\frac{1.04}{(.51)(1.98)} < \frac{\sigma_1^2}{\sigma_2^2} < \frac{(1.04)(1.98)}{.51}$$

or

$$1.03 < \frac{\sigma_1^2}{\sigma_2^2} < 4.04.$$

Thus we estimate the reduction in the variance of the amount of impurities to be as large as 4.04 to 1 or as small as 1.03 to 1. The actual reduction would likely be somewhere between these two extremes. This would suggest that the adjustment is quite effective in reducing the variation in the amount of impurities in the chemical.

EXERCISES

9.41 The stability of measurements of the characteristics of a manufactured product is important in maintaining product quality. In fact, it is sometimes better to possess small variation in the measured value of some important characteristic of a product and have the process mean slightly off target than to suffer wide variation with a mean value that perfectly fits requirements. The latter situation may produce a higher percentage of defective product than the former. A manufacturer of light bulbs suspected that one of his production lines was producing bulbs with a high variation in length of life. To test this theory, he compared the lengths of life of $n = 50$ bulbs randomly sampled from the suspect line and $n = 50$ from a line that seemed to be "in control." The sample means and variances for the two samples were as follows:

"Suspect Line"	Line "in Control"
$\bar{y}_1 = 1{,}520$	$\bar{y}_2 = 1{,}476$
$s_1^2 = 92{,}000$	$s_2^2 = 37{,}000$

Do the data provide sufficient evidence to indicate that bulbs produced by the "suspect line" possess a larger variance in length of life than those produced by the line that is assumed to be in control? Use $\alpha = .05$.

9.42 Use the method of Section 9.7 to obtain a 90 percent confidence interval for the variance ratio, Exercise 9.41.

9.43 A personnel officer, planning to use a Student's *t* test to compare the mean number of monthly absences for two categories of employees, noticed a possible difficulty. The variation in the numbers of absences per month seemed to differ for the two groups. As a check, the personnel officer randomly selected five months and counted the number of absences for each group. The data are shown in the table.

Category *A*	Category *B*
20	37
14	29
19	51
22	40
25	26

(a) About which assumption necessary for use of the *t* test was the personnel officer concerned?

(b) Do the data provide sufficient evidence to indicate that the variances differ for the populations of absences for the two employee categories? Test with $\alpha = .10$ and interpret the results of the test.

9.44 A pharmaceutical manufacturer purchases a particular material from two different suppliers. The mean level of impurities in the raw material is approximately the same for both suppliers but the manufacturer is concerned about the variability of the impurities from shipment to shipment. If the level of impurities tends to vary excessively for one source of supply, it could affect the quality of the pharmaceutical product. To compare the variation in percentage impurities for the two suppliers, the manufacturer selects ten shipments from each of the two suppliers and measures the percentage of impurities in the raw material for each shipment. The sample means and variances are shown in the table.

Supplier *A*	Supplier *B*
$\bar{y}_1 = 1.89$	$\bar{y}_2 = 1.85$
$s_1^2 = .273$	$s_2^2 = .094$
$n_1 = 10$	$n_2 = 10$

(a) Do the data provide sufficient evidence to indicate a difference in the variability of the shipment impurity levels for the two suppliers? Test using $\alpha = .10$. Based on the results of your test, what recommendation would you make to the pharmaceutical manufacturer?

(b) Find a 90 percent confidence interval for σ_B^2 and interpret your results.

9.45 The blood cholesterol values for randomly selected patients were compared for two diets, one low fat and the other normal. The sample means, variances, and sample sizes are as follows:

Low Fat	Normal
$\bar{y}_1 = 170$	$\bar{y}_2 = 196$
$s_1^2 = 198$	$s_2^2 = 435$
$n_1 = 19$	$n_2 = 24$

(a) Do the data provide sufficient evidence to indicate a difference in variation for the populations of patients from which the two samples were drawn? Use $\alpha = .10$.
(b) Why is the answer to part (a) of importance if we wish to determine whether the low-fat diet is effective in lowering blood cholesterol?

9.8
ASSUMPTIONS

As noted earlier, the tests and confidence intervals based on the Student's *t*, the chi-square, and the *F* statistic require that the data satisfy specific assumptions in order that the error probabilities (for the tests) and the confidence coefficients (for the confidence intervals) be equal to the values that we have specified. For example, if the assumptions are violated by selecting a sample from a nonnormal population, and the data are used to construct a 95 percent confidence interval for μ, the actual confidence coefficient might (unknown to us) be equal to .85 instead of .95. The assumptions are summarized next for your convenience.

Assumptions

1. For all tests and confidence intervals described in this chapter, it is assumed that samples are randomly selected from normally distributed populations.
2. When two samples are selected, we assume that they are selected in an independent manner except in the case of the paired-difference experiment.
3. For tests or confidence intervals concerning the difference between two population means, μ_1 and μ_2, based on independent random samples, we assume that $\sigma_1^2 = \sigma_2^2$.

In a practical sampling situation, you never know everything about the probability distribution of the sampled population. If you did, there would be no need for sampling or statistics. Second, it is highly unlikely that a population would possess, exactly, the characteristics described above. Consequently, to be useful, the inferential methods described in this chapter must give good inferences when moderate departures from the assumptions are present. For example, if the population possesses a mound-shaped distribution that is nearly normal, we would like a 95 percent confidence interval constructed for μ to be one with a confidence coefficient close to .95. Similarly, if we conduct a *t* test of the null hypothesis, $\mu_1 = \mu_2$, based on independent random samples from normal populations where σ_1^2 and σ_2^2 are not exactly equal, we want the probability of incorrectly rejecting the null hypothesis, α, to be approximately equal to the value we used in locating the rejection region.

A statistical method that is insensitive to departures from the assumptions upon which the method is based is said to be robust. The *t* tests are quite robust to moderate departures from normality. In contrast, the chi-square and *F* tests are sensitive to departures from normality. The t test for comparing two means is moderately robust to departures from the assumption $\sigma_1^2 = \sigma_2^2$ when $n_1 = n_2$. However, the test becomes sensitive to departures from this assumption as n_1 becomes large relative to n_2 (or vice versa).

If you are concerned that your data do not satisfy the assumptions prescribed for one of the statistical methods described in this chapter, you may be able to use a nonparametric statistical method to make your inference. These methods, which require few or no asumptions about the nature of the population probability distributions, are particularly useful for testing hypotheses and some nonparametric methods have been developed for estimating population parameters. Tests of hypotheses concerning the location of a population distribution or a test for the equivalence of two population distributions are presented in Chapter 14. If you can select relatively large samples, you need not worry about nonparametric estimation techniques. Then you can usually use a large-sample estimation procedure similar to those presented in Chapter 8.

9.9

SUMMARY

It is important to note that the t, χ^2, and F statistics employed in the small-sample statistical methods discussed in the preceding sections are based upon the assumption that the sampled populations possess a normal probability distribution. This requirement will be adequately satisfied for many types of experimental measurements.

You will observe the very close relationship connecting Student's t and the z statistic and, therefore, the similarity of the methods for testing hypotheses and the construction of confidence intervals. The χ^2 and F statistics employed in making inferences concerning population variances do not, of course, follow this pattern, but the reasoning employed in the construction of the statistical tests and confidence intervals is identical for all the methods we have presented.

A summary of the confidence intervals and statistical tests described in Chapters 8 and 9 is presented in Appendix I.

Tips on Problem Solving: To help you decide whether the techniques of this chapter are appropriate for the solution of a problem, ask yourself the following questions:

1. Does the problem imply that an inference should be made about a population mean or the difference between two means? Are the samples small, say $n < 30$? If the answers to both questions are "yes," you may be able to use one of the methods of Sections 9.3, 9.4, or 9.5. If the sample sizes are large $n \geq 30$, you can use the methods of Chapter 8. In practice, you would also need to verify that the assumptions underlying each procedure are satisfied. Are the population distributions nearly normal and have the sampling procedures conformed to those prescribed for the statistical method?

2. When comparing population means, were the observations from the two populations selected in a paired manner? If they were, you must use the paired-difference analysis of Section 9.5. If the samples were selected independently and in a random manner, use the methods of Section 9.4.

3. Is data variation the primary objective of the problem? If it is, you may be required to make an inference about a population variance, σ^2 (Section 9.6), or to compare two population variances, σ_1^2 and σ_2^2 (Section 9.7).

(*Note:* The tips on problem solving following Section 8.14 will also be helpful in solving problems in this chapter.)

REFERENCES

Barr, A. J., J. H. Goodnight, J. P. Sall, and J. T. Helwig, *A User's Guide to SAS 76.* Raleigh, N.C.: SAS Institute, Inc., 1976.

Brown, M. B., ed., *Biomedical Computer Programs.* Berkeley, Calif.: University of California, 1977.

Nie, N., C. H. Hull, J. G. Jenkins, K. Steinbrenner, and D. H. Bent, *Statistical Package for the Social Sciences,* 2nd ed. New York: McGraw-Hill Book Company, 1975.

Snedecor, G. W., and W. G. Cochran, *Statistical Methods,* 6th ed. Ames, Iowa: The Iowa State University Press, 1967.

Steel, R. G. D., and J. H. Torrie, *Principles and Procedures of Statistics.* New York: McGraw-Hill Book Company, 1960.

SUPPLEMENTARY EXERCISES

9.46 What assumptions are made when Student's t test is employed to test an hypothesis concerning a population mean?

9.47 A chemical process has produced, on the average, 800 tons of chemical per day. The daily yields for the past week are 785, 805, 790, 793, and 802 tons. Do these data indicate that the average yield is less than 800 tons and hence that something is wrong with the process? Test at the 5 percent level of significance.

9.48 Find a 90 percent confidence interval for the mean yield in Exercise 9.47.

9.49 Refer to Exercises 9.47 and 9.48. How large should the sample be in order that the width of the confidence interval be reduced to approximately 5 tons?

9.50 The mean and standard deviation for a sample of 19 measurements were found to equal 24.7 and 1.8, respectively. Find a 98 percent confidence interval for the mean of the population.

9.51 A manufacturer can tolerate a small amount (.05 milligrams per liter) of impurities in a raw material needed for manufacturing its product. Because the laboratory test for the impurities is subject to experimental error, the manufacturer tests each batch 10 times. Assume that the mean value of the experimental error is 0 and hence that the mean value of the 10 test readings is an unbiased estimate of the true amount of the impurities in the batch. For a particular batch of the raw material, the mean of the 10 test readings is .058 milligrams per liter (mg/l), with a standard deviation of .012 mg/l. Do the data provide sufficient evidence to indicate that the amount of impurities in the batch exceeds .05 mg/l?

9.52 A coin-operated soft-drink machine was designed to discharge, on the average, 7 ounces of beverage per cup. To test the machine, 10 cupfuls of beverage were drawn from the machine and measured. The mean and standard deviation of the 10 measurements were 7.1 and .12 ounces, respectively. Do these data present sufficient evidence to indicate that the mean discharge differs from 7 ounces? Test at the 10 percent level of significance.

9.53 Find a 90 percent confidence interval for the mean discharge in Exercise 9.52.

9.54 The main stem growth measured for a sample of 17 four-year-old red pine trees produced a mean and standard deviation equal to 11.3 and 3.4 inches, respectively. Find a 90 percent confidence interval for the mean growth of a population of four-year-old red pine trees subjected to similar environmental conditions.

9.55 In a general chemistry experiment, it is desired to determine the amount (in milliliters) of sodium (NaOH) solution needed to neutralize 1 gram of a specified acid. This will be an exact amount, but when run in the laboratory, variation will occur as the result of experimental error. Three titrations are made using phenolphthalein as an indicator of the neutrality of the solution (pH equals 7 for a neutral solution). The three volumes of sodium hydroxide required to attain a pH of 7 in each of the three titrations are as follows: 82.10, 75.75, and 75.44 milliliters. Use a 90 percent confidence interval to estimate the mean number of milliliters required to neutralize 1 gram of the acid.

9.56 What assumptions are made about the populations from which independent random samples are obtained when utilizing the t distribution in making small-sample inferences concerning the difference in population means?

9.57 Two random samples, each containing 11 measurements, were drawn from normal populations possessing means μ_1 and μ_2, respectively, and a common variance, σ^2. The sample means and variances are as follows:

Population I	Population II
$\bar{y}_1 = 60.4$	$\bar{y}_2 = 65.3$
$s_1^2 = 31.40$	$s_2^2 = 44.82$

Do the data present sufficient evidence to indicate a difference between the population means? Test at the $\alpha = .10$ level of significance.

9.58 Find a 90 percent confidence interval for the difference between the population means in Exercise 9.57.

9.59 The temperature of operation of two paint-drying ovens associated with two manufacturing production lines was recorded for 20 days. (Pairing was ignored.) The means and variances of the two samples are

$$\bar{y}_1 = 164, \quad \bar{y}_2 = 168,$$
$$s_1^2 = 81, \quad s_2^2 = 172.$$

Do the data present sufficient evidence to indicate a difference in temperature variability for the two ovens? Test the hypothesis that $\sigma_1^2 = \sigma_2^2$ at the $\alpha = .10$ level of significance.

9.60 A production plant has two extremely complex fabricating systems, with one being twice the age of the other. Both systems are checked, lubricated, and maintained once every two weeks. The number of finished products fabricated daily by each of the systems is recorded for 30 working days. The results are given in the table. Do these data present sufficient evidence to conclude that the variability in daily production warrants increased maintenance of the older fabricating system? Use a 5 percent level of significance.

New System	Old System
$\bar{y}_1 = 246$	$\bar{y}_2 = 240$
$s_1 = 15.6$	$s_2 = 28.2$

9.61 Four sets of identical twins (pairs *A*, *B*, *C*, and *D*) were selected at random from a population of identical twins. One child was taken at random from each pair to form an "experimental group." These four children were sent to school. The other four children were kept at home as a control group. At the end of the school year the following IQ scores were obtained:

Pair	Experimental Group	Control Group
A	110	111
B	125	120
C	139	128
D	142	135

Does this evidence justify the conclusion that lack of school experience has a depressing effect on IQ scores? Use $\alpha = .10$.

9.62 A comparison of reaction times for two different stimuli in a psychological word-association experiment produced the following results when applied to a random sample of 16 people:

Stimulus	Reaction Time (sec)							
1	1	3	2	1	2	1	3	2
2	4	2	3	3	1	2	3	3

Do the data present sufficient evidence to indicate a difference in mean reaction time for the two stimuli? Test at the $\alpha = .05$ level of significance.

9.63 Refer to Exercise 9.62. Suppose that the word-association experiment had been conducted using 8 people as blocks and making a comparison of reaction time within each person; that is, each person would be subjected to both stimuli in a random order. The data for the experiment are as follows:

Person	Reaction Time (sec)	
	Stimulus 1	Stimulus 2
1	3	4
2	1	2
3	1	3
4	2	1
5	1	2
6	2	3
7	3	3
8	1	3

Do the data present sufficient evidence to indicate a difference in mean reaction time for the two stimuli? Test at the $\alpha = .05$ level of significance.

9.64 Obtain a 90 percent confidence interval for $(\mu_1 - \mu_2)$ in Exercise 9.62.

9.65 Obtain a 95 percent confidence interval for $(\mu_1 - \mu_2)$ in Exercise 9.63.

9.66 Analyze the data in Exercise 9.63 as though the experiment had been conducted in an unpaired manner. Calculate a 95 percent confidence interval for $(\mu_1 - \mu_2)$ and compare with the answer to Exercise 9.65. Does it appear that blocking increased the amount of information available in the experiment?

9.67 The following data give readings in foot-pounds of the impact strength on two kinds of packaging material. Determine whether there is evidence of a difference in mean strength between the two kinds of material. Test at the $\alpha = .10$ level of significance.

A	*B*
1.25	.89
1.16	1.01
1.33	.97
1.15	.95
1.23	.94
1.20	1.02
1.32	.98
1.28	1.06
1.21	.98
$\Sigma y_i = 11.13$	$\Sigma y_i = 8.80$
$\bar{y}_1 = 1.237$	$\bar{y}_2 = .978$
$\Sigma y_i^2 = 13.7973$	$\Sigma y_i^2 = 8.6240$

9.68 Would the amount of information extracted from the data in Exercise 9.67 be increased by pairing successive observations and analyzing the differences? Calculate 90 percent confidence intervals for $(\mu_1 - \mu_2)$ for the two methods of analysis (unpaired and paired) and compare the widths of the intervals.

9.69 When should one employ a paired-difference analysis in making inferences concerning the difference between two means?

9.70 An experiment was conducted to compare the density of cakes prepared from two different cake mixes, *A* and *B*. Six cake pans received batter *A* and six received batter *B*. Expecting a variation in oven temperature, the experimenter placed an *A* and a B side by side at six different locations within the oven. The six paired observations are as follows:

Mix	Density (oz/in.^{3})					
A	.135	.102	.098	.141	.131	.144
B	.129	.120	.112	.152	.135	.163

Do the data present sufficient evidence to indicate a difference in the average density for cakes prepared using the two types of batter? Test at the $\alpha = .05$ level of significance.

9.71 Place a 95 percent confidence interval on the difference between the average densities for the two mixes in Exercise 9.70.

9.72 The 1977 interim emission standards for hydrocarbons (HC) and carbon monoxide (CO) in automobile exhaust systems is 1.5 grams per mile (HC) and 15 grams per mile (CO). Analysis of the exhaust for a random sample of six automobiles of a particular 1977 model gave the following (HC) readings:

1.37, 1.44, 1.28, 1.51, 1.39, 1.32.

(a) Do the tests on the six automobiles provide sufficient evidence to indicate that the mean fleet automobile output of HC for this model is less than the 1.5 grams per mile limit? Test using $\alpha = .05$.

(b) Find a 90 percent confidence interval for the mean HC output for the fleet.

9.73 Two plastics, each produced by a different process, were tested for ultimate strength. The following measurements represent breaking load in units of 1,000 pounds per square inch:

Plastic 1	Plastic 2
15.3	21.2
18.7	22.4
22.3	18.3
17.6	19.3
19.1	17.1
14.8	27.7

Do the data present sufficient evidence to indicate a difference between the mean ultimate strengths for the two plastics?

9.74 Refer to Exercise 9.73. Find a 90 percent confidence interval for the difference between the means, $(\mu_1 - \mu_2)$.

9.75 Refer to Exercise 9.73. How many observations would be required to estimate $(\mu_1 - \mu_2)$ correct to within 500 pounds per square inch with a probability of .90?

9.76 Refer to Exercise 9.47. Find a 90 percent confidence interval for σ^2, the variance of the population of daily yields.

9.77 A manufacturer of a machine to package soap powder claimed that the machine could load cartons at a given weight with a range of no more than two-fifths of an ounce. The mean and variance of a sample of eight 3-pound boxes were found to equal 3.1 and .018, respectively. Test the hypothesis that the variance of the population of weight measurements is $\sigma^2 = .01$ against the alternative, $\sigma^2 > .01$. Use an $\alpha = .05$ level of significance.

9.78 Find a 90 percent confidence interval for σ^2 in Exercise 9.77.

9.79 Under what assumptions may the F distribution be used in making inferences about the ratio of population variances?

9.80 A dairy is in the market for a new bottle-filling machine and is considering models *A* and *B* manufactured by company *X* and company *Y*, respectively. If ruggedness, cost, and convenience are comparable in the two models, the deciding factor is the variability of fills (the model producing fills with the smaller variance being preferred). Let σ_1^2 and σ_2^2 be the fill variances for models *A* and *B*, respectively, and consider various tests of the null hypothesis $H_0: \sigma_1^2 = \sigma_2^2$. Obtaining samples of fills from the two machines and utilizing the test statistic s_1^2/s_2^2, one could set up as the rejection regions an upper-tail area, a lower-tail area, or a two-tailed area of the *F* distribution depending on his point of view. Which type of rejection region would be most favored by the following persons:
(a) The manager of the dairy? Why?
(b) A salesman for company *X*? Why?
(c) A salesman for company *Y*? Why?

9.81 The closing prices of two common stocks were recorded for a period of 15 days. The means and variances are

$$\bar{y}_1 = 40.33, \quad \bar{y}_2 = 42.54,$$
$$s_1^2 = 1.54, \quad s_2^2 = 2.96.$$

Do these data present sufficient evidence to indicate a difference in variability of the two stocks for the populations associated with the two samples?

9.82 Place a 90 percent confidence interval on the ratio of the two population variances in Exercise 9.81.

9.83 Place a 98 percent confidence interval on the population variance in Exercise 9.47.

9.84 Refer to Exercise 9.80. Wishing to demonstrate that the variability of fills is less for model *A* than for model *B*, a salesman for company *X* acquired a sample of 30 fills from a machine of model *A* and a sample of 10 fills from a machine of model *B*. The sample variances were $s_1^2 = .027$ and $s_2^2 = .065$, respectively. Does this result provide statistical support at the .05 level of significance for the salesman's claim?

9.85 A chemical manufacturer claims that the purity of his product never varies more than 2 percent. Five batches were tested and gave purity readings of 98.2, 97.1, 98.9, 97.7, 97.9 percent. Do the data provide sufficient evidence to contradict the manufacturer's claim? (*Hint:* To be generous, let a range of 2 percent equal 4σ.)

9.86 Refer to Exercise 9.85. Find a 90 percent confidence interval for σ^2.

9.87 An educational psychologist wished to compare two methods, *A* and *B*, for teaching arithmetic. Ten pairs of students having the same IQ level of achievement in arithmetic were selected to participate in the experiment. From each pair, one student was randomly assigned to method *A* and the second to method *B*. Each student was tested at the end of a 4-week period. The following achievement scores were recorded:

Pair	Method *A*	Method *B*
1	36	35
2	37	35
3	41	40
4	42	41
5	36	36
6	35	34
7	42	40
8	33	31
9	40	39
10	38	37

(a) Do the data provide sufficient evidence to indicate that the mean achievement scores differ for the two methods? (Use $\alpha = .05$.)
(b) Estimate the mean difference in achievement scores using a 98 percent confidence interval.

9.88 A cannery prints "weight 16 ounces" on its label. The quality-control supervisor selects 9 cans at random and weighs them. He finds $\bar{y} = 15.7$ and $s = .5$. Do the data present sufficient evidence to indicate that the weight is less than that claimed on the label? (Use $\alpha = .05$.)

9.89 A psychologist wishes to verify that a certain drug increases the reaction time to a given stimulus. The following reaction times in tenths of a second were recorded before and after injection of the drug for each of four subjects:

	Reaction Time	
Subject	Before	After
1	7	13
2	2	3
3	12	18
4	12	13

Test at the 5 percent level of significance to determine whether the drug significantly increases reaction time.

9.90 A physician compared the blood pressures of four patients before and after treatment by a drug. The blood pressures are as follows:

	Blood Pressure	
Patient	Before Drug	After Drug
1	158	135
2	169	146
3	154	143
4	155	141

Estimate the mean drop in blood pressure using a 90 percent confidence interval.

9.91 A car dealer decided to compare the mean monthly sales of two salespersons, call them *A* and *B*. Because the strength of sales varies with the season and with people's opinions about the economy, the car dealer decided to make the comparison on a monthly basis. The data shown in the table at the top of page 322 give the monthly sales (to the nearest thousands of dollars) for the two salespersons.
(a) Do the data provide sufficient evidence to indicate a difference in mean sales for the two salespersons? Test with $\alpha = .05$.
(b) Find a 95 percent confidence interval for $\mu_A - \mu_B$ and interpret your results.

9.92 PCBs are found not only in the milk of pregnant women (Exercise 8.1) and in clams (Exercise 9.4); concentrations of PCBs have been found to be dangerously high in some game birds along the Coosa River in Georgia (*Environment News,* January 1977). The FDA considers concentrations higher than 5 parts per million (ppm) dangerous for human consumption. Six wild ducks tested in the Coosa area (four woodcocks and two mallards) showed the following PCB concentrations:

28, 26, 11, 7.8, 11.5, 11.5.

Month	Salesperson *A*	Salesperson *B*
January	130	105
February	141	109
March	163	147
April	176	159
May	147	150
June	160	134
July	145	123
August	129	130
September	104	91
October	139	124
November	163	141
December	151	147

If these ducks can be regarded as a random sample of ducks from the Coosa area, estimate the mean concentration per duck using a 90 percent confidence interval. Interpret the interval. What assumptions must you make in order that your test be valid?

9.93 How much combustion efficiency should a homeowner expect from an oil furnace? The EPA (*Environment News,* January 1977) states that 80 percent or above is excellent, 75 to 79 is good, 70 to 74 is fair, and below 70 percent is poor. A home heating contractor who sells two makes of oil heaters (call them *A* and *B*) decided to compare their mean efficiencies. An analysis was made of the efficiencies for 8 heaters of type *A* and 6 of type *B*. The efficiency ratings in percentages, for the 14 heaters are shown in the table.

Type *A*	Type *B*
72	78
78	76
73	81
69	74
75	82
74	75
69	
75	

(a) Do the data provide sufficient evidence to indicate a difference in mean efficiencies for the two makes of home heaters?
(b) Find a 90 percent confidence interval for $(\mu_A - \mu_B)$ and interpret the result.

9.94 At a time when energy conservation is so important, some scientists think we should give closer scrutiny to the cost (in energy) of producing various forms of food. One recent study compares the mean amount of oil required to produce one acre of different types of crops. For example, suppose that we wish to compare the mean amount of oil required to produce one acre of corn versus one acre of cauliflower. The readings in barrels of oil per acre, based on 20-acre plots, seven for each crop, are shown in the table. Use these data to find a 90 percent confidence interval for the difference in the mean amount of oil required to produce these two crops.

Corn	Cauliflower
5.6	15.9
7.1	13.4
4.5	17.6
6.0	16.8
7.9	15.8
4.8	16.3
5.7	17.1

9.95 The effect of alcohol consumption on the body appears to be much greater at high altitudes than at sea level. To test this theory, a scientist randomly selects 12 subjects and randomly divides them into two groups of 6 each. One group is transported to an altitude of 12,000 feet and each subject ingests a drink containing 100 cc of alcohol. The second group of 6 receives the same drink at sea level. After two hours, the amount of alcohol in the blood (grams per 100 cc) for each subject is measured. The data are shown in the table. Do the data provide sufficient evidence to support the theory that retention of alcohol in the blood is greater at high altitude? Test with $\alpha = .10$.

Sea Level	12,000 feet
.07	.13
.10	.17
.09	.15
.12	.14
.09	.10
.13	.14

EXPERIENCES WITH REAL DATA

Design an experiment to compare the means of two populations using either independent random samples selected from the two populations, Section 9.4, or a paired comparison as described in Section 9.5. Select sampling situations for which the t statistic, Section 9.4 or 9.5, would be appropriate. Conduct a statistical test to determine whether your data provide sufficient evidence to indicate a difference in the population means.

To provide some suggestions, utilize experimental data that you personally can collect in a biology, chemistry, physics, psychology, or geology laboratory course. Or you might utilize data collected from educational or sociological populations. As an illustration, you might wish to compare the mean cost of a bag of groceries purchased from two sections of a city, say from a high- and a low-income area. Identify 20 or 30 items that might constitute the purchases available in stores located in both areas of the city. Randomly select n_1 supermarkets from area 1 and n_2 from area 2. Record the prices of each item on the grocery list for each store in the sample and total the cost of the groceries for each store. Compare the mean prices of the grocery list for supermarkets in the two city areas.

Calculation of $\bar{y}$ and s^2 can most easily be accomplished on an electronic desk calculator, but you can use packaged programs and an electronic computer to perform the calculations. Useful packaged programs are available in the Biomed (Biomedical Programs), the SAS (Statistical Analysis Systems), and the SPSS (Statistical Package for the Social Sciences) program libraries (see the references).

CHAPTER OBJECTIVES

GENERAL OBJECTIVE Chapters 8 and 9 presented methods for making inferences about population means based on large random samples (Chapter 8) and small random samples (Chapter 9). The object of this chapter is to extend this methodology to consider the case in which the mean value of y is related to another variable, call it x. By making simultaneous observations on y and the x variable, we can use information contained in the x measurements to estimate the mean value of y and to predict particular values of y for preassigned values of x. This chapter will be devoted to the case where y is a linear function of one predictor variable x. The general case, where y is related to one or more predictor variables, say $x_1, x_2, \ldots, x_k$, will be discussed in Section 10.9.

SPECIFIC OBJECTIVES

1. To give practical illustrations of the types of problems that can be solved using the techniques of linear regression and correlation. *Sections 10.1 through 10.9*
2. To distinguish between deterministic and probabilistic models and to identify their advantages and limitations, and to present a linear probabilistic model for relating a response y to a single independent variable x. *Section 10.2*
3. To explain how the method of least squares can be used to fit a linear probabilistic model to data. *Section 10.3*
4. To provide a method for determining whether x contributes information for the prediction of y. *Sections 10.4, 10.5*
5. To present a confidence interval for estimating the mean value of y for a given value of x. *Section 10.6*
6. To give a prediction interval for predicting a particular value of y for a given value of x. *Section 10.7*
7. To present measures of the strength of the linear relationship between y and x—the simple linear coefficient of correlation and the coefficient of determination. *Sections 10.8, 10.9*
8. To familiarize you with a typical computer printout for an interesting application of regression analysis. *Section 10.10*
9. To emphasize the assumptions you must make in order for the estimation and test procedures described in this chapter to be valid. *Section 10.11*

10 LINEAR REGRESSION AND CORRELATION

10.1
INTRODUCTION

An estimation problem of more than casual interest to high school seniors, freshmen entering college, their parents, and a university administration concerns the expected academic achievement of a particular student after he has enrolled in a university. For example, we might wish to estimate a student's grade-point average at the end of the freshman year *before* the student has been accepted or enrolled in the university. At first glance this would seem to be a difficult task.

The statistical approach to this problem is, in many respects, a formalization of the procedure we might follow intuitively. If data were available giving the high school academic grades, psychological and sociological information, as well as the grades attained at the end of the college freshman year for a large number of students, we might categorize the students into groups possessing similar characteristics. Certainly, highly motivated students who have had a high rank in their high school class, have graduated from a high school with known superior academic standards, and so on, should achieve, on the average, a high grade-point average at the end of the college freshman year. On the other hand, students who lack proper motivation, who achieved only moderate success in high school, would not be expected, on the average, to do as well. Carrying this line of thought to the ultimate and idealistic extreme, we would expect the grade-point average of a student to be a *function* of the many variables that define the characteristics, psychological and physical, of the individual as well as those that define the environment, academic and social, to which he will be exposed. Ideally, we would like to possess a mathematical equation that would relate a student's grade-point average to all these independent variables so that it could be used for prediction.

You will observe that the problem we have defined is of a very general nature. We are interested in a random variable, y, that is related to a number of independent variables, $x_1, x_2, x_3, \ldots$. The variable, y, for our example, would be the student's grade-point average, and the independent variables might be

x_1 = rank in high school class,
x_2 = score on a mathematics achievement test,
x_3 = score on a verbal achievement test,

and so on. The ultimate objective would be to measure $x_1, x_2, x_3, \ldots$ for a particular student, substitute these values into the prediction equation, and thereby predict the student's grade-point average. To accomplish this end, we must first locate the related variables $x_1, x_2, x_3, \ldots$ and obtain a measure of the strength of their relationship to y. Then we must construct a good prediction equation that will express y as a function of the selected independent variables.

Practical examples of our prediction problem are very numerous in business, industry, and the sciences. The stockbroker wishes to predict stock market behavior as a function of a number of "key indices" which are observable and serve as the independent variables, $x_1, x_2, x_3, \ldots$. The manager of a manufacturing plant would like to relate yield of a chemical to a number of process variables. He would then use the prediction equation to find settings for the controllable process variables that would provide the maximum yield of the chemical. The personnel director of a corporation, like the admissions director of a university, wishes to test and measure individual characteristics so that he may hire the person best suited for a particular job. The biologist would like to relate body characteristics to the amounts of various glandular secretions. The political scientist may wish to relate success in a political campaign to the characteristics of a candidate, his opposition, and various campaign issues and promotional techniques. Certainly, all these prediction problems are, in many respects, one and the same.

In this chapter we shall be primarily concerned with the *reasoning* involved in acquiring a prediction equation based upon one or more independent variables. Thus we will restrict our attention to the simple problem of predicting y as a *linear* function of a *single* variable and observe that the solution for the multivariable problem, for example, predicting student grade-point average, will consist of a generalization of our technique. We shall show you how to fit a simple linear model to a set of data, a process called a regression analysis, and shall show you how to use the model for estimation and prediction. The methodology for finding the multivariable predictor, called a multiple regression analysis, is fairly complex, as will later be apparent to you, and is omitted from our discussion.

10.2
A SIMPLE LINEAR PROBABILISTIC MODEL

We will introduce our topic by considering the problem of predicting a student's final grade in a college freshman calculus course based upon his score on a mathematics achievement test administered prior to college entrance. As noted in Section 10.1, we wish to determine whether the achievement test is really worthwhile—whether the achievement test score is related to a student's grade in calculus—and, in addition, we wish to obtain an equation that may be useful for prediction purposes. The evidence, presented in Table 10.1, represents a sample of the achievement test scores and calculus grades for 10 college freshmen. Hopefully, the 10 students represent a random sample drawn from the population of freshmen who have already entered the university or will do so in the immediate future.

Our initial approach to the analysis of the data of Table 10.1 would be to plot the data as points on a graph, representing a student's calculus grade as y and the corresponding achievement test score as x. The graph is shown in Figure 10.1. You will quickly observe that y appears to increase as x increases. Do you think this arrangement of the points could have occurred due to chance even if x and y were unrelated?

One method of obtaining a prediction equation relating y to x would be to place a ruler on the graph and move it about until it seems to pass through the points and provide what we might regard as the "best fit" to the data. Indeed, if we were to draw a line through the points, it would appear that our prediction problem were solved. Certainly, we may now use the graph to predict a student's calculus grade as a function of his score on the mathematics achievement test.

TABLE 10.1
Mathematics achievement test scores and final calculus grades for college freshmen

Student	Mathematics Achievement Test Score	Final Calculus Grade
1	39	65
2	43	78
3	21	52
4	64	82
5	57	92
6	47	89
7	28	73
8	75	98
9	34	56
10	52	75

Furthermore, we note that we have chosen a *mathematical model* that expresses the supposed functional relation between y and x.

You should recall several facts concerning the graphing of mathematical functions. First, the mathematical equation of a straight line is

$$y = \beta_0 + \beta_1 x,$$

where β_0 is the y intercept and β_1 is the slope of the line. Second, the line that we may graph corresponding to any linear equation is unique. Each equation will correspond to only one line and vice versa. Thus, when we draw a line through

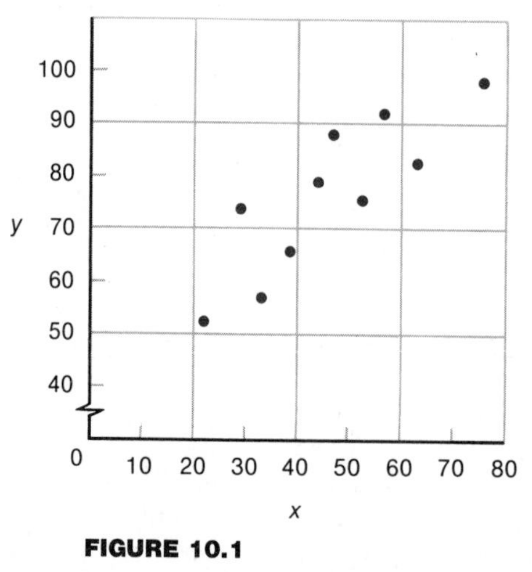

FIGURE 10.1
Plot of the data in Table 10.1

the points, we have automatically chosen a mathematical equation

$$y = \beta_0 + \beta_1 x,$$

where β_0 and β_1 have unique numerical values.

The linear model,

$$y = \beta_0 + \beta_1 x,$$

is said to be a deterministic mathematical model because when a value of x is substituted into the equation the value of y is determined and no allowance is made for error. Fitting a straight line through a set of points by eye produces a deterministic model. Many other examples of deterministic mathematical models may be found by leafing through the pages of elementary chemistry, physics, or engineering textbooks.

Deterministic models are quite suitable for explaining physical phenomena and predicting when the error of prediction is negligible for practical purposes. Thus, Newton's Law, which expresses the relation between the force, F, imparted by a moving body with mass, m, and acceleration, a, given by the deterministic model

$$F = ma,$$

predicts force with very little error for most practical applications. "Very little" is, of course, a relative concept. An error of .1 inch in forming an I-beam for a bridge is extremely small but would be impossibly large in the manufacture of parts for a wristwatch. Thus, in many physical situations the error of prediction cannot be ignored. Indeed, consistent with our stated philosophy, we would be hesitant to place much confidence in a prediction unaccompanied by a measure of its goodness. For this reason, a visual choice of a line to relate the calculus grade and achievement test score would be of limited utility.

In contrast to the deterministic model, we might employ a probabilistic mathematical model to explain a physical phenomenon. As we might suspect, probabilistic mathematical models contain one or more random elements with specified probability distributions. For our example, we shall relate the calculus score to the achievement test score by the equation

$$y = \beta_0 + \beta_1 x + \epsilon,$$

where ϵ is assumed to be a random variable with expected value equal to zero and variance equal to σ^2. In addition, we shall assume that any pair, ϵ_i and ϵ_j, corresponding to two observations, y_i and y_j, are independent. In other words, we assume that the *average* or expected value of y is linearly related to x and that observed values of y will deviate above and below this line by a random amount, ϵ. Furthermore, we have assumed that the distribution of errors about the line will be identically the same, regardless of the value of x, and that any pair of errors will be independent of one another. The assumed line, giving the expected value of y for a given value of x, is indicated in Figure 10.2 (next page). The probability distribution of the random error, ϵ, is shown for several values of x.

In Section 10.3, we shall consider the problem of finding the prediction equation, or regression line as it is commonly known in statistics.

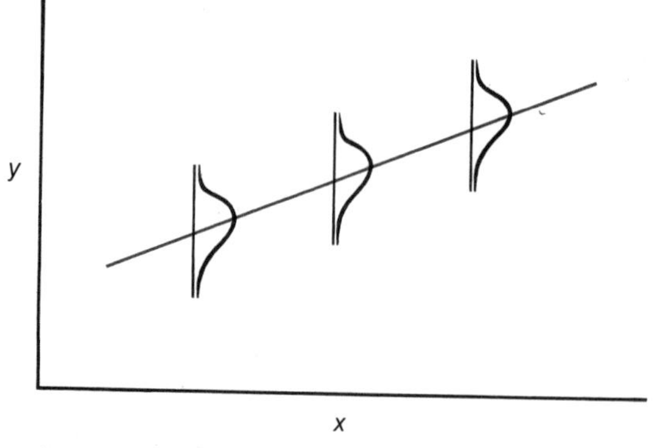

FIGURE 10.2
Linear probabilistic model

EXERCISES

10.1 Graph the line corresponding to the equation $y = 3x + 2$ by graphing the points corresponding to $x = 0$, 1, and 2. Give the y intercept and slope for the line.

10.2 Graph the line corresponding to the equation, $2y = x + 4$. Give the y intercept and slope for the line.

10.3 What is the difference between deterministic and probabilistic mathematical models?

10.3 THE METHOD OF LEAST SQUARES

The statistical procedure for finding the "best-fitting" straight line for a set of points would seem, in many respects, a formalization of the procedure employed when we fit a line by eye. For instance, when we visually fit a line to a set of data, we move the ruler until we think that we have minimized the deviations of the points from the prospective line. If we denote the predicted value of y obtained from the fitted line as $\hat{y}$, the prediction equation will be

$$\hat{y} = \hat{\beta}_0 + \hat{\beta}_1 x,$$

where $\hat{\beta}_0$ and $\hat{\beta}_1$ represent estimates of the true β_0 and β_1. This line for the data of Table 10.1 is shown in Figure 10.3. The vertical lines drawn from the prediction line to each point represent the deviations of the points from the predicted value of y. Thus the deviation of the ith point is

$$y_i - \hat{y}_i, \qquad \text{where} \qquad \hat{y}_i = \hat{\beta}_0 + \hat{\beta}_1 x_i.$$

Having decided that in some manner or other we will attempt to minimize the deviations of the points in choosing the best-fitting line, we must now define what we mean by "best." That is, we wish to define a criterion for "best fit" that will seem intuitively reasonable, which is objective, and which under certain conditions will give the best prediction of y for a given value of x.

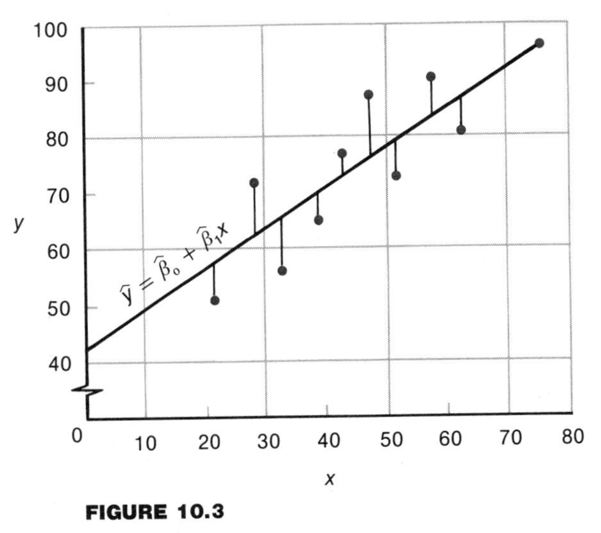

FIGURE 10.3
Linear prediction equation

We shall employ a criterion of goodness that is known as the principle of least squares and which may be stated as follows. Choose as the "best-fitting" line the line that minimizes the sum of squares of the deviations of the observed values of y from those predicted.* Expressed mathematically, we wish to choose values for $\hat{\beta}_0$ and $\hat{\beta}_1$ that minimize

$$\text{SSE} = \sum_{i=1}^{n} (y_i - \hat{y}_i)^2.$$

The symbol SSE represents the sum of squares of deviations or, as commonly called, the sum of squares for error.

Substituting for $\hat{y}_i$ in SSE, we obtain

$$\text{SSE} = \sum_{i=1}^{n} [y_i - (\hat{\beta}_0 + \hat{\beta}_1 x_i)]^2.$$

The method for finding the numerical values of $\hat{\beta}_0$ and $\hat{\beta}_1$ that minimize SSE utilizes the differential calculus and hence is beyond the scope of this text. We simply state that the least-squares solutions for $\hat{\beta}_0$ and $\hat{\beta}_1$ are given by the following formulas:

Least-Squares Estimators of β_0 and β_1

$$\hat{\beta}_1 = \frac{\text{SS}_{xy}}{\text{SS}_x} \quad \text{and} \quad \hat{\beta}_0 = \bar{y} - \hat{\beta}_1 \bar{x},$$

*The deviations of the points about the least-squares line satisfy another property. The sum of the deviations equals 0, i.e., $\sum_{i=1}^{n} (y_i - \hat{y}_i) = 0$ (proof omitted). Many other lines can be found that satisfy this property.

where

$$SS_x = \sum_{i=1}^{n} (x_i - \overline{x})^2 = \sum_{i=1}^{n} x_i^2 - \frac{\left(\sum_{i=1}^{n} x_i\right)^2}{n}$$

and

$$SS_{xy} = \sum_{i=1}^{n} (x_i - \overline{x})(y_i - \overline{y}) = \sum_{i=1}^{n} x_i y_i - \frac{\left(\sum_{i=1}^{n} x_i\right)\left(\sum_{i=1}^{n} y_i\right)}{n}.$$

Note that SS_x (sum of squares for x) is computed using the familiar shortcut formula for calculating sums of squares of deviations that was presented in Chapter 3. SS_{xy} is computed using a very similar formula and hence should be easy to remember. Once $\widehat{\beta}_0$ and $\widehat{\beta}_1$ have been computed, substitute their values into the equation of a line to obtain the least-squares prediction equation,

$$\widehat{y} = \widehat{\beta}_0 + \widehat{\beta}_1 x.$$

There is one important point to note here. Rounding errors can greatly affect the answer you obtain in calculating SS_x and SS_{xy}. If you must round a number, it is recommended that you carry six or more significant figures in the calculations. (Note also that in working exercises, rounding errors might cause some slight discrepancies between your answers and the answers given in the back of the text.)

EXAMPLE 10.1

Obtain the least-squares prediction line for the data of Table 10.1.

Solution The calculation of $\widehat{\beta}_0$ and $\widehat{\beta}_1$ for the data of Table 10.1 is simplified by the use of Table 10.2.

Substituting the appropriate sums from Table 10.2 into the least-squares equations, we obtain

$$SS_x = \sum_{i=1}^{n} x_i^2 - \frac{\left(\sum_{i=1}^{n} x_i\right)^2}{n} = 23{,}634 - \frac{(460)^2}{10}$$

$$= 2{,}474$$

$$SS_{xy} = \sum_{i=1}^{n} x_i y_i - \frac{\left(\sum_{i=1}^{n} x_i\right)\left(\sum_{i=1}^{n} y_i\right)}{n} = 36{,}854 - \frac{(460)(760)}{10}$$

$$= 1{,}894$$

$$\overline{y} = \frac{\sum_{i=1}^{n} y_i}{n} = \frac{760}{10} = 76$$

and

$$\overline{x} = \frac{\sum_{i=1}^{n} x_i}{n} = \frac{460}{10} = 46$$

TABLE 10.2

Calculations for the data, Table 10.1

	y_i	x_i	x_i^2	x_iy_i	y_i^2
	65	39	1,521	2,535	4,225
	78	43	1,849	3,354	6,084
	52	21	441	1,092	2,704
	82	64	4,096	5,248	6,724
	92	57	3,249	5,244	8,464
	89	47	2,209	4,183	7,921
	73	28	784	2,044	5,329
	98	75	5,625	7,350	9,604
	56	34	1,156	1,904	3,136
	75	52	2,704	3,900	5,625
Sum	760	460	23,634	36,854	59,816

Hence

$$\widehat{\beta}_1 = \frac{\text{SS}_{xy}}{\text{SS}_x} = \frac{1,894}{2,474} = .765562 \approx .77$$

and

$$\widehat{\beta}_0 = \bar{y} - \widehat{\beta}_1\bar{x} = 76 - (.765562)(46) = 40.7841 \approx 40.78.$$

Then, according to the principle of least squares, the best-fitting straight line relating the calculus grade to the achievement test score is

$$\widehat{y} = \widehat{\beta}_0 + \widehat{\beta}_1 x$$

or

$$\widehat{y} = 40.78 + .77x.$$

The graph of this equation is shown in Figure 10.3. Note that the y intercept, 40.78, is the value of $\widehat{y}$ when $x = 0$. The slope of the line, .77, gives the estimated change in y for a one-unit change in x.

We may now predict y for a given value of x by referring to Figure 10.3 or by substituting into the prediction equation. For example, if a student scored $x = 50$ on the achievement test, the student's predicted calculus grade would be

$$\widehat{y} = \widehat{\beta}_0 + \widehat{\beta}_1 x = 40.78 + (.77)(50) = 79.28.$$

Our next obvious step would be to place a bound upon our error of estimation. We shall consider this and related problems in succeeding sections.

Tips on Problem Solving

1. Be careful of rounding errors. Carry at least six significant figures in computing sums of squares of deviations or the sums of squares of cross products of deviations.

2. Always plot the data points and graph your least-squares line. If the line does not provide a reasonable fit to the data points, you may have committed an error in your calculations.

EXERCISES

10.4 Given five points whose coordinates are:

x	-2	-1	0	1	2
y	1	2	2	3	5

(a) Find the least-squares line for the data.
(b) As a rough check on the calculations in part (a), plot the five points and graph the line. Does the line appear to provide a good fit to the data points?

10.5 Given the points whose coordinates are:

x	1	2	3	4	5	6
y	4.7	3.9	3.7	3.0	2.6	1.9

(a) Find the least-squares line for the data.
(b) As a rough check on the calculations in part (a), plot the six points and graph the line. Does the line appear to provide a good fit to the data points?

10.6 The murder rate per 100,000 population increased in a city over a 6-year period as follows:

Year, x	1	2	3	4	5	6
Murder Rate, y	8	11	16	19	25	29

(a) Find the least-squares line relating y to x.
(b) As a check on your calculations, plot the six points and graph the line.

10.7 The 1977 EPA gas mileage guide gives the engine size and miles per gallon (mpg) ratings for eleven gasoline-fueled subcompact cars shown in the table. The engine sizes are in total cubic inches of cylinder volume.

Car	Cylinder Volume, x	mpg (Combined), y
VW Rabbit	388	34
Mazda 808	312	38
Toyota Corolla	284	41
Pinto	560	30
Chevette	340	33
Vega	560	28
Dodge Colt	392	35
Subaru	388	32
Toyota Celica	536	26
Datsun B-210	340	34
Buick Opel	444	27

(a) Plot the data points on graph paper.
(b) Find the least-squares line for the data.
(c) Graph the least-squares line to see how well it fits the data.
(d) Use the least-squares line to estimate the mean miles per gallon for a subcompact automobile which has 500 cubic inches of engine volume. (*Note:* We shall find a confidence interval for this mean in Section 10.6.)

10.4

CALCULATING s^2, AN ESTIMATOR OF σ^2

Recall that we constructed a probabilistic model for y in Section 10.2,

$$y = \beta_0 + \beta_1 x + \epsilon,$$

where ϵ is a random error with mean value equal to 0 and variance equal to σ^2. Thus each observed value of y is subject to a random error, ϵ, which will enter into the computations of $\hat{\beta}_0$ and $\hat{\beta}_1$ and will introduce errors in these estimates. Further, if we use the least-squares line,

$$\hat{y} = \hat{\beta}_0 + \hat{\beta}_1 x,$$

to predict some future value of y, the random errors will affect the error of prediction. Consequently, the variability of the random errors, measured by σ^2, plays an important role when estimating or predicting using the least-squares line.

The first step toward acquiring a bound on a prediction error requires that we estimate σ^2, the variance of the random error, ϵ. For this purpose it would seem reasonable to use SSE, the sum of squares of deviations (sum of squares for error) about the predicted line. Indeed, it can be shown that

$$\hat{\sigma}^2 = s^2 = \frac{\text{SSE}}{n - 2}$$

provides a good estimator for σ^2 which will be unbiased and be based upon $(n - 2)$ degrees of freedom.

The sum of squares of deviations, SSE, may be calculated directly by using the prediction equation to calculate $\hat{y}$ for each point, then calculating the deviations $(y_i - \hat{y}_i)$, and finally calculating

$$\text{SSE} = \sum_{i=1}^{n} (y_i - \hat{y}_i)^2.$$

This tends to be a very tedious procedure and is rather poor from a computational point of view because the numerous subtractions tend to introduce computational rounding errors. An easier and computationally better procedure is to use the formula

$$\text{SSE} = \text{SS}_y - \hat{\beta}_1 \text{SS}_{xy},$$

where

$$\text{SS}_y = \sum_{i=1}^{n} (y_i - \bar{y})^2 = \sum_{i=1}^{n} y_i^2 - \frac{\left(\sum_{i=1}^{n} y_i\right)^2}{n}$$

and

$$SS_{xy} = \sum_{i=1}^{n} x_i y_i - \frac{\left(\sum_{i=1}^{n} x_i\right)\left(\sum_{i=1}^{n} y_i\right)}{n}.$$

Note that SS_{xy} was used in the calculation of $\hat{\beta}_1$ and hence has already been computed.

EXAMPLE 10.2

Calculate an estimate of σ^2 for the data of Table 10.1.

Solution SS_{xy} and $\hat{\beta}_1$ were computed in Example 10.1. There we found $SS_{xy} = 1{,}894$ and $\hat{\beta}_1 = .765562$. To find SS_y, we need the values of $\sum_{i=1}^{n} y_i$ and $\sum_{i=1}^{n} y_i^2$ given in Table 10.2. Substituting, we have

$$SS_y = \sum_{i=1}^{n} y_i^2 - \frac{\left(\sum_{i=1}^{n} y_i\right)^2}{n}$$

$$= 59{,}816 - \frac{(760)^2}{10} = 2{,}056.$$

Then,

$$\begin{aligned} SSE &= SS_y - \hat{\beta}_1 SS_{xy} \\ &= 2{,}056 - (.765562)(1{,}894) \\ &= 606.03. \end{aligned}$$

Then, since the number of data points is $n = 10$,

$$s^2 = \frac{SSE}{n-2} = \frac{606.03}{8} = 75.754.$$

How can you interpret these values of SSE and s^2? Turn to Figure 10.3 and note the deviations of the $n = 10$ points from the least-squares line (shown as the vertical line segments between the points and the line). SSE $= 606.03$ is equal to the sum of squares of the numerical values of these deviations. This quantity is then used to calculate $s^2 = 75.754$ and $s = \sqrt{75.754} = 8.70$, estimates of σ^2 and σ.

The practical interpretation that can be given to s ultimately rests on the meaning of σ. Since σ measures the spread of the y values about the line of means $E(y) = \beta_0 + \beta_1 x$ (see Figure 10.2), we would expect (from the Empirical Rule) approximately 95 percent of the y values to fall within 2σ of that line. Since we do not know σ, $2s$ provides an approximate value for the half width of this interval. Now return to Figure 10.3 and note the location of the data points about the least-squares line. Since we used the $n = 10$ data points to fit the least-squares line, you would not be too surprised to find that most of the points fall within $2s = 2(8.7) = 17.40$ of the line. If you check Figure 10.3, you will see that all 10 points fall within $2s$ of the least-squares line. (You will find that, in general, most of the data points used to fit the least-squares line will fall within $2s$ of the line. This provides you with a rough check for your calculated value of s.)

But s will play a much more important role in this chapter than the application described. As mentioned at the beginning of this section, the less the variability of the y values about the line of means (i.e., the smaller the value of σ), the closer the least-squares line will be to the line of means. Consequently, s will play an important role in evaluating the goodness of all of the inferential methods described in this chapter.

Tips on Problem Solving

1. To reduce rounding error, always carry at least six significant figures when calculating SS_y and SSE. You can round when you obtain the answer for SSE if you desire.
2. As a check on your calculated value of s, remember that s measures the spread of the points about the least-squares line. Therefore, you would expect (by the Empirical Rule) most of the points to fall within $2s$ of the least-squares line. For example, if the points appear to fall in a band roughly equal to 4 units in width on the scale of the y variable and if your calculated value of s is 10, your value of s is too large. You have made an error. For example, perhaps you forgot to divide SSE by $(n - 2)$.

EXERCISES

10.8 Calculate SSE and s^2 for the data, Exercise 10.4.

10.9 Calculate SSE and s^2 for the data, Exercise 10.5.

10.10 Calculate SSE and s^2 for the data, Exercise 10.6.

10.11 Calculate SSE and s^2 for the data, Exercise 10.7.

10.5

INFERENCES CONCERNING THE SLOPE OF THE LINE, β_1

The initial inference desired in studying the relationship between y and x concerns the existence of the relationship. Does x contribute information for the prediction of y? That is, do the data present sufficient evidence to indicate that y increases (or decreases) linearly as x increases over the region of observation? Or, is it quite probable that the points would fall on the graph in a manner similar to that observed in Figure 10.1 when y and x are completely unrelated?

The practical question we pose concerns the value of β_1, which is the average change in y for a one-unit change in x. Stating that y does not increase (or decrease) linearly as x increases is equivalent to saying that $\beta_1 = 0$. Thus, we would wish to test an hypothesis that $\beta_1 = 0$ against the alternative that $\beta_1 \neq 0$. As we might suspect, the estimator, $\widehat{\beta}_1$, is extremely useful in constructing a test statistic to test this hypothesis. Therefore, we wish to examine the distribution of estimates, $\widehat{\beta}_1$, that would be obtained when samples, each containing n points, are repeatedly drawn from the population of interest. If we assume that the random error, ϵ, is normally distributed, in addition to the previously stated assumptions, it can be shown that both $\widehat{\beta}_0$ and $\widehat{\beta}_1$ will be normally distributed in repeated sampling and that the expected value and variance of $\widehat{\beta}_1$ will be

$$E(\widehat{\beta}_1) = \beta_1,$$

$$\sigma^2_{\widehat{\beta}_1} = \frac{\sigma^2}{\text{SS}_x}.$$

Thus, $\widehat{\beta}_1$ is an unbiased estimator of β_1, we know its standard deviation, and hence we can construct a z statistic in the manner described in Section 8.12. Then,

$$z = \frac{\widehat{\beta}_1 - \beta_1}{\sigma_{\widehat{\beta}_1}} = \frac{\widehat{\beta}_1 - \beta_1}{\sigma / \sqrt{\text{SS}_x}}$$

would possess a standardized normal distribution in repeated sampling. Since the actual value of σ^2 is unknown, we would wish to obtain the estimated standard deviation of $\widehat{\beta}_1$, which is $\dfrac{s}{\sqrt{\text{SS}_x}}$. Substituting s for σ in z, we obtain, as in Chapter 9, a test statistic,

$$t = \frac{\widehat{\beta}_1 - \beta_1}{s / \sqrt{\text{SS}_x}} = \frac{\widehat{\beta}_1 - \beta_1}{s} \sqrt{\text{SS}_x},$$

which can be shown to follow a Student's t distribution in repeated sampling with $(n - 2)$ degrees of freedom. Note that the number of degrees of freedom associated with s^2 determines the number of degrees of freedom associated with t. Thus we observe that the test of an hypothesis that β_1 equals some particular numerical value, say β_{10}, is the familiar t test encountered in Chapter 9.

Test of an Hypothesis Concerning the Slope of a Line

Null Hypothesis: H_0: $\beta_1 = \beta_{10}$.

Alternative Hypothesis: Specified by the experimenter, depending on the values of β_1 which he wishes to detect.

Test Statistic: $$t = \frac{\widehat{\beta}_1 - \beta_{10}}{s} \sqrt{\text{SS}_x}.$$

Rejection Region: See the critical values of t, Table 4, Appendix II, for $(n - 2)$ degrees of freedom.

EXAMPLE 10.3

Use the data of Table 10.1 to determine whether a linear relationship exists between a freshman's mathematics achievement test score, x, and his or her final calculus grade, y.

Solution We wish to test the null hypothesis

$$H_0: \beta_1 = 0$$

for the calculus grade–achievement test score data in Table 10.1. The test statistic will be

$$t = \frac{\hat{\beta}_1 - 0}{s} \sqrt{SS_x}$$

and, if we choose $\alpha = .05$, we will reject H_0 when $t > 2.306$ or $t < -2.306$. The critical value of t is obtained from the t table using $(n - 2) = 8$ degrees of freedom. Substituting into the test statistic, we obtain

$$t = \frac{\hat{\beta}_1}{s} \sqrt{SS_x} = \frac{.765562}{8.70} \sqrt{2{,}474}$$

or

$$t = 4.377.$$

Observing that the test statistic exceeds the critical value of t, we will reject the null hypothesis, $\beta_1 = 0$, and conclude that there is evidence to indicate that the calculus final grade is linearly related to the achievement test score.

Once we have decided that x and y are linearly related, we would be interested in examining this relationship in detail. If x increases by one unit, what is the predicted change in y and how much confidence can be placed in the estimate? In other words, we require an estimate of the slope β_1. You will not be surprised to observe a continuity in the procedures of Chapters 9 and 10. That is, the confidence interval for β_1, with confidence coefficient $(1 - \alpha)$, can be shown to be

$$\hat{\beta}_1 \pm t_{\alpha/2} \text{ (estimated } \sigma_{\hat{\beta}_1})$$

or

A Confidence Interval for β_1

$$\hat{\beta}_1 \pm \frac{t_{\alpha/2} s}{\sqrt{SS_x}},$$

where $t_{\alpha/2}$ is based on $(n - 2)$ degrees of freedom.

EXAMPLE 10.4

Find a 95 percent confidence interval for β_1 based upon the data of Table 10.1.

Solution The 95 percent confidence interval for β_1, based upon the data of Table 10.1, is

$$\hat{\beta}_1 \pm \frac{t_{.025} s}{\sqrt{SS_x}}.$$

Substituting, we obtain

$$.77 \pm \frac{(2.306)(8.70)}{\sqrt{2,474}}$$

or

$$.77 \pm .40.$$

Several points concerning the interpretation of our results deserve particular attention. As we have noted, β_1 is the slope of the assumed line over the region of observation and indicates the *linear* change in y for a one-unit change in x. Even though we do not reject the null hypothesis, $\beta_1 = 0$, x and y may be related. In the first place, we must be concerned with the probability of committing a type II error, that is, accepting the null hypothesis that the slope β_1 equals 0 when this hypothesis is false. Second, it is possible that x and y might be *perfectly* related in a curvilinear but not in a linear manner. For example, Figure 10.4 depicts a curvilinear relationship between y and x over the domain of x, $a \leq x \leq f$. We note that a straight line would provide a good predictor of y if fitted over a small interval in the x domain, say $b \leq x \leq c$. The resulting line would be line 1. On the other hand, if we attempt to fit a line over the region $c \leq x \leq d$, β_1 will equal zero and the best fit to the data will be the horizontal line 2. This would occur even though *all* the points fell perfectly on the curve and y and x possessed a functional relation as defined in Section 10.2. Thus, we must take care in drawing conclusions if we do not find evidence to indicate that β_1 differs from zero. Perhaps we have chosen the wrong type of probabilistic model for the physical situation.

Note that these comments contain a second implication. If the data provide values of x in an interval $b \leq x \leq c$, then the calculated prediction equation is appropriate only over this region. You can see that extrapolation in predicting y for values of x outside of the region, $b \leq x \leq c$, for the situation indicated in Figure 10.4 would result in a serious prediction error.

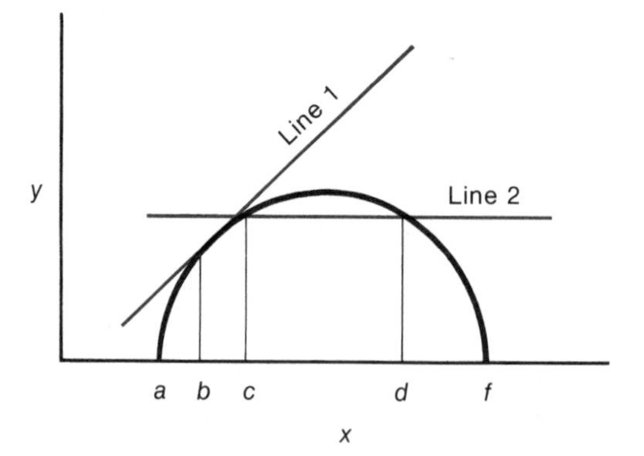

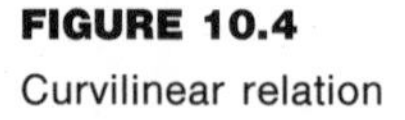
FIGURE 10.4
Curvilinear relation

Finally, if the data present sufficient evidence to indicate that β_1 differs from zero, we do not conclude that the true relationship between y and x is linear. Undoubtedly, y is a function of a number of variables which demonstrate their existence to a greater or lesser degree in terms of the random error ϵ which appears in the model. This, of course, is why we have been obliged to use a probabilistic model in the first place. Large errors of prediction imply either curvatures in the true relation between y and x, the presence of other important variables that do not appear in the model, or, as most often is the case, both. All we can say is that we have evidence to indicate that y changes as x changes and that we may obtain a better prediction of y using x and the linear predictor than simply using $\bar{y}$ and ignoring x. Note that this *does not* imply a *causal* relationship between x and y. A third variable may have caused the change in both x and y, producing the relationship that we have observed.

EXERCISES

10.12 Do the data, Exercise 10.4, present sufficient evidence to indicate that y and x are linearly related? (Test the hypothesis that $\beta_1 = 0$; use $\alpha = .05$.)

10.13 Find a 90 percent confidence interval for the slope of the line in Exercise 10.12. Interpret this interval estimate.

10.14 Do the data, Exercise 10.5, present sufficient evidence to indicate that y and x are linearly related? (Test the hypothesis that $\beta_1 = 0$; use $\alpha = .05$.)

10.15 Find a 95 percent confidence interval for the slope of the line in Exercise 10.14. Interpret this interval estimate.

10.16 Do the data, Exercise 10.6, present sufficient evidence to indicate that y and x are linearly related? (Test the hypothesis that $\beta_1 = 0$; use $\alpha = .05$.)

10.17 Find a 90 percent confidence interval for the slope of the line in Exercise 10.16. Give a practical interpretation to this interval estimate.

10.18 Do the data, Exercise 10.7, present sufficient evidence to indicate that y and x are linearly related? (Test the hypothesis that $\beta_1 = 0$; use $\alpha = .05$.)

10.19 Find a 95 percent confidence interval for the slope of the line in Exercise 10.18. Give a practical interpretation to this interval estimate.

10.20 A study was conducted to determine the effects of sleep deprivation on subjects' ability to solve simple problems. The amount of sleep deprivation varied over 8, 12, 16, 20, and 24 hours without sleep. A total of 10 subjects participated in the study, 2 at each sleep-deprivation level. After his specified sleep-deprivation period, each subject was administered a set of simple addition problems and the number of errors recorded. The following results were obtained:

Number of Errors, y	8, 6	6, 10	8, 14	14, 12	16, 12
Number of Hours Without Sleep, x	8	12	16	20	24

(a) Find the least-squares line appropriate to these data.
(b) Plot the points and graph the least-squares line as a check on your calculations.
(c) Calculate s^2.

10.21 Do the data in Exercise 10.20 present sufficient evidence to indicate that number of errors is linearly related to number of hours without sleep? (Test using $\alpha = .05$.) Would you expect the relation between y and x to be linear if x were varied over a wider range (say $x = 4$ to $x = 48$)?

10.22 Find a 95 percent confidence interval for the slope of the line in Exercise 10.20. Give a practical interpretation to this interval estimate.

10.23 A marketing research experiment was conducted to study the relationship between the length of time necessary for a buyer to reach a decision and the number of alternative package designs of a product presented. Brand names were eliminated from the packages to reduce the effects of brand preferences. The buyers made their selections using the manufacturer's product descriptions on the packages as the only buying guide. The length of time necessary to reach a decision is recorded for 15 participants in the marketing research study.

Length of Decision Time, y (sec)	5, 8, 8, 7, 9	7, 9, 8, 9, 10	10, 11, 10, 12, 9
Number of Alternatives, x	2	3	4

(a) Find the least-squares line appropriate for these data.
(b) Plot the points and graph the line as a check on your calculations.
(c) Calculate s^2.
(d) Do the data present sufficient evidence to indicate that the length of decision time is linearly related to the number of alternative package designs? (Test at the $\alpha = .05$ level of significance.)

10.24 Is the per capita consumption of cheese growing in the United States? A cheese importer would tell you that it depends upon the type. The data shown in the table give the per capita consumption of two types of cheeses, Swiss and the combination of Dutch cheeses, Edam and Gouda, for the period 1965 to 1976.*

		Per Capita Consumption (Pounds)	
Year	x = Year − 1970	Swiss	Edam and Gouda
1965	−5	.73	.07
1966	−4	.80	.10
1967	−3	.81	.10
1968	−2	.93	.15
1969	−1	.85	.10
1970	0	.90	.11
1971	1	.95	.10
1972	2	1.08	.11
1973	3	1.08	.12
1974	4	1.21	.11
1975	5	1.12	.11
1976	6	1.28	.11

(a) Let y represent the per capita consumption of Swiss cheese. Find a least-squares line appropriate for the data.

* *Source: Dairy Situation,* Economic Research Service, U.S. Department of Agriculture, September 1977.

(b) Plot the points and graph the line as a check on your calculations.
(c) Calculate s^2.
(d) Do the data provide sufficient increase to indicate that the mean annual change β_1 in the per capita consumption of Swiss cheese differs from 0? Test using $\alpha = .10$.
(e) Find a 90 percent confidence interval for the mean annual change in the consumption rate. Interpret this interval estimate.
(f) If you were a Swiss cheese importer, how much of an increase would you expect in the mean per capita consumption of Swiss cheese in 1978 over 1977?

10.25 Refer to Exercise 10.24 and perform the same type of data analysis for the per capita consumption measurements for the Dutch (Edam and Gouda) cheese.

10.6

ESTIMATING THE EXPECTED VALUE OF y FOR A GIVEN VALUE OF x

In Chapters 8 and 9 we studied methods for estimating a population mean, μ, and encountered numerous practical applications of these methods in the examples and exercises. Now let us consider a generalization of this problem.

Estimating the mean value of y for a given value of x [that is, estimating $E(y|x)$] can be a very important practical problem. If a corporation's profit, y, is linearly related to advertising expenditures, x, the corporation would wish to estimate the mean profit for a given expenditure, x. Similarly, a research physician might wish to estimate the mean response of a human to a specific drug dosage, x, and an educator might wish to know the mean grade expected of calculus students who acquired a mathematics achievement test score of $x = 50$. Let us see how our least-squares prediction equation can be employed to obtain these estimates.

Assume that x and y are linearly related according to the probabilistic model defined in Section 10.2 and therefore that $E(y|x) = \beta_0 + \beta_1 x$ represents the expected value of y for a given value of x. Since the fitted line

$$\hat{y} = \hat{\beta}_0 + \hat{\beta}_1 x$$

attempts to estimate the true linear relation (that is, we estimate β_0 and β_1), then $\hat{y}$ would be used to estimate the *expected* value of y as well as a *particular* value of y for a given value of x. It would seem quite reasonable to assume that the errors of estimation and prediction would differ for these two cases. In this situation, we consider the estimation of the expected value of y for a given value of x.

Observe that two lines are drawn in Figure 10.5. The first line represents the line of means for the true relationship,

$$E(y|x) = \beta_0 + \beta_1 x,$$

and the second is the fitted prediction equation,

$$\hat{y} = \hat{\beta}_0 + \hat{\beta}_1 x.$$

You can see that the error of estimating the expected value of y when $x = x_p$ will be the deviation between the two lines above the point x_p and that this

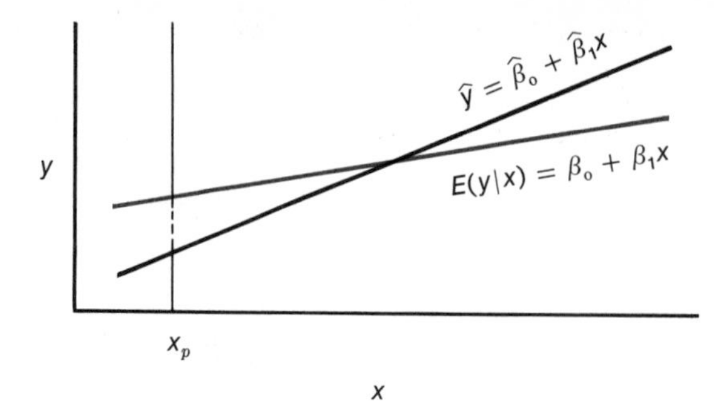

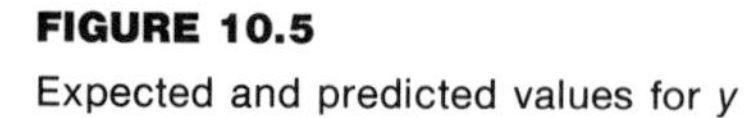

FIGURE 10.5

Expected and predicted values for y

error will increase as we move to the end points of the interval over which x has been measured. It can be shown that the predicted value,

$$\hat{y} = \hat{\beta}_0 + \hat{\beta}_1 x,$$

is an unbiased estimator of $E(y|x)$, that is, $E(\hat{y}) = \beta_0 + \beta_1 x$, and that it will be normally distributed with variance

$$\sigma_{\hat{y}}^2 = \sigma^2 \left[\frac{1}{n} + \frac{(x_p - \bar{x})^2}{SS_x}\right].$$

The corresponding estimated variance of $\hat{y}$ would use s^2 to replace σ^2 in the above expression.

The results outlined above may be used to test an hypothesis concerning the mean or expected value of y for a given value of x, say x_p. (This, of course, would also enable us to test an hypothesis concerning the y intercept, β_0, which is the special case where $x_p = 0$.) The null hypothesis would be

$$H_0\colon E(y|x = x_p) = E_0,$$

where E_0 is the hypothesized numerical value of $E(y)$ when $x = x_p$. Once again, it can be shown that the quantity

$$t = \frac{\hat{y} - E_0}{\text{estimated } \sigma_{\hat{y}}} = \frac{\hat{y} - E_0}{s\sqrt{\dfrac{1}{n} + \dfrac{(x_p - \bar{x})^2}{SS_x}}}$$

follows a Student's t distribution in repeated sampling with $(n - 2)$ degrees of freedom. Thus the statistical test is conducted in exactly the same manner as the other t tests previously discussed.

A Test Concerning the Expected Value of y

Null Hypothesis: H_0: $E(y|x = x_p) = E_0$.

Alternative Hypothesis: Specified by the experimenter depending on the values of $E(y|x)$ which he or she wishes to detect.

Test Statistic:
$$t = \frac{\hat{y} - E_0}{s\sqrt{\dfrac{1}{n} + \dfrac{(x_p - \bar{x})^2}{SS_x}}}.$$

Rejection Region: See the critical values of t, Table 4, Appendix II, for $(n - 2)$ degrees of freedom.

The corresponding confidence interval, with confidence coefficient $(1 - \alpha)$, for the expected value of y, given $x = x_p$, is

A Confidence Interval for $E(y|x)$

$$\hat{y} \pm t_{\alpha/2} s \sqrt{\frac{1}{n} + \frac{(x_p - \bar{x})^2}{SS_x}}.$$

where $t_{\alpha/2}$ is based on $(n - 2)$ degrees of freedom.

EXAMPLE 10.5

Find a 95 percent confidence interval for the expected value of y, the final calculus grade, given the mathematics achievement test score is $x = 50$.

Solution To estimate the mean calculus grade for students whose achievement test score was $x_p = 50$, we would use

$$\hat{y} = \hat{\beta}_0 + \hat{\beta}_1 x_p,$$

to calculate $\hat{y}$, the estimate of $E(y|x = 50)$. Then,

$$\hat{y} = 40.78 + (.77)(50) = 79.28.$$

The formula for the 95 percent confidence interval would be

$$\hat{y} \pm t_{.025} s \sqrt{\frac{1}{n} + \frac{(x_p - \bar{x})^2}{SS_x}}.$$

Substituting into this expression, we find the 95 percent confidence interval for the expected (mean) calculus grade, given an achievement test score of 50, is

$$79.28 \pm (2.306)(8.70)\sqrt{\frac{1}{10} + \frac{(50 - 46)^2}{2{,}474}}$$

or

$$79.28 \pm 6.55.$$

Thus we estimate that the mean calculus grade for the population of students acquiring mathematics achievement test scores of $x = 50$ will fall in the interval 79.28 ± 6.55, or 72.73 to 85.83.

EXERCISES

10.26 Refer to Exercise 10.4. Estimate the expected value of y given $x = 1$ using a 90 percent confidence interval.

10.27 Refer to Exercise 10.7. Find a 90 percent confidence interval for the mean number of miles per gallon (EPA combined) that you would expect to obtain from a subcompact with a 500-cubic-inch engine. Interpret the interval.

10.28 Refer to Exercise 10.20. Estimate the mean number of errors corresponding to $x = 20$ hours of sleep deprivation. Use a 95 percent confidence interval.

10.29 In manufacturing an antibiotic, the yield is a function of time. Data collected show that a process yielded the following pounds of antibiotic for the time periods shown:

Time, x (days)	1	2	3	4	5	6
Yield, y	23	31	40	46	52	63

(a) For various reasons, it is convenient to schedule production using a 4-day cycle. Estimate the mean yield for the amount of antibiotic produced over a 4-day period. Use a 95 percent confidence interval.

(b) In a practical situation, the yield for a zero length of time must be 0. Explain why the least-squares line does not go through the origin. Should it?

10.30 If you try to rent an apartment or buy a house, you will find that real estate representatives establish apartment rents and house prices on the basis of the square footage of the heated floor space. The data in the table give the square footages and sales prices of $n = 12$ houses randomly selected from those sold in a small city.

Square Feet x	Price y
1,460	$38,700
2,108	59,300
1,743	51,400
1,499	41,100
1,864	52,400
2,391	64,900
1,977	55,400
1,610	47,000
1,530	42,400
1,759	48,200
1,821	54,300
2,216	61,700

(a) Estimate the mean increase in the price for an increase of one square foot for houses sold in the city. Use a 90 percent confidence interval. Interpret your estimate.

(b) Suppose that you are a real estate salesperson and you desire an estimate of the mean sales price of houses with a total of 2,000 square feet of heated space. Use a 95 percent confidence interval and interpret your estimate.

(c) Calculate the price per square foot for each house and then calculate the sample mean. Why is this estimate of the mean cost per square foot not equal to the answer in part (a)? Should it? Explain.

10.7

PREDICTING A PARTICULAR VALUE OF y FOR A GIVEN VALUE OF x

Although the expected value of y for a particular value of x is of interest for our example, Table 10.1, we are primarily interested in *using* the prediction equation, $\hat{y} = \hat{\beta}_0 + \hat{\beta}_1 x$, based upon our observed data to predict the final calculus grade for some prospective student selected from the population of interest. That is, we want to use the prediction equation obtained for the 10 measurements, Table 10.1, to predict the final calculus grade for a new student selected from the population. If the student's achievement test score was x_p, we intuitively see that the error of prediction (the deviation between $\hat{y}$ and the actual grade, y, that the student will obtain) is composed of two elements. Since the student's grade will equal

$$y = \beta_0 + \beta_1 x_p + \epsilon,$$

$(y - \hat{y})$ will equal the deviation between $\hat{y}$ and the expected value of y, described in Section 10.6 (and shown in Figure 10.5), *plus* the random amount ϵ which represents the deviation of the student's grade from the expected value (see Figure 10.6). Thus the variability in the error for predicting a single value of y will exceed the variability for estimating the expected value of y.

It can be shown that the variance of the error of predicting a particular value of y when $x = x_p$, that is, $(y - \hat{y})$, is

$$\sigma^2_{\text{error}} = \sigma^2\left[1 + \frac{1}{n} + \frac{(x_p - \bar{x})^2}{SS_x}\right].$$

When n is very large, the second and third terms in the brackets will become small and the variance of the prediction error will approach σ^2. These results may be used to construct the following prediction interval for y, given $x = x_p$. The confidence coefficient for the prediction interval is $(1 - \alpha)$.

A Prediction Interval for y

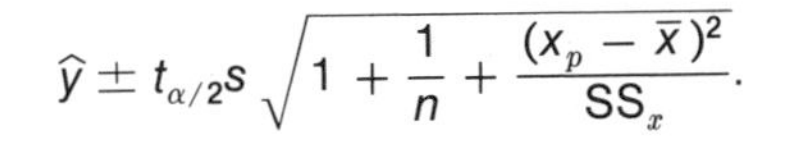

$$\hat{y} \pm t_{\alpha/2}s\sqrt{1 + \frac{1}{n} + \frac{(x_p - \bar{x})^2}{SS_x}}.$$

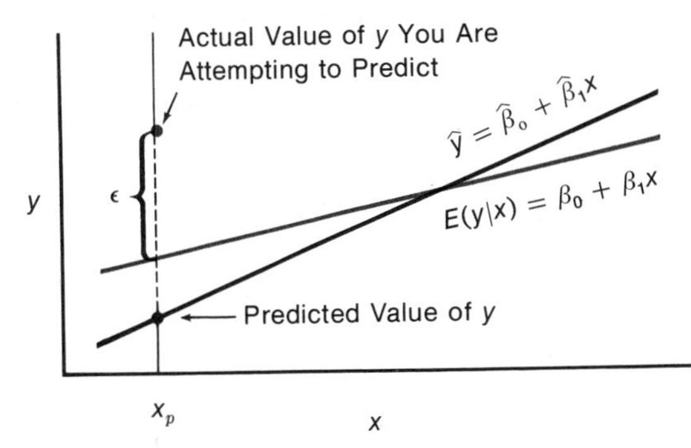

FIGURE 10.6
Error in predicting a particular value of y

EXAMPLE 10.6

Predict the final calculus grade for some new student who scored $x = 50$ on the mathematics achievement test.

Solution For example, if a prospective student scored $x_p = 50$ on the achievement test, we would predict that his final calculus grade would be

$$79.28 \pm (2.306)(8.70)\sqrt{1 + \frac{1}{10} + \frac{(50 - 46)^2}{2{,}474}}$$

or

$$79.28 \pm 21.10.$$

Note that in a practical situation we would likely possess the grades and achievement test scores for many more than the $n = 10$ students indicated in Table 10.1 and that this would reduce somewhat the width of the prediction interval.

Again, note the distinction between the confidence interval for $E(y|x)$ discussed in Section 10.6 and the prediction interval presented in this section. $E(y|x)$ is a mean, a parameter of a population of y values, and y is a random

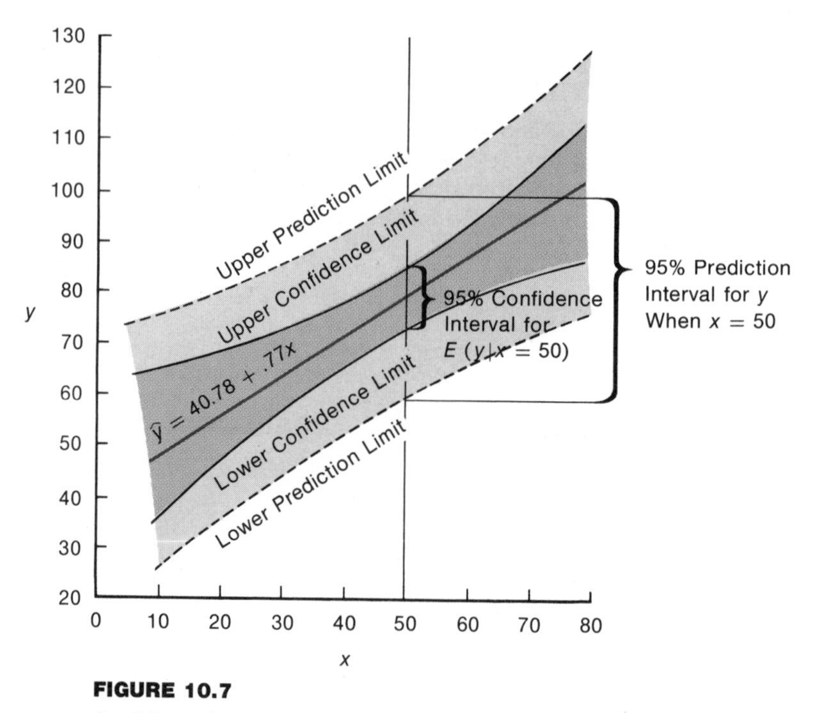

FIGURE 10.7

Confidence intervals for $E(y|x)$ and prediction intervals for y based on data of Table 10.1

variable that oscillates in a random manner about $E(y|x)$. The mean value of y when $x = 50$ is vastly different from some value of y, chosen at random from the set of all y values for which $x = 50$. To make this distinction when making inferences, we always *estimate* the value of a parameter and *predict* the value of a random variable. As noted in our earlier discussion and as shown in Figures 10.5 and 10.6, the error of predicting y is different from the error of estimating $E(y|x)$. This is evident in the difference in widths of the two prediction and confidence intervals.

A graph of the confidence interval for $E(y|x)$ and the prediction interval for a particular value of y for the data of Table 10.1 is shown in Figure 10.7. The plot of the confidence interval is shown by solid lines, the prediction interval is identified by dashed lines. Note how the widths of the intervals increase as you move to the right or left of $\bar{x} = 46$. Particularly, see the confidence interval and prediction interval for $x = 50$ which were calculated in Examples 10.5 and 10.6.

EXERCISES

10.31 Refer to Exercise 10.4. Suppose that a value of y is sampled at some future time when $x = 1$. Find a 90 percent prediction interval for y. Compare with the confidence interval on $E(y|x = 1)$ for Exercise 10.26.

10.32 Refer to Exercise 10.29. Suppose that the production process is operated for a 4-day period at some particular time in the future. Find a 95 percent prediction interval for the yield. Compare with the confidence interval on the mean yield for $x = 4$ days.

10.33 Refer to Exercise 10.24. Find a 95 percent prediction interval for the per capita Swiss cheese consumption in the United States for the year 1977. Interpret your prediction interval. Are there any qualifications that you would place on your prediction? Explain.

10.34 Refer to Exercise 10.30. Suppose that a house containing 1,780 square feet of heated floor space is offered for sale. Give a 90 percent prediction interval for the price at which the house will sell. Interpret this prediction.

10.35 In an effort to increase the production in the semiarid regions of the world, it is necessary that we be able to monitor the amount of moisture in the soil. Two methods are available. One utilizes mathematical formulas, temperatures, rainfalls, and evaporation rates to arrive at a computed value of moisture content. The other relies on direct measurement of the weight of soil before and after evaporation of the soil moisture. A comparison of the theoretically computed moisture content (in percent of dry weight of soil) with the measured moisture content is shown below for 9 soil samples randomly selected from a semiarid region.

Computed Soil Moisture, x	10.2	13.6	9.7	7.3	13.1	15.6	9.0	10.4	12.1
Measured Soil Moisture, y	11.8	15.4	9.2	8.0	13.8	16.1	8.4	10.2	10.5

(a) Fit a least-squares line to the data.

(b) If the computed soil moisture reading at some future time was 10.0, find a 90 percent prediction interval for the actual measured soil moisture.

10.36 Most sophomore physics students are required to conduct an experiment verifying Hooke's Law. Hooke's Law states that when a force is applied to a body that is long in comparison to its cross-sectional area, the change y in its length is proportional to the force x, i.e.,

$$y = \beta_1 x$$

where β_1 is a constant of proportionality. The results of an actual physics student's laboratory experiment are shown in the table. Six lengths of steel wire, .34 millimeters (mm) in diameter and 2 meters (m) long, were used to obtain the 6 force-length change measurements.

Force x (kg)	Change in Length y (mm)
29.4	4.25
39.2	5.25
49.0	6.50
58.8	7.85
68.6	8.75
78.4	10.00

(a) Fit the model, $y = \beta_0 + \beta_1 x + \epsilon$, to the data using the method of least squares.
(b) Plot the points and graph the line as a check on your calculations.
(c) Find a 95 percent confidence interval for the slope of the line.
(d) According to Hooke's Law, the line should pass through the point (0, 0), i.e., β_0 should equal 0. Test the hypothesis that $E(y) = 0$ when $x = 0$.
(e) Predict the elongation of a 2-m length of wire when a force of 55 kg is applied. Use a 95 percent prediction interval.

10.8

A COEFFICIENT OF CORRELATION

It is sometimes desirable to obtain an indicator of the strength of the linear relationship between two variables, y and x, which will be independent of their respective scales of measurement. We will call this a measure of the linear correlation between y and x.

The measure of linear correlation commonly used in statistics is called the Pearson product moment coefficient of correlation. This quantity, denoted by the symbol r, is defined as follows:

Pearson Product Moment Coefficient of Correlation

$$r = \frac{SS_{xy}}{\sqrt{SS_x SS_y}}$$

We shall show you how to compute the Pearson product moment coefficient of correlation for the grade-point data, Table 10.1, and then we shall explain how it measures the strength of the relationship between y and x.

EXAMPLE 10.7

Calculate the coefficient of correlation for the calculus grade–achievement test score data of Table 10.1.

Solution The coefficient of correlation for the calculus grade–achievement test score data, Table 10.1, may be obtained by using the formula for r and the quantities

$$\text{SS}_{xy} = 1{,}894$$
$$\text{SS}_x = 2{,}474$$

and

$$\text{SS}_y = 2{,}056,$$

which were computed previously.

Then,

$$r = \frac{\text{SS}_{xy}}{\sqrt{\text{SS}_x \text{SS}_y}} = \frac{1{,}894}{\sqrt{2{,}474(2{,}056)}} = .84.$$

A study of the coefficient of correlation, r, yields rather interesting results and explains the reason for its selection as a measure of linear correlation. We note that the denominators used in calculating r and $\widehat{\beta}_1$ will always be positive, since they both involve sums of squares of numbers. Since the numerator used in calculating r is identical to the numerator of the formula for the slope, $\widehat{\beta}_1$, the coefficient of correlation, r, will assume exactly the same sign as $\widehat{\beta}_1$ and will equal zero when $\widehat{\beta}_1 = 0$. Thus $r = 0$ implies no linear correlation between y and x. A positive value for r will imply that the line slopes upward to the right; a negative value indicates that it slopes downward to the right (see Figure 10.8).

The interpretation of nonzero values of r may be obtained by comparing the errors of prediction for the prediction equation

$$\widehat{y} = \widehat{\beta}_0 + \widehat{\beta}_1 x$$

with the predictor of y, $\bar{y}$, that would be employed if x were ignored. Figure 10.9(a) and (b) shows the lines $\widehat{y} = \widehat{\beta}_0 + \widehat{\beta}_1 x$ and $\widehat{y} = \bar{y}$ fit to the same set of data. Certainly, if x is of any value in predicting y, then SSE, the sum of squares of deviations of y about the linear model, should be less than the sum of squares about the predictor, $\bar{y}$, which would be

$$\text{SS}_y = \sum_{i=1}^{n} (y_i - \bar{y})^2.$$

Indeed, we see that SSE can *never* be larger than

$$\text{SS}_y = \sum_{i=1}^{n} (y_i - \bar{y})^2$$

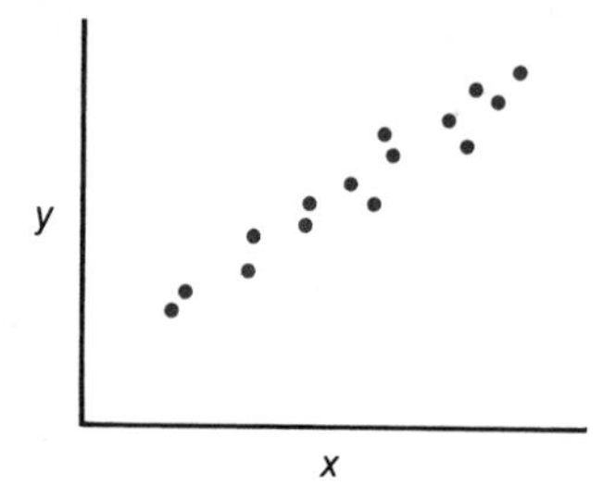

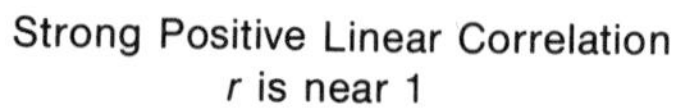
Strong Positive Linear Correlation
r is near 1

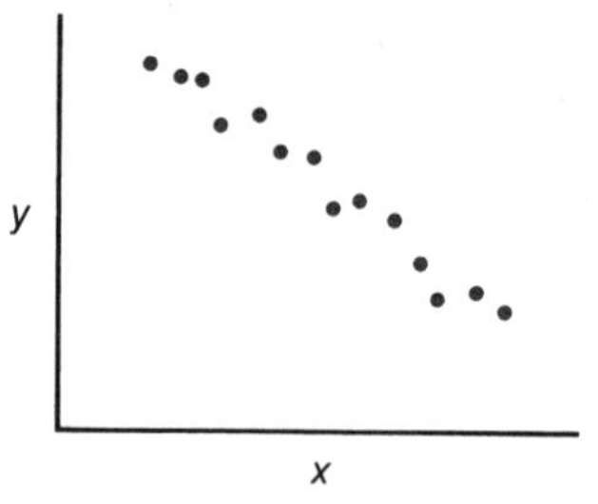

Strong Negative Linear Correlation
r is near -1

No Apparent Linear Correlation
r is near 0

Curvilinear, but Not Linear Correlation
r is near 0

FIGURE 10.8
Some typical scatter diagrams with approximate values of r

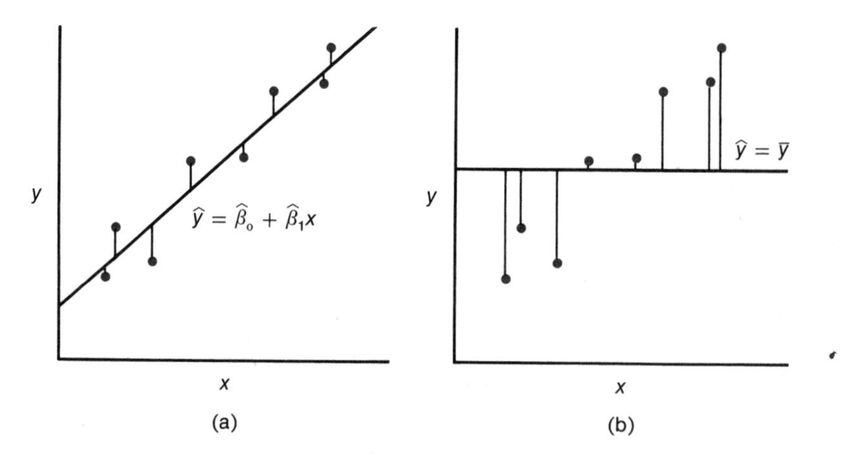

FIGURE 10.9
Two models fit to the same data

because

$$\text{SSE} = \text{SS}_y - \hat{\beta}_1 \text{SS}_{xy} = \text{SS}_y - \left(\frac{\text{SS}_{xy}}{\text{SS}_x}\right) \text{SS}_{xy}$$

$$= \text{SS}_y - \frac{(\text{SS}_{xy})^2}{\text{SS}_x}.$$

Therefore, SSE is equal to SS_y minus a positive quantity.

Furthermore, with the aid of a bit of algebraic manipulation, you can show that

$$r^2 = 1 - \frac{\text{SSE}}{\text{SS}_y}$$

$$= \frac{\text{SS}_y - \text{SSE}}{\text{SS}_y}.$$

In other words, r^2 will lie in the interval

$$0 \leq r^2 \leq 1$$

and r will equal $+1$ or -1 only when all the points fall exactly on the fitted line, that is, when SSE equals zero. Actually, we see that

The Coefficient of Determination

$$r^2 = \frac{\text{SS}_y - \text{SSE}}{\text{SS}_y} = \frac{\sum_{i=1}^{n} (y_i - \bar{y})^2 - \text{SSE}}{\sum_{i=1}^{n} (y_i - \bar{y})^2}$$

is equal to the ratio of the reduction in the sum of squares of deviations obtained by using the linear model to the total sum of squares of deviations about the sample mean $\bar{y}$, which would be the predictor of y if x were ignored. Thus r^2, called the coefficient of determination, would seem to give a more meaningful interpretation of the strength of the relation between y and x than would the correlation coefficient, r.

You will observe that the sample correlation coefficient, r, is an estimator of a population correlation coefficient, ρ, which would be obtained if the coefficient of correlation were calculated using all the points in the population. A discussion of a test of an hypothesis concerning the value of ρ is omitted as well as a bound on the error of estimation. Ordinarily, we would be interested in testing the null hypothesis that $\rho = 0$ and, since this is algebraically equivalent to testing the hypothesis that $\beta_1 = 0$, we have already considered this problem. If the evidence in the sample suggests that y and x are related, it would seem that we would redirect our attention to the ultimate objective of our data analysis, using the prediction equation to obtain interval estimates for $E(y|x)$ and prediction intervals for y.

While r gives a rather nice measure of the goodness of fit of the least-squares line to the fitted data, its use in making inferences concerning ρ would seem to be of dubious practical value in many situations. It would seem unlikely that a phenomenon y, observed in the physical sciences, and especially the social sciences, would be a function of a single variable. Thus the correlation coefficient between a student's predicted grade-point average and any one variable likely would be quite small and of questionable value. A larger reduction in SSE could possibly be obtained by constructing a predictor of y based upon a set of variables $x_1, x_2, \ldots$.

One further reminder concerning the interpretation of r is worthwhile. It is not uncommon for researchers in some fields to speak proudly of sample correlation coefficients, r, in the neighborhood of .5 (and, in some cases, as low as .1) as indicative of a "relation" between y and x. Certainly, even if these values were *accurate* estimates of ρ, only a very weak relation would be indicated. A value $r = .5$ would imply that the use of x in predicting y reduced the sum of squares of deviations about the prediction line by only $r^2 = .25$ or 25 percent. A correlation coefficient, $r = .1$, would imply only an $r^2 = .01$ or 1 percent reduction in the total sum of squares of deviations that could be explained by x.

If the linear coefficients of correlation between y and each of two variables, x_1 and x_2, were calculated to be .4 and .5, respectively, it does not follow that a predictor using both variables would account for a $[(.4)^2 + (.5)^2] = .41$ or a 41 percent reduction in the sum of squares of deviations. Actually, x_1 and x_2 might be highly correlated and therefore contribute the same information for the prediction of y.

Finally, we remind you that r is a measure of *linear* correlation and that x and y could be perfectly related by some curvilinear function when the observed value of r is equal to zero.

EXERCISES

10.37 How does the coefficient of correlation measure the strength of the linear relationship between two variables, y and x?

10.38 Describe the significance of the algebraic sign and the magnitude of r.

10.39 What value does r assume if all the sample points fall on the same straight line and if
(a) the line has positive slope?
(b) the line has negative slope?

10.40 An experiment was conducted in a supermarket to observe the relation between the amount of display space allotted to a brand of coffee (brand A) and its weekly sales. The amount of space allotted to brand A was varied over 3-, 6-, and 9-square-feet displays in a random manner over 12 weeks while the space allotted to competing brands was maintained at a constant 3 square feet for each. The following data were observed:

Weekly Sales, y (dollars)	526	421	581	630	412	560	434	443	590	570	346	672
Space Allotted, x (ft^2)	6	3	6	9	3	9	6	3	9	6	3	9

(a) Find the least-squares line appropriate for the data.
(b) Calculate r and r^2. Interpret.
(c) Find a 90 percent confidence interval for the mean weekly sales given that 6 square feet is allotted for display.
(d) Use a 90 percent prediction interval to predict the weekly sales at some time in the future if 6 square feet is allotted for display.
(e) By what percentage was the sum of squares of deviations reduced by using the least-squares predictor, $\hat{y} = \hat{\beta}_0 + \hat{\beta}_1 x$, rather than $\bar{y}$ as a predictor of y for these data?
(f) Would you expect the relation between y and x to be linear if x were varied over a wider range (say $x = 1$ to $x = 30$)?

10.41 For the following data:

x	1	2	3	4	5	6
y	7	5	5	3	2	0

(a) Find the least-squares line.
(b) Sketch the least-squares line on graph paper and plot the six points.
(c) Calculate the simple coefficient of correlation, r, and interpret.
(d) By what percentage was the sum of squares of deviations reduced by using the least-squares predictor, $\hat{y} = \hat{\beta}_0 + \hat{\beta}_1 x$, rather than $\bar{y}$ as a predictor of y?

10.42 Reverse the slope of the line, Exercise 10.41, by reordering the y observations:

x	1	2	3	4	5	6
y	0	2	3	5	5	7

Repeat the steps of Exercise 10.41. Notice the change in sign of r and the relation between the values of r^2 of Exercises 10.41 and 10.42.

10.43 Geothermal power could be an important source of energy in future years. Since the amount of energy contained in a pound of water is a function of its temperature, you might wonder whether water obtained from deeper wells contains more energy per pound. The data in the table is reproduced from an article on geothermal systems by A. J. Ellis.*

Location of Well	Average (Max.) Drill Hole Depth (m)	Average (Max.) Temperature (°C)
El Tateo, Chile	650	230
Ahuachapan, El Salvador	1,000	230
Namafjall, Iceland	1,000	250
Larderello (region), Italy	600	200
Matsukawa, Japan	1,000	220
Cerro Prieto, Mexico	800	300
Wairakei, New Zealand	800	230
Kizildere, Turkey	700	190
The Geysers, U.S.	1,500	250

*A. J. Ellis, "Geothermal Systems," *Amer. Scientist,* September–October 1975.

(a) Find the least-squares line.
(b) Sketch the least-squares line on graph paper and plot the nine points.
(c) Calculate the simple coefficient of correlation, r, and interpret it.
(d) By what percentage was the sum of squares of deviations reduced by using the least-squares predictor, $\hat{y} = \hat{\beta}_0 + \hat{\beta}_1 x$, rather than $\bar{y}$?

10.9

A MULTIVARIABLE PREDICTOR

A prediction equation based upon a number of variables, $x_1, x_2, \ldots, x_k$, could be obtained by the method of least squares in exactly the same manner as that employed for the simple linear model. For example, we might wish to fit the model

$$y = \beta_0 + \beta_1 x_1 + \beta_2 x_2 + \beta_3 x_3 + \epsilon,$$

where y = student grade-point average at the end of the freshman year, x_1 = rank in high school class divided by the number in class, x_2 = score on a mathematics achievement test, x_3 = score on a verbal and written achievement test, to data on the achievement of freshmen college students. (Note that we could add other variables as well as the squares, cubes, and cross products of x_1, x_2, and x_3.)

We would require a random sample of n freshmen selected from the population of interest and would record the values of y, x_1, x_2, and x_3, which, for each student, could be regarded as coordinates of a point in four-dimensional space. Then, ideally, we would like to possess a multidimensional "ruler" (in our case, a plane) that we could visually move about among the n points until the deviations of the observed values of y from the predicted values of y would in some sense be a minimum. Although we cannot graph points in four dimensions, the student readily recognizes that this device is provided by the method of least squares, which, mathematically, performs the task for us.

The sum of squares of deviations of the observed values of y from the fitted model would be

$$\text{SSE} = \sum_{i=1}^{n} (y_i - \hat{y}_i)^2$$

$$= \sum_{i=1}^{n} [y_i - (\hat{\beta}_0 + \hat{\beta}_1 x_{1i} + \hat{\beta}_2 x_{2i} + \hat{\beta}_3 x_{3i})]^2,$$

where $\hat{y} = \hat{\beta}_0 + \hat{\beta}_1 x_1 + \hat{\beta}_2 x_2 + \hat{\beta}_3 x_3$ is the fitted model and $\hat{\beta}_0$, $\hat{\beta}_1$, $\hat{\beta}_2$, and $\hat{\beta}_3$ are estimates of the model parameters. We would then use the calculus to find the estimates, $\hat{\beta}_0$, $\hat{\beta}_1$, $\hat{\beta}_2$, and $\hat{\beta}_3$, that make SSE a minimum. The estimates, as for the simple linear model, would be obtained as the solution of a set of four simultaneous linear equations known as the least-square equations.

Although the reasoning employed in obtaining the least-squares multivariable predictor is identical to the procedure developed for the simple linear model, a description of the least-squares equations, their solution, and related methods of inference becomes quite complex. These topics are omitted from this elementary discussion. Our primary objective in this chapter has been to indicate the generality and usefulness of the method of least squares as well as its role in prediction.

10.10 CASE STUDY: A COMPUTER ANALYSIS OF WHOOPING CRANE SIGHTINGS

The calculations associated with a multiple (or even a simple) regression analysis are so complex and time consuming, they are often performed on a computer. Since you may wish to use a computer to solve some of the problems in this chapter, we shall show you a computer printout of a regression analysis for an interesting set of data. This will help to familiarize you with the format of a computer printout and will give you a real-life application of regression analysis.

Our regression analysis will pertain to the sightings of whooping cranes, an endangered and protected species that is near extinction. The whooping crane, a native of North America, is the tallest living bird on the continent. According to wildlife officials, in 1941 there were only 21 known whooping cranes in the wilds and this number had dwindled to 15 by 1948. It is believed that the number of whooping cranes living, both in the wilds and in captivity, in 1977 is 100.

The birds, which winter in known wildlife refuges in the southwestern and south central United States, are monitored individually each year. The data we shall analyze are contained in an article by R. S. Miller and D. B. Botkin, which gives the number of whooping cranes sighted in the Arkansas National Wildlife Refuge from 1938 to 1972.* Miller and Botkin use the data to develop a prediction equation to forecast the number of whooping crane sightings in the Arkansas National Wildlife Refuge for 1975, 1980, etc. We shall use only a portion of their data, the number of whooping crane sightings at two-year intervals, for the period 1952 to 1972, and shall perform a computer regression analysis of the data. We shall then use our prediction equation to forecast the number of sightings in 1980 and shall compare our forecast with the forecast of Miller and Botkin.

The data for the 1952-to-1972 period are shown in Table 10.3. To simplify the numbers involved, we let $x = \text{year} - 1950$. The annual number of sightings appear in the "Count" column of Table 10.3.

The printouts for computer multiple regression analyses of data vary from one computer program to another but they are similar. Once you become familiar with the printout for one computer regression program, you will be able to quickly adapt this knowledge to others. The whooping crane data were analyzed using a SAS (Statistical Analysis System)† multiple regression computer program pack-

*Miller, R. S., and D. B. Botkin, "Endangered Species: Models and Predictions," *Amer. Scientist,* *62,* no. 2 (1974).
†See references.

TABLE 10.3

Whooping crane sightings, Arkansas National Wildlife Refuge, 1952 to 1972

Year	x = Year − 1950	Count y
1952	2	21
1954	4	21
1956	6	24
1958	8	32
1960	10	36
1962	12	32
1964	14	42
1966	16	43
1968	18	50
1970	20	57
1972	22	51

age. The printout is shown in Table 10.4. Many of the numbers appearing in the printout appear in the format used for a multiple regression analysis (the analysis for a multivariable linear model) and hence are not relevant to our discussion. The pertinent items in the printout are boxed and numbered individually or in groups. We will discuss each of these groups and show you how to extract the pertinent numbers from the printout.

1. Least-squares estimates of β_0 and β_1. The estimates are shown in the bottom left corner of the printout. Thus, to the nearest hundredth,

$$\hat{\beta}_0 = 15.36, \qquad \hat{\beta}_1 = 1.82,$$

and the prediction equation is

$$\hat{y} = 15.36 + 1.82x \qquad \text{where } x = \text{year} - 1950.$$

2. The degrees of freedom for s^2, the values of SSE, s^2, and s. At the top left of the printout, you will notice that the first four columns are marked "SOURCE," "DF," "SUM OF SQUARES," and "MEAN SQUARE." Go to the row of the table marked "ERROR" and proceed to the right. You will find d.f. = 9 in

TABLE 10.4

SAS computer printout for the regression analysis of whooping crane sightings

STATISTICAL ANALYSIS SYSTEM

GENERAL LINEAR MODELS PROCEDURE

DEPENDENT VARIABLE: COUNT

SOURCE	DF	SUM OF SQUARES	MEAN SQUARE	F VALUE	PR > F	(5) R-SQUARE	C.V.
MODEL	1	1454.54545455	1454.54545455	126.98	0.0001	0.933816	9.1025
ERROR	9	103.09090909	11.45454545		STD DEV		COUNT MEAN
CORRECTED TOTAL	10	1557.63636364			3.38445645		37.18181818

SOURCE	DF	TYPE I SS	F VALUE	PR > F	DF	TYPE IV SS	F VALUE	PR > F
YEAR	1	1454.54545455	126.98	0.0001	1	1454.54545455	126.98	0.0001

PARAMETER	(1) ESTIMATE	(3) T FOR H0: PARAMETER = 0	PR > |T|	(4) STD ERROR OF ESTIMATE
INTERCEPT	15.36363636	7.02	0.0001	2.18862574
YEAR	1.81818182	11.27	0.0001	0.16134763

the DF column, SSE = 103.0909 . . . in the SUM OF SQUARES column, and $s^2 = 11.4545$ . . . in the MEAN SQUARE column. The value of s is shown to be 3.384. . . .

3. Computed value of t for the test, H_0: $\beta_1 = 0$. The computed value of t for the t test, Section 10.5,

$$t = \frac{\widehat{\beta}_1}{s_{\widehat{\beta}_1}}$$

is given at the bottom of the third column under "T FOR H0: PARAMETER = 0." The first number under this heading, 7.02, gives the computed value of t to test the null hypothesis that the intercept, β_0, equals 0. Since this test is of little use to us, we ignore this value of t. The second number in the column (in the YEAR row) gives the value of t to test H_0: $\beta_1 = 0$. The computed value of t for this test is $t = 11.27$. Clearly, this value of t is so large, we reject the null hypothesis, $\beta_1 = 0$, and conclude $\beta_1 \neq 0$.

If you want the significance level for this test, it is given to the right of the t value under the column headed "PR > |T|." This column gives the probability of observing a value of t greater than or equal to the observed value, 11.27 (or less than $t = -11.27$). This value, .0001, is the significance level for the test.

4. Confidence interval for β_1. A confidence interval for β_1 is not given directly in the printout but it can be easily calculated. From Section 10.5, the confidence interval for β_1 is

$$\widehat{\beta}_1 \pm t_{\alpha/2}(\text{estimated standard deviation of } \widehat{\beta}_1)$$

or

$$\widehat{\beta}_1 \pm t_{\alpha/2}\, s_{\widehat{\beta}_1}.$$

The quantity $t_{\alpha/2}$ is given in Table 4, Appendix II. For d.f. = 9 and $\alpha = .05$, $t_{.025} = 2.262$. The value of $s_{\widehat{\beta}_1}$ is given in the last line of the printout under STD ERROR OF ESTIMATE. Thus, $s_{\widehat{\beta}_1} = .1613$ . . . . Substituting these values and the value of $\widehat{\beta}_1$ into the formula for the confidence interval,

$$\widehat{\beta}_1 \pm t_{\alpha/2}\, s_{\widehat{\beta}_1},$$
$$1.82 \pm (2.262)(.161),$$

or

$$1.82 \pm .36.$$

Therefore, the 95 percent confidence interval for β_1 is 1.46 to 2.18. Since β_1 is the mean increase (or decrease) in the crane population over a two-year period, we estimate this increase to fall in the interval 1.46 to 2.18 cranes per two-year period.

5. The coefficient of determination, r^2. The value of r^2 is given at the top right of the printout under R-SQUARE. Thus $r^2 = .9338$ . . . . The coefficient of correlation is not given in the printout but it will take the same sign (positive) as β_1 and can easily be calculated. Thus,

$$r = \sqrt{r^2} = \sqrt{.9338}$$

or

$$r = +.97.$$

Miller and Botkin do not state the type of probabilistic model they fit to their data but they give the values that they forecast for the number of cranes that will be observed in 1975, 1980, etc. and give the standard deviations for their prediction. Particularly, they forecast 70 cranes for 1980. Let us see how a forecast obtained from our least-squares prediction equation compares with Miller and Botkin's forecast.

In 1980,

$$x = \text{year} - 1950 = 30.$$

Then the predicted value of y for 1980 is

$$\begin{aligned}\hat{y} &= 15.36 + 1.82x \\ &= 15.36 + 1.82(30)\end{aligned}$$

or

$$\hat{y} = 69.96.$$

You will notice that this value is very close to the forecast of 70 cranes given by Miller and Botkin.

How good are these forecasts? Miller and Botkin give a value of 15.6 which they indicate by the symbol s.d. We will use the method of Section 10.7 to find a 95 percent prediction interval for our forecast. From Section 10.7, a 95 percent prediction interval for y is

$$\hat{y} \pm t_{.025} s \sqrt{1 + \frac{1}{n} + \frac{(x_p - \bar{x})^2}{SS_x}}$$

Since $\bar{x}$ and SS_x are not given explicitly in the printout, we can calculate them from Table 10.3. You can verify that

$$\sum_{i=1}^{11} x_i = 132 \qquad \text{and} \qquad \sum_{i=1}^{11} x_i^2 = 2{,}024.$$

Then

$$\bar{x} = \frac{\sum_{i=1}^{11} x_i}{n} = \frac{132}{11} = 12$$

and

$$SS_x = \sum_{i=1}^{11} x_i^2 - \frac{\left(\sum_{i=1}^{11} x_i\right)^2}{n} = 2{,}024 - \frac{(132)^2}{11} = 440.$$

Since we wish to predict the whooping crane population for $x = 30$ (the year 1980), we set $x_p = 30$. Substituting this value along with s, $t_{.025}$, n, $\bar{x}$, and SS_x into the formula for the prediction interval, we obtain

$$\bar{y} \pm t_{.025} s \sqrt{1 + \frac{1}{n} + \frac{(x_p - \bar{x})^2}{SS_x}}$$

$$69.96 \pm (2.262)(3.38) \sqrt{1 + \frac{1}{11} + \frac{(30 - 12)^2}{440}}$$

or

$$69.96 \pm 10.33.$$

As noted before, our least-squares prediction equation forecasts the same number of whooping crane sightings as forecast by Miller and Botkin. We give an error of prediction of 10.33. Clearly, this forecast must be viewed with caution because, as noted earlier, it is dangerous to predict y for values of x that fall outside the range of the x values used in fitting the model. Some very basic change in the system could occur between now and 1980 that would change the form of the relationship between x and y. Some new predator might appear on the scene, some pesticide might reduce their population, or efforts to clean up our environment might cause the whooping crane population to rapidly increase. Nevertheless, we have done our best. And in any case, all forecasting models, both deterministic and probabilistic, are subject to this risk.

10.11 ASSUMPTIONS

The assumptions for a regression analysis are as follows:

Assumptions for a Regression Analysis

1. The response y can be represented by the probabilistic model,

$$y = \beta_0 + \beta_1 x + \epsilon.$$

2. x is measured without error.
3. ϵ is a random variable such that, for a given value of x,

$$E(\epsilon) = 0$$
$$\sigma_\epsilon^2 = \sigma^2,$$

and all pairs, ϵ_i, ϵ_j, are independent in a probabilistic sense.
4. ϵ possesses a normal probability distribution.

At first glance, you might forget assumption 1 or fail to catch its significance. Models, deterministic or otherwise, are, as the name implies, only models for real relationships that occur in nature. Consequently, there will always be some error (slight if we are careful) due to the fact that there is curvature in the response curve. If you have obtained a good fit to the data, then the only difficulty that inherent lack of fit can cause is some problem if you attempt to use the model to predict y for some value of x outside the range of values used to fit the least-squares equation. Of course, this will always occur if x is time and you attempt to forecast y at some point in the future. But then this problem occurs with any kind of model for future time predictions. Consequently, you make the forecast but keep the limitations in mind.

The assumption that the variance of the ϵ's (and consequently y) is constant and equal to σ^2 will not be true for some types of data. Similarly, if x is time, say year, it is possible that y values measured over adjacent years will tend to be dependent (an overly large value of y in 1974 might signal a large value of y in 1975). Substantial departure from either of these assumptions will affect the confidence coefficients and significance levels associated with interval estimates and tests described in this chapter.

Like unequal variances and correlation of the random errors, if the normality assumption (4) is not satisfied, the confidence coefficients and significance levels for interval estimates and tests will not be what we expect them to be. But modest departures from normality will not seriously disturb these values.

10.12

SUMMARY

Although it was not stressed, you will observe that the prediction of a particular value of a random variable, y, was considered for the most elementary situation in Chapters 8 and 9. Thus, if we possessed no information concerning variables related to y, the sole information available for predicting y would be provided by its probability distribution. As we noted in Chapter 5, the probability that y would fall between two specific values, say y_1 and y_2, would equal the area under the probability distribution curve over the interval $y_1 \leq y \leq y_2$. And, if we were to select randomly one member of the population, we would most likely choose μ, or some other measure of central tendency, as the most likely value of y to be observed. Thus, we would wish to estimate μ, and this of course was considered in Chapters 8 and 9.

Chapter 10 is concerned with the problem of predicting y when auxiliary information is available on other variables, say $x_1, x_2, x_3, \ldots$, which are related to y and hence assist in its prediction. Primarily, we have concentrated on the problem of predicting y as a linear function of a single variable, x, which provides the simplest extension of the prediction problem beyond that considered in Chapters 8 and 9.

REFERENCES

Anderson, R. L., and T. A. Bancroft, *Statistical Theory in Research*. New York: McGraw-Hill Book Company, 1952.

Barr, A. J., J. H. Goodnight, J. P. Sall, and J. T. Helwig, *A User's Guide to SAS 76*. SAS Institute, P.O. Box 10066, Raleigh, N.C. 27605.

Brown, M. B., ed., *Biomedical Computer Programs*. Berkeley, Calif.: University of California, 1977.

Chapman, D. G., and R. A. Schaufele, *Elementary Probability Models and Statistical Inference*. New York: John Wiley & Sons, Inc., 1970.

Draper, N. R., and H. Smith, *Applied Regression Analysis*. New York: John Wiley & Sons, Inc., 1966.

Freund, J. E., *Mathematical Statistics,* 2nd ed. Englewood Cliffs, N.J.: Prentice-Hall, Inc., 1971.

Kendall, M. G., and A. Stuart, *The Advanced Theory of Statistics,* Vol. 2. New York: Hafner Press, 1961.

Mendenhall, W., *An Introduction to Linear Models and the Design and Analysis of Experiments*. Belmont, Calif.: Wadsworth Publishing Company, Inc., 1968.

Mendenhall, W., and R. L. Scheaffer, *Mathematical Statistics with Applications*. North Scituate, Mass.: Duxbury Press, 1973.

Nie, N., C. H. Hull, J. G. Jenkins, K. Steinbrenner, and D. H. Bent, *Statistical Package for the Social Sciences*, 2nd ed. New York: McGraw-Hill Book Company, 1975.

SUPPLEMENTARY EXERCISES

10.44 Graph the line corresponding to the equation $y = 2x + 1$ by locating points corresponding to $x = 0$, 1, and 2.

10.45 Given the linear equation $2x + 3y + 6 = 0$:
(a) Give the y intercept and slope for the line.
(b) Graph the line corresponding to the equation.

10.46 Follow the instructions given in Exercise 10.45 for the linear equation: $2x - 3y - 5 = 0$.

10.47 Follow the instructions given in Exercise 10.45 for the linear equation $x/y = 1/2$.

10.48 Given five points whose coordinates are:

y	0	0	1	1	3
x	−2	−1	0	1	2

(a) Find the least-squares line for the data.
(b) As a check on the calculations in (a), plot the five points and graph the line.

10.49 Given the following data for corresponding values of two variables, y and x:

y	2	1.5	1	2.5	2.5	4	5
x	−3	−2	−1	0	1	2	3

follow the instructions of Exercise 10.48.

10.50 Calculate s^2 for the data in Exercise 10.48.

10.51 Calculate s^2 for the data in Exercise 10.49.

10.52 Do the data in Exercise 10.48 present sufficient evidence to indicate that y and x are linearly related? (Test the hypothesis that $\beta_1 = 0$, using $\alpha = .05$.)

10.53 Do the data in Exercise 10.49 present sufficient evidence to indicate that y and x are linearly related? (Test the hypothesis that $\beta_1 = 0$, using $\alpha = .05$.)

10.54 For what configurations of sample points will s^2 be zero?

10.55 For what parameter is s^2 an unbiased estimator? Explain how this parameter enters into the description of the probabilistic model $y = \beta_0 + \beta_1 x + \epsilon$.

10.56 Find a 90 percent confidence interval for the slope of the line in Exercise 10.48.

10.57 Find a 95 percent confidence interval for the slope of the line in Exercise 10.49.

10.58 Refer to Exercise 10.48. Obtain a 90 percent confidence interval for the expected value of y when $x = 1$.

10.59 Refer to Exercise 10.49. Obtain a 95 percent confidence interval for the expected value of y when $x = -1$.

10.60 Refer to Exercise 10.49. Given that $x = 2$, find an interval estimate for a particular value of y. Use a confidence coefficient equal to .90.

10.61 Calculate the coefficient of correlation, r, for the data in Exercise 10.48. What is the significance of this particular value of r?

10.62 Calculate the coefficient of correlation for the data in Exercise 10.49.

10.63 By what percentage was the sum of squares of deviations reduced by using the least-squares predictor, $\hat{y} = \hat{\beta}_0 + \hat{\beta}_1 x$, rather than $\bar{y}$ as a predictor of y for the data in Exercise 10.49?

10.64 An experiment was conducted to observe the effect of an increase in temperature on the potency of an antibiotic. Three 1-ounce portions of the antibiotic were stored for equal lengths of time at each of the following temperatures: 30°, 50°, 70°, and 90°. The potency readings observed at the temperature of the experimental period were:

Potency Readings, y	38, 43, 29	32, 26, 33	19, 27, 23	14, 19, 21
Temperature, x	30°	50°	70°	90°

(a) Find the least-squares line appropriate for this data.
(b) Plot the points and graph the line as a check on your calculations.
(c) Calculate s^2.

10.65 Refer to Exercise 10.64. Estimate the change in potency for a one-unit change in temperature. Use a 90 percent confidence interval.

10.66 Refer to Exercise 10.64. Estimate the mean potency corresponding to a temperature of 50°. Use a 90 percent confidence interval.

10.67 Refer to Exercise 10.64. Suppose that a batch of the antibiotic were stored at 50° for the same length of time as the experimental period. Predict the potency of the batch at the end of the storage period. Use a 90 percent confidence interval.

10.68 Calculate the coefficient of correlation for the data in Exercise 10.64. Interpret the results.

10.69 A psychological experiment was conducted to study the relationship between the length of time necessary for a human being to reach a decision and the number of alternatives presented. The questions presented to the participants required a classification of an object into two or more classes, similar to the situation that one might encounter in grading potatoes. Five individuals classified one item each for a two-class, two-decision situation. Five each were also allotted to three-class and four-class categories. The length of time necessary to reach a decision is recorded for the 15 participants.

Length of Reaction Time, y (sec)	1, 3, 3, 2, 4	2, 4, 3, 4, 5	5, 6, 5, 7, 4
Number of Alternatives, x	2	3	4

(a) Find the least-squares line appropriate for this data.
(b) Plot the points and graph the line as a check on your calculations.
(c) Calculate s^2.

10.70 Do the data in Exercise 10.69 present sufficient evidence to indicate that the length of reaction time is linearly related to the number of alternatives? (Test at the $\alpha = .05$ level of significance.)

10.71 A comparison of 12 student grade-point averages at the end of the college freshman year with corresponding scores on an IQ test produced the following results:

G.P.A., y	2.1	2.2	3.1	2.3	3.4	2.9	2.9	2.7	2.1	1.7	3.3	3.5
IQ Score, x	116	129	123	121	131	134	126	122	114	118	132	129

(a) Find the least-squares prediction equation appropriate for the data.
(b) Graph the points and the least-squares line as a check on your calculations.
(c) Calculate s^2.

10.72 Do the data in Exercise 10.71 present sufficient evidence to indicate that x is useful in predicting y? (Test by use of $\alpha = .05$.)

10.73 Calculate the coefficient of correlation for the data in Exercise 10.71.

10.74 Refer to Exercise 10.71. Obtain a 90 percent confidence interval for the expected grade-point average given an IQ score equal to 120.

10.75 Use the least-squares equation in Exercise 10.71 to predict the grade-point average of a particular student whose IQ score is equal to 120. Use a 90 percent prediction interval.

10.76 An experiment was conducted to investigate the effect of a training program on the length of time for a typical male college student to complete the 100-yard dash. Nine students were placed in the program. The reduction y in time to complete the 100-yard dash was measured for three students at the end of 2 weeks, for three at the end of 4 weeks, and for three at the end of 6 weeks of training. The data are shown below:

Reduction in Time, y (seconds)	1.6, .8, 1.0	2.1, 1.6, 2.5	3.8, 2.7, 3.1
Length of Training, x (weeks)	2	4	6

(a) Find the least-squares line for these data.
(b) Estimate the mean reduction in running time after 4 weeks of training. Use a 90 percent confidence interval.

10.77 Refer to Exercise 10.76. Suppose that only three students had been employed in the experiment and that the reduction in running time was measured for each student at the end of 2, 4, and 6 weeks. Would the assumptions required for the confidence interval, Exercise 10.76(b), be satisfied? Explain.

10.78 The following coded data represent the chemical yield, y, for different settings of temperature, x:

x	−2	−1	0	1	2
y	4	3	3	2	1

(a) Calculate the least-squares line for this data.
(b) As a check on (a) plot the observed points, (x, y), and graph the fitted line, $\hat{y}$.
(c) Calculate SSE and s for the data.

10.79 Refer to Exercise 10.78. Do the data present sufficient evidence to indicate a *linear* relationship between x and y? Test at the $\alpha = .05$ level.

10.80 Refer to Exercise 10.78. Estimate the true value of β_1 using a 95 percent confidence interval.

10.81 Refer to Exercise 10.78.
(a) Predict a particular value of y for $x = 1$ using a 90 percent prediction interval.
(b) If you were to estimate the expected value of y for $x = 1$, would the bounds of error be larger or smaller? (Assume that the confidence coefficient is .90.)

10.82 Refer to Exercise 10.78.
(a) Calculate r, the coefficient of correlation.
(b) By what percentage has the sums of squares for error been reducing by using the linear predictor, $\hat{y}$, rather than $\bar{y}$?

10.83 In the accompanying table, x is the tensile force applied to a steel specimen in thousands of pounds and y is the resulting elongation in thousandths of an inch.

x	1	2	3	4	5
y	2	4	5	6	8

(a) Assuming the regression of y on x to be linear, find the least-squares line for the data.
(b) Plot the points and graph the line found in (a) as a check on your calculations.
(c) Calculate SSE and s for the data.

10.84 Refer to Exercise 10.83. Use a 95 percent prediction interval to predict the elongation if the experiment is to be run one more time at $x = 4$.

10.85 Refer to Exercise 10.83. Construct a 90 percent confidence interval for the mean change in elongation of the specimen per thousand pounds of tensile stress.

10.86 Refer to Exercise 10.83. If a force of 0 pounds is applied, the resulting elongation should be 0 units. Perform a test of an hypothesis to see if the line is consistent with the preceding statement. Use $\alpha = .05$.

10.87 Refer to Exercise 10.83.
(a) Find the coefficient of correlation, r, for the above set of data.
(b) By what percentage has the sum of squares of error been reduced by using the linear predictor, $\hat{y}$, rather than $\bar{y}$?

10.88 Suppose that the following data were collected on emphysema patients: the number of years the patient smoked and inhaled (x) and a physician's subjective evaluation of the extent of lung damage (y). The latter variable is measured on a scale of 0 to 100. Measurements taken on 10 patients are as follows:

Patient	Years Smoking, x	Lung Damage, y
1	25	55
2	36	60
3	22	50
4	15	30
5	48	75
6	39	70
7	42	70
8	31	55
9	28	30
10	33	35

(a) Calculate the coefficient of correlation, r, between years smoking (x) and lung damage (y).

(b) Calculate the coefficient of determination, r^2. Interpret r^2.
(c) Fit a least-squares line to the data. Graph the line and plot the data points. Compare with your computed least-squares line and your computed values of r and r^2.

10.89 Some varieties of nematodes, round worms that live in the soil and frequently are so small as to be invisible to the naked eye, feed upon the roots of lawn grasses and other plants. This pest, which is particularly troublesome in warm climates, can be treated by the application of nematicides. Data collected on the percent kill of nematodes for various rates of application (dosages given in pounds per acre of active ingredient) are as follows:

Rate of Application, x	2	3	4	5
Percent Kill, y	50, 56, 48	63, 69, 71	86, 82, 76	94, 99, 97

(a) Calculate the coefficient of correlation, r, between rates of application (x) and percent kill (y).
(b) Calculate the coefficient of determination, r^2, and interpret.
(c) Fit a least-squares line for the data.
(d) Suppose that you wish to estimate the mean percent kill for an application of 4 pounds of the nematicide per acre. Do the data satisfy the assumptions that are required for the confidence intervals of Section 10.6?

10.90 If you play tennis, you know that tennis rackets vary in their physical characteristics. The data shown in the table give measures of bending stiffness and twisting stiffness, as measured by engineering tests, for twelve tennis rackets.*

Racket	Bending Stiffness x	Twisting Stiffness y
Dunlop Maxply Fort	419	227
Garcia 240	407	231
Bancroft Bjorn Borg	363	200
Wilson Jack Kramer	360	211
Davis Classic	257	182
Spalding Smasher III	622	304
Yonex T-7500	424	384
Prince	359	194
Wilson T-4000	346	158
Yamaha YFG-30	556	225
Head Competition II	474	305
Adidas Adistar	441	235

(a) If a racket possesses bending stiffness, is it also likely to possess twisting stiffness? Do the data provide evidence that x and y are correlated? (*Hint:* Test H_0: $\beta_1 = 0$. As indicated in the text, this test is equivalent to testing the null hypothesis that $\rho = 0$.)
(b) Calculate the coefficient of determination, r^2, and interpret its value.

**Source:* R. Gillen, ed., "Equipment Preview 1976: How to Pick the Right Racquet for Your Game," *Tennis USA,* February 1976, pp. 87–91, 99.

10.91 Soybean meal production, a major source of protein, varies with the weather, rainfall, and the production of competing products. The data in the table show the annual U.S production (in 100,000 tons) for the years 1960 to 1977.*

Year	Year − 1960 x	Soybean Production y
1960	0	9.45
1961	1	10.3
1962	2	11.1
1963	3	10.6
1964	4	11.3
1965	5	12.9
1966	6	13.5
1967	7	13.7
1968	8	14.6
1969	9	17.6
1970	10	18.0
1971	11	17.0
1972	12	16.7
1973	13	19.7
1974	14	16.7
1975	15	20.8
1976	16	18.5
1977	17	20.1

(a) Fit a least-squares line to the data.
(b) Forecast the U.S. soybean meal production for 1978 using a 90 percent prediction interval.
(c) Notice that you have forecast a value of y outside the range of the x values used to develop the prediction equation. How might this affect the interpretation of your prediction interval?

10.92 Drivers suspected of drinking may face a new type of breath analyzer, a Vacu-Sampler. A. H. Principe reports on this new device and gives data linking Vacu-Sampler

Breathanalyzer y	Vacu-Sampler x
.150	.154
.100	.085
.090	.079
.140	.144
.080	.078
.110	.078
.120	.097
.100	.088
.090	.082
.090	.072
.090	.080
.090	.092
.080	.067
.080	.077
.060	.053

**Source: Fats and Oils Situation*, Economic Research Service, U.S. Department of Agriculture, October 1977.

readings to those of the device currently in use, the Breathanalyzer.* The data shown in the table give corresponding Breathanalyzer and Vacu-Sampler readings for 15 drinkers.

(a) Find the least-squares line relating Breathanalyzer readings, y, to corresponding readings on the Vacu-Sampler, x.
(b) Plot the data and graph the least-squares line.
(c) Do the data provide sufficient evidence to indicate that Breathanalyzer readings are linearly related to Vacu-Sampler readings?
(d) Suppose that a person's breath was analyzed using a Vacu-Sampler and gave a reading of .010. Predict the corresponding reading that would be recorded for a Breathanalyzer using a 90 percent prediction interval.

EXPERIENCES WITH REAL DATA

Conduct a study to determine whether a correlation exists between a student's quantitative ability and his or her verbal ability, where these two quantities are measured by their respective Student Aptitude Test (SAT) scores or similar tests used by your college or university.

The data for this study are available in the office of your college or university registrar. Visit the registrar's office and request the SAT verbal score, x, and quantitative score, y, for $n = 100$ students.

1. Fit a least-squares line to the data relating y and x.
2. Calculate r^2 and the linear coefficient of correlation, r. Describe the strength of the relationship between y and x.
3. Do the data provide evidence to indicate a linear correlation between y and x? (*Hint:* Recall that a test of the hypothesis, $\rho = 0$, is equivalent to testing the hypothesis that $\beta_1 = 0$.)

What have you learned about the relationship between SAT verbal and quantitative scores?

Computations associated with this study are acquired most easily using either an electronic desk calculator or an electronic computer. Packaged computer programs are available to fit a general linear model (described in Section 10.9) to a set of data using the method of least squares. Since our model, $E(y) = \beta_0 + \beta_1 x$, is a special case of the general linear model, these programs are suitable for the computations required for the aptitude data. Particularly good programs are those in the Biomed (Biomedical Programs), the SAS (Statistical Analysis System), and the SPSS (Statistical Package for the Social Sciences) program libraries (see the references).

* *Source:* A. H. Principe, "The Vacu-Sampler: A New Device for the Encapsulation of Breath and Other Gaseous Samples," *J. Police Sci. and Adm.*, 2, no. 4 (1974).

CHAPTER OBJECTIVES

GENERAL OBJECTIVE Many types of surveys and experiments result in the observation of qualitative rather than quantitative response variables. That is, the responses can be classified but not quantified. Consequently, data from these experiments will be the number of response observations falling in each of the data classes included in the experiment. The objective of this chapter is to present methods for analyzing count (or enumerative) data.

SPECIFIC OBJECTIVES

1. To present the properties of the multinomial experiment. Except for minor exceptions, which are noted, all the techniques presented in this chapter will assume that the sampling satisfies the properties of a multinomial experiment. *Section 11.1*
2. To present the chi-square test, which will be used to test hypotheses concerning the class (or cell) probabilities. *Section 11.2*
3. To present a test of an hypothesis concerning specific values of the cell probabilities. *Section 11.3*
4. To present a method for testing for dependence between two methods of classification. *Sections 11.4, 11.5*
5. To suggest other applications of the general statistical test of Section 11.2. *Section 11.6*
6. To emphasize the assumptions you must make in order for the test procedures described in this chapter to be valid. *Section 11.7*

11

ANALYSIS OF ENUMERATIVE DATA

11.1

A DESCRIPTION OF THE EXPERIMENT

Many experiments, particularly in the social sciences, result in enumerative (or count) data. For instance, the classification of people into five income brackets would result in an enumeration or count corresponding to each of the five income classes. Or, we might be interested in studying the reaction of a mouse to a particular stimulus in a psychological experiment. If a mouse will react in one of three ways when the stimulus is applied and if a large number of mice were subjected to the stimulus, the experiment would yield three counts, indicating the number of mice falling in each of the reaction classes. Similarly, a traffic study might require a count and classification of the type of motor vehicles using a section of highway. An industrial process manufactures items that fall into one of three quality classes: acceptable, seconds, and rejects. A student of the arts might classify paintings in one of k categories according to style and period in order to study trends in style over time. We might wish to classify ideas in a philosophical study or style in the field of literature. The results of an advertising campaign would yield count data indicating a classification of consumer reaction. Indeed, many observations in the physical sciences are not amenable to measurement on a continuous scale and hence result in enumerative or classificatory data.

The illustrations in the preceding paragraph exhibit, to a reasonable degree of approximation, the following characteristics which define a multinomial experiment:

1. The experiment consists of n identical trials.
2. The outcome of each trial falls into one of k classes or cells.
3. The probability that the outcome of a single trial will fall in a particular cell, say cell i, is p_i ($i = 1, 2, \ldots, k$), and remains the same from trial to trial. Note that

$$p_1 + p_2 + p_3 + \cdots + p_k = 1.$$

4. The trials are independent.
5. We are interested in $n_1, n_2, n_3, \ldots, n_k$, where n_i ($i = 1, 2, \ldots, k$) is equal to the number of trials in which the outcome falls in cell i. Note that $n_1 + n_2 + n_3 + \cdots + n_k = n$.

The above experiment is analogous to tossing n balls at k boxes, where each ball must fall in one of the boxes. The boxes are arranged such that the probability that a ball will fall in a box varies from box to box but remains the same for a particular box in repeated tosses. Finally, the balls are tossed in such a way that the trials are independent. At the conclusion of the experiment, we observe n_1 balls in the first box, n_2 in the second, . . . , and n_k in the kth. The total number of balls is equal to

$$\sum_{i=1}^{k} n_i = n.$$

Note the similarity between the binomial and multinomial experiments and, in particular, that the binomial experiment represents the special case for the multinomial experiment when $k = 2$. The two cell probabilities, p and q, of the binomial experiment are replaced by the k cell probabilities, $p_1, p_2, \ldots, p_k$, of the multinomial experiment. The objective of this chapter is to make inferences about the cell probabilities, $p_1, p_2, \ldots, p_k$. The inferences will be expressed in terms of a statistical test of an hypothesis concerning their specific numerical values or their relationship, one to another.

If we were to proceed as in Chapter 6, we would derive the probability of the observed sample $(n_1, n_2, \ldots, n_k)$ for use in calculating the probability of the type I and type II errors associated with a statistical test. Fortunately, we have been relieved of this chore by the British statistician Karl Pearson, who proposed a very useful test statistic for testing hypotheses concerning $p_1, p_2, \ldots, p_k$, and gave its approximate probability distribution in repeated sampling.

11.2 THE CHI-SQUARE TEST

Suppose that $n = 100$ balls were tossed at the cells and that we knew that p_1 was equal to .1. How many balls would be expected to fall in the first cell? Referring to Chapter 6 and utilizing knowledge of the binomial experiment, we would calculate

$$E(n_1) = np_1 = 100(.1) = 10.$$

In like manner, the expected number falling in the remaining cells may be calculated using the formula

$$E(n_i) = np_i, \qquad i = 1, 2, \ldots, k.$$

Now suppose that we hypothesize values for $p_1, p_2, \ldots, p_k$ and calculate the expected value for each cell. Certainly, if our hypothesis is true, the cell counts, n_i, should not deviate greatly from their expected values, np_i ($i = 1, 2, \ldots, k$). Hence it would seem intuitively reasonable to use a test statistic involving the k deviations.

$$(n_i - np_i) \qquad i = 1, 2, \ldots, k.$$

In 1900, Karl Pearson proposed the following test statistic, which is a function of the squares of the deviations of the observed counts from their expected values, weighted by the reciprocals of their expected values:

$$X^2 = \sum_{i=1}^{k} \frac{[n_i - E(n_i)]^2}{E(n_i)}$$

$$= \sum_{i=1}^{k} \frac{[n_i - np_i]^2}{np_i}.$$

Although the mathematical proof is beyond the scope of this text, it can be shown that when n is large, X^2 will possess, approximately, a chi-square probability distribution in repeated sampling. Experience has shown that the cell counts, n_i, should not be too small in order that the chi-square distribution provide an adequate approximation to the distribution of X^2. As a rule of thumb, we will require that all expected cell counts equal or exceed five, although Cochran (see the references) has noted that this value can be as low as one for some situations.

You will recall the use of the chi-square probability distribution for testing an hypothesis concerning a population variance, σ^2, in Section 9.6. Particularly, we stated that the shape of the chi-square distribution would vary depending upon the number of degrees of freedom associated with s^2, and we discussed the use of Table 5, Appendix II, which presents the critical values of χ^2 corresponding to various right-hand tail areas of the distribution. Therefore, we must know which χ^2 distribution to use—that is, the number of degrees of freedom—in approximating the distribution of X^2, and we must know whether to use a one-tailed or two-tailed test in locating the rejection region for the test. The latter problem may be solved directly. Since large deviations of the observed cell counts from those expected would tend to contradict the null hypothesis concerning the cell probabilities $p_1, p_2, \ldots, p_k$, we would reject the null hypothesis when X^2 is large and employ a one-tailed statistical test using the upper tail values of χ^2 to locate the rejection region.

The determination of the appropriate number of degrees of freedom to be employed for the test can be rather difficult and therefore will be specified for the physical applications described in the following sections. In addition, we will state the principle involved (which is fundamental to the mathematical proof of the approximation) so that the reader may understand why the number of degrees of freedom changes with various applications. This states that the appropriate number of degrees of freedom will equal the number of cells, k, less one degree of freedom for each independent linear restriction placed upon the observed cell counts. For example, one linear restriction is always present because the sum of the cell counts must equal n; that is,

$$n_1 + n_2 + n_3 + \cdots + n_k = n.$$

Other restrictions will be introduced for some applications because of the necessity for estimating unknown parameters required in the calculation of the expected cell frequencies or because of the method in which the sample is collected. These will become apparent as we consider various practical examples.

11.3

A TEST OF AN HYPOTHESIS CONCERNING SPECIFIED CELL PROBABILITIES

The simplest hypothesis concerning the cell probabilities would be one that specifies numerical values for each. For example, suppose that we were to consider a simple extension of the rat experiment discussed in Exercise 6.33.

One or more rats proceed down a ramp to one of three doors. We wish to test the hypothesis that the rats have no preference concerning the choice of a door and therefore that

$$H_0: p_1 = p_2 = p_3 = 1/3,$$

where p_i is the probability that a rat will choose door i, $i = 1, 2$, or 3.

Suppose that the rat was sent down the ramp $n = 90$ times and that the three observed cell frequencies were $n_1 = 23$, $n_2 = 36$, and $n_3 = 31$. The expected cell frequency would be the same for each cell: $E(n_i) = np_i = 90(1/3) = 30$. The observed and expected cell frequencies are presented in Table 11.1. Noting the discrepancy between the observed and expected cell frequency, we would wonder whether the data present sufficient evidence to warrant rejection of the hypothesis of no preference.

The chi-square test statistic for our example will possess $(k - 1) = 2$ degrees of freedom since the only linear restriction on the cell frequencies is that

$$n_1 + n_2 + \cdots + n_k = n,$$

or, for our example,

$$n_1 + n_2 + n_3 = 90.$$

Therefore, if we choose $\alpha = .05$, we would reject the null hypothesis when $X^2 > 5.99147$ (see Table 5, Appendix II).

Substituting into the formula for X^2, we obtain

$$X^2 = \sum_{i=1}^{k} \frac{[n_i - E(n_i)]^2}{E(n_i)} = \sum_{i=1}^{k} \frac{[n_i - np_i]^2}{np_i}$$

$$= \frac{(23 - 30)^2}{30} + \frac{(36 - 30)^2}{30} + \frac{(31 - 30)^2}{30}$$

$$= 2.87.$$

Since X^2 is less than the tabulated critical value of χ^2, 5.991, the null hypothesis is not rejected and we conclude that the data do not present sufficient evidence to indicate that the rat has a preference for a particular door.

TABLE 11.1

Observed and expected cell counts for the rat experiment

	Door		
	1	2	3
Observed cell frequency	$n_1 = 23$	$n_2 = 36$	$n_3 = 31$
Expected cell frequency	30	30	30

EXERCISES

11.1 List the characteristics of a multinomial experiment.

11.2 A city expressway utilizing four lanes in each direction was studied to see whether drivers preferred to drive on the inside lanes. A total of 1,000 automobiles were observed during the heavy early morning traffic and their respective lanes recorded. The results were as follows:

Lane	1	2	3	4
Observed Count	294	276	238	192

Do the data present sufficient evidence to indicate that some lanes are preferred over others? (Test the hypothesis that $p_1 = p_2 = p_3 = p_4 = 1/4$ using $\alpha = .05$.)

11.3 The Mendelian theory states that the number of peas of a certain type falling into the classifications round and yellow, wrinkled and yellow, round and green, and wrinkled and green should be in the ratio 9:3:3:1. Suppose that 100 such peas revealed 56, 19, 17, and 8 in the respective classes. Do these data disagree with the Mendelian theory? Use $\alpha = .05$.

11.4 Medical statistics show that deaths due to four major diseases, call them *A*, *B*, *C*, and *D*, account for 15, 21, 18, and 14 percent, respectively, of all nonaccidental deaths. A study of the causes of 308 nonaccidental deaths at a hospital serving primarily indigent black patients gave the following counts of patients dying of diseases *A*, *B*, *C*, and *D*.

Disease	Number of Deaths
A	43
B	76
C	85
D	21
Other	83
	308

Do these data provide sufficient evidence to indicate that the proportions of people dying of diseases *A*, *B*, *C*, and *D* at this hospital differ from the proportions accumulated for the population at large?

11.5 You may recall the exercise in Chapter 8 (Exercise 8.107) that reported on research that linked the prevalence of schizophrenia with birth during particular months of the year. Suppose you are working on a similar problem and you suspect a linkage between a disease observed in later life with month of birth. You have records of 400 cases of the disease and you classify them according to month of birth. The data appear in the table. Do the data present sufficient evidence to indicate that the proportion of cases of the disease per month varies from month to month? Test with $\alpha = .10$.

Month	Jan	Feb	Mar	Apr	May	June	July	Aug	Sept	Oct	Nov	Dec
No. of Births	38	31	42	46	28	31	24	29	33	36	27	35

11.6 Officials in a particular community are seeking a federal program which they hope will boost local income levels. As justification, the city claims that its local income distribution differs substantially from the national distribution and that incomes tend to be lower than expected. A random sample of 2,000 family incomes were classified and compared with the corresponding national percentages. They are shown in the table. Do the data provide sufficient evidence to indicate that the distribution of family incomes within the city differs from the national distribution? Test with $\alpha = .05$.

Income	National Percentages	City Salary Class Frequency
more than $50,000	2	27
$25,000 to $50,000	16	193
$20,000 to $25,000	13	234
$15,000 to $20,000	19	322
$10,000 to $15,000	20	568
$5,000 to $10,000	19	482
below $5,000	11	174
Total	100	2000

11.4

CONTINGENCY TABLES

A problem frequently encountered in the analysis of count data concerns the independence of two methods of classification of observed events. For example, we might wish to classify defects found on furniture produced in a manufacturing plant, first, according to the type of defect and, second, according to the production shift. Ostensibly, we wish to investigate a contingency—a dependence between the two classifications. Do the proportions of various types of defects vary from shift to shift?

A total of $n = 309$ furniture defects was recorded and the defects were classified according to one of four types: A, B, C, or D. At the same time, each piece of furniture was identified according to the production shift in which it was manufactured. These counts are presented in Table 11.2, which is known as a contingency table. (*Note:* Numbers in parentheses are the expected cell frequencies.)

Let p_A equal the unconditional probability that a defect will be of type A. Similarly, define p_B, p_C, and p_D as the probabilities of observing the three other types of defects. Then these probabilities, which we will call the column probabilities of Table 11.2, will satisfy the requirement

$$p_A + p_B + p_C + p_D = 1.$$

TABLE 11.2
Contingency table

	Type of Defect				
Shift	*A*	*B*	*C*	*D*	Total
1	15 (22.51)	21 (20.99)	45 (38.94)	13 (11.56)	94
2	26 (22.99)	31 (21.44)	34 (39.77)	5 (11.81)	96
3	33 (28.50)	17 (26.57)	49 (49.29)	20 (14.63)	119
Total	74	69	128	38	309

In like manner, let p_i ($i = 1, 2,$ or 3) equal the row probability that a defect will have occurred on shift i, where

$$p_1 + p_2 + p_3 = 1.$$

Then, if the two classifications are independent of each other, a cell probability will equal the product of its respective row and column probabilities in accordance with the multiplicative law of probability. For example, the probability that a particular defect will occur on shift 1 and be of type A is $(p_1)(p_A)$. Thus, we observe that the numerical values of the cell probabilities are unspecified in the problem under consideration. The null hypothesis specifies only that each cell probability will equal the product of its respective row and column probabilities and therefore imply independence of the two classifications.

The analysis of the data obtained from a contingency table differs from the problem discussed in Section 11.3 because we must *estimate* the row and column probabilities to estimate the expected cell frequencies.

If proper estimates of the cell probabilities are obtained, the estimated expected cell frequencies may be substituted for the $E(n_i)$ in X^2, and X^2 will continue to possess a distribution in repeated sampling that is approximated by the chi-square probability distribution. The proof of this statement as well as a discussion of the methods for obtaining the estimates are beyond the scope of this text. Fortunately, the procedures for obtaining the estimates, known as the method of maximum likelihood and the method of minimum chi-square, yield estimates that are intuitively obvious for our relatively simple applications.

It can be shown that the maximum likelihood estimator of a column probability will equal the column total divided by $n = 309$. If we denote the total of column j as c_j, then

$$\hat{p}_A = \frac{c_1}{n} = \frac{74}{309},$$

$$\hat{p}_B = \frac{c_2}{n} = \frac{69}{309},$$

$$\hat{p}_C = \frac{c_3}{n} = \frac{128}{309},$$

$$\hat{p}_D = \frac{c_4}{n} = \frac{38}{309}.$$

Likewise, the row probabilities p_1, p_2, and p_3 may be estimated using the row totals r_1, r_2, and r_3:

$$\widehat{p}_1 = \frac{r_1}{n} = \frac{94}{309},$$

$$\widehat{p}_2 = \frac{r_2}{n} = \frac{96}{309},$$

$$\widehat{p}_3 = \frac{r_3}{n} = \frac{119}{309}.$$

Denote the observed frequency of the cell in row i and column j of the contingency table as n_{ij}. Then the estimated expected value of n_{11} will be

$$\widehat{E}(n_{11}) = n[\widehat{p}_1\widehat{p}_A] = n\left(\frac{r_1}{n}\right)\left(\frac{c_1}{n}\right)$$

$$= \frac{r_1 c_1}{n},$$

where $(\widehat{p}_1\widehat{p}_A)$ is the estimated cell probability. Likewise, we may find the estimated expected value for any other cell, say $\widehat{E}(n_{23})$,

$$\widehat{E}(n_{23}) = n[\widehat{p}_2\widehat{p}_3] = n\left(\frac{r_2}{n}\right)\left(\frac{c_3}{n}\right)$$

$$= \frac{r_2 c_3}{n}.$$

In other words, we observe that the estimated expected value of the observed cell frequency, n_{ij}, for a contingency table is equal to the product of its respective row and column totals divided by the total frequency; that is,

Estimated Expected Cell Frequency

$$\widehat{E}(n_{ij}) = \frac{r_i c_j}{n}.$$

The estimated expected cell frequencies for our example are shown in parentheses in Table 11.2.

We may now use the expected and observed cell frequencies shown in Table 11.2 to calculate the value of the test statistic:

$$X^2 = \sum_{j=1}^{4}\sum_{i=1}^{3} \frac{[n_{ij} - \widehat{E}(n_{ij})]^2}{\widehat{E}(n_{ij})}$$

$$= \frac{(15 - 22.51)^2}{22.51} + \frac{(26 - 22.99)^2}{22.99} + \cdots + \frac{(20 - 14.63)^2}{14.63}$$

$$= 19.18.$$

The only remaining obstacle involves the determination of the appropriate number of degrees of freedom associated with the test statistic. We will give this as a rule that we will attempt to justify. The degrees of freedom associated with a contingency table possessing r rows and c columns will always equal $(r-1)(c-1)$. Thus, for our example, we will compare X^2 with the critical value of χ^2 with $(r-1)(c-1) = (3-1)(4-1) = 6$ degrees of freedom.

You will recall that the number of degrees of freedom associated with the X^2 statistic will equal the number of cells (in this case, $k = rc$) less one degree of freedom for each independent linear restriction placed upon the observed cell frequencies. The total number of cells for the data of Table 11.2 is $k = 12$. From this we subtract one degree of freedom because the sum of the observed cell frequencies must equal n; that is,

$$n_{11} + n_{12} + \cdots + n_{34} = 309.$$

In addition, we used the cell frequencies to estimate three of the four column probabilities. Note that the estimate of the fourth column probability will be determined once we have estimated p_A, p_B, and p_C because

$$p_A + p_B + p_C + p_D = 1.$$

Thus, we lose $(c-1) = 3$ degrees of freedom for estimating the column probabilities.

Finally, we used the cell frequencies to estimate $(r-1) = 2$ row probabilities and, therefore, we lose $(r-1) = 2$ additional degrees of freedom. The total number of degrees of freedom remaining will be

$$\text{d.f.} = 12 - 1 - 3 - 2 = 6.$$

And, in general, we see that the total number of degrees of freedom associated with an $r \times c$ contingency table will be

$$\text{d.f.} = rc - 1 - (c-1) - (r-1) = rc - c - r + 1$$

or

$$\text{d.f.} = (r-1)(c-1).$$

Therefore, if we use $\alpha = .05$, we will reject the null hypothesis that the two classifications are independent if $X^2 > 12.5916$. Since the value of the test statistic, $X^2 = 19.18$, exceeds the critical value of χ^2, we will reject the null hypothesis. The data present sufficient evidence to indicate that the proportion of the various types of defects varies from shift to shift. A study of the production operations for the three shifts would likely reveal the cause.

EXAMPLE 11.1

A survey was conducted to evaluate the effectiveness of a new flu vaccine which had been administered in a small community. The vaccine was provided free of charge in a two-shot sequence over a period of two weeks to those wishing to avail themselves of it. Some people received the two-shot sequence, some appeared only for the first shot, and others received neither.

A survey of 1,000 local local inhabitants in the following spring provided the information shown in Table 11.3. Do the data present sufficient evidence to indicate that the vaccine was successful in reducing the number of flu cases in the community?

TABLE 11.3
Data tabulation for Example 11.1

	No Vaccine	One Shot	Two Shots	Total
Flu	24 (14.4)	9 (5.0)	13 (26.6)	46
No flu	289 (298.6)	100 (104.0)	565 (551.4)	954
Total	313	109	578	1,000

Solution The question stated above asks whether the data provide sufficient evidence to indicate a dependence between the vaccine classification and the occurrence or nonoccurrence of flu. We therefore analyze the data as a contingency table.

The estimated expected cell frequencies may be calculated using the appropriate row and column totals,

$$\widehat{E}(n_{ij}) = \frac{r_i c_j}{n}.$$

Thus,

$$\widehat{E}(n_{11}) = \frac{r_1 c_1}{n} = \frac{46(313)}{1,000} = 14.4,$$

$$\widehat{E}(n_{12}) = \frac{r_1 c_2}{n} = \frac{46(109)}{1,000} = 5.0,$$

$$\vdots$$

$$\widehat{E}(n_{23}) = \frac{r_2 c_3}{n} = \frac{954(578)}{1,000} = 551.4.$$

These values are shown in parentheses in Table 11.3.

The value of the test statistic, X^2, will now be computed and compared with the critical value of χ^2 possessing $(r - 1)(c - 1) = (1)(2) = 2$ degrees of freedom. Then, for $\alpha = .05$, we will reject the null hypothesis when $X^2 > 5.99147$. Substituting into the formula for X^2, we obtain

$$X^2 = \frac{(24 - 14.4)^2}{14.4} + \frac{(289 - 298.6)^2}{298.6} + \cdots + \frac{(565 - 551.4)^2}{551.4}$$
$$= 17.35.$$

Observing that X^2 falls in the rejection region, we reject the null hypothesis of independence of the two classifications. A comparison of the percentage incidence of flu for each of the three categories would suggest that those receiving the two-shot sequence were less susceptible to the disease. Further analysis of the data could be obtained by deleting one of the three categories, the second column, for example, to compare the effect of the vaccine with that of no vaccine. This could be done by using a 2×2 contingency table or treating the two categories as two binomial populations and using the methods of Section 8.9. Or, we might wish to analyze the data by comparing the results of the two-shot vaccine sequence with those of the combined no vaccine–one shot group. That is, we would combine the first two columns of the 2×3 table into one.

EXERCISES

11.7 A recent complaint of blacks and minorities has resulted in at least one lawsuit. According to the complaint, culturally biased IQ tests result in many black students being assigned to EMR (educable mentally retarded) classes. Typical of data in support of this contention would be the following. A school district has an enrollment of 6,000 elementary and junior high school students of which 1,640 are black and 4,360 are white. A total of 280 students are enrolled in EMR classes and of these 163 are black. The data are given in the table. Does it appear that student assignment to EMR classes is dependent on race? Test with $\alpha = .01$.

School Class Assignment	Race		Total
	Black	White	
EMR	163	117	280
non-EMR	1,477	4,243	5,720
Total	1,640	4,360	6,000

11.8 An experiment was conducted to investigate the effect of general hospital experience on the attitudes of physicians toward lower-class people. A random sample of 50 physicians who had just completed 4 weeks of service in a general hospital were categorized according to their concern for lower-class people before and after their general hospital experience. The data are shown below. Do the data provide sufficient evidence to indicate a change in "concern" due to the general hospital experience?

Concern Before Experience in General Hospital	After Experience in a General Hospital		Total
	High	Low	
Low	27	5	32
High	9	9	18

11.9 A survey of 477 hospitals by the American College of Surgeons' Commission on Cancer produced the data in the table, which classifies women with liver tumors into 6 classes.* The women are classified according to whether they used oral contraceptives and according to type of liver tumor. Do the data provide sufficient evidence

** Source:* J. Vana, G. Murphy, B. Aronoff, and H. Baker, "Primary Liver Tumors and Oral Contraceptives," *J. Amer. Med. Assoc.*, *238*, no. 20 (1977).

to indicate a dependence between type of tumor and whether or not they used oral contraceptives?

	Type of Tumor		
	Benign	Malignant	Total
Contraceptive users	138	49	187
Nonusers	39	41	80
Use not known	35	76	111
Total	212	166	378

11.10 An analysis of accident data was made to determine the distribution of numbers of fatal accidents for automobiles of three sizes. The data for 346 accidents are as follows:

	Size of Auto		
	Small	Medium	Large
Fatal	67	26	16
Not fatal	128	63	46

Do the data indicate that the frequency of fatal accidents is dependent on the size of automobiles?

11.11 A group of 306 people were interviewed to determine their opinion concerning a particular current American foreign-policy issue. At the same time their political affiliation was recorded. The data are as follows:

	Approve of Policy	Do Not Approve of Policy	No Opinion
Republicans	114	53	17
Democrats	87	27	8

Do the data present sufficient evidence to indicate a dependence between party affiliation and the opinion expressed for the sampled population?

11.5

$r \times c$ TABLES WITH FIXED ROW OR COLUMN TOTALS

In the previous section we have described the analysis of an $r \times c$ contingency table using examples which, for all practical purposes, fit the multinomial experiment described in Section 11.1. Although the methods of collecting data in many surveys may adhere to the requirements of a multinomial experiment, other methods do not. For example, we might not wish to sample randomly the population described in Example 11.1 because we might find that, owing to chance, one category is completely missing. People who have received no flu shots might fail to appear in the sample. Thus we might decide beforehand to interview a specified number of people in each column category, thereby fixing the column totals in advance. Although these restrictions tend to disturb somewhat our visualization of the experiment in the multinomial context, they have no effect on the analysis of the data. As long as we wish to test the hypothesis of independence of two classifications and none of the row or column probabilities are specified in advance, we may analyze the data as an $r \times c$ contingency table. It can be shown that the resulting X^2 will possess a probability distribution in repeated sampling that is approximated by a chi-square distribution with $(r - 1)(c - 1)$ degrees of freedom.

To illustrate, suppose that we wish to test an hypothesis concerning the equivalence of four binomial populations as indicated in the following example.

EXAMPLE 11.2

A survey of voter sentiment was conducted in four midcity political wards to compare the fraction of voters favoring candidate A. Random samples of 200 voters were polled in each of the four wards with results as shown in Table 11.4. Do the data present sufficient evidence to indicate that the fractions of voters favoring candidate A differ in the four wards?

Solution You will observe that the test of an hypothesis concerning the equivalence of the parameters of the four binomial populations corresponding to the four wards is identical to an hypothesis implying independence of the row and column classifications. Thus, if we denote the fraction of voters favoring A as p and hypothesize that p is the same for all four wards, we imply that the first- and second-row probabilities are equal to p and $(1 - p)$, respectively. The probability that a member of the sample of $n = 800$ voters falls in a particular ward will equal one-fourth, since this was fixed in advance. Then the cell probabilities for the table would be obtained by multiplication of the appropriate row and column probabilities

TABLE 11.4

Data tabulation for Example 11.2

	Ward				
	1	2	3	4	Total
Favor *A*	76 (59)	53 (59)	59 (59)	48 (59)	236
Do not favor *A*	124 (141)	147 (141)	141 (141)	152 (141)	564
Total	200	200	200	200	800

under the null hypothesis and be equivalent to a test of independence of the two classifications.

The estimated expected cell frequencies, calculated using the row and column totals, appear in parentheses in Table 11.4. We see that

$$X^2 = \sum_{j=1}^{4} \sum_{i=1}^{2} \frac{[n_{ij} - \hat{E}(n_{ij})]^2}{\hat{E}(n_{ij})}$$

$$= \frac{(76 - 59)^2}{59} + \frac{(124 - 141)^2}{141} + \cdots + \frac{(152 - 141)^2}{141}$$

$$= 10.72.$$

The critical value of χ^2 for $\alpha = .05$ and $(r - 1)(c - 1) = (1)(3) = 3$ degrees of freedom is 7.81473. Since X^2 exceeds this critical value, we reject the null hypothesis and conclude that the fraction of voters favoring candidate A is not the same for all four wards.

EXERCISES

11.12 According to a study conducted at the University of Florida Health Center, orange juice may provide protection against colds (*Orlando Sentinel Star,* September 25, 1975). Twenty-four volunteers were selected for the experiment. Twelve subjects were given dummy pills (placebos) and twelve drank a specified daily amount of orange juice for a three-week period. The cold virus was introduced into all 24 subjects by means of nose drops. Two of the 12 drinking orange juice contracted the cold in comparison with 9 of the 12 on the placebos. Do these data provide sufficient evidence to indicate a dependence between drinking orange juice and contracting colds? Test at the $\alpha = .10$ level.

11.13 A study to determine the effectiveness of a drug (serum) for arthritis resulted in the comparison of two groups each consisting of 200 arthritic patients. One group was inoculated with the serum, the other received a placebo (an inoculation that appears to contain serum but actually is nonactive). After a period of time, each person in the study was asked to state whether his arthritic condition was improved. The following results were observed:

	Treated	Untreated
Improved	117	74
Not improved	83	126

Do these data present sufficient evidence to indicate that the serum was effective in improving the condition of arthritic patients?
(a) Test by means of the X^2 statistic. Use $\alpha = .05$.
(b) Test by use of the z test in Section 8.12.

11.14 A particular poultry disease is thought to be noncommunicable. To test this theory, 30,000 chickens were randomly partitioned into three groups of 10,000. One group had no contact with diseased chickens, one had moderate contact, and the third had heavy contact. After a 6-month period, the following data were collected on the number of diseased chickens in each group of 10,000. Do the data provide sufficient evidence to indicate a dependence between the amount of contact between diseased and nondiseased fowls and the incidence of the disease?

	No Contact	Moderate Contact	Heavy Contact
Number of diseased chickens	87	89	124
Number of nondiseased chickens	9,913	9,911	9,876
Total	10,000	10,000	10,000

11.15 In a recent article, H. W. Menard writes about manganese nodules, a mineral-rich concoction found abundantly on the deep-sea floor.* In one portion of his report, Menard provides data relating the magnetic age of the earth's crust to "the probability of finding manganese nodules." The data shown in the table give the number of samples of the earth's core and the percentage of those that contain manganese nodules for each of a set of magnetic-crust ages. Do the data provide sufficient evidence to indicate that the probability of finding manganese nodules in the deep-sea earth's crust is dependent upon the magnetic-age classification? Test with $\alpha = .05$.

Age	Number of Samples	Percentage with Nodules
Miocene–recent	389	5.9
Oligocene	140	17.9
Eocene	214	16.4
Paleocene	84	21.4
Late Cretaceous	247	21.1
Early and Middle Cretaceous	1,120	14.2
Jurassic	99	11.0

11.16 A study of the purchase decisions for three stock portfolio managers, *A*, *B*, and *C*, was conducted to compare the rates of stock purchases that resulted in profits over a time period that was less than or equal to one year. One hundred randomly selected purchases obtained for each of the managers gave the following results:

	Manager		
	A	*B*	*C*
Purchases that resulted in a profit	63	71	55
Purchases that resulted in no profit	37	29	45
Total	100	100	100

*H. W. Menard, "Time, Chance and the Origin of Manganese Nodules," *Amer. Scientist*, September–October 1976.

Do the data provide evidence of differences among the rates of successful purchases for the three managers?

11.17 A survey was conducted to study the relationship between lung disease and air pollution. Four areas were chosen for the survey, two cities that were frequently plagued with smog and two nonurban areas in states that possessed low air-pollution counts. Only adult permanent residents of the area were included in the study. Random samples of 400 adult permanent residents from each area gave the following results:

Area	City *A*	City *B*	Nonurban Area 1	Nonurban Area 2
Number with Lung Disease	34	42	21	18

(a) Do the data provide sufficient evidence to indicate a difference in the proportions with lung disease for the four locations?
(b) Should cigarette smokers have been excluded from the samples? How would this affect inferences drawn from the data?

11.6

OTHER APPLICATIONS

The applications of the chi-square test in analyzing enumerative data described in Sections 11.3, 11.4, and 11.5 represent only a few of the interesting classificatory problems which may be approximated by the multinomial experiment and for which our method of analysis is appropriate. By and large, these applications are complicated to a greater or lesser degree because the numerical values of the cell probabilities are unspecified and hence require the estimation of one or more population parameters. Then, as in Sections 11.4 and 11.5, we can estimate the cell probabilities. Although we omit the mechanics of the statistical tests, several additional applications of the chi-square test are worth mention as a matter of interest.

For example, suppose that we wish to test an hypothesis stating that a population possesses a normal probability distribution. The cells of a sample frequency histogram (for example, Figure 3.2) would correspond to the k cells of the multinomial experiment and the observed cell frequencies would be the number of measurements falling in each cell of the histogram. Given the hypothesized normal probability distribution for the population, we could use the areas under the normal curve to calculate the theoretical cell probabilities and hence the expected cell frequencies. The difficulty arises when μ and σ are unspecified for the normal population and these parameters must be estimated to obtain the estimated cell probabilities. This difficulty can, of course, be surmounted.

The construction of a two-way table to investigate dependency between two classifications can be extended to three or more classifications. For example, if we wish to test the mutual independence of three classifications, we would employ a three-dimensional "table" or rectangular parallelepiped. The reasoning and methodology associated with the analysis of both the two- and three-way tables are identical, although the analysis of the three-way table is a bit more complex.

A third and interesting application of our methodology would be its use in the investigation of the rate of change of a multinomial (or binomial) population as a function of time. For example, we might study the decision-making ability of a human (or any animal) as he is subjected to an educational program and tested over time. If, for instance, he is tested at prescribed intervals of time and the test is of the yes or no type yielding a number of correct answers, y, that would follow a binomial probability distribution, we would be interested in the behavior of the probability of a correct response, p, as a function of time. If the number of correct responses was recorded for c time periods, the data would fall in a $2 \times c$ table similar to that in Example 11.2 (Section 11.5). We would then be interested in testing the hypothesis that p is equal to a constant, that is, that no learning has occurred, and we would then proceed to more interesting hypotheses to determine whether the data present sufficient evidence to indicate a gradual (say, linear) change over time as opposed to an abrupt change at some point in time. The procedures we have described could be extended to decisions involving more than two alternatives.

You will observe that our learning example is common to business, to industry, and to many other fields, including the social sciences. For example, we might wish to study the rate of consumer acceptance of a new product for various types of advertising campaigns as a function of the length of time that the campaign has been in effect. Or, we might wish to study the trend in the lot fraction defective in a manufacturing process as a function of time. Both of these examples, as well as many others, require a study of the behavior of a binomial (or multinomial) process as a function of time.

The examples that we have just described are intended to suggest the relatively broad application of the chi-square analysis of enumerative data, a fact that should be borne in mind by the experimenter concerned with this type of data. The statistical test employing X^2 as a test statistic is often called a "goodness-of-fit" test. Its application for some of these examples requires care in the determination of the appropriate estimates and the number of degrees of freedom for X^2, which, for some of these problems, may be rather complex.

11.7 ASSUMPTIONS

The following assumptions must be satisfied if X^2 is to possess approximately a chi-square distribution and, consequently, if the tests described in this chapter are to be valid:

Assumptions

1. The cell counts, $n_1, n_2, \ldots, n_k$, satisfy the conditions of a multinomial experiment (or a set of multinomial experiments created by restrictions on row or column totals).
2. The expected values of all cell counts should equal or exceed 5.

It is absolutely essential that assumption 1 be satisfied. The chi-square goodness-of-fit tests, of which these tests are special cases, compare observed frequencies with expected frequencies and apply only to data generated by a multinomial experiment.

The larger the sample size n, the closer the chi-square distribution will approximate the distribution of X^2. We have stated in assumption 2 that n must be large enough so that all the expected cell frequencies will be equal to 5 or more. This is a safe figure. Actually, the expected cell frequencies can be smaller for some tests than for others. For information on the minimum expected cell frequencies for specific goodness-of-fit tests, see the paper by Cochran listed in the references.

11.8

SUMMARY

The preceding material has been concerned with a test of an hypothesis regarding the cell probabilities associated with a multinomial experiment. When the number of observations, n, is large, the test statistic, X^2, can be shown to possess, approximately, a chi-square probability distribution in repeated sampling, the number of degrees of freedom being dependent upon the particular application. In general we assume that n is large and that the minimum expected cell frequency is equal to or is greater than five.

Several words of caution concerning the use of the X^2 statistic as a method of analyzing enumerative-type data are appropriate. The determination of the correct number of degrees of freedom associated with the X^2 statistic is very important in locating the rejection region. If the number is incorrectly specified, erroneous conclusions might result. Also, note that nonrejection of the null hypothesis does not imply that it should be accepted. We would have difficulty in stating a meaningful alternative hypothesis for many practical applications and, therefore, would lack knowledge of the probability of making a type II error. For example, we hypothesize that the two classifications of a contingency table are independent. A specific alternative would have to specify some measure of dependence, which may or may not possess practical significance to the experimenter. Finally, if parameters are missing and the expected cell frequencies must be estimated, the estimators of missing parameters should be of a particular type in order that the test be valid. In other words, the application of the chi-square test for other than the simple applications outlined in Sections 11.3, 11.4, and 11.5 will require experience beyond the scope of this introductory presentation of the subject.

REFERENCES

Anderson, R. L., and T. A. Bancroft, *Statistical Theory in Research*. New York: McGraw-Hill Book Company, 1952.

Barr, A. J., J. H. Goodnight, J. P. Sall, and J. T. Helwig, *A User's Guide to SAS 76*. Raleigh, N.C.: SAS Institute, Inc., 1976.

Brown, M. B., ed., *Biomedical Computer Programs.* Berkeley, Calif.: University of California, 1977.

Chapman, D. G., and R. A. Schaufele, *Elementary Probability Models and Statistical Inference.* New York: John Wiley & Sons, Inc., 1970.

Cochran, W. G., "The χ^2 Test of Goodness of Fit." *Ann. Math. Stat., 23* (1952), pp. 315–345.

Dixon, W. J., and F. J. Massey, Jr., *Introduction to Statistical Analysis,* 3rd ed. New York: McGraw-Hill Book Company, 1969.

Kendall, M. G., and A. Stuart, *The Advanced Theory of Statistics,* Vol. 2. New York: Hafner Press, 1961.

Nie, N., C. H. Hull, J. G. Jenkins, K. Steinbrenner, and D. H. Bent, *Statistical Package for the Social Sciences,* 2nd ed. New York: McGraw-Hill Book Company, 1975.

SUPPLEMENTARY EXERCISES

11.18 A die was rolled 600 times, with the following results:

Observed Number	1	2	3	4	5	6
Frequency	89	113	98	104	117	79

Do these data present sufficient evidence to indicate that the die is unbalanced? Test using $\alpha = .05$.

11.19 After inspecting the data in Exercise 11.18, one might wish to test the hypothesis that the probability of a "6" is 1/6 against the alternative that this probability is less than 1/6.

(a) Carry out the above test using $\alpha = .05$.

(b) What tenet of good statistical practice is violated in the test of part (a)?

11.20 A manufacturer of buttons wished to determine whether the fraction of defective buttons produced by three machines varied from machine to machine. Samples of 400 buttons were selected from each of the three machines and the number of defectives counted for each sample. The results are as follows:

Machine Number	1	2	3
Number of Defectives	16	24	9

Do these data present sufficient evidence to indicate that the fraction of defective buttons varies from machine to machine? Test by use of $\alpha = .05$.

11.21 A survey was conducted by an auto repairman to determine whether various auto ills were dependent upon the make of the auto. His survey, restricted to this year's model, produced the results shown in the table. Do these data present sufficient evidence to indicate a dependency between auto makes and type of repair for these new-model cars? Note that the repairman was not utilizing all the information available when he conducted his survey. In conducting a study of this type, what other factors should be recorded?

Make	Type of Repair		
	Electrical	Fuel Supply	Other
A	17	19	7
B	14	7	9
C	6	21	12
D	33	44	19
E	7	9	6

11.22 A manufacturer of floor polish conducted a consumer-preference experiment to see whether a new floor polish, *A*, was superior to those produced by four of his competitors. A sample of 100 housewives viewed five patches of flooring that had received the five polishes and each indicated the patch that she considered superior in appearance. The lighting, background, etc., were approximately the same for all five patches. The results of the survey are as follows:

Polish	*A*	*B*	*C*	*D*	*E*
Frequency	27	17	15	22	19

Do these data present sufficient evidence to indicate a preference for one or more of the polished patches of floor over the others? If one were to reject the hypothesis of "no preference" for this experiment, would this imply that polish *A* is superior to the others? Can you suggest a better method of conducting the experiment?

11.23 A sociologist conducted a survey to determine whether the incidence of various types of crime varied from one part of a particular city to another. The city was partitioned into three regions and the crimes classified as homicide, auto theft, grand larceny, petty larceny, and others. An analysis of 1,599 cases produced the following results:

City Region	Type of Crime				
	Homicide	Auto Theft	Grand Larceny (Neglecting Auto Theft)	Petty Larceny	Other
1	12	239	191	122	47
2	17	163	278	201	54
3	7	98	109	44	17

Do these data present sufficient evidence to indicate that the occurrence of various types of crime is dependent upon city region?

11.24 A survey was conducted to investigate interest of middle-aged adults in physical-fitness programs in Rhode Island, Colorado, California, and Florida. The objective of the investigation was to determine whether adult participation in physical-fitness programs varies from one region of the United States to another. A random sample of people was interviewed in each state and the following data were recorded.

	Rhode Island	Colorado	California	Florida
Participate	46	63	108	121
Do not participate	149	178	192	179

Do the data indicate a difference in adult participation in physical-fitness programs from one state to another?

11.25 The governor of each state was polled to determine his opinion concerning a particular domestic policy issue. At the same time, his party affiliation was recorded. The data are given here. If we assume that the 50 governors represent a random sample of political leaders throughout the nation, do the data present sufficient evidence to indicate a dependence between party affiliation and the opinion expressed on the domestic policy issue? (Use $\alpha = .05$.)

	Approve of Policy	Do Not Approve	No Opinion
Republican	18	5	5
Democrat	8	8	6

11.26 A carpet company was interested in comparing the fraction of new-home builders favoring carpet over other floor coverings for homes in three different areas of a city. The objective was to decide how to allocate sales effort to the areas. A survey was conducted and the data are as follows:

	Areas		
	1	2	3
Carpet	69	126	16
Other Material	78	99	27

Do the data indicate a difference in the percentage favoring carpet from one region of the city to another?

11.27 Refer to Exercise 11.26. Estimate the difference in the fractions of new-home builders who favor carpet between areas 1 and 2. Use a 95 percent confidence interval.

11.28 A survey was conducted to determine student, faculty, and administration attitudes on a new university parking policy. The distribution of those favoring or opposed to the policy is shown below.

	Student	Faculty	Administration
Favor	252	107	43
Opposed	139	81	40

Do the data provide sufficient evidence to indicate that attitudes regarding the parking policy are independent of student, faculty, or administration status?

11.29 The chi-square test used in Exercise 11.13 is equivalent to the two-tailed z test of Section 8.12 provided α is the same for the two tests. Show algebraically that the chi-square test statistic X^2 is the square of the test statistic z for the equivalent test.

11.30 It is often not clear whether all properties of a binomial experiment are actually met in a given application. A "goodness-of-fit" test is desirable for such cases. Suppose that an experiment consisting of four trials was repeated 100 times. The number of repetitions on which a given number of successes was obtained is recorded in the following table:

Possible Results (Number of Successes)	0	1	2	3	4
Number of Times Obtained	11	17	42	21	9

Estimate p (assuming that the experiment was binomial), obtain estimates of the expected cell frequencies, and test for goodness of fit. To determine the appropriate number of degrees of freedom for X^2, note that p was estimated by a linear combination of the frequencies obtained.

11.31 A survey of student opinion concerning a resolution presented to the student council was studied to determine whether the resulting opinion was independent of fraternity and sorority affiliation. Two hundred students were interviewed, with the following results:

	In Favor	Opposed	Undecided
Fraternity	37	16	5
Sorority	30	22	8
No affiliation	32	44	6

Do these data present sufficient evidence to indicate that student opinion concerning the resolution was dependent upon affiliation with fraternities or sororities? Test using $\alpha = .05$.

11.32 The responses for the data in Exercise 11.31 were reclassified according to whether the student was male or female.

	In Favor	Opposed	Undecided
Female	39	46	9
Male	60	36	10

Do the data present sufficient evidence to indicate that student reaction to the resolution varied for the various opinion categories depending upon whether the student was male or female?

11.33 A radio station conducted a survey to study the relationship between the number of radios per household and family income. The survey, based upon $n = 1{,}000$ interviews, produced the results shown in the table. Do the data present sufficient evidence to indicate that the number of radios per household is dependent upon family income? Test at the $\alpha = .10$ level of significance.

Number of Radios per Household	Family Income			
	Less Than $4,000	$4,000–7,000	$7,000–10,000	More Than $10,000
1	126	362	129	78
2	29	138	82	56

11.34 A trucking company claims that 30 percent of its trucks carry produce, 45 percent carry local freight, and 25 percent carry the bottled produce from neighboring vineyards. A group of investors who are interested in purchasing the trucking company decides to check this claim. The group randomly samples the bills of lading of 100 trucks and checks their cargos. The survey produced the following counts on the number of trucks carrying specific cargos:

Produce: 21
Local Freight: 39
Wine: 40

Do these data disagree with the trucking company's claim? Test by using $\alpha = .10$.

11.35 A problem that sometimes occurs during surgical operations is the occurrence of infections during blood transfusions. An experiment was conducted to determine whether the injection of antibodies reduced the probability of infection. An examination of the records of 138 patients produced the data shown in the accompanying table. Do the data provide sufficient evidence to indicate that injections of antibodies affect the likelihood of transfusional infections? Test by using $\alpha = .10$.

	Infection	No Infection
Antibody	4	78
No antibody	11	45

11.36 By tradition U.S. labor unions have been content to leave the management of the company to the managers and corporate executives. But in Europe worker participation in management decision making is an accepted idea and one that is continually spreading. To study the effect of worker satisfaction with worker participation in managerial decision making, 100 workers were interviewed in each of two separate West German manufacturing plants. One plant had active worker participation in managerial decision making; the other did not. Each selected worker was asked whether he or she generally approved of the managerial decisions made within the firm. The results of the interviews are shown in the table.

	Participative Decision Making	No Participative Decision Making
Generally approve of the firm's decisions	73	51
Do not approve of the firm's decisions	27	49

(a) Do the data provide sufficient evidence to indicate that approval or disapproval of management's decisions depends on whether workers participate in decision making? Test by using the X^2 test statistic. Use $\alpha = .05$.

(b) Do these data support the hypothesis that workers in a firm with participative decision making more generally approve of the firm's managerial decisions than those employed by firms without participative decision making? Test by using the z test presented in Section 8.12. This problem requires a one-tailed test. Why?

11.37 A new study reported in the *New England Journal of Medicine* (December 8, 1977) suggests that aspirin can protect male surgery patients from forming postsurgical blood clots in their veins. Of 23 men who received 4 aspirin tablets per day, only 4 developed blood clots in comparison with 14 of 25 men who took dummy pills (placebos) intended to look like aspirin but which in fact were inert. Do these data provide sufficient evidence to indicate a link between the use of aspirin and the frequency of formation of postsurgical blood clots?

11.38 An occupant-traffic study was conducted to aid in the remodeling of an office building that contains three entrances. The choice of entrance was recorded for a sample of 200 persons entering the building. Do the data shown in the table indicate that there is a difference in preference for the three entrances? Find a 90 percent confidence interval for the proportion of persons favoring entrance 1.

	Entrance		
	1	2	3
Number entering	83	61	56

11.39 The computer, which at its advent was expected to play its primary role in scientific computation, has become essential in banking, in the recording of personal data, in merchandising, and in many areas of our daily lives. Along with this growth in computer applications have come numerous cases of computer abuse, financial fraud, theft of information, etc. The data in the accompanying table give four different types of computer abuse that were reported and verified for the years 1970 to 1973.* The frequency of computer abuses would be expected to increase as the years go by unless safeguards are found to prevent their occurrence. But are the proportions of the four types of abuses changing over time? Test by using $\alpha = .10$.

Year	Financial Fraud	Theft of Information or Property	Unauthorized Use	Vandalism	Total
1970	7	5	9	8	29
1971	22	18	6	6	52
1972	12	15	6	12	45
1973	21	15	16	9	61
Total	62	53	37	35	187

11.40 Exercise 8.104 gave the national statistics shown in the accompanying table on admission to medical colleges, classified according to major, for the 1976–1977

* *Source:* D. B. Parker, S. Nycum, and S. S. Oura, *Computer Abuse*. Menlo Park, Calif.: Stanford Research Institute, 1973.

academic year. Do the data provide sufficient evidence to indicate that the proportion of applicants admitted to medical schools is dependent upon the major field of study? Explain.

Major	Number of Admissions	Percentage of Total of Major Applications Admitted
biology	15,488	36
chemistry	4,813	45
zoology	3,025	35
premed	2,243	38
biochemistry	1,387	50
interdisciplinary	172	52

11.41 As part of a study of small-town police, Galliher and colleagues questioned police in towns of varying sizes to determine what types of people they watched most carefully because of their potential for criminal activity.* Particularly, they were interested in determining whether the "most watched" type of person depended upon the size of the town. They interviewed police in 224 small towns and they obtained the data shown in the table. Do the data provide sufficient information to indicate a dependence between the type of persons most watched and town size? Test by using $\alpha = .05$.

	Town Size		
Type of Persons Most Watched	Under 5,000	5,000–10,000	10,000–25,000
known criminals	21	26	25
young people	39	15	20
strangers and other suspicious persons	13	5	11
bar crowds	11	2	2
others	10	11	13
Total	94	59	71

EXPERIENCES WITH REAL DATA

Sociologists and educators periodically point to the changing attitudes of college students, not only within an educational institution over a period of time but also between college entering classes. For example, students during the 1950s have been labeled complacent and likely to accept the pronouncements of those in authority without question;

*J. F. Galliher, L. P. Donavan, and D. L. Adams, "Small Town Police: Trouble, Tasks and Publics," *J. Police Sci. and Adm.*, *3*, no. 1 (1975).

students of the 1960s were supposed to be the questioning, ever-seeking idealists. (These statements are subject to question!) And some of my colleagues profess now to see a distinct difference between entering and senior classes in their attitude toward study. All these pronouncements are subject to debate and that leads us to a class problem associated with real data.

A current social question concerns the interpretation that should be given to equal rights and opportunity for minority groups. For example, *should equal rights and opportunity imply that each human, regardless of race, religion, or sex, be accorded an equal opportunity for job employment and/or entrance to graduate or professional schools? Or, does it mean that society should redress the wrongs of the past and accord members of minority groups a priority status in seeking jobs and/or educational opportunities?* On this particular issue, the attitudes of entering freshmen may very well differ from those of seniors, who will soon be entering the job market or competing for admission to professional schools.

Conduct a survey to determine whether there is a dependence between college class and the response to the italicized questions above. The 12 categories of the study will appear as follows:

	Freshmen	Sophomores	Juniors	Seniors
Favor equal opportunity for each person				
Favor priority for minority groups				
No opinion, or neither of the above				

Use the student directory to randomly select 100 students from each college class. Contact each student by telephone to ascertain his or her response to the questions. Analyze your data and determine whether the data provide sufficient evidence to indicate a dependence between college class and student attitude to the survey questions.

Because seniors face the problem of admission to professional or graduate schools, the competition for a limited number of good jobs, and imminent competition for advancement, their attitude regarding the meaning of "equal opportunity" may differ from others. Collapse the 3×4 categorization of the data to a 3×2 categorization as follows:

	Seniors	Others
Favor equal opportunity for each person		
Favor priority for minority groups		
No opinion, or neither of the above		

Do the data provide sufficient evidence to indicate that the opinions of seniors differ from the others?

The calculations for this experiment can easily be accomplished on an electronic desk calculator, but you can use packaged programs and an electronic computer to perform the calculations. Useful packaged programs are available in the Biomed (Biomedical Programs), the SAS (Statistical Analysis Systems), and the SPSS (Statistical Package for the Social Sciences) program libraries (see the references).

CHAPTER OBJECTIVES

GENERAL OBJECTIVE In the preceding chapters, we studied the mechanisms and the concepts involved in making inferences about a population based on information contained in a sample. In this chapter we shall demonstrate to you how certain factors affect the quantity of information contained in a sample and how you can use this information to minimize the cost of sampling.

SPECIFIC OBJECTIVES

1. To explain how to measure the quantity of information in a sample that pertains to a specific parameter. *Section 12.1*
2. To identify the factors that affect the quantity of information in an experiment. *Section 12.1*
3. To familiarize you with the terminology employed in the design of experiments. *Section 12.2*
4. To explain how to use random-number tables to select random samples. *Section 12.3*
5. To explain how the selection of treatments (volume-increasing experimental designs) increases the amount of information in an experiment. *Section 12.4*
6. To explain how blocking can be used to remove sources of random variation (noise-reducing experimental designs) and thereby increase the amount of information in an experiment. *Section 12.5*

12

CONSIDERATIONS IN DESIGNING EXPERIMENTS

12.1

THE FACTORS AFFECTING THE INFORMATION IN A SAMPLE

The information in a sample that is available to make an inference about a population parameter can be measured by the width (or half-width) of the confidence interval that could be constructed from the sample data. Thus, a large sample confidence interval for a population mean is

$$\bar{y} \pm 2\frac{\sigma}{\sqrt{n}}.$$

The widths of almost all the commonly employed confidence intervals are, like the confidence interval for a population mean, dependent on the population variance, σ^2, and the sample size, n. The less variation in the population, measured by σ^2, the smaller will be the confidence interval. Similarly, the width of the confidence interval will decrease as n increases. This interesting phenomenon would lead us to believe that two factors affect the quantity of information in a sample pertinent to a parameter, namely the variation of the data and the sample size, n. We shall find this deduction to be slightly oversimplified but essentially true.

A strong similarity exists between the audio theory of communication and the theory of statistics. Both are concerned with the transmission of a message (signal) from one point to another and, consequently, both are theories of information. For example, the telephone engineer is responsible for transmitting a verbal message that might originate in New York City and be received in New Orleans. Or equivalently, a speaker may wish to communicate with a large and noisy audience. If static or background noise is sizable for either example, the receiver may acquire only a sample of the complete signal, and from this partial information he must infer the nature of the complete message. Similarly, scientific experimentation is conducted to verify certain theories about natural phenomena, or simply to explore some aspect of nature, and hopefully to deduce—either exactly or with a good approximation—the relationship of certain natural variables. Thus one might think of experimentation as the communication between nature and a scientist. The message about the natural phenomenon is contained, in garbled form, in the experimenter's sample data. Imperfections in his measuring instruments, nonhomogeneity of experimental material, and many other factors contribute background noise (or static) that tends to obscure nature's signal and cause the observed response to vary in a random manner. For both the communications engineer and the statistician, two factors affect the quantity of information in an experiment: the magnitude of the background noise (or variation) and the volume of the signal. The greater the noise or, equivalently, the variation, the less information will be contained in the sample. Likewise, the louder the signal, the greater the amplification will be and hence it is more likely that the message will penetrate the noise and be received.

The design of experiments is a very broad subject concerned with methods of sampling to reduce the variation in an experiment, to amplify nature's signal, and thereby to acquire a specified quantity of information at minimum cost. Despite the complexity of the subject, some of the important considerations in the design of good experiments can be easily understood and should be presented to the beginner. We take these considerations as our objective in the succeeding discussion.

12.2

THE PHYSICAL PROCESS OF DESIGNING AN EXPERIMENT

We commence our discussion of experimental design by clarifying terminology and then, through examples, by identifying the steps that one must take in designing an experiment.

DEFINITION

The objects upon which measurements are taken are called experimental units.

If an experimenter subjects a set of $n = 10$ rats to a stimulus and measures the response of each, a rat is the experimental unit. The collection of $n = 10$ measurements is a sample. Similarly, if a set of $n = 10$ items is selected from a list of hospital supplies in an inventory audit, each item is an experimental unit. The observation made on each experimental unit is the dollar value of the item actually in stock, and the set of 10 measurements constitutes a sample.

What one does to the experimental units that makes them differ from one population to another is called a treatment. Thus, one might wish to study the density of a specific kind of cake when baked at $x = 350°F$, $x = 400°F$, and $x = 450°F$ in a given oven. An experimental unit would be a single mix of batter in the oven at a given point in time. The three temperatures, $x = 350°$, $400°$, and $450°$, would represent three treatments. The millions and millions of cakes that conceptually *could* be baked at 350°F would generate a population of densities, and one could similarly generate populations corresponding to 400°F and 450°F. The objective of the experiment would be to compare the cake density, y, for the three populations. Or, we might wish to study the effect of temperature of baking, x, on cake density by fitting a linear or curvilinear model to the data points as described in Chapter 10.

In another experiment we might wish to compare tire wear for two manufacturers, A and B. Each tire–wheel combination tested at a particular time would represent an experimental unit, and each of the two manufacturers would represent a treatment. Note that one does not physically treat the tires to make them different. They receive two different treatments by the very fact that they are manufactured by two different companies in different locations and in different factories.

As a third example, consider an experiment conducted to investigate the effect of various amounts of nitrogen and phosphate on the yield of a variety of corn. An experimental unit would be a specified acreage, say 1 acre, of corn. A treatment would be a fixed number of pounds of nitrogen, x_1, and phosphate, x_2, applied to a given acre of corn. For example, one treatment might be to use $x_1 = 100$ pounds of nitrogen per acre and $x_2 = 200$ pounds of phosphate. A second treatment could be $x_1 = 150$ and $x_2 = 100$. Note that the experimenter could experiment with different amounts (x_1, x_2) of nitrogen and phosphate and that each combination would represent a treatment.

Most experiments involve a study of the effect of one or more independent variables on a response.

DEFINITION

Independent experimental variables are called factors.

Factors can be quantitative or qualitative.

DEFINITION

A quantitative factor is one that can take values corresponding to points on a real line. Factors that are not quantitative are said to be qualitative.

Oven temperature, pounds of nitrogen, and pounds of phosphate are examples of quantitative factors. In contrast, manufacturers, types of drugs, or physical locations are factors that cannot be quantified and are called qualitative.

DEFINITION

The intensity setting of a factor is called a level.

Thus the three temperatures, 350°, 400°, and 450°, represent three levels of the quantitative factor "oven temperature." Similarly, the two treatments, manufacturer *A* and manufacturer *B*, represent two levels of the qualitative factor, "manufacturers." Note that a third or fourth tire manufacturer could have been included in the tire-wear experiment and resulted in either three or four levels of the factor "manufacturers."

We noted previously that what one does to experimental units that makes them differ from one population to another is called a treatment. Since every treatment implies a combination of one or more factor levels, we have a more precise definition for the term.

DEFINITION

A treatment is a specific combination of factor levels.

The experiment may involve only a single factor such as temperature in the baking experiment. Or, it could be composed of combinations of levels of two (or more) factors as for the corn-fertilizing experiment. Each combination would represent a treatment. One of the early steps in the design of an experiment is the selection of factors to be studied and a decision regarding the combinations of levels (treatments) to be employed in the experiment.

The design of an experiment implies one final problem after selecting the factor combinations (treatments) to be employed in an experiment. One must decide how the treatments should be assigned to the experimental units. Should the treatments be randomly assigned to the experimental units or should some semirandom pattern be employed? For example, should the tires corresponding to manufacturers *A* and *B* be randomly assigned to all the automobile wheels, or should one each of tire types *A* and *B* be assigned to the rear wheels of each car?

The foregoing discussion suggests that the design of an experiment involves four steps:

Steps Employed in Designing an Experiment

1. The selection of factors to be included in the experiment and a specification of the population parameter(s) of interest.
2. Deciding how much information is desired pertinent to the parameter(s) of interest. (For example, how accurately do you wish to estimate the parameters?)
3. The selection of the treatments to be employed in the experiment (combination of factor levels) and deciding on the number of experimental units to be assigned to each.
4. Deciding on the manner in which the treatments should be applied to the experimental units.

Steps 3 and 4 correspond to the two factors that affect the quantity of information in an experiment. First, how one selects the treatments (combinations of factor levels) and the number of experimental units assigned to each affects the intensity of nature's signal pertinent to the population parameter(s) of interest to the experimenter. Second, the method of assigning the treatments to the experimental units affects the background noise or, equivalently, the variation of the experimental units. We will examine each of these assertions in detail in Sections 12.4 and 12.5 after digressing briefly in Section 12.3 to consider the implications of random sampling and how to draw a random sample.

12.3

RANDOM SAMPLING AND THE COMPLETELY RANDOMIZED DESIGN

Random sampling—that is, giving every possible sample in a population an equal probability of selection—has two purposes. First, it avoids the possibility of bias introduced by a nonrandom selection of sample elements. For example, a sample selection of voters from telephone directories in 1936 indicated a clear win for Landon. However, the sample did not represent a random selection from the whole population of eligible voters, because the majority of telephone users in

1936 were Republicans. As another example, suppose that we sample to determine the percent of homeowners favoring the construction of a new city park and modify our original random sample by ignoring owners who are not at home. The result may yield a biased response because those at home will likely have children and may be more inclined to favor the new park. The second purpose of random sampling is to provide a probabilistic basis for the selection of a sample. That is, it treats the selection of a sample as an experiment (Chapter 4), enabling the statistician to calculate the probability of an observed sample and to use this probability in making inferences. Thus, we learned that under fairly general conditions, the mean, $\bar{y}$, of a random sample of n elements will possess a probability distribution that is approximately normal when n is large (the Central Limit Theorem). Fundamental to the proof of the Central Limit Theorem is the assumption of random sampling. Similarly, the confidence intervals and tests of hypotheses based on Student's t, Chapter 9, required the assumption of random sampling.

The random selection of independent samples to compare two or more populations is the simplest type of experimental design. The populations differ because we have applied different treatments (factor combinations) and we now wish to consider step 4 in the design of an experiment—deciding how to apply the treatments to the experimental units, or equivalently, how to select samples.

DEFINITION

The selection of independent random samples from p populations is called a completely randomized design.

The comparison of five brands of aspirin, *A, B, C, D,* and *E,* by randomly selecting 100 pills from the production of each manufacturer would be a completely randomized design with $n_A = n_B = \cdots = n_E = 100$. Similarly, we might wish to compare five teaching techniques, *A, B, C, D,* and *E,* using 25 students in each class ($n_A = n_B = \cdots = n_E = 25$). The populations corresponding to *A, B, C, D,* and *E* are nonexistent (that is, they are conceptual) because students taught by the five techniques either do not exist or they are unavailable. Consequently, random samples are obtained for the five conceptual populations by randomly selecting 125 students of the type envisioned for the study and then *randomly* assigning 25 students to each of the five teaching techniques. The scheme will yield independent random samples of 25 students subjected to each technique and will result in a completely randomized design.

As you might surmise, not all designs are completely randomized, but all good experiments rely on randomization to some extent. For example, the paired-difference experiment of Chapter 9 implied the application of the treatments *A* and *B* to pairs of experimental units that were nearly homogeneous. Then *A* and *B* were *randomly* assigned, one each, to the experimental units in the pair. This design, which uses sampling that is partially, but not completely, random, was employed to reduce the experimental error (reduce the background noise) and, thereby, to increase the information in the experiment. We will comment further on noise-reducing designs in Section 12.5.

Before concluding this section on randomization, let us consider a method

TABLE 12.1
Portion of a table of random numbers

15574	35026	98924
45045	36933	28630
03225	78812	50856
88292	26053	21121

for selecting a random sample or randomly assigning experimental units to a set of treatments.

The simplest and most reliable way to select a random sample of n elements from a large population is to employ a table of random numbers such as that shown in Table 13, Appendix II. Random-number tables are constructed so that integers occur randomly and with equal frequency. For example, suppose that the population contains $N = 1{,}000$ elements. Number the elements in sequence, 1 to 1,000. Then turn to a table of random numbers such as the excerpt shown in Table 12.1.

Select n of the random numbers in order. The population elements to be included in the random sample will be given by the first three digits of the random numbers (unless the first four digits are 1,000). Thus, if $n = 5$, we would include elements numbered 155, 450, 32, 882, and 350. So as not to use the same sequence of random numbers over and over again, the experimenter should select different starting points in Table 13 to begin the selection of random numbers for different samples.

The random assignment of 40 experimental units, 10 to each of four treatments, A, B, C, and D, is equally easy using the random-number table. Number the experimental units 1 to 40. Select a sequence of random numbers and refer only to the first two digits since $n = 40$. Discard random numbers greater than 40. Then assign the experimental units associated with the first 10 numbers that appear to A, those associated with the second 10 to B, and so on. To illustrate, refer to Table 12.1. Starting in the first column and moving top to bottom, we acquire the numbers 15 and 3 (discard 45 and 88 because no experimental units possess those numbers). This procedure would be continued until all integers, 1 to 40, were selected. The resulting numbers would occur in random order.

The analysis of data for a completely randomized design is treated in Section 13.3.

EXERCISES

12.1 Suppose that a telephone company wishes to select a random sample of $n = 20$ (we select this small number to simplify this exercise) out of 7,000 customers for a survey of customer attitudes concerning service. If the customers are numbered for identification purposes, indicate the customers whom you will include in your sample. Use the random-number table and explain how you selected your sample.

12.2 A small city contains 20,000 voters. Use the random-number table to identify the voters to be included in a random sample of $n = 15$.

12.3 Twelve people are each to receive a vaccination in random order. Use the random-number tables to select the ordering.

12.4 If a survey of a newspaper's readership is obtained by requesting readers to respond to a questionnaire published in the newspaper, are the resulting responses likely to give a random sample of readership opinion? Explain.

12.5 A random sample of the public opinion for a small town was obtained by selecting every tenth person to pass by the busiest corner in the downtown area. Will this sample have the characteristics of a random sample selected from the towns citizenry? Explain.

12.6 Suppose that you decide to conduct a telephone public opinion survey, randomly sampling members from the telephone directory. The survey is conducted from 9 A.M. to 5 P.M. Will the resulting responses represent a random sample of adult public opinion in the community? Explain.

12.4 VOLUME-INCREASING EXPERIMENTAL DESIGNS

You will recall that designing to increase the volume of a signal, that is, the information pertinent to one or more population parameters, depends on the selection of treatments and the number of experimental units assigned to each (step 3 in designing an experiment). The treatments, or combinations of factor levels, identify points at which one or more response measurements will be made and indicate the general location in which the experimenter is focusing his attention. As we shall see, some designs contain more information concerning specific population parameters than others for the same number of observations. Other very costly experiments contain no information concerning certain population parameters. And no single design is best in acquiring information concerning all types of population parameters. Indeed, the problem of finding the best design for focusing information on a specific population parameter has been solved in only a few specific cases.

The purpose of this section is not to present a general theory or find the best selection of factor-level combinations for a given experiment but rather to present a few examples to illustrate the principles involved. The optimal design providing the maximum amount of information pertinent to the parameter(s) of interest (for a fixed sample size, n) will be given for the following two examples.

The simplest example of an information-focusing experiment is the problem of estimating the difference between a pair of population means, $(\mu_1 - \mu_2)$, based on independent random samples. In this instance, the two treatments have already been selected and the question concerns the allocation of experimental units to the two samples. If the experimenter plans to invest money sufficient to sample a total of n experimental units, how many units should he select from populations 1 and 2, say n_1 and n_2 $(n_1 + n_2 = n)$, respectively, so as to maximize the information in the data pertinent to $(\mu_1 - \mu_2)$? Thus if $n = 10$, should he select $n_1 = n_2 = 5$ observations from each population or would an allocation of $n_1 = 4$ and $n_2 = 6$ be better?

Recall that the estimator of $(\mu_1 - \mu_2)$, $(\bar{y}_1 - \bar{y}_2)$, has a standard deviation

$$\sigma_{(\bar{y}_1 - \bar{y}_2)} = \sqrt{\frac{\sigma_1^2}{n_1} + \frac{\sigma_2^2}{n_2}}.$$

The smaller $\sigma_{(\bar{y}_1 - \bar{y}_2)}$, the smaller will be the corresponding error of estimation and the greater will be the quantity of information in the sample pertinent to $(\mu_1 - \mu_2)$. If, as we frequently assume, $\sigma_1^2 = \sigma_2^2 = \sigma^2$, then

$$\sigma_{(\bar{y}_1 - \bar{y}_2)} = \sigma\sqrt{\frac{1}{n_1} + \frac{1}{n_2}}.$$

You can verify that for a fixed total number $(n = n_1 + n_2)$ of observations, this quantity is a minimum for $n_1 = n_2$. Consequently, the sample contains a maximum of information on $(\mu_1 - \mu_2)$ when the n experimental units are equally divided between the two treatments. We illustrate (but do not prove) our point with an example.

EXAMPLE 12.1

Calculate the standard deviation for $(\bar{y}_1 - \bar{y}_2)$ when $\sigma_1^2 = \sigma_2^2 = \sigma^2$ and
(a) $n = 10$, $n_1 = n_2 = 5$.
(b) $n = 10$, $n_1 = 4$, $n_2 = 6$.

Solution

$$\sigma_{(\bar{y}_1 - \bar{y}_2)} = \sigma\sqrt{\frac{1}{n_1} + \frac{1}{n_2}}.$$

(a) When $n_1 = n_2 = 5$,

$$\sigma_{(\bar{y}_1 - \bar{y}_2)} = \sigma\sqrt{\frac{1}{5} + \frac{1}{5}} = \sigma\sqrt{.4}.$$

(b) When $n_1 = 4$, $n_2 = 6$,

$$\sigma_{(\bar{y}_1 - \bar{y}_2)} = \sigma\sqrt{\frac{1}{4} + \frac{1}{6}} = \sigma\sqrt{.42}.$$

Note that $\sigma_{(\bar{y}_1 - \bar{y}_2)}$ is smaller for an equal allocation, $n_1 = n_2 = 5$.

Equal allocation of experimental units to the two treatments is *not* best when $\sigma_1^2 \neq \sigma_2^2$. Then, the allocation of the n experimental units to give the maximum information concerning $(\mu_1 - \mu_2)$ assigns values to n_1 and n_2 that are proportional to σ_1 and σ_2, respectively. Proof of this statement is omitted.

As a second example, consider the problem of fitting a straight line through a set of n points using the least-squares method of Chapter 10 (see Figure 12.1). Further, suppose that we are primarily interested in the slope of the line, β_1, in the linear model

$$y = \beta_0 + \beta_1 x + \varepsilon.$$

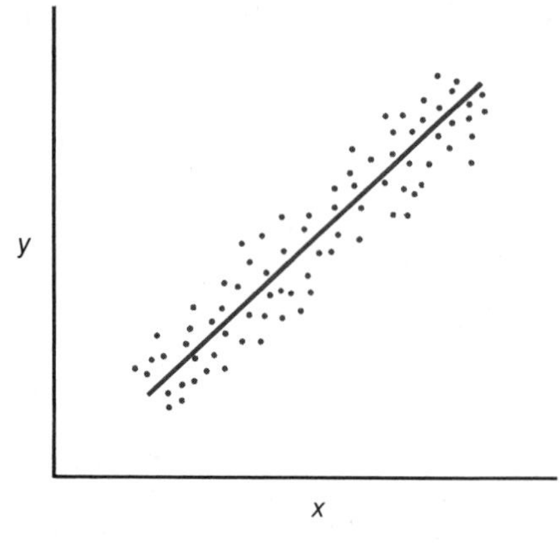

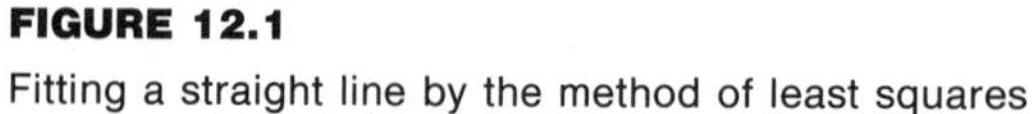
FIGURE 12.1
Fitting a straight line by the method of least squares

If we have the option of selecting the n values of x for which y will be observed, which values of x will maximize the quantity of information on β_1? Thus we have one single quantitative factor, x, and have the problem of deciding on the levels to employ $(x_1, x_2, \ldots, x_n)$ as well as the number of observations to be taken at each.

A strong lead to the best design for fitting a straight line can be achieved by viewing Figure 12.2(a) and (b). Suppose that y was linearly related to x and generated data similar to that shown in Figure 12.2(a) for the interval $x_1 < x < x_2$. Note the approximate range of variation for a given value of x. Now suppose that instead of the wide range for x employed in Figure 12.2(a), the experimenter selected data from the same population but over the very narrow range $x_3 < x < x_4$, as shown in Figure 12.2(b). The variation in y, given x, is the same as for Figure 12.2(a). Which distribution of data points would provide the

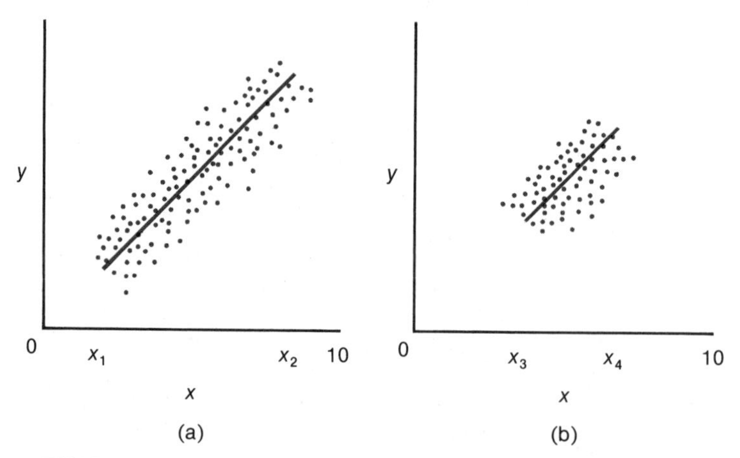

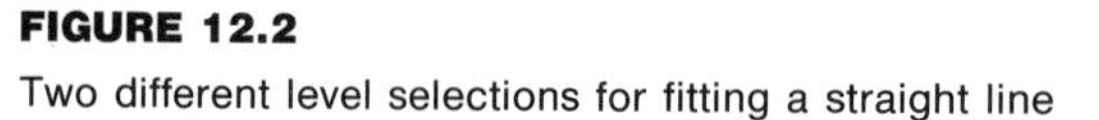
FIGURE 12.2
Two different level selections for fitting a straight line

greater amount of information concerning the slope of the line, β_1? You might guess (correctly) that the best estimate of slope will occur when the levels of x are selected farther apart as shown in Figure 12.2(a). The data for 12.2(b) could yield a very inaccurate estimate of the slope and, as a matter of fact, might leave a question as to whether the slope is positive or negative.

The best design for estimating the slope, β_1, can be determined by considering the standard deviation of $\widehat{\beta}_1$,

$$\sigma_{\widehat{\beta}_1} = \frac{\sigma}{\sqrt{\sum_{i=1}^{n}(x_i - \bar{x})^2}}.$$

The larger the sum of squares of deviations of $x_1, x_2, \ldots, x_n$ about their mean, the smaller will be the standard deviation of $\widehat{\beta}_1$. The experimenter will usually have some experimental region, say $x_1 < x < x_2$, over which he wishes to observe y and this range will frequently be selected prior to experimentation. Then, the smallest value for $\sigma_{\widehat{\beta}_1}$ will occur when the n data points are equally divided, with half located at the lower boundary of the region, x_1, and half at the upper boundary, x_2. (Proof is omitted.) Thus, if an experimenter wishes to fit a line using $n = 10$ data points in the interval $2 < x < 6$, he should select five data points at $x = 2$ and five at $x = 6$. Before concluding discussion of this example, you should note that observing all values of y at only two values of x will not provide information on curvature of the response curve in case the assumption of linearity in the relation of y and x is incorrect. It is frequently safer to select a few points (as few as one or two) somewhere near the middle of the experimental region to detect curvature if it should be present (see Figure 12.3).

To summarize, we have given optimal designs (factor-level combinations and the allocation of experimental units per combination) for comparing a pair of means and fitting a straight line. These two simple designs illustrate the manner in which information in an experiment can be increased or decreased by the selection of the factor-level combinations that represent treatments and by changing the allocation of observations to treatments. Thus, we have demonstrated that factor-level selection and the allocation of experimental units to treatments can greatly affect the information in an experiment pertinent to a

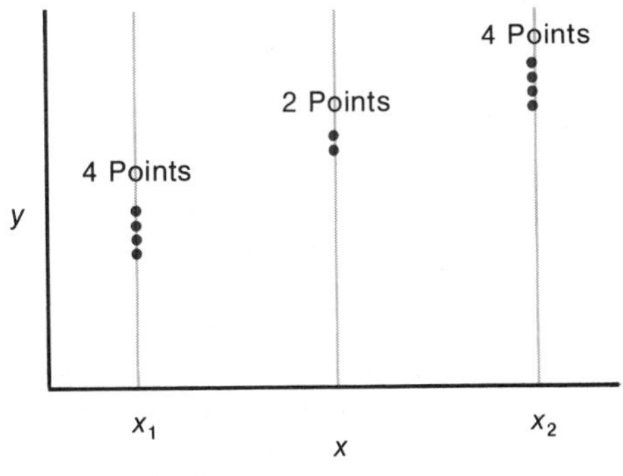

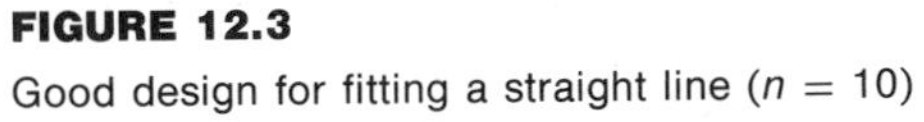

FIGURE 12.3

Good design for fitting a straight line ($n = 10$)

particular population parameter and, thereby, can amplify nature's signal. Thus step 3, Section 12.2, is an important consideration in the design of good experiments.

EXERCISES

12.7 Suppose that one wishes to compare the means for two populations. If $\sigma_1^2 = \sigma_2^2$, what is the optimal allocation of n_1 and n_2 ($n_1 + n_2 = n$) for a fixed sample size n? Does this operation employ the principle of noise reduction or signal amplification?

12.8 Refer to Exercise 12.7. Suppose that $\sigma_1^2 = 9$, $\sigma_2^2 = 25$, and $n = 90$. What allocation of $n = 90$ to the two samples will result in the maximum amount of information on $(\mu_1 - \mu_2)$?

12.9 Refer to Exercise 12.8. Suppose that we allocate $n_1 = n_2$ observations to each sample. How large must n_1 and n_2 be in order to obtain the same amount of information as that implied by the solution to Exercise 12.8?

12.10 Suppose that you wish to estimate the slope of a line using the method of least squares, that you can afford only 12 experimental points, and that the assumptions of Section 10.2 are satisfied. How can you select the locations of the points (the settings of the independent variable, x) so as to minimize the variance of $\hat{\beta}_1$? Note that this selection of settings accomplishes an increase in "volume" without increasing the sample size. Essentially, it shifts information within the design so that it focuses on the slope, β_1.

12.5 NOISE-REDUCING EXPERIMENTAL DESIGNS

Noise-reducing experimental designs increase the information in an experiment by decreasing the background noise (variation) caused by uncontrolled nuisance variables. Thus, by serving as filters to screen out undesirable noise, they permit nature's signal to be received more clearly. Reduction of noise can be accomplished in step 4 of the design of an experiment, namely, in the method for assigning treatments to the experimental units.

Designing for noise reduction is based on the single principle of making all comparisons of treatments within relatively homogeneous groups of experimental units. The more homogeneous the experimental units, the easier it is to detect a difference in treatments. Because noise-reducing, or filtering, designs work with blocks of relatively homogeneous experimental units, they are called block designs.

The paired-difference experiment, Section 9.5, employed the blocking principle to achieve a very sizable reduction in noise and a large increase in information pertinent to the difference between treatment means. The objective of the experiment was to make an inference about the difference in mean wear for two types of automobile tires, A and B. Five tires of each type were to be mounted on the rear wheels of five automobiles subjected to a large and specified number of miles of driving and the wear recorded for each. The experimenter realized that tire wear might vary substantially from automobile to automobile, owing primarily to the difference in drivers and the completely different sequence of starts and stops to which each car would be subjected over the test distance. In contrast, the abrasions applied to the rear wheels of the same car would be relatively equivalent. Thus, he decided to block his design on automobiles, assigning an A and a B tire type in a random manner to the rear wheels of each automobile. This eliminated auto-to-auto variability in comparing treatments and produced approximately 400 times as much information concerning $(\mu_A - \mu_B)$ in comparison with a completely randomized design employing the same number of tires. This fact was made evident by comparing the width of the confidence intervals based on the data generated by the paired and unpaired analyses.

The paired-difference design is a special case of the most elementary noise-reducing design, the randomized block design. The randomized block design is employed to compare the means of any number of treatments, say p, within relatively homogeneous blocks of p experimental units. Any number of blocks may be employed, but each treatment must appear exactly once in each block. Thus the paired-difference experiment is a randomized block design containing $p = 2$ treatments.

DEFINITION

A randomized block design containing *b* blocks and *p* treatments consists of *b* blocks of *p* experimental units each. The treatments are randomly assigned to the units in each block, with each treatment appearing exactly once in every block.

The difference between a randomized block design and the completely randomized design can be demonstrated by considering an experiment designed to compare subject reaction to a set of four stimuli (treatments) in a stimulus–response psychological experiment. We shall denote the treatments as T_1, T_2, T_3, and T_4.

Suppose that eight subjects are to be randomly assigned to each of the four treatments. Random assignment of subjects to treatments (or vice versa) randomly distributes errors due to person-to-person variability to the four treatments and yields four samples that are, for all practical purposes, random and independent. This would be a completely randomized experimental design.

The experimental error associated with a completely randomized design is composed of a number of components. Some of these are due to the difference between subjects, to the failure of repeated measurements within a subject to be identical (owing to variations in physical and psychological conditions), to the failure of the experimenter to administer a given stimulus with exactly the same intensity in repeated measurements, and, finally, to errors of measurement. Reduction of any of these causes of error will increase the information in the experiment.

The subject-to-subject variation in the above experiment can be eliminated by using subjects as blocks. Thus each subject would receive each of the four treatments assigned in a random sequence. The resulting randomized block design would appear as in Figure 12.4. Now only eight subjects are required to obtain eight response measurements per treatment. Note that each treatment occurs exactly once in each block.

The word "randomization" in the name of the design implies that the treatments are randomly assigned within a block. For our experiment, position in the block would pertain to the position in the sequence when assigning the stimuli to a given subject over time. The purpose of the randomization (that is, position in the block) is to eliminate bias caused by fatigue or learning.

Blocks may represent time, location, or experimental material. Thus, if three treatments are to be compared and there is a suspected trend in the mean response over time, a substantial part of the time–trend variation may be removed by blocking. All three treatments would be randomly applied to experimental units in one small block of time. This procedure would be repeated in succeeding blocks of time until the required amount of data is collected. A comparison of the sale of competitive products in supermarkets should be made within supermarkets, thus using the supermarkets as blocks and removing market-to-market variability. Animal experiments in agriculture and medicine often utilize animal litters as blocks, applying all the treatments, one each, to animals within a litter. Because of heredity, animals within a litter are more homogeneous than those between litters. This type of blocking removes the litter-to-litter variation. The analysis of data generated by a randomized block design is discussed in Section 13.8.

The randomized block design is only one of many types of block designs. Blocking in two directions can be accomplished using a Latin square design. Suppose that the subjects of the preceding example became fatigued as the stimuli were applied so that the last stimulus always produced a lower response than the first. If this trend (and consequent lack of homogeneity of the experimental units in a block) were true for all subjects, a Latin square design would be appropriate. The design would be constructed as shown in Figure 12.5. Each stimulus is applied once to each subject and occurs exactly once in each position of the order of presentation. Thus all four stimuli occur in each row and in each column of the 4×4 configuration. The resulting design is called a 4×4 Latin square. A Latin square design for three treatments will require a 3×3 configuration and, in general, p treatments will require a $p \times p$ array of experimental units. If more observations are desired per treatment, the experimenter would utilize several Latin square configurations in one experiment. In the example above, it would require running two Latin squares to obtain eight observations per

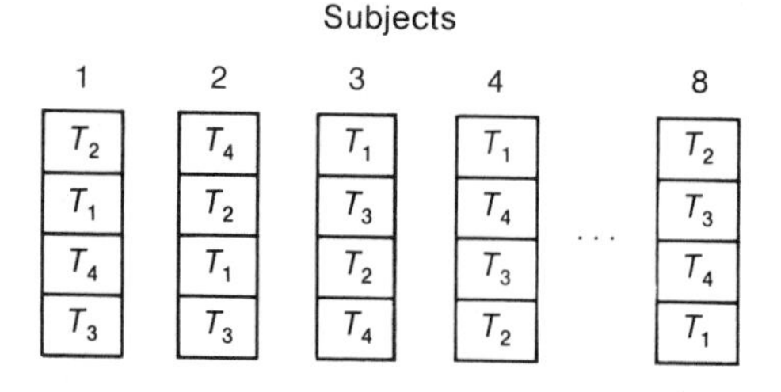

FIGURE 12.4

Randomized block design

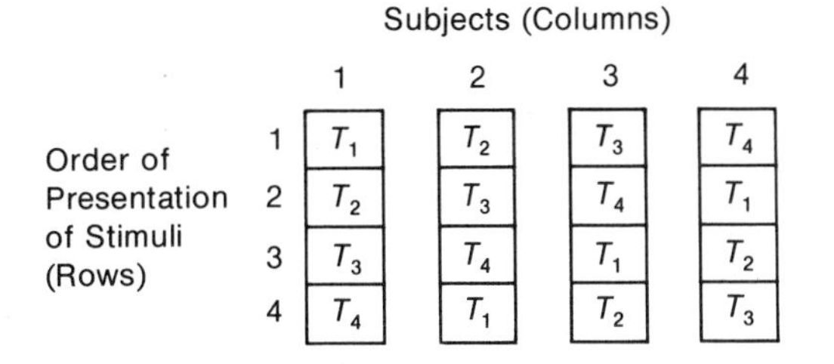

FIGURE 12.5
Latin square design

treatment. The experiment would then contain the same number of observations per treatment as the randomized block design, Figure 12.4.

A comparison of means for any pair of stimuli would eliminate the effect of subject-to-subject variation but would, in addition, eliminate the effect of the fatigue trend within each stimulus because each treatment was applied in each position of the stimuli time administering sequence. Consequently, the effect of trend would be canceled in comparing the means.

A more extensive discussion of block designs and their analyses is contained in the references. The objective of this section is to make the reader aware of the existence of block designs, how they work, and how they can produce substantial increases in the quantity of information in an experiment by reducing nuisance variation.

EXERCISES

12.11 Consider the tire-wear experiment of Section 12.5. Why do we regard "automobiles" as blocks rather than levels of a factor?

12.12 Could an independent variable be a factor in one experiment and a nuisance variable (a noise contributor) in another?

12.13 Suppose that you intended to compare the mean responses of humans to five drugs that were not expected to produce a long-lasting effect that would carry over from one part of the experiment to another. Each drug is to be administered to each of seven people. The intent is to make a comparison of drug effects within each person.
(a) What type of experimental design is to be used and how is it supposed to increase the quantity of information in the experiment?
(b) Use the random-number table to obtain an ordering of drug applications within the seven people.

12.14 An experiment was conducted to compare the mean yield (in soybeans) of soybean plants subjected to six different fertilizer combinations. Six plots 50 feet by 50 feet square were selected at each of eight different locations (farms), and the plots at each farm were located as near as possible to each other. The six fertilizer combinations were randomly assigned to the plots at each of the eight farms, the plots were seeded with soybeans using the same rate of planting for all plots, and the crop was harvested for each plot at the end of the growing season.
(a) Identify the experimental units.
(b) Identify the treatments.
(c) What type of experimental design was employed?

(d) What principle was employed to increase the quantity of information in the experiment?

12.15 Describe an experimental situation for which a randomized block design would be appropriate.

12.16 What is a Latin square design? What is the objective of the design? Describe an experimental situation for which a Latin square design would be appropriate.

12.6 SUMMARY

The objective of this chapter is to identify the factors that affect the quantity of information in an experiment and to use this knowledge to design better experiments. The subject, design of experiments, is very broad and is certainly not susceptible to condensation into a single chapter in an introductory text. In contrast, the philosophy underlying design, the methods for varying information in an experiment, and desirable strategies for design are easily explained and comprise the objective of Chapter 12.

Two factors affect the quantity of information in an experiment—the volume of nature's signal and the magnitude of variation caused by uncontrolled variables. The volume of information pertinent to a parameter of interest depends on the selection of factor-level combinations (treatments) to be included in the experiment and on the allocation of the total number of experimental units to the treatments. This choice determines the focus of attention of the experimenter.

The second method for increasing the information in an experiment concerns the method for assigning treatments to the experimental units. Blocking, comparing treatments within relatively homogeneous blocks of experimental material, can be used to eliminate block-to-block variation when comparing treatments. As such, it serves as a filter to reduce the unwanted variation that tends to obscure nature's signal.

Thus the selection of factors and the selection of factor levels are important considerations in shifting information in an experiment to amplify the information on a population parameter. The use of blocking in assigning treatments to experimental units reduces the noise created by uncontrolled variables and, consequently, increases the information in an experiment.

The analysis of some elementary experimental designs is given in Chapter 13. A more extensive treatment of the design and analysis of experiments is a course in itself. The reader interested in exploring this subject is directed to the references.

REFERENCES

Cochran, W. G., and G. M. Cox, *Experimental Designs,* 2nd ed. New York: John Wiley & Sons, Inc., 1957.

Hicks, C. R., *Fundamental Concepts in the Design of Experiments*. New York: Holt, Rinehart and Winston, Inc., 1964.

Mendenhall, W., *An Introduction to Linear Models and the Design and Analysis of Experiments.* Belmont, Calif.: Wadsworth Publishing Company, Inc., 1968.

Scheaffer, R., W. Mendenhall, and L. Ott, *Elementary Survey Sampling,* 2nd ed. North Scituate, Mass.: Duxbury Press, 1979.

Winer, B. J., *Statistical Principles in Experimental Design,* 2nd ed. New York: McGraw-Hill Book Company, 1971.

SUPPLEMENTARY EXERCISES

12.17 How can one measure the quantity of information in a sample pertinent to a specific population parameter?

12.18 Give the two factors that affect the quantity of information in an experiment.

12.19 What is a random sample?

12.20 Give two reasons for the use of random samples.

12.21 A political analyst wishes to select a sample of $n = 20$ people from a population of 2,000. Use the random-number table to identify the people to be included in the sample.

12.22 Two drugs, *A* and *B*, are to be applied to five rats each. Suppose that the rats are numbered from 1 to 10. Use the random-number table to randomly assign the rats to the two treatments.

12.23 Refer to Exercise 12.22, and suppose that the experiment involved three drugs, *A, B,* and *C,* with five rats assigned to each. Use the random table to randomly assign the 15 rats to the three treatments.

12.24 A population contains 50,000 voters. Use the random-number table to identify the voters to be included in a random sample of $n = 15$.

12.25 What is a factor?

12.26 State the steps involved in designing an experiment.

12.27 If one were to design an experiment, what part of the design procedure would result in signal amplification?

12.28 Refer to Exercise 12.27. What part of the design procedure would result in noise reduction?

12.29 Complete the assignment of treatments for the following 3 × 3 Latin square design:

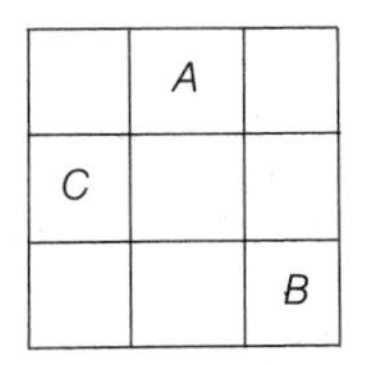

12.30 Suppose that one wishes to study the effect of the stimulant digitalis on the blood pressure of rats over a dosage range of $x = 2$ to $x = 5$ units. The response is expected to be linear over the region. Six rats are available for the experiment, and each rat can receive only one dose. What dosages of digitalis should be employed in the experiment and how many rats should be run at each dosage to maximize the quantity of information on β_1, the slope of the regression line? Which aspect of design—noise reduction or signal amplification—is implied in this experiment?

12.31 Refer to Exercise 12.30. Consider two methods for selecting the dosages. Method 1 assigns three rats to the dosage $x = 2$ and three rats to $x = 5$. Method 2 equally spaces the dosages between $x = 2$ and $x = 5$ ($x = 2, 2.6, 3.2, 3.8, 4.4, 5.0$). Suppose that σ is known and that the relationship between $E(y)$ and x is truly linear (see Chapter 10). How much larger will be the confidence interval for the slope (β_1) for method 2 in comparison with method 1? Approximately how many observations would be required to obtain the same confidence interval as obtained by the optimal assignment of method 1?

12.32 Refer to Exercise 12.30. Why might it be advisable to assign one or two points at $x = 3.5$?

12.33 An experiment is to be conducted to compare the effect of digitalis on the contraction of the heart muscle of a rat. The experiment is conducted by removing the heart from a live rat, slicing the heart into thin layers, and treating the layers to a dosage of digitalis. The muscle contraction is then measured. If four dosages (*A, B, C,* and *D*) are to be employed, what advantage might be derived by applying *A, B, C,* and *D* to a slice of tissue from each rat heart? What principle of design is illustrated by this example?

12.34 Describe the factors that affect the quantity of information in an experiment and the design procedures that control these factors.

12.35 When one statistic tends to support a theory and a second statistic, based on a completely different survey, tends to refute it, the reason may not be the occurrence of a highly improbable event. The apparent disagreement of statistics may be due to the fact that the researchers are sampling different populations. For example, researchers at the Florida Technological University conducted a survey to determine the effect of Florida's new mandatory three-year sentence for gun-related crimes.* To determine the effectiveness of the law, the researchers conducted a survey of 277 prisoners convicted of the use of handguns and found that 73 percent said that they would continue to use handguns upon release from prison. They therefore concluded that the law was ineffective. In response, the sponsor of the law pointed out that Florida Department of Criminal Law Enforcement statistics show a 30 percent drop in gun-related crimes since the law was enacted. Which of the two populations is the more pertinent in deciding whether the law was effective in reducing the use of handguns in crimes? Discuss the two statistics and explain their relevance to the researchers objective.

EXPERIENCES WITH REAL DATA

Select three competitive consumer products that can be visually evaluated. For example, you might compare the appeal of three different package designs for a consumer product (soap, food, etc.). Or you might wish to compare the appeal of automobile designs, house designs, or works of art. Assume that a single response will be obtained when a person rates one of the objects on a scale of 1 to 10.

Design an experiment to collect the data. Identify the experimental units, treatments, and any blocking that might be involved in the design. Explain how your design will acquire more information for a given expenditure than a completely randomized design.

* *Source:* "Senator Says Gun Research Is Wrong," *Gainesville Sun,* December 23, 1977.

CHAPTER OBJECTIVES

GENERAL OBJECTIVE Methods for comparing two population means, based on independent random samples and on a paired-difference experiment, were presented in Chapter 9. Chapter 13 extends these analyses to the comparison of any number of population means using a technique called an analysis of variance. In this chapter we shall explain the logic of an analysis of variance and give the analyses for three experimental designs.

SPECIFIC OBJECTIVES

1. To explain the logic of an analysis of variance. *Section 13.2*
2. To present the analysis of variance for a comparison of two treatment means and to relate the resulting F test to the t test of Chapter 9. *Section 13.2*
3. To give the analysis of variance for comparing two or more population means by using a completely randomized design. *Sections 13.3, 13.4, 13.5, 13.6, 13.7*
4. To give the analysis of variance for a randomized block design and to explain how to interpret the results of the analysis. *Sections 13.8, 13.9, 13.10*
5. To give the analysis of variance for a Latin square design and to explain how to interpret the results of the analysis. *Sections 13.11, 13.12, 13.13*
6. To familiarize you with typical computer printouts for analyses of variance. *Sections 13.7, 13.10, 13.13*
7. To explain how to select sample sizes to obtain a specific amount of information concerning the difference between a pair of treatment means. *Section 13.14*
8. To explain how blocking increases the information in an experiment. *Sections 13.8, 13.9, 13.11, 13.12, 13.15*
9. To summarize the assumptions upon which the tests and confidence intervals of an analysis of variance are based and to explain what happens and what to do when they are not satisfied. *Section 13.16*

13 THE ANALYSIS OF VARIANCE

13.1

INTRODUCTION

Most experiments involve a study of the effect of one or more independent variables on a response. In Chapter 12 we learned that a response, y, can be affected by two types of independent variables, quantitative and qualitative. Those independent variables that can be controlled in an experiment are called factors.

The analysis of data generated by a multivariable experiment requires identification of the independent variables in the experiment. These not only will be factors (controlled independent variables) but also will be directions of blocking. If one studies the wear for three types of tires, *A*, *B*, and *C*, on each of four automobiles, "tire types" is a factor representing a single qualitative variable at three levels. Automobiles are blocks and represent a single qualitative variable at four levels. The response for a Latin square design depends upon the factors that represent treatments, but it also is affected by two qualitative independent block variables, "rows" and "columns."

It is not possible to present a comprehensive treatment of the analysis of multivariable experiments in a single chapter of an introductory text. However, it is possible to introduce the reasoning upon which one method of analysis, the analysis of variance, is based and to show how the technique is applied to a few common experimental designs.

The application of noise reduction and signal amplification to the design of experiments was illustrated in Chapter 12. In particular, the completely randomized and the randomized block designs were shown to be generalizations of simple designs for the unpaired and paired comparisons of means discussed in Chapter 9. Treatments correspond to combinations of factor levels and identify the different populations of interest to the experimenter. Chapter 13 presents an introduction to the analysis of variance and gives methods for the analysis of the completely randomized, the randomized block, and the Latin square designs.

13.2

THE ANALYSIS OF VARIANCE

The methodology for the analysis of experiments involving several independent variables can best be explained in terms of the linear probabilistic model of Chapter 10. Although elementary and unified, this approach is not susceptible to the condensation necessary for inclusion in an elementary text. Instead, we shall attempt an intuitive discussion using a procedure known as the analysis of variance. Actually, the two approaches are connected and the analysis of variance can easily be explained in a general way in terms of the linear model. The interested reader can consult an introductory text on this subject by Mendenhall (1968), which is listed in the references.

As the name implies, the analysis-of-variance procedure attempts to analyze the variation of a response and to assign portions of this variation to each of a set of independent variables. The reasoning is that response variables vary only because of variation in a set of unknown independent variables. Since the experimenter will rarely, if ever, include all the variables affecting the response in his experiment, random variation in the response is observed even though all independent variables considered are held constant. The objective of the analysis of variance is to locate important independent variables in a study and to determine how they interact and affect the response.

The rationale underlying the analysis of variance can be indicated best with a symbolical discussion. The actual analysis of variance—that is, "how to do it"—can be illustrated with an example.

You will recall that the variability of a set of n measurements is proportional to the sum of squares of deviations, $SS_y = \sum_{i=1}^{n} (y_i - \bar{y})^2$, and that this quantity is used to calculate the sample variance. The analysis of variance partitions SS_y, called the total sum of squares of deviations, into parts, each of which is attributed to one of the independent variables in the experiment, plus a remainder that is associated with random error. This may be shown diagrammatically as indicated in Figure 13.1 for three independent variables.

If a multivariable linear regression model were written for the response y, as suggested in Section 10.9, the portion of the total sum of squares of deviations assigned to error would be the sum of squares of deviations of the y values about their respective predicted values obtained from the prediction equation, $\hat{y}$. You will recall that this quantity, represented by the sum of squares of deviations of the y values about a straight line, was denoted as SSE in Chapter 10.

For the cases that we consider, and when the response is unrelated to the independent variables, it can be shown that each of the pieces of the total sum of squares of deviations, divided by an appropriate constant, provides an independent and unbiased estimator of σ^2, the variance of the experimental error. When a variable is highly related to the response, its portion (called the "sum of

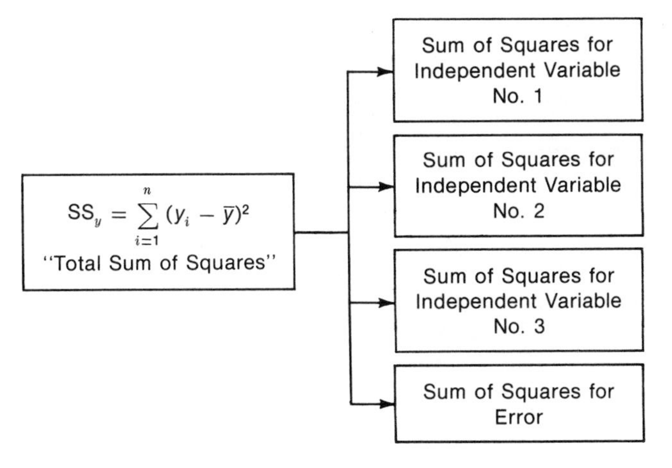

FIGURE 13.1
Partitioning of the total sum of squares of deviations

squares" for the variable) will be inflated. This condition can be detected by comparing the estimate of σ^2 for a particular independent variable with that obtained from SSE using an F test (see Section 9.7). If the estimate for the independent variable is significantly larger, the F test will reject an hypothesis of "no effect for the independent variable" and produce evidence to indicate a relation to the response.

The logic behind an analysis of variance can be illustrated by considering a familiar example, the comparison of two population means for an unpaired experiment (completely randomized design), which was analyzed in Chapter 9 using a Student's t statistic. We shall commence by giving a graphic and intuitive explanation of the procedure.

Suppose that we have selected random samples of five observations each from two populations, I and II, and that the y values are plotted as shown in Figure 13.2. Note that the $n_1 = 5$ observations from population I lie to the left; the $n_2 = 5$ observations from population II lie to the right. The sample means, $\bar{y}_1$ and $\bar{y}_2$, are shown as horizontal lines in the figure and the deviations of the y values about their respective means are the vertical line segments. Now examine Figure 13.2. Do you think the data provide sufficient evidence to indicate a difference between the two population means? Before we explain, let us look at another figure.

The same two sets of five points are plotted in Figure 13.3 except that the relative distance between the two sets is greater in (b) than in (a) and even greater in (c). Therefore, the distance between $\bar{y}_1$ and $\bar{y}_2$ increases as you move from Figure 13.3(a) to (b) and then to (c), but the relative variation within each set is held constant.

Now view the three plots, (a), (b), and (c), and decide which situation, (a), (b), or (c), provides the greatest evidence to indicate a difference between μ_1 and μ_2. We think you will choose Figure 13.3(c) because that plot shows the

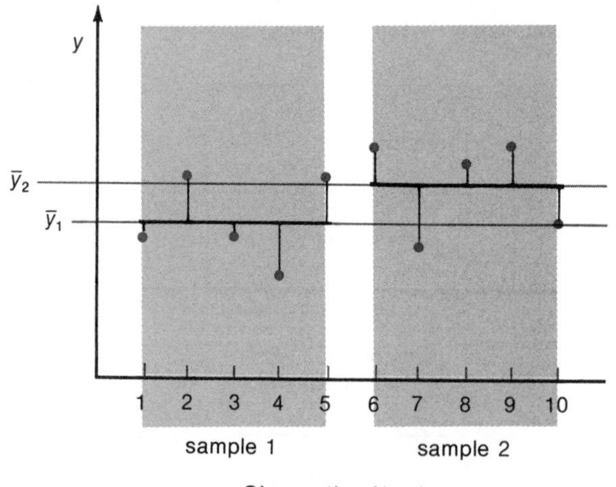

FIGURE 13.2

Graphic portrayal of the deviations of the y values about their means

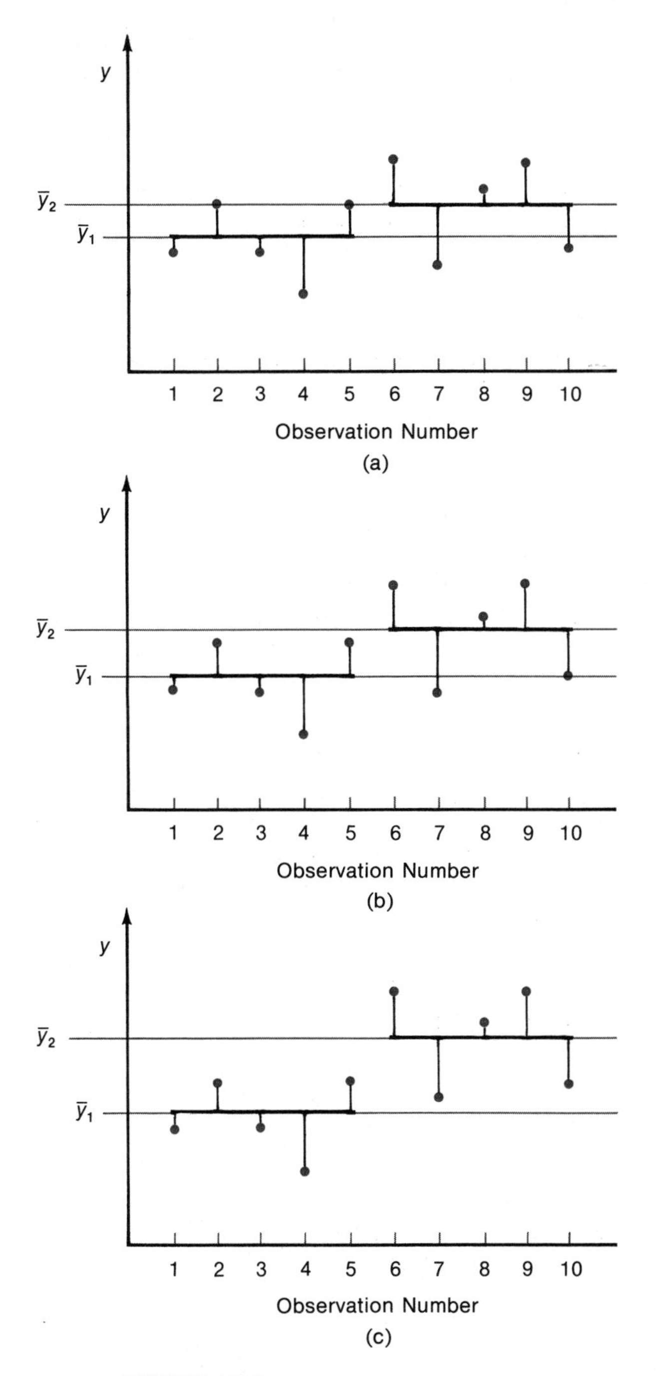

FIGURE 13.3

Three fictitious sets of measurements, $n_1 = n_2 = 5$ (the relative positions of points within each set is held constant)

greatest difference between sample means in comparison with the variation of the points about their respective sample means. This latter variation was held constant for all three plots.

Note that the population means actually may differ for Figure 13.3(a), but this fact would not be apparent, because the variation of the points about their respective sample means is too large in comparison with the difference between $\bar{y}_1$ and $\bar{y}_2$. Figure 13.4 shows the same difference between sample means as for Figure 13.3(a), but the variation within samples has been reduced. Now it appears that a difference does exist between μ_1 and μ_2.

Now let us leave our intuitive discussion and consider the two sample comparison of means for sample sizes n_1 and n_2. Particularly, we shall want to see how the total sum of squares of deviations can be partitioned into portions corresponding to the difference between the means and another to the variation within the two samples. The total sum of squares of deviations of all $(n_1 + n_2)$ y values about the general mean is

$$\text{Total SS} = \sum_{i=1}^{2} \sum_{j=1}^{n_i} (y_{ij} - \bar{y})^2,$$

where $\bar{y}$ is the average of all $(n_1 + n_2)$ observations contained in the two samples. Then with a bit of algebra you can show that

$$\text{Total SS} = \sum_{i=1}^{2} \sum_{j=1}^{n_i} (y_{ij} - \bar{y})^2 = \underbrace{\sum_{i=1}^{2} n_i(\bar{y}_i - \bar{y})^2}_{\text{SST}} + \underbrace{\sum_{i=1}^{2} \sum_{j=1}^{n_i} (y_{ij} - \bar{y}_i)^2}_{\text{SSE}}$$

where $\bar{y}_i$ is the average of the observations in the ith sample, $i = 1, 2$. The first quantity to the right of the equal sign, called the sum of squares for treatments

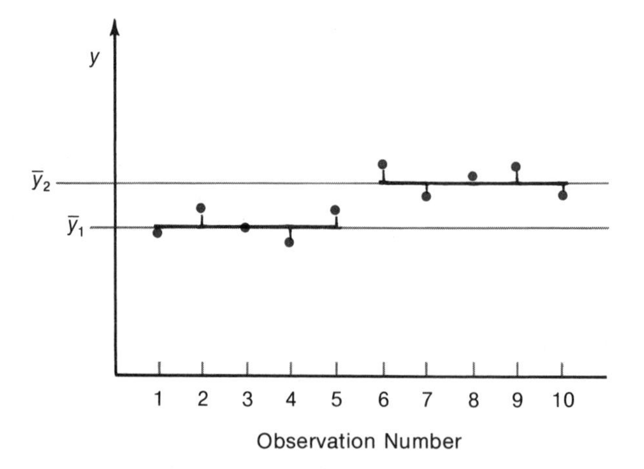

FIGURE 13.4

Small amount of within-sample variation in comparison with the difference between sample means

and denoted by the symbol SST, can be shown (with a bit of algebra) to equal

$$\text{SST} = \frac{n_1 n_2}{n_1 + n_2}(\bar{y}_1 - \bar{y}_2)^2.$$

Thus SST, which increases as the difference between $\bar{y}_1$ and $\bar{y}_2$ increases, measures the variation between the sample means. Consequently, SST would be larger for the data of Figure 13.3(c) than for Figure 13.3(a).

The second quantity to the right of the equal sign is the familiar pooled sum of squares of deviations computed in the t test of Section 9.4. It is the sum of the sum of squares of deviations of the y values about their respective sample means. This pooled sum of squares measures within sample variation, a variation that is usually attributed to experimental error, and is consequently called "sum of squares for error" (denoted by the symbol SSE).

The quantities SST and SSE measure the two kinds of variation that we viewed in the graphic representation of Figure 13.3, the variation between the means of the samples and the variation of the observations within samples. The greater the variation between means (the larger SST) in comparison with the variation within samples (SSE), the greater the weight of evidence to indicate a difference between μ_1 and μ_2. How large is large? When will SST be large enough (relative to SSE) to indicate a real difference between μ_1 and μ_2?

As indicated in Chapter 9,

$$s^2 = \text{MSE} = \frac{\text{SSE}}{n_1 + n_2 - 2}$$

provides an unbiased estimator of σ^2. (In the language of analysis of variance, s^2 is usually denoted as MSE, meaning "mean square for error.") Also, when the null hypothesis is true (that is, $\mu_1 = \mu_2$), SST divided by an appropriate number of degrees of freedom yields a second unbiased estimator of σ^2, which we shall denote as MST. For this example, the number of degrees of freedom for MST is equal to 1.

When the null hypothesis is true (that is, $\mu_1 = \mu_2$), MSE (the mean square for error) and MST (the mean square for treatments) will estimate the same quantity and should be "roughly" of the same magnitude. When the null hypothesis is false and $\mu_1 \neq \mu_2$, MST will probably be larger than MSE.

The preceding discussion, along with a review of the variance ratio, Section 9.7, suggests the use of

$$\frac{\text{MST}}{\text{MSE}}$$

as a test statistic to test the hypothesis $\mu_1 = \mu_2$ against the alternative, $\mu_1 \neq \mu_2$. Indeed, when both populations are normally distributed, it can be shown that MST and MSE are independent in a probabilistic sense and that

$$F = \frac{\text{MST}}{\text{MSE}}$$

possesses the F probability distribution of Section 9.7. Disagreement with the null hypothesis is indicated by a large value of F, and hence the rejection region for a given α will be

$$F \geq F_{\alpha}.$$

Thus the analysis-of-variance test results in a one-tailed F test. The degrees of freedom for the F will be those associated with MST and MSE, which we will denote as ν_1 and ν_2, respectively. Although we have not indicated how one determines ν_1 and ν_2, in general, $\nu_1 = 1$ and $\nu_2 = (n_1 + n_2 - 2)$ for the two-sample experiment described.

EXAMPLE 13.1

The coded values for the measure of elasticity in plastic, prepared by two different processes, are given below for samples of six drawn randomly from each of the two processes.

	Process A	Process B
	6.1	9.1
	7.1	8.2
	7.8	8.6
	6.9	6.9
	7.6	7.5
	8.2	7.9
Total	43.7	48.2

Do the data present sufficient evidence to indicate a difference in mean elasticity for the two processes?

Solution Although the Student's t could be used as the test statistic for this example, we shall use the analysis-of-variance F test.

The two desired sums of squares of deviations* are

$$\begin{aligned} \text{SST} &= n_1 \sum_{i=1}^{2} (\bar{y}_i - \bar{y})^2 \\ &= \frac{n_1}{2}(\bar{y}_1 - \bar{y}_2)^2 = \frac{6}{2}\left(\frac{43.7}{6} - \frac{48.2}{6}\right)^2 \\ &= 1.6875 \end{aligned}$$

and
$$\begin{aligned} \text{SSE} &= \sum_{i=1}^{2}\sum_{j=1}^{6} (y_{ij} - \bar{y}_i)^2 = \sum_{j=1}^{6} (y_{1j} - \bar{y}_1)^2 + \sum_{j=1}^{6} (y_{2j} - \bar{y}_2)^2 \\ &= 5.8617. \end{aligned}$$

(The reader may verify that SSE is the pooled sum of squares of the deviations for the two samples discussed in Section 9.4, p. 288.) The mean squares for treatment and error are, respectively,

$$\text{MST} = \frac{\text{SST}}{1} = 1.6875,$$

$$s^2 = \text{MSE} = \frac{\text{SSE}}{2n_1 - 2} = \frac{5.8617}{10} = .5862.$$

* *Note:* The formula for SST applies to the special case where $n_1 = n_2$.

To test the null hypothesis $\mu_1 = \mu_2$, we compute the test statistic

$$F = \frac{\text{MST}}{\text{MSE}} = \frac{1.6875}{.5862} = 2.88.$$

The critical value of the F statistic for $\alpha = .05$ is 4.96. Although the mean square for treatments is almost three times as large as the mean square for error, it is not large enough to reject the null hypothesis. Consequently, there is not sufficient evidence to indicate a difference between μ_1 and μ_2.

As noted, the purpose of the preceding example was to illustrate the computations involved in a simple analysis of variance. The F test for comparing two means is equivalent to a Student's t test, because an F statistic with one degree of freedom in the numerator is equal to t^2 (proof omitted). Had the t test been used for Example 13.1, we would have found $t = -1.6967$, which we see satisfies the relationship, $t^2 = (-1.6967)^2 = 2.88 = F$. Similarly, you can see that the relationship, $t^2 = F$, holds for the critical values. The square of $t_{.025} = 2.228$ (used for the two-tailed test with $\alpha = .05$ and $\nu = 10$ degrees of freedom) is equal to $F_{.05} = 4.96$. Since each value of F corresponds to two values of t, one positive and one negative, the F test with one degree of freedom in the numerator always corresponds to a two-tailed t test.

Of what value is the Total SS? The answer is that it provides an easy way to compute SSE. Since the Total SS partitions into SST and SSE, that is,

$$\text{Total SS} = \text{SST} + \text{SSE},$$

then

$$\text{SSE} = \text{Total SS} - \text{SST}.$$

Both the Total SS and SST are easy to compute. Hence one can easily find SSE by substituting into the expression above. For this example

$$\begin{aligned}\text{Total SS} = \sum_{i=1}^{2}\sum_{j=1}^{6}(y_{ij} - \bar{y})^2 &= \sum_{i=1}^{2}\sum_{j=1}^{6} y_{ij}^2 - \frac{\left(\sum_{i=1}^{2}\sum_{j=1}^{6} y_{ij}\right)^2}{12} \\ &= (\text{sum of squares of all } y \text{ values}) - \text{CM} \\ &= 711.35 - \frac{(91.9)^2}{12} = 7.5492.\end{aligned}$$

(the term CM denotes "correction for the mean"). Then

$$\begin{aligned}\text{SSE} &= \text{Total SS} - \text{SST} \\ &= 7.5492 - 1.6875 \\ &= 5.8617.\end{aligned}$$

This is exactly the same value obtained by the tedious computation and pooling of the sums of squares of deviations from the individual samples.

EXERCISE

13.1 Analyze the data of Exercise 9.62 by the procedure outlined in Section 13.2. Determine whether there is evidence of a difference in mean reaction times for the two stimuli. Test at the $\alpha = .05$ level of significance. If you have not worked Exercise 9.62, analyze the same data using Student's t test (as described in Exercise 9.62). Note that both methods lead to the same conclusion and that the computed values of F and t, for the two methods, are related. That is, $F = t^2$ (this will hold true only for the comparison of *two* population means).

13.3 A COMPARISON OF MORE THAN TWO MEANS

An analysis of variance to detect a difference in a set of more than two population means is a simple generalization of the analysis of variance of Section 13.2. You will recall (Chapter 12) that the random selection of independent samples from p populations is known as a completely randomized experimental design.

Assume that independent random samples have been drawn from p normal populations with means $\mu_1, \mu_2, \ldots, \mu_p$, respectively, and variance σ^2. Thus, all populations are assumed to possess equal variances. And, to be completely general, we will allow the sample sizes to be unequal and let n_i, $i = 1, 2, \ldots, p$, be the number in the sample drawn from the ith population. The total number of observations in the experiment will be $n = n_1 + n_2 + \cdots + n_p$.

Let y_{ij} denote the measured response on the jth experimental unit in the ith sample and let T_i and $\bar{T}_i$ represent the total and mean, respectively, for the observations in the ith sample. (The modification in the symbols for sample totals and averages will simplify the computing formulas for the sums of squares.) Then, as in the analysis of variance involving two means,

$$\text{Total SS} = \text{SST} + \text{SSE}$$

(proof deferred to Section 13.4), where

$$\text{Total SS} = \sum_{i=1}^{p}\sum_{j=1}^{n_i}(y_{ij} - \bar{y})^2 = \sum_{i=1}^{p}\sum_{j=1}^{n_i} y_{ij}^2 - \text{CM}$$

$$= (\text{sum of squares of all } y \text{ values}) - \text{CM}$$

$$\text{CM} = \frac{(\text{total of all observations})^2}{n} = \frac{\left(\sum_{i=1}^{p}\sum_{j=1}^{n_i} y_{ij}\right)^2}{n} = n\bar{y}^2$$

$$\text{SST} = \sum_{i=1}^{p} n_i(\bar{T}_i - \bar{y})^2 = \sum_{i=1}^{p}\frac{T_i^2}{n_i} - \text{CM}$$

$= $ (sum of squares of the treatment totals with each square divided by the number of observations in that particular total) $-$ CM

$$\text{SSE} = \text{Total SS} - \text{SST}.$$

Although the easy way to compute SSE is by subtraction, as just shown, it is interesting to note that SSE is the pooled sum of squares for all p samples and is equal to

$$\text{SSE} = \sum_{i=1}^{p} \sum_{j=1}^{n_i} (y_{ij} - \bar{T}_i)^2.$$

The unbiased estimator of σ^2 based on $(n_1 + n_2 + \cdots + n_p - p)$ degrees of freedom is

$$s^2 = \text{MSE} = \frac{\text{SSE}}{n_1 + n_2 + \cdots + n_p - p}.$$

The mean square for treatments will possess $(p - 1)$ degrees of freedom, that is, one less than the number of means, and

$$\text{MST} = \frac{\text{SST}}{p - 1}.$$

To test the null hypothesis

$$H_0\colon \mu_1 = \mu_2 = \cdots = \mu_p$$

against the alternative that at least one of the equalities does not hold, MST is compared with MSE using the F statistic based upon $\nu_1 = p - 1$ and $\nu_2 = \left(\sum_{i=1}^{p} n_i - p\right) = n - p$ degrees of freedom. The null hypothesis will be rejected if

$$F = \frac{\text{MST}}{\text{MSE}} > F_\alpha,$$

where F_α is the critical value of F, based on $(p - 1)$ and $(n - p)$ degrees of freedom, for probability of a type I error, α.

Intuitively, the greater the difference between the observed treatment means, $\bar{T}_1, \bar{T}_2, \ldots, \bar{T}_p$, the greater will be the evidence to indicate a difference between their corresponding population means. It can be seen from the formula for SST that SST $= 0$ when all the observed treatment means are identical because then $\bar{T}_1 = \bar{T}_2 = \cdots = \bar{T}_p = \bar{y}$, and the deviations appearing in SST, $(\bar{T}_i - \bar{y})$, $i = 1, 2, \ldots, p$, will equal zero. As the treatment means get farther apart, the deviations $(\bar{T}_i - \bar{y})$ will increase in absolute value and SST will increase in magnitude. Consequently, the larger the value of SST, the greater will be the weight of evidence favoring a rejection of the null hypothesis. This same line of reasoning will apply to the F tests employed in the analysis of variance for all designed experiments.

The test is summarized as follows:

F Test for Comparing *p* Population Means

$H_0\colon \mu_1 = \mu_2 = \cdots = \mu_p$.
H_a: One or more pairs of population means differ.

Test Statistic: $F = \dfrac{\text{MST}}{\text{MSE}}$

where F is based on $\nu_1 = (p - 1)$ and $\nu_2 = (n - p)$ degrees of freedom.

Rejection Region: Reject if $F > F_\alpha$, where F_α lies in the upper tail of the F distribution (with $\nu_1 = p - 1$ and $\nu_2 = n - p$) and satisfies the expression $P(F > F_\alpha) = \alpha$.

The assumptions underlying the analysis-of-variance F tests should receive particular attention. The samples are assumed to have been randomly selected from the p populations in an independent manner. The populations are assumed to be normally distributed with equal variances, σ^2, and means $\mu_1, \mu_2, \ldots, \mu_p$. Moderate departures from these assumptions will not seriously affect the properties of the test. This is particularly true of the normality assumption.

EXAMPLE 13.2

Four groups of students were subjected to different teaching techniques and tested at the end of a specified period of time. Because of dropouts in the experimental groups (sickness, transfers, etc.), the number of students varied from group to group. Do the following data present sufficient evidence to indicate a difference in the mean achievement for the four teaching techniques?

	Techniques			
	1	2	3	4
	65	75	59	94
	87	69	78	89
	73	83	67	80
	79	81	62	88
	81	72	83	
	69	79	76	
		90		
T_i	454	549	425	351
$\bar{T}_i$	75.67	78.43	70.83	87.75

Solution

$$\text{CM} = \frac{\left(\sum_{i=1}^{4}\sum_{j=1}^{n_i} y_{ij}\right)^2}{n} = \frac{(\text{total of all observations})^2}{n}$$

$$= \frac{(1{,}779)^2}{23} = 137{,}601.8,$$

$$\text{Total SS} = \sum_{i=1}^{4}\sum_{j=1}^{n_i} y_{ij}^2 - \text{CM}$$

$$= (\text{sum of squares of all } y \text{ values}) - \text{CM}$$

$$= (65)^2 + (87)^2 + (73)^2 + \cdots + (88)^2 - \text{CM}$$

$$= 139{,}511 - 137{,}601.8 = 1{,}909.2,$$

$$\begin{aligned}
\text{SST} &= \sum_{i=1}^{4} \frac{T_i^2}{n_i} - \text{CM} \\
&= \text{(sum of squares of the treatment totals with each square divided by the number of observations in that particular total)} - \text{CM} \\
&= \frac{(454)^2}{6} + \frac{(549)^2}{7} + \frac{(425)^2}{6} + \frac{(351)^2}{4} - \text{CM} \\
&= 138{,}314.4 - 137{,}601.8 \\
&= 712.6, \\
\text{SSE} &= \text{Total SS} - \text{SST} = 1{,}196.6.
\end{aligned}$$

The mean squares for treatment and error are

$$\text{MST} = \frac{\text{SST}}{p-1} = \frac{712.6}{3} = 237.5,$$

$$\text{MSE} = \frac{\text{SSE}}{n_1 + n_2 + \cdots + n_p - p} = \frac{\text{SSE}}{n-p} = \frac{1{,}196.6}{19} = 63.0.$$

The test statistic for testing the hypothesis, $\mu_1 = \mu_2 = \mu_3 = \mu_4$, is

$$F = \frac{\text{MST}}{\text{MSE}} = \frac{237.5}{63.0} = 3.77,$$

where

$$\nu_1 = (p-1) = 3, \qquad \nu_2 = \sum_{i=1}^{p} n_i - p = 19.$$

The critical value of F for $\alpha = .05$ is $F_{.05} = 3.13$. Since the computed value of F exceeds $F_{.05}$, we reject the null hypothesis and conclude that the evidence is sufficient to indicate a difference in mean achievement for the four teaching techniques.

You may feel that the above conclusion could have been made on the basis of visual observation of the treatment means. It is not difficult to construct a set of data that will lead the "visual" decision maker to erroneous results.

13.4

PROOF OF ADDITIVITY OF THE SUMS OF SQUARES (OPTIONAL)

The proof that

$$\text{Total SS} = \text{SST} + \text{SSE}$$

for the completely randomized design is presented in this section for the benefit of the interested reader. It may be omitted without loss of continuity.

The proof utilizes the three summation theorems of Chapter 2 and the device of adding and subtracting $\bar{T}_i$ within the expression for the Total SS. Thus

$$\text{Total SS} = \sum_{i=1}^{p}\sum_{j=1}^{n_i}(y_{ij} - \bar{y})^2$$

$$= \sum_{i=1}^{p}\sum_{j=1}^{n_i}(y_{ij} - \bar{T}_i + \bar{T}_i - \bar{y})^2 = \sum_{i=1}^{p}\sum_{j=1}^{n_i}[(y_{ij} - \bar{T}_i) + (\bar{T}_i - \bar{y})]^2$$

$$= \sum_{i=1}^{p}\sum_{j=1}^{n_i}[(y_{ij} - \bar{T}_i)^2 + 2(y_{ij} - \bar{T}_i)(\bar{T}_i - \bar{y}) + (\bar{T}_i - \bar{y})^2].$$

Summing first over j, we obtain

$$\text{Total SS} = \sum_{i=1}^{p}\left[\sum_{j=1}^{n_i}(y_{ij} - \bar{T}_i)^2 + 2(\bar{T}_i - \bar{y})\sum_{j=1}^{n_i}(y_{ij} - \bar{T}_i) + n_i(\bar{T}_i - \bar{y})^2\right],$$

where

$$\sum_{j=1}^{n_i}(y_{ij} - \bar{T}_i) = T_i - n_i\bar{T}_i = T_i - T_i = 0.$$

Consequently, the middle term in the expression for the Total SS is equal to zero. Then, summing over i, we obtain

$$\text{Total SS} = \sum_{i=1}^{p}\sum_{j=1}^{n_i}(y_{ij} - \bar{T}_i)^2 + \sum_{i=1}^{p} n_i(\bar{T}_i - \bar{y})^2,$$

The first expression on the right is SSE, the pooled sum of squares of deviations of the sample measurements about their respective means. The second is the formula for SST. Therefore,

$$\text{Total SS} = \text{SSE} + \text{SST}.$$

Proof of the additivity of the analysis-of-variance sums of squares for other experimental designs can be obtained in a similar manner. The proofs are omitted.

13.5

AN ANALYSIS-OF-VARIANCE TABLE FOR A COMPLETELY RANDOMIZED DESIGN

The calculations of the analysis of variance are usually displayed in an analysis-of-variance (ANOVA or AOV) table. The table for the design of Section 13.3 involving p treatment means is shown in Table 13.1. Column 1 shows the source of each sum of squares of deviations; column 2 gives the respective degrees of freedom; columns 3 and 4 give the corresponding sums of squares and mean squares, respectively. A calculated value of F, comparing MST and

TABLE 13.1
ANOVA table for a comparison of means

Source	d.f.	SS	MS	F
Treatments	$p-1$	SST	MST = SST/$(p-1)$	MST/MSE
Error	$n-p$	SSE	MSE = SSE/$(n-p)$	
Total	$n-1$	$\sum_{i=1}^{p}\sum_{j=1}^{n_i}(y_{ij}-\bar{y})^2$		

MSE, is usually shown in column 5. Note that the degrees of freedom and sums of squares add to their respective totals.

The ANOVA table for Example 13.2, shown in Table 13.2, gives a compact presentation of the appropriate computed quantities for the analysis of variance.

TABLE 13.2
ANOVA table for Example 13.2

Source	d.f.	SS	MS	F
Treatments	3	712.6	237.5	3.77
Error	19	1196.6	63.0	
Total	22	1909.2		

13.6

ESTIMATION FOR THE COMPLETELY RANDOMIZED DESIGN

Confidence intervals for a single treatment mean and the difference between a pair of treatment means, Section 13.3, are identical to those given in Chapter 9. The confidence interval for the mean of treatment i or the difference between treatments i and j are, respectively,

Completely Randomized Design: 100(1 − α) Percent Confidence Intervals for:

A Single Treatment Mean: $\bar{T}_i \pm t_{\alpha/2}s/\sqrt{n_i}$.

The Difference Between Two Treatment Means:

$$(\bar{T}_i - \bar{T}_j) \pm t_{\alpha/2}\, s\sqrt{\frac{1}{n_i} + \frac{1}{n_j}},$$

where

$$s = \sqrt{s^2} = \sqrt{\text{MSE}} = \sqrt{\frac{\text{SSE}}{n_1 + n_2 + \cdots + n_p - p}}$$

$$n = n_1 + n_2 + \cdots + n_p$$

and $t_{\alpha/2}$ is based upon $(n - p)$ degrees of freedom.

Note that the confidence intervals stated above are appropriate for single treatment means or a comparison of a pair of means selected prior to observation of the data. The stated confidence coefficients are based on random sampling. If one were to look at the data and always compare the largest and smallest sample means, the assumption of randomness would be disturbed. This is because the difference between the largest and smallest sample means you would expect to be larger than for a pair selected at random.

EXAMPLE 13.3

Find a 95 percent confidence interval for the mean score for teaching technique 1, Example 13.2.

Solution The 95 percent confidence interval for the mean score is

$$\bar{T}_1 \pm \frac{t_{.025}s}{\sqrt{n_1}}$$

or

$$75.67 \pm \frac{(2.093)(7.94)}{\sqrt{6}}$$

or

$$75.67 \pm 6.78.$$

Thus we infer that the interval 75.67 ± 6.78 or 68.89 to 82.45 encloses the mean score for students subjected to teaching technique 1.

EXAMPLE 13.4

Find a 95 percent confidence interval for the difference in mean score for teaching techniques 1 and 4, Example 13.2.

Solution The 95 percent confidence interval for $(\mu_1 - \mu_4)$ is

$$(\bar{T}_1 - \bar{T}_4) \pm t_{\alpha/2}s\sqrt{\frac{1}{n_1} + \frac{1}{n_4}}$$

or $$(75.67 - 87.75) \pm (2.093)(7.94)\sqrt{\frac{1}{6} + \frac{1}{4}}$$

or $$-12.08 \pm 10.73.$$

Thus we estimate that the interval -12.08 ± 10.73 or -22.81 to -1.35 encloses the difference in mean scores for teaching techniques 1 and 4.

Because all points in the interval are negative, we infer that μ_4 is larger than μ_1. Also, note that the variability in scores within students is rather large. Consequently, the sample sizes should be increased if the experimenter wishes to reduce the width of the confidence interval.

Tips on Problem Solving: The following suggestions apply to all the analyses of variance in this chapter.

1. When calculating sums of squares, be certain to carry at least six significant figures before performing subtractions.
2. Remember, sums of squares can never be negative. If you obtain a negative sum of squares, you have made a mistake in arithmetic.
3. Always check your analysis-of-variance table to make certain that the degrees of freedom sum to the total degrees of freedom, $n - 1$, and that the sums of squares sum to the Total SS.

13.7

A COMPUTER PRINTOUT FOR A COMPLETELY RANDOMIZED DESIGN

Computer packages for analyses of variance are readily available and are very similar. The purpose of this section is to familiarize you with a typical printout for the analysis of variance for a completely randomized design in case you would like to perform your computations on a computer. As for the regression analysis, Section 10.10, we will use the SAS (Statistical Analysis System) computer package (listed in the references).

So that you will more readily identify the elements of a SAS printout, we have given the analysis of variance for Example 13.2. The SAS printout is shown in Table 13.3. You will note that it is broken into two parts, both shaded. The upper table breaks the Total sum of squares (called CORRECTED TOTAL) into two sources, MODEL and ERROR. You can see that the numbers appearing in the ERROR row (df, SS, MS) of Table 13.3 are identical (except for rounding) to the numbers appearing in the "Error" row of the ANOVA table, Table 13.2. The MODEL row in the printout, Table 13.3, corresponds to all effects other than error. In this case, there is only one other source of variation, namely treatments. Consequently, the row identified as MODEL gives the values of df, SST, MST and F. You can see that these numbers are identical to the numbers that appear in the "Treatments" row of the ANOVA table, Table 13.2.

The lower shaded table gives a breakdown of the MODEL source of variance into its component sources. In this particular case, there is only one source, TREATMENTS. Consequently, this line repeats the value of SST (712.58 . . .), gives the computed F value (3.77) for a test of the null hypothesis, "no difference between treatment means," and gives the probability of observing

TABLE 13.3

A SAS computer printout for Example 13.2

STATISTICAL ANALYSIS SYSTEM

ANALYSIS OF VARIANCE PROCEDURE

DEPENDENT VARIABLE: Y

SOURCE	DF	SUM OF SQUARES	MEAN SQUARE	F VALUE	PR > F	R-SQUARE	C.V.
MODEL	3	712.58643892	237.52881297	3.77	0.0280	0.373235	10.2602
ERROR	19	1196.63095238	62.98057644		STD DEV		Y MEAN
CORRECTED TOTAL	22	1909.21739130			7.93603027		77.34782609

SOURCE	DF	ANOVA SS	F VALUE	PR > F
TRTMENTS	3	712.58643892	3.77	0.0280

a value of F as large or larger than 3.77, given the null hypothesis is true. This probability, 0.0280, is the significance level for the test.

The value of s, $s = 7.936\ldots$, is given at the right of the printout. This is useful in calculating confidence intervals. We omit discussion of the last two columns of the printout because they are not relevant to the analysis of variance that we have given in Section 13.5.

EXERCISES

13.2 A clinical psychologist wished to compare three methods for reducing hostility levels in university students. A certain psychological test (HLT) was used to measure the degree of hostility. High scores on this test were taken to indicate great hostility. Eleven students obtaining high and nearly equal scores were used in the experiment. Five were selected at random from among the 11 problem cases and treated by method A. Three were taken at random from the remaining six students and treated by method B. The other three students were treated by method C. All treatments continued throughout a semester. Each student was given the HLT test again at the end of the semester, with the following results:

Method	Scores on the HLT Tests				
A	73	83	76	68	80
B	54	74	71		
C	79	95	87		

(a) Perform an analysis of variance for this experiment.

(b) Do the data provide sufficient evidence to indicate a difference in mean student response for the three methods after treatment?

13.3 Refer to Exercise 13.2. Let μ_A and μ_B, respectively, denote the mean scores at the end of the semester for the populations of extremely hostile students who are treated throughout that semester by method A and method B.

(a) Find a 95 percent confidence interval for μ_A.

(b) Find a 95 percent confidence interval for μ_B.

(c) Find a 95 percent confidence interval for $\mu_A - \mu_B$.
(d) Is it correct to claim that the confidence intervals found in parts (a), (b), and (c) are jointly valid?

13.4 Three groups of fourth graders were randomly selected and assigned, one group each, to three different physical exercise programs to determine whether the programs were effective in increasing the childrens' abilities to throw an object. Twenty-eight students were involved in the experiment, 10 in a control group (no exercise) and 9 each assigned to two different exercise regimes, each of which lasted four weeks. The velocity at which a child was able to throw a test ball was measured before and after the four weeks of exercise and the gain (or loss) in velocity *y* (in feet per second) was recorded. A table of mean gains for the three groups and a partially completed analysis-of-variance table for the data are as shown.

Sample Means

Control	Exercise Regime *A*	Exercise Regime *B*
−1.34	.32	3.69

ANOVA Table

Source	d.f.	SS	MS
Exercise regimes	—	64.31	—
Error	—	—	—
Total	—	402.33	

(a) Fill in the missing numbers in the analysis-of-variance table.
(b) Do the data provide sufficient evidence to indicate a difference in population means for the three groups? Explain the implications of the test results.
(c) Find a 95 percent confidence interval for the difference in mean gain between children in the control group versus those on exercise regime *B*.
(d) Find a 95 percent confidence interval for the mean gain for children on exercise regime *B*.

13.5 An experiment was conducted to compare the price of a loaf of bread (a particular brand) at four city locations. Four stores were randomly sampled in locations 1, 2, and 3 but only two were selected from location 4 (only two carried the brand). Note that a completely randomized design was employed. Conduct an analysis of variance for the data.

Location	Prices (cents)			
1	59	63	65	61
2	58	61	64	63
3	54	59	55	58
4	69	70		

(a) Do the data provide sufficient evidence to indicate a difference in mean price of the bread in stores located in the four areas of the city?
(b) Suppose that prior to seeing the data, we wished to compare the mean prices between locations 1 and 4. Estimate the difference in means using a 95 percent confidence interval.

13.6 An experiment was conducted to compare the effectiveness of three training programs, *A*, *B*, and *C*, in training assemblers of a piece of electronic equipment. Fifteen employees were randomly assigned, 5 each, to the three programs. After completion of the courses, each person was required to assemble four pieces of the equipment and the average length of time required to complete the assembly was recorded. Due to resignation from the company, only 4 employees completed program *A* and only 3 completed *B*. The data are shown in the accompanying table. A SAS computer printout of the analysis of variance for the data is also shown below. Use the information in the printout to answer the following questions:

Training Program	Average Assembly Time (min)				
A	59	64	57	62	
B	52	58	54		
C	58	65	71	63	64

```
                              STATISTICAL ANALYSIS SYSTEM
                            ANALYSIS OF VARIANCE PROCEDURE
DEPENDENT VARIABLE: Y
SOURCE              DF   SUM OF SQUARES      MEAN SQUARE   F VALUE      PR > F   R-SQUARE          C.V.
MODEL                2     170.45000000      85.22500000      5.70      0.0251   0.559005        6.3802
ERROR                9     134.46666667      14.94074074               STD DEV                  Y MEAN
CORRECTED TOTAL     11     304.91666667                               3.86532544            60.58333333

SOURCE              DF          ANOVA SS   F VALUE    PR > F
TRTMENTS             2      170.45000000      5.70    0.0251
```

(a) Do the data provide sufficient evidence to indicate a difference in mean assembly time for people trained by the three programs?
(b) Find a 90 percent confidence interval for the difference in mean assembly time between persons trained by programs *A* and *B*.
(c) Find a 90 percent confidence interval for the mean assembly time for persons trained in program *A*.
(d) Do you think the data will satisfy (approximately) the assumption that they have been selected from normal populations? Why?

13.7 An ecological study was conducted to compare rate of growth of vegetation at four swampy undeveloped sites and to determine the cause of any differences that might be observed. Part of the study involved the measurement of leaf lengths of a particular plant species at a preselected date in May. Six plants were randomly selected at each of the four sites to be used in the comparison. The following data represent the mean leaf length per plant, in centimeters, for a random sample of 10 leaves per plant.

Location	Mean Leaf Length (cm)					
1	5.7	6.3	6.1	6.0	5.8	6.2
2	6.2	5.3	5.7	6.0	5.2	5.5
3	5.4	5.0	6.0	5.6	4.9	5.2
4	3.7	3.2	3.9	4.0	3.5	3.6

(a) You will recall that the test and estimation procedures for an analysis of variance require that the observations be selected from normally distributed (at least, roughly so) populations. Why might you feel reasonably confident that your data satisfy this assumption?
(b) Do the data provide sufficient evidence to indicate a difference in mean leaf length among the four locations?
(c) Suppose, prior to seeing our data, that we decided to compare the mean leaf

length between locations 1 and 4. Test the null hypothesis that $\mu_1 = \mu_4$ against the alternative that $\mu_1 \neq \mu_4$.

(d) Refer to part (c). Find a 90 percent confidence interval for $(\mu_1 - \mu_4)$.

(e) Rather than use an analysis-of-variance F test, it would seem simpler to examine one's data, select the two locations that have the smallest and largest sample mean lengths, and then compare these two means using a Student's t test. If there is evidence to indicate a difference in these means, there is clearly evidence of a difference among the four. (If you were to use this logic, there would be no need for the analysis-of-variance F test). Explain why this procedure is invalid.

13.8 Water samples were taken at four different locations in a river to determine whether the quantity of dissolved oxygen, a measure of water pollution, varied from one location to another. Locations 1 and 2 were selected above an industrial plant, one near the shore and the other in midstream; location 3 was adjacent to the industrial water discharge for the plant, and location 4 was slightly downriver in midstream. Live water specimens were randomly selected at each location, but one specimen, corresponding to location 4, was lost in the laboratory. The data are shown below (the greater the pollution, the lower will be the dissolved oxygen readings).

(a) Do the data provide sufficient evidence to indicate a difference in mean dissolved oxygen content for the four locations?

(b) Compare the mean dissolved oxygen content in midstream above the plant with the mean content adjacent to the plant (location 2 versus location 3). Use a 95 percent confidence interval.

Location	Mean Dissolved Oxygen Content				
1	5.9	6.1	6.3	6.1	6.0
2	6.3	6.6	6.4	6.4	6.5
3	4.8	4.3	5.0	4.7	5.1
4	6.0	6.2	6.1	5.8	

13.9 An experiment was conducted to examine the effect of age on heart rate when a subject is subjected to a specific amount of exercise. Ten male subjects were randomly selected from four age groups, 10–19, 20–39, 40–59, and 60–69. Each subject walked a treadmill at a fixed grade for a period of 12 minutes and the increase in heart rate, the difference before and after exercise, was recorded (in beats per minute). These data are shown in the table.

	Age			
	10–19	20–39	40–59	60–69
	29	24	37	28
	33	27	25	29
	26	33	22	34
	27	31	33	36
	39	21	28	21
	35	28	26	20
	33	24	30	25
	29	34	34	24
	36	21	27	33
	22	32	33	32
Total	309	275	295	282

(a) Do the data provide sufficient evidence to indicate a difference in mean increase in heart rate among the four age groups? Test by using $\alpha = .05$.
(b) Find a 90 percent confidence interval for the difference in mean increase in heart rate between the 10–19-years age group and the 60–69-years age group.
(c) Find a 90 percent confidence interval for the mean increase in heart rate for the 20–39-years age group.
(d) Approximately how many people would you need in each group if you wished to be able to estimate a group mean correct to within 2 beats per minute with probability equal .95?

13.8

THE ANALYSIS OF VARIANCE FOR A RANDOMIZED BLOCK DESIGN

The method for constructing a randomized block design was presented in Section 12.5.

The randomized block design implies the presence of two qualitative independent variables, "blocks" and "treatments." Consequently, the total sum of squares of deviations of the response measurements about their mean may be partitioned into three parts: the sums of squares for blocks, treatments, and error.

Denote the total and mean of all observations in block i as B_i and $\bar{B}_i$, respectively. Similarly, let T_j and $\bar{T}_j$ represent the total and the mean for all observations receiving treatment j. Then, for a randomized block design involving b blocks and p treatments,

$$\text{Total SS} = \text{SSB} + \text{SST} + \text{SSE},$$

$$\text{Total SS} = \sum_{i=1}^{b}\sum_{j=1}^{p}(y_{ij} - \bar{y})^2 = \sum_{i=1}^{b}\sum_{j=1}^{p} y_{ij}^2 - \text{CM},$$

$$= (\text{sum of squares of all } y \text{ values}) - \text{CM}$$

$$\text{SSB} = p\sum_{i=1}^{b}(\bar{B}_i - \bar{y})^2 = \frac{\sum_{i=1}^{b} B_i^2}{p} - \text{CM},$$

$$= \frac{\text{sum of squares of all block totals}}{\text{number of observations in a single total}} - \text{CM}$$

$$\text{SST} = b\sum_{j=1}^{p}(\bar{T}_j - \bar{y})^2 = \frac{\sum_{j=1}^{p} T_j^2}{b} - \text{CM},$$

$$= \frac{\text{sum of squares of all treatment totals}}{\text{number of observations in a single total}} - \text{CM}$$

where

$$\bar{y} = (\text{average of all } n = bp \text{ observations}) = \frac{\sum_{i=1}^{b}\sum_{j=1}^{p} y_{ij}}{n}$$

TABLE 13.4
ANOVA table for a randomized block design

Source	d.f.	SS	MS	F
Blocks	$b - 1$	SSB	SSB/$(b - 1)$	MSB/MSE
Treatments	$p - 1$	SST	SST/$(p - 1)$	MST/MSE
Error	$n - b - p + 1$	SSE	MSE	
Total	$n - 1$	Total SS		

and

$$\text{CM} = \frac{(\text{total of all observations})^2}{n} = \frac{\left(\sum_{i=1}^{b}\sum_{j=1}^{p} y_{ij}\right)^2}{n}.$$

The analysis of variance for the randomized block design is presented in Table 13.4. The degrees of freedom associated with each sum of squares is shown in the second column. Mean squares are calculated by dividing the sums of squares by their respective degrees of freedom.

To test the null hypothesis "there is no difference in treatment means," we use the F statistic,

$$F = \frac{\text{MST}}{\text{MSE}}$$

and reject if $F > F_\alpha$ based on $\nu_1 = (p - 1)$ and $\nu_2 = (n - b - p + 1)$ degrees of freedom.

Blocking not only reduces the experiment error, it also provides an opportunity to see whether evidence exists to indicate a difference in the mean response for blocks. Under the null hypothesis that there is no difference in mean response for blocks, MSB provides an unbiased estimator for σ^2 based on $(b - 1)$ degrees of freedom. Where a real difference exists in the block means, MSB will tend to be inflated in comparison with MSE. Therefore,

$$F = \frac{\text{MSB}}{\text{MSE}}$$

can be used as a test statistic to test for differences among block means. As in the test for treatments, the rejection region for the test will be

$$F > F_\alpha,$$

based on $\nu_1 = b - 1$ and $\nu_2 = n - b - p + 1$ degrees of freedom.

F Test for Comparing *p* Treatments, Using a Randomized Block Design

H_0: The population treatments means are equal.

H_a: One or more pairs of population treatment means differ.

Test Statistic: $F = \dfrac{\text{MST}}{\text{MSE}},$

where F is based on $\nu_1 = (p - 1)$ and $\nu_2 = (n - b - p + 1)$ degrees of freedom.

Rejection Region: Reject if $F > F_\alpha$.

EXAMPLE 13.5

A stimulus–response experiment involving three treatments was laid out in a randomized block design using four subjects. The response was the length of time to reaction measured in seconds. The data (treatment identification numbers are circled) are as follows:

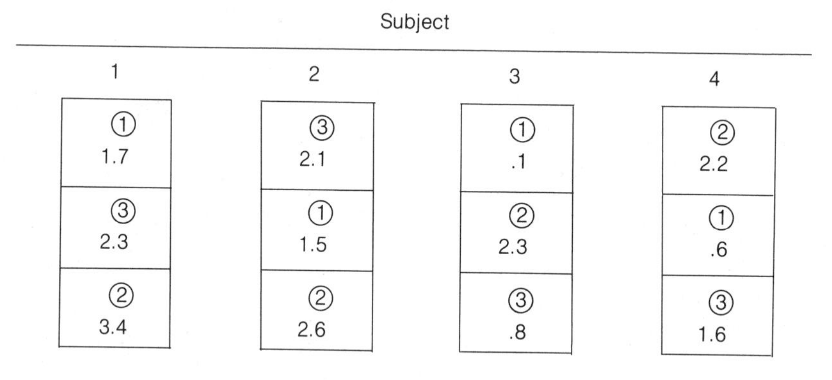

Subject 1	Subject 2	Subject 3	Subject 4
① 1.7	③ 2.1	① .1	② 2.2
③ 2.3	① 1.5	② 2.3	① .6
② 3.4	② 2.6	③ .8	③ 1.6

Do the data present sufficient evidence to indicate a difference in the mean response for stimuli (treatments)? Subjects?

Solution The treatment and block totals are as follows: $T_1 = 3.9$, $T_2 = 10.5$, $T_3 = 6.8$, $B_1 = 7.4$, $B_2 = 6.2$, $B_3 = 3.2$, $B_4 = 4.4$.

The sums of squares for the analysis of variance are shown individually below, and jointly in the analysis-of-variance table. Thus

$$\text{CM} = \frac{(\text{Total})^2}{n} = \frac{(21.2)^2}{12} = 37.4533,$$

$$\begin{aligned}
\text{Total SS} &= \sum_{i=1}^{4}\sum_{j=1}^{3}(y_{ij} - \bar{y})^2 = \sum_{i=1}^{4}\sum_{j=1}^{3} y_{ij}^2 - \text{CM} \\
&= (\text{sum of squares of all } y \text{ values}) - \text{CM} \\
&= (1.7)^2 + (2.3)^2 + \cdots + (1.6)^2 - \text{CM} \\
&= 46.8600 - 37.4533 = 9.4067,
\end{aligned}$$

$$\begin{aligned}
\text{SSB} &= \frac{\sum_{i=1}^{4} B_i^2}{3} - \text{CM} \\
&= \frac{\text{sum of squares of all block totals}}{\text{number of observations in a single total}} - \text{CM}
\end{aligned}$$

$$= \frac{(7.4)^2 + (6.2)^2 + (3.2)^2 + (4.4)^2}{3} - \text{CM}$$

$$= 40.9333 - 37.4533 = 3.4800,$$

$$\text{SST} = \frac{\sum_{j=1}^{3} T_j^2}{4} - \text{CM}$$

$$= \frac{\text{sum of squares of all treatment totals}}{\text{number of observations in a single total}} - \text{CM}$$

$$= \frac{(3.9)^2 + (10.5)^2 + (6.8)^2}{4} - \text{CM}$$

$$= 42.9250 - 37.4533 = 5.4717,$$

$$\text{SSE} = \text{Total SS} - \text{SSB} - \text{SST}$$

$$= 9.4069 - 3.4800 - 5.4717 = .4550.$$

The analysis-of-variance table for Example 13.5 is shown in Table 13.5. We use the ratio of mean-square treatments to mean-square error to test an hypothesis of no difference in the expected response for treatments. Thus,

$$F = \frac{\text{MST}}{\text{MSE}} = \frac{2.74}{.076} = 36.05.$$

The critical value of the F statistic ($\alpha = .05$) for $\nu_1 = 2$ and $\nu_2 = 6$ degrees of freedom is $F_{.05} = 5.14$. Since the computed value of F exceeds the critical value, there is sufficient evidence to reject the null hypothesis and conclude that a real difference does exist in the expected response for the four stimuli.

A similar test may be conducted for the null hypothesis that no difference exists in the mean response for subjects. Rejection of this hypothesis would imply that subject-to-subject variability does exist, and that blocking is desirable. The computed value of F based on $\nu_1 = 3$ and $\nu_2 = 6$ degrees of freedom is

$$F = \frac{\text{MSB}}{\text{MSE}} = \frac{1.16}{.076} = 15.26.$$

TABLE 13.5

ANOVA table for Example 13.5

Source	d.f.	SS	MS	F
Blocks	3	3.4767	1.16	15.26
Treatments	2	5.4767	2.74	36.05
Error	6	.4533	.076	
Total	11	9.4067		

Since this value of F exceeds the corresponding tabulated critical value, $F_{.05} = 4.76$, we reject the null hypothesis and conclude that a real difference exists in the expected response in the group of subjects.

13.9

ESTIMATION FOR THE RANDOMIZED BLOCK DESIGN

The confidence interval for the difference between a pair of means is exactly the same as for the completely randomized design, Section 13.5. It is

Randomized Block Design: A 100(1 $-$ α) Percent Confidence Interval for the Difference Between Two Treatment Means

$$(\bar{T}_i - \bar{T}_j) \pm t_{\alpha/2}\, s\sqrt{\frac{2}{b}},$$

where $n_i = n_j = b$, the number of observations contained in a treatment mean and $t_{\alpha/2}$ is based upon $(n - b - p + 1)$ degrees of freedom.

The difference between the confidence intervals for the completely randomized and the randomized block designs is that s, appearing in the expression above, will tend to be smaller than for the completely randomized design.

Similarly, one may construct a $100(1 - \alpha)$ percent confidence interval for the difference between a pair of block means. Each block contains p observations corresponding to the p treatments. Therefore, the confidence interval is

$$(\bar{B}_i - \bar{B}_j) \pm t_{\alpha/2}\, s\sqrt{\frac{2}{p}},$$

where $t_{\alpha/2}$ is based upon $(n - b - p + 1)$ degrees of freedom.

EXAMPLE 13.6

Construct a 95 percent confidence interval for the difference between treatments 1 and 2, Example 13.5.

Solution The confidence interval for the difference in mean response for a pair of treatments is

$$(\bar{T}_i - \bar{T}_j) \pm t_{\alpha/2}\, s\sqrt{\frac{2}{b}},$$

where for our example, $t_{.025}$ is based upon 6 degrees of freedom. Then $t_{.025} = 2.447$ and $s = \sqrt{\text{MSE}} = \sqrt{.076} = .28$. Substituting these values along with the values of $\bar{T}_1$ and $\bar{T}_2$, we have

$$(.98 - 2.63) \pm (2.447)(.28)\sqrt{\frac{2}{4}}$$

or

$$-1.65 \pm .48.$$

Thus we infer that the interval $-1.65 \pm .48$ or -2.13 to -1.17, encloses the difference in mean reaction times for stimuli 1 and 2. Because all points in the interval are negative, we conclude that the mean time to react for stimulus 2 is larger than for stimulus 1.

Tips on Problem Solving: Be careful of this point: Unless the blocks have been randomly selected from a population of blocks, you cannot obtain a confidence interval for a single treatment mean. This is because the sample treatment mean is biased by the positive and negative effects that the blocks have on the response.

The same comment applies to the Latin square design of Section 13.11. Further discussion of this topic can be found in the references.

13.10

A COMPUTER PRINTOUT FOR A RANDOMIZED BLOCK DESIGN

The SAS computer printout for the randomized block data, Example 13.5, is shown in Table 13.6. As for the printout of the analysis of variance for a completely randomized design, Table 13.3, we have shaded two portions of the printout.

The upper table gives a breakdown of the Total SS into two parts, one corresponding to MODEL and the second to ERROR. The numbers appearing in the row corresponding to ERROR are identical (except for rounding error) to the numbers in the "Error" row of the ANOVA table, Table 13.5. The row corresponding to MODEL gives the totals of d.f. and SS corresponding to BLOCKS and TREATMENTS. A breakdown of these totals is given in the lower table. There

TABLE 13.6

A SAS computer printout for Example 13.5

STATISTICAL ANALYSIS SYSTEM

ANALYSIS OF VARIANCE PROCEDURE

DEPENDENT VARIABLE: Y

SOURCE	DF	SUM OF SQUARES	MEAN SQUARE	F VALUE	PR > F	R-SQUARE	C.V.
MODEL	5	8.95166667	1.79033333	23.61	0.0007	0.951630	15.5875
ERROR	6	0.45500000	0.07583333		STD DEV		Y MEAN
CORRECTED TOTAL	11	9.40666667			0.27537853		1.76666667

SOURCE	DF	ANOVA SS	F VALUE	PR > F
BLOCKS	3	3.48000000	15.30	0.0032
TRTMENTS	2	5.47166667	36.08	0.0005

you see that the degrees of freedom and sums of squares for BLOCKS and TREATMENTS correspond to the quantities given in the ANOVA table, Table 13.5.

Notice that MSB and MST are not given in either table but that they can easily be calculated from the sums of squares if they are desired. The lower table gives the F values and significance levels for tests of the null hypotheses of "no difference between block means" ($F = 15.30$) and "no difference between treatment means" ($F = 36.08$). The value of s, $s = .275\ldots$, given in the sixth column, can be used to calculate confidence intervals for the difference between any pair of treatment or block means. The slight discrepancies between the numbers in the ANOVA table, Table 13.5, and those in the computer printout, Table 13.6, are due to rounding errors.

EXERCISES

13.10 A study was conducted to compare automobile gasoline mileage for three brands of gasoline, *A*, *B*, and *C*. Four automobiles, all of the same make and model, were employed in the experiment and each gasoline brand was tested in each automobile. Using each brand within the same automobile has the effect of eliminating (blocking out) automobile-to-automobile variability. The data, in miles per gallon, are as follows:

Gasoline Brand	Automobile			
	1	2	3	4
A	15.7	17.0	17.3	16.1
B	17.2	18.1	17.9	17.7
C	16.1	17.5	16.8	17.8

(a) Do the data provide sufficient evidence to indicate a difference in mean mileage per gallon for the three gasolines?

(b) Is there evidence of a difference in mean mileage for the four automobiles?

(c) Suppose that *prior to looking at the data,* we had decided to compare the mean mileage per gallon for gasoline brands *A* and *B*. Find a 90 percent confidence interval for this difference.

13.11 An experiment was conducted to compare the effect of four different chemicals, *A*, *B*, *C*, and *D*, in producing water resistance in textiles. A strip of material, randomly selected from a bolt, was cut into four pieces and the pieces were randomly assigned to receive one of the four chemicals, *A*, *B*, *C*, or *D*. This process was replicated three times, thus producing a randomized block design. The design, with moisture-resistance measurements, is as shown (low readings indicate low moisture penetration). A SAS computer printout of the analysis of variance for the data is also presented. Use the information in the printout to answer the following questions:

(a) Do the data provide sufficient evidence to indicate a difference in the mean moisture penetration for fabric treated with the four chemicals?

(b) Do the data provide evidence to indicate that blocking increased the amount of information in the experiment?

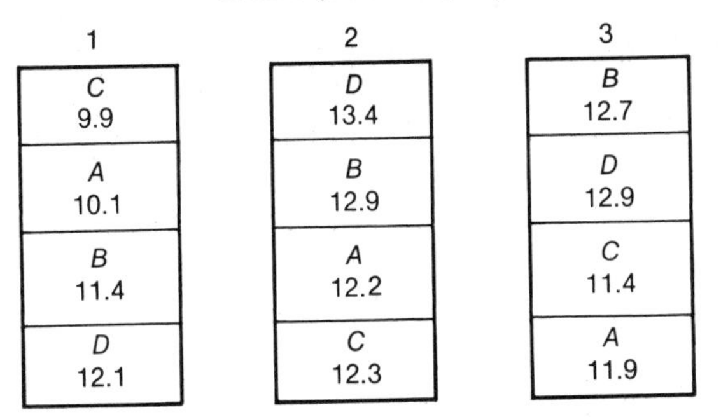

STATISTICAL ANALYSIS SYSTEM

ANALYSIS OF VARIANCE PROCEDURE

DEPENDENT VARIABLE: Y

SOURCE	DF	SUM OF SQUARES	MEAN SQUARE	F VALUE	PR > F	R-SQUARE	C.V.
MODEL	5	12.37166667	2.47433333	27.75	0.0004	0.958549	2.5023
ERROR	6	0.53500000	0.08916667		STD DEV		Y MEAN
CORRECTED TOTAL	11	12.90666667			0.29860788		11.93333333

SOURCE	DF	ANOVA SS	F VALUE	PR > F
BLOCKS	2	7.17166667	40.21	0.0003
TRTMENTS	3	5.20000000	19.44	0.0017

(c) Find a 95 percent confidence interval for the difference in mean moisture penetration for fabrics treated by chemicals A and D. Interpret the interval.

13.12 An experiment was conducted to determine the effect of three methods of soil preparation on the first-year growth of slash pine seedlings. Four locations (state forest lands) were selected and each location was divided into three plots. Since it was felt that soil fertility within a location was more homogeneous than between locations, a randomized block design was employed using locations as blocks. The methods of soil preparation were A (no preparation), B (light fertilization), and C (burning). Each soil preparation was randomly applied to a plot within each location. On each plot the same number of seedlings were planted and the observation recorded was the average first-year growth of the seedlings on each plot.

Soil Preparation	Location			
	1	2	3	4
A	11	13	16	10
B	15	17	20	12
C	10	15	13	10

(a) Conduct an analysis of variance. Do the data provide sufficient evidence to indicate a difference in the mean growth for the three soil preparations?
(b) Is there evidence to indicate a difference in mean rates of growth for the four locations?
(c) Use a 90 percent confidence interval to estimate the difference in mean growth for methods A and B.

13.13 A study was conducted to compare the effect of three levels of digitalis on the level of calcium in the heart muscle of dogs. A description of the actual experimental procedure is omitted, but it is sufficient to note that the general level of calcium uptake varies from one animal to another so that comparison of digitalis levels (treatments) had to be blocked on heart muscles. That is, the tissue for a heart muscle was regarded as a block and comparisons of the three treatments were made within a given muscle. The calcium uptakes for the three levels of digitalis, *A*, *B*, and *C*, were compared based on the heart muscle of four dogs. The results are as follows:

Dogs

1	2	3	4
A 1342	*C* 1698	*B* 1296	*A* 1150
B 1608	*B* 1387	*A* 1029	*C* 1579
C 1881	*A* 1140	*C* 1549	*B* 1319

(a) Calculate the sums of squares for this experiment and construct an analysis-of-variance table.
(b) How many degrees of freedom are associated with SSE?
(c) Do the data present sufficient evidence to indicate a difference in the mean uptake of calcium for the three levels of digitalis?
(d) Do the data indicate a difference in the mean uptake in calcium for the four heart muscles?
(e) Give the standard deviation of the difference between the mean calcium uptake for two levels of digitalis.
(f) Find a 95 percent confidence interval for the difference in mean response between treatments *A* and *B*.

13.14 An experiment was conducted to investigate the toxic effect of three chemicals, *A*, *B*, and *C*, on the skin of rats. One-inch squares of skin were treated with the chemicals and then scored from 0 to 10 depending on the degree of irritation. Three adjacent 1-inch squares were marked on the backs of eight rats and each of the three chemicals was applied to each rat. Thus the experiment was blocked on rats to eliminate the variation in skin sensitivity from rat to rat. The data are as follows:

Rats

1	2	3	4	5	6	7	8
B 5	*A* 9	*A* 6	*C* 6	*B* 8	*C* 5	*C* 5	*B* 7
A 6	*C* 4	*B* 9	*B* 8	*C* 8	*A* 5	*B* 7	*A* 6
C 3	*B* 9	*C* 3	*A* 5	*A* 7	*B* 7	*A* 6	*C* 7

(a) Do the data provide sufficient evidence to indicate a difference in the toxic effect of the three chemicals?
(b) Estimate the difference in mean score for chemicals *A* and *B* using a 95 percent confidence interval.

13.15 A building contractor employs three construction engineers, *A*, *B*, and *C*, to estimate and bid on jobs. To determine whether one tends to be a more conservative (or liberal) estimator than the others, the contractor selects four projected construction jobs and has each estimator independently estimate the cost (dollars per square foot) of each job. The data are shown in the table.

Estimator (Treatments)	Construction Job (Blocks)				Total
	1	2	3	4	
A	35.10	34.50	29.25	31.60	130.45
B	37.45	34.60	33.10	34.40	139.55
C	36.30	35.10	32.45	32.90	136.75
Total	108.85	104.20	94.80	98.90	406.75

(a) Do the data provide sufficient evidence to indicate a difference in the mean building costs estimated by the three estimators? Test by using $\alpha = .05$.

(b) Find a 90 percent confidence interval for the difference in the mean of the estimates produced by estimators *A* and *B*. Interpret the interval.

(c) Do the data support the contention that the mean estimate of the cost per square foot varies from job to job?

13.11

THE ANALYSIS OF VARIANCE FOR A LATIN SQUARE DESIGN

The method for constructing a Latin square design for comparing p treatments is presented in Section 12.5. The purpose of the design is to remove unwanted variation as might occur in the mechanized application of icing to cakes on a conveyor belt. Variation in the thickness of icing could occur across the belt due to the variation in pressure at the applicator nozzles. Similarly, the thickness of icing could vary somewhat along the length of the belt due to variations in the consistency of the icing supplied to the machine. Now suppose that we wish to compare three different types of cake mixes, *A*, *B*, and *C*, that result in different porosities which affect absorption of the icing into the cakes. Then the thickness of the resulting icing, y, could be compared for the three treatments (mixes) by employing a 3×3 Latin square design. Each mix would appear in each column (across the conveyor belt) and in each row as one proceeds down the belt. The design configuration is shown in Figure 13.5.

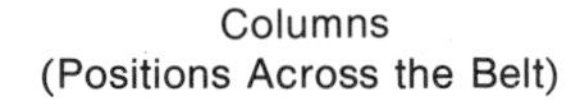

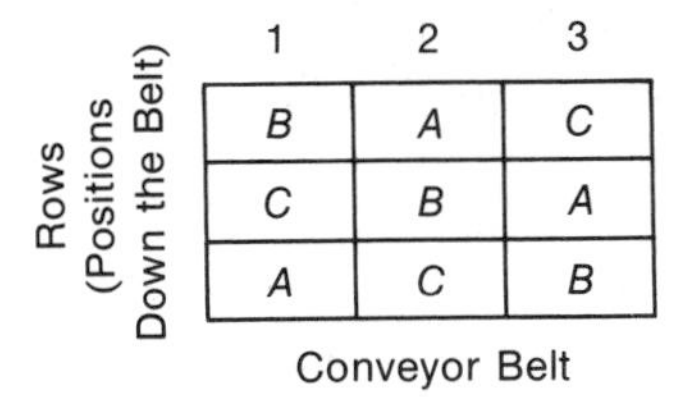

FIGURE 13.5
A 3×3 Latin square design

TABLE 13.7

ANOVA table for a Latin square design

Source	d.f.	SS	MS	F
Rows	$p-1$	SSR	SSR/$(p-1)$	MSR/MSE
Columns	$p-1$	SSC	SSC/$(p-1)$	MSC/MSE
Treatments	$p-1$	SST	SST/$(p-1)$	MST/MSE
Error	$n-3p+2$	SSE	MSE	
Total	$n-1$	Total SS		

The three independent variables in a Latin square design are "rows," "columns," and treatments. All are qualitative variables, although the treatments could be levels of a single quantitative factor or combinations of levels for two or more factors. Thus the total variation in an analysis of variance can be partitioned into four parts, one each corresponding to the variation in rows, columns, treatments, and experimental error. The analysis-of-variance table for a $p \times p$ Latin square design is shown in Table 13.7. As for previous designs, the four sums of squares add to the total sum of squares of deviations, Total SS.

The formulas for computing the total sum of squares, SSR, SSC, and SST, are identical to the corresponding formulas given for the randomized block design. Let y_{ij} denote an observation in row i and column j. Then

$$\text{Total SS} = \sum_{j=1}^{p}\sum_{i=1}^{p}(y_{ij} - \bar{y})^2 = \sum_{j=1}^{p}\sum_{i=1}^{p} y_{ij}^2 - \text{CM},$$
$$= (\text{sum of squares of all } y \text{ values}) - \text{CM},$$

where

$$\text{CM} = \frac{(\text{total of all observations})^2}{n} = \frac{\left(\sum_{j=1}^{p}\sum_{i=1}^{p} y_{ij}\right)^2}{n}.$$

Similarly, let R_i and $\bar{R}_i$, C_j and $\bar{C}_j$, and T_k and $\bar{T}_k$ represent the total and average for all observations in row i, column j, and treatment k, respectively. Then all sums of squares for rows, columns, and treatments are

$$\text{SSR} = p\sum_{i=1}^{p}(\bar{R}_i - \bar{y})^2 = \frac{\sum_{i=1}^{p} R_i^2}{p} - \text{CM}$$
$$= \frac{\text{sum of squares of all row totals}}{\text{number of observations in a single total}} - \text{CM},$$

$$\text{SSC} = p\sum_{j=1}^{p}(\bar{C}_j - \bar{y})^2 = \frac{\sum_{j=1}^{p} C_j^2}{p} - \text{CM}$$
$$= \frac{\text{sum of squares of all column totals}}{\text{number of observations in a single total}} - \text{CM},$$

$$\text{SST} = p\sum_{k=1}^{p}(\bar{T}_k - \bar{y})^2 = \frac{\sum_{k=1}^{p} T_k^2}{p} - \text{CM}$$

$$= \frac{\text{sum of squares of all treatment totals}}{\text{number of observations in a single total}} - \text{CM}.$$

The sums of squares for error, SSE, can be obtained by subtraction. Thus

$$\text{Total SS} = \text{SSR} + \text{SSC} + \text{SST} + \text{SSE}.$$

Hence

$$\text{SSE} = \text{Total SS} - \text{SSR} - \text{SSC} - \text{SST}.$$

The mean squares corresponding to rows, columns, and treatments can be obtained by dividing the respective sum of squares by $(p - 1)$. Thus, for example,

$$\text{MST} = \frac{\text{SST}}{(p-1)}.$$

The hypothesis, "no difference in mean response for treatments," is tested using the F statistic when

$$F = \frac{\text{MST}}{\text{MSE}}.$$

Similarly, the F statistic can be used to test an hypothesis of no difference between rows (or columns) by using the ratio of MSR (or MSC) to MSE.

F Test for Comparing *p* Treatments, Using a Latin Square Design

H_0: The population treatment means are equal.

H_a: One or more pairs of population means differ.

Test Statistic: $F = \frac{\text{MST}}{\text{MSE}}$

where F is based on $\nu_1 = (p - 1)$ and $\nu_2 = (n - 3p + 2)$ degrees of freedom.

Rejection Region: Reject if $F > F_\alpha$.

EXAMPLE 13.7

An experiment was conducted to investigate the difference in mean time to assemble four different electronic devices, T_1, T_2, T_3, and T_4. Two sources of unwanted variation affect the response: the variation between people and the effect of fatigue if a person assembles a series of the devices over time. Consequently, four assemblers were selected and each assembled all four of the devices, T_1, T_2, T_3, and T_4, in the following Latin square design (the observed responses, in minutes, are shown in the cells):

Rows (Position in Assembly Sequence)	Columns (Assemblers)				Total
	1	2	3	4	
1	T_3 44	T_1 41	T_2 30	T_4 40	155
2	T_2 41	T_3 42	T_4 49	T_1 49	181
3	T_1 59	T_4 41	T_3 59	T_2 34	193
4	T_4 58	T_2 37	T_1 53	T_3 59	207
Total	202	161	191	182	736

Do the data provide sufficient evidence to indicate a difference in mean time to assemble the four devices? A difference in the mean time to assemble for people? Is there evidence of a fatigue factor (a difference in mean response for positions in the assembly sequence)?

Solution The totals for rows, columns, and treatments are:

	i			
	1	2	3	4
R_i	155	181	193	207
C_i	202	161	191	182
T_i	202	142	204	188

$$\text{CM} = \frac{(736)^2}{16} = \frac{541{,}696}{16} = 33{,}856.0.$$

Then,

$$\text{Total SS} = \sum_{j=1}^{4}\sum_{i=1}^{4} y_{ij}^2 - \text{CM} = 35{,}186.0 - 33{,}856.0 = 1{,}330.0,$$

$$\text{SSR} = \frac{\sum_{i=1}^{4} R_i^2}{4} - \text{CM}$$

$$= \frac{(155)^2 + (181)^2 + (193)^2 + (207)^2}{4} - \text{CM}$$

$$= 34{,}221.0 - 33{,}856.0 = 365.0,$$

$$\text{SSC} = \frac{\sum_{j=1}^{4} C_j^2}{4} - \text{CM}$$

$$= \frac{(202)^2 + (161)^2 + (191)^2 + (182)^2}{4} - \text{CM}$$

$$= 34{,}082.5 - 33{,}856.0 = 226.5,$$

$$\text{SST} = \frac{\sum_{k=1}^{4} T_k^2}{4} - \text{CM}$$

$$= \frac{(202)^2 + (142)^2 + (204)^2 + (188)^2}{4} - \text{CM}$$

$$= 34{,}482.0 - 33{,}856.0 = 626.0.$$

Finally,

$$\text{SSE} = \text{Total SS} - \text{SSR} - \text{SSC} - \text{SST}$$

$$= 1{,}330.0 - 365.0 - 226.5 - 626.0$$

$$= 112.5.$$

The analysis-of-variance table for the example is shown in Table 13.8. Note that the mean squares were obtained by dividing the sums of squares by their respective degrees of freedom.

TABLE 13.8
ANOVA table for Example 13.7

Source	d.f.	SS	MS	*F*
Rows	3	365	121.67	6.49
Columns	3	226.5	75.50	4.03
Treatments	3	626	208.67	11.13
Error	6	112.5	18.75	
Total	15	1,330.0		

All the computed F statistics are based on $\nu_1 = 3$ and $\nu_2 = 6$ degrees of freedom. The corresponding tabulated critical value is $F_{.05} = 4.76$. A comparison of the computed F statistics with the tabulated value indicates that the computed F's for both rows and treatments exceed the critical value. Thus the data provide sufficient evidence to indicate a difference in the mean time to assemble the four devices. The data also show a difference in mean time to assemble for rows, or equivalently, positions in the sequence of assembly. Thus it appears that fatigue, boredom, or some other factor increases the mean time to assemble as the length of employment increases.

13.12

ESTIMATION FOR THE LATIN SQUARE DESIGN

A confidence interval for the difference between a pair of treatment means, with confidence coefficient $(1 - \alpha)$, is obtained in the same manner as for the completely randomized and the randomized block designs. Since each treatment mean will contain p observations, $n_1 = n_2 = p$, the standard deviation of the difference in a pair of means is

$$\sigma_{\bar{T}_i - \bar{T}_j} = \sigma\sqrt{\frac{1}{p} + \frac{1}{p}}.$$

The $100(1 - \alpha)$ percent confidence interval is

Latin Square Design: A 100(1 − α) Percent Confidence Interval for the Difference Between Two Treatment Means

$$(\bar{T}_i - \bar{T}_j) \pm t_{\alpha/2}\, s\sqrt{\frac{2}{p}},$$

where $t_{\alpha/2}$ has $n - 3p + 2$ degrees of freedom.

Corresponding confidence intervals for the difference between a pair of row or column means are, respectively,

$$(\bar{R}_i - \bar{R}_j) \pm t_{\alpha/2}\, s\sqrt{\frac{2}{p}} \quad \text{and} \quad (\bar{C}_i - \bar{C}_j) \pm t_{\alpha/2}\, s\sqrt{\frac{2}{p}}.$$

EXAMPLE 13.8

Refer to Example 13.7. Estimate the difference in mean response between treatments 1 and 2 using a 95 percent confidence interval.

Solution From the analysis-of-variance table, Example 13.7, observe that $s^2 = \text{MSE} = 18.75$ is based on six degrees of freedom. Consequently, the tabulated value, $t_{\alpha/2}$, with six degrees of freedom is $t_{.025} = 2.447$. Then the 95 percent confidence interval for the difference between the treatment means, $(\mu_1 - \mu_2)$, is

$$(\bar{T}_1 - \bar{T}_2) \pm t_{\alpha/2}\, s\sqrt{\frac{2}{p}}$$

or

$$(50.5 - 35.5) \pm (2.447)(4.33)\sqrt{\frac{2}{4}}$$

or

$$15 \pm 7.49.$$

Thus we estimate the difference in mean time to assemble the two devices to be between 7.51 and 22.49 minutes.

13.13

A COMPUTER PRINTOUT FOR A LATIN SQUARE DESIGN

Now that you are familiar with the SAS computer printouts for the analyses of variance for the completely random and the randomized block designs (Sections 13.7 and 13.10), you will be able to identify the appropriate quantities in the SAS printout for the analysis of variance of the data for the Latin square design, Example 13.7. This printout is shown in Table 13.9.

The upper table (shaded) shown in the printout gives a breakdown of the Total SS into two parts, the first corresponding to MODEL and the second to ERROR. Note that the numbers appearing in the ERROR row are identical to those in the "Error" row of the ANOVA, Table 13.8. The lower table of the printout gives a breakdown of the MODEL degrees of freedom and sum of squares into portions corresponding to ROWS, COLUMNS, and TREATMENTS. Note that the degrees of freedom and sums of squares in this table correspond to those of the ANOVA, Table 13.8. As in the printout for the randomized block design, Section 13.10, the lower table does not give the mean squares, MSR, MSC, and MST, but they are not needed because the computed F values, necessary for the F tests, along with their significance levels, are shown. Note that they correspond with the values given in the ANOVA, Table 13.8. The value of s, $s = 4.33\ldots$, given in the sixth column of the printout can be used to calculate confidence intervals for the difference between any pair of treatment, row, or column means.

TABLE 13.9

SAS computer printout for Example 13.7

STATISTICAL ANALYSIS SYSTEM

ANALYSIS OF VARIANCE PROCEDURE

DEPENDENT VARIABLE: Y

SOURCE	DF	SUM OF SQUARES	MEAN SQUARE	F VALUE	PR > F	R-SQUARE	C.V.
MODEL	9	1217.50000000	135.27777778	7.21	0.0129	0.915414	9.4133
ERROR	6	112.50000000	18.75000000		STD DEV		Y MEAN
CORRECTED TOTAL	15	1330.00000000			4.33012702		46.00000000

SOURCE	DF	ANOVA SS	F VALUE	PR > F
ROWS	3	365.00000000	6.49	0.0259
COLUMNS	3	226.50000000	4.03	0.0692
TRTMENTS	3	626.00000000	11.13	0.0073

EXERCISES

13.16 Give the analysis of variance for the following 3×3 Latin square design. Find a 95 percent confidence interval for the difference in mean response for treatments A and C; assume that this is a preplanned comparison.

Rows	Columns 1	2	3
1	*B* 12	*A* 7	*C* 17
2	*C* 10	*B* 7	*A* 4
3	*A* 2	*C* 8	*B* 12

13.17 Paper machines distribute a thin mixture of wood fibers and water to a wide wire mesh belt that is traveling at a very high speed. Thus it is conceivable that the distribution of fibers, thickness, porosity, etc., will vary across the belt as well as along the belt and produce corresponding variations in the strength of the final paper product. A paper company designed an experiment to compare the strength of four coatings intended to improve the appearance of packaging paper. Because the uncoated paper strength could vary down as well as across a roll (because of the reasons described above), the coatings were applied in a Latin square design. The strength measurements for the four coatings, *A, B, C,* and *D,* were as follows:

Position Down the Roll	Position Across the Roll 1	2	3	4
1	*B* 12.4	*A* 10.4	*C* 13.1	*D* 11.8
2	*C* 13.4	*B* 12.4	*D* 11.8	*A* 10.9
3	*A* 10.5	*D* 11.4	*B* 12.3	*C* 12.9
4	*D* 11.4	*C* 13.3	*A* 10.7	*B* 12.0

(a) Do the data present sufficient evidence to indicate a difference in mean strength for paper treated with the four paper coatings?
(b) Suppose that prior to seeing our data, we wished to compare the mean strength between paper coated with coatings *A* and *C*. Estimate the difference using a 95 percent confidence interval.
(c) Do the data provide sufficient evidence to indicate a difference in mean strength for locations across the roll?
(d) Down the roll?

13.18 The voltage within an electronic system was to be investigated for four different conditions, *A, B, C,* and *D,* but unfortunately the voltage measurements were affected by the line voltage within the laboratory. Since the line voltage assumed the same pattern within each day due to the usage patterns of various industrial firms in the area and since the general level of line voltage varied from day to day, a Latin square design was employed. The data are shown below:

Time Period Within a Day	Days			
	1	2	3	4
1	*A* 116	*B* 108	*C* 126	*D* 112
2	*C* 111	*D* 124	*A* 122	*B* 121
3	*B* 120	*C* 115	*D* 126	*A* 109
4	*D* 118	*A* 125	*B* 116	*C* 127

(a) Calculate the appropriate sums of squares and construct an analysis-of-variance table for this data.
(b) Do the data present sufficient evidence to indicate a difference between treatments?
(c) Estimate the difference in mean responses for voltage conditions *A* and *B*. Use a 90 percent confidence interval.

3.19 A 4 × 4 Latin square design was employed to investigate the gasoline consumption at a fixed horsepower for four different brands of gasoline, *A, B, C,* and *D*. Four engines of the same design and specifications were employed in the experiment and each gasoline was tested in each of the engines. The tests, one gasoline per engine, were conducted on four different days. The data are shown in the table. A SAS computer printout for the analysis of variance is also shown. Use the information in the printout to answer the following questions:

Rows (Days)	Columns (Engines)			
	1	2	3	4
1	*D* 19.5	*B* 16.8	*A* 19.8	*C* 19.2
2	*B* 18.0	*A* 17.9	*C* 17.9	*D* 17.7
3	*A* 18.1	*C* 21.0	*D* 17.5	*B* 17.2
4	*C* 20.1	*D* 19.2	*B* 17.0	*A* 18.5

(a) Do the data provide sufficient evidence to indicate a difference in mean gasoline consumption for the four brands of gasoline? Test by using $\alpha = .05$.

STATISTICAL ANALYSIS SYSTEM

ANALYSIS OF VARIANCE PROCEDURE

DEPENDENT VARIABLE: Y

SOURCE	DF	SUM OF SQUARES	MEAN SQUARE	F VALUE	PR > F	R-SQUARE	C.V.
MODEL	9	14.99750000	1.66638889	1.42	0.3465	0.679927	5.8754
ERROR	6	7.06000000	1.17666667		STD DEV		Y MEAN
CORRECTED TOTAL	15	22.05750000			1.08474267		18.46250000

SOURCE	DF	ANOVA SS	F VALUE	PR > F
ROWS	3	2.13250000	0.60	0.6360
COLUMNS	3	2.20250000	0.62	0.6252
TRTMENTS	3	10.66250000	3.02	0.1156

(b) Do the data provide sufficient evidence to indicate that blocking on engines increased the amount of information in the experiment concerning mean gasoline consumption for the four brands of gasoline? Explain.

(c) Do the data provide sufficient evidence to indicate that blocking on days increased the amount of information in the experiment concerning the mean gasoline consumption for the four brands of gasoline? Explain.

(d) If you conducted a similar experiment in the future, should you block on engines and days? Explain.

(e) Find a 90 percent confidence interval for the difference in mean gasoline consumption between brands A and C. Interpret the interval.

13.14

SELECTING THE SAMPLE SIZE

Selecting the sample size for the completely randomized or the randomized block design is an extension of the procedure of Section 8.10. We confine our attention to the case of equal sample sizes, $n_1 = n_2 = \cdots = n_p$, for the p treatments of the completely randomized design. The number of observations per treatment is equal to b for the randomized block design. Thus the problem is to select n_1 or b for these two designs so as to purchase a specified quantity of information.

The selection of sample size follows a similar procedure for both designs; hence we will outline a general method. First the experimenter must decide on the parameter (or parameters) of major interest. Usually, he will wish to compare a pair of treatment means. Second, he must specify a bound on the error of estimation that he is willing to tolerate. Once determined, he need only select n_i (the number of observations in a treatment mean) or, correspondingly, b (the number of blocks in the randomized block design) that will reduce the half-width of the confidence interval for the parameter so that it is less than or equal to the specified bound on the error of estimation. It should be emphasized that the sample-size solution will *always* be an approximation, since σ is unknown and s is unknown until the sample is acquired. The best available value will be used for s in order to produce an approximate solution. We will illustrate the procedure with an example.

EXAMPLE 13.9

A completely randomized design is to be conducted to compare five teaching techniques on classes of equal size. Estimation of the difference in mean response to an achievement test is desired correct to within 30 test-score points, with probability equal to .95. It is expected that the test scores for a given teaching technique will possess a range approximately equal to 240. Find the approximate number of observations required for each sample in order to acquire the specified information.

Solution The confidence interval for the difference between a pair of treatment means is

$$(\bar{T}_i - \bar{T}_j) \pm t_{\alpha/2}\, s \sqrt{\frac{1}{n_i} + \frac{1}{n_j}}.$$

Therefore, we will wish to select n_i and n_j so that

$$t_{\alpha/2}\, s \sqrt{\frac{1}{n_i} + \frac{1}{n_j}} \leq 30.$$

The value of σ is unknown and s is a random variable. However, an approximate solution for $n_i = n_j$ can be obtained by guessing s to be roughly equal to one-fourth of the range. Thus $s \approx (240)/4 = 60$. The value of $t_{\alpha/2}$ will be based upon $(n_1 + n_2 + \cdots + n_5 - 5)$ degrees of freedom, and for even moderate values of n_i, $t_{.025}$ will approximately equal 2. Then

$$t_{.025}\, s \sqrt{\frac{1}{n_i} + \frac{1}{n_j}} \approx 2(60) \sqrt{\frac{2}{n_i}} \leq 30$$

or

$$n_i \approx 32, \qquad i = 1, 2, \ldots, 5.$$

EXAMPLE 13.10

An experiment is to be conducted to compare the toxic effect of three chemicals on the skin of rats. The resistance to the chemicals was expected to vary substantially from rat to rat. Therefore, all three chemicals were to be tested on each rat, thereby blocking out rat-to-rat variability.

The standard deviation of the experimental error was unknown, but prior experimentation involving several applications of a given chemical on the same rat suggested a range of response measurements equal to 5 units.

Find a value for b such that the error of estimating the difference between a pair of treatment means is less than one unit, with probability equal to .95.

Solution A very approximate value for s would be one-fourth the range, or $s \approx 1.25$. Then we wish to select b so that

$$t_{.025}\, s \sqrt{\frac{1}{b} + \frac{1}{b}} = t_{.025}\, s \sqrt{\frac{2}{b}} \leq 1.$$

Since $t_{.025}$ will depend upon the degrees of freedom associated with s^2, which will be $(n - b - p + 1)$, we will guess $t_{.025} \approx 2$. Then

$$2(1.25) \sqrt{\frac{2}{b}} \leq 1$$

or

$$b \approx 13, \qquad i = 1, 2, 3.$$

Thus approximately 13 rats will be required to obtain the required information.

The degrees of freedom associated with s^2 will be 24 based on this solution. Therefore, the guessed value of t would seem to be adequate for this approximate solution.

The sample-size solutions for Examples 13.9 and 13.10 are very approximate, and are intended to provide only a rough estimate of approximate size and consequent cost of the experiment. The experimenter will obtain information on σ as the data are being collected and can recalculate a better approximation to n as he or she proceeds.

Selecting the sample size for Latin square designs follows essentially the same procedure as for the completely randomized and the randomized block designs. The only difference is that one must decide on the number of $p \times p$ Latin squares that will be needed to acquire the desired information. We omit discussion of this topic because we have not shown how to conduct an analysis of variance for more than one $p \times p$ Latin square. The reader interested in this topic should consult the references.

EXERCISES

13.20 Refer to Exercise 13.7. How many plants would have to be selected and analyzed for each location if we wish to estimate the difference in mean leaf length for two locations with an error of less than .3 and with a probability approximately equal to .95?

13.21 Refer to Exercise 13.14. Approximately how many rats would be required to estimate the difference in mean scores for two chemicals correct to within 1 unit?

13.22 Refer to Exercise 13.13. Approximately how many replications would be required for each treatment (observations per treatment) in order that the error of estimating the difference in mean response for a pair of treatments be less than 20 (with probability equal to .95)? Assume that additional observations would be made within a randomized block design.

13.23 A manufacturer wished to compare the mean daily production for four production lines, A, B, C, and D, which manufacture the same product. Since daily production is affected by the quality of raw materials and labor associated with supply to the lines, it was decided to make the comparison of line productivity by making comparisons *within* each day (thereby eliminating day to day variations associated with supply).

(a) What type of experimental design was the manufacturer using to collect his data?

(b) Suppose that the mean daily yield per line is in the neighborhood of 1,000 units and that an earlier analysis of the differences in daily production, recorded each day over a long period of time, suggests that the variance of $(y_1 - y_2)$, the difference in daily yields for production lines A and B, is equal to 5,000. For how many days must the daily production be recorded for the three production lines to estimate the difference in mean production between any pair of lines correct to within 25 units?

13.24 Refer to Exercise 13.11. Suppose you wish to estimate the difference in mean moisture absorption between fabrics treated by two different chemicals correct to within .2 units, with probability equal to .95. Approximately how many bolts of fabric (blocks) should be used for the experiment?

13.25 An experiment is to be conducted to compare the mean amount of effluents discharged by three chemical plants. Suppose that the variance of the effluent measurements at all three plants is approximately equal to .03. If you wish to estimate the mean amount of effluents (pounds/gallon) at a plant correct to within .05 lb/gal, with probability near .95, approximately how many effluent specimens would have to be randomly selected at a single point?

13.15

SOME CAUTIONARY COMMENTS ON BLOCKING

Two points need to be emphasized when using block designs. First, you should not use a randomized block design to investigate the effects of two factors on a response (letting treatments be the levels of one factor and blocks be the levels of the second factor). If the two factors do not affect the response independently of each other—that is, the effect of one factor on the response depends upon the level of the other factor (called "factor interaction")—a "randomized block" analysis of variance could lead to very erroneous conclusions about the relationship between the factors and the response. Similarly, you should not use a Latin square design to investigate the effects of three factors on a response. Remember, the object of block designs is to block out the variation of nuisance variables (blocks) and thereby reduce SSE. Information on designs for multivariable experiments can be found in any standard text on the design of experiments (see the references).

The second point is that blocking is not always beneficial. Just as we noted for the paired-difference experiment (a randomized block design with two treatments), blocking produces a gain in information if the between-block variation is larger than the within-block variation. Then blocking removes this larger source of variation from SSE and s^2 assumes a smaller value (as does the population variance σ^2) because of the design. At the same time, you lose information because blocking reduces the number of degrees of freedom associated with SSE (and s^2). (To see this, compare the ANOVA tables for the completely randomized design, Table 13.1, and the randomized block design, Table 13.3.) Consequently, if blocking is to be beneficial, the gain in information due to the elimination of block variation must outweigh the loss due to a reduction in the number of degrees of freedom associated with SSE. Unless blocking leaves you with only a small number of degrees of freedom for SSE, the loss in degrees of freedom causes only a small reduction in information in the experiment. Consequently, if you have a reason for suspecting that there is block-to-block variation, it will usually pay to block.

13.16

ASSUMPTIONS

The assumptions about the probability distribution of the random response, y, are the same for the analysis of variance as for the regression models of Chapter 10. These are as follows:

Assumptions

1. For any treatment or block combination, y is normally distributed with variance equal to σ^2.
2. The observations have been selected independently so that any pair of y values, y_i and y_j, are independent in a probabilistic sense for all i and j $(i \neq j)$.

In a practical situation, you can never be certain that the assumptions are satisfied but you will often have a fairly good idea whether the assumptions are reasonable for your data. To illustrate, it has been mentioned that the inferential methods of Chapters 9, 10, and 13 are not seriously affected by moderate departures from the assumptions of normality, but you would want the probability distribution of y at least to be mound-shaped. So if y is a discrete random variable that only assumes three values, say $y = 10, 11, 12$, then it is unreasonable to assume that the probability distribution of y is approximately normal!

The assumption of a constant variance for y for the various experimental conditions should be approximately satisfied, although violation of even this assumption is not too serious if the sample sizes for the various experimental conditions are equal. However, suppose the response is binomial, say the proportion $\hat{p}$ of people who favor a particular food product. We know that the variance of a proportion (Section 6.4) is

$$\sigma^2_{\hat{p}} = \frac{pq}{n}, \qquad \text{where } q = 1 - p,$$

and therefore that the variance is dependent on p. The variance of the response is a function of the mean response and will change from one experimental setting to another, and the assumptions of the analysis of variance have been violated. A similar situation occurs when the response measurements are Poisson data (mentioned in the exercises, Chapter 5). If the response possesses a Poisson probability distribution, the variance of the response will equal the mean. Consequently, Poisson response data also violate the analysis-of-variance assumptions.

Many kinds of data are not measurable and hence are unsuitable for an analysis of variance. For example, many responses cannot be measured but can be ranked. Consumer preference studies yield data of this type: you know you prefer product A to B and you prefer B to C, but you have difficulty assigning an exact value to the strength of your preference.

So what do you do when the assumptions of an analysis of variance are not satisfied? For example, suppose the variances of the responses for various experimental conditions are not equivalent. This situation can be remedied sometimes by transforming the response measurements. That is, instead of using the original response measurements, we might use their square roots, logarithms, or some other function of the response, y. Transformations intended to stabilize the variance of the response have been found to make the probability distributions of the transformed responses more nearly normal. See the references for discussions of these topics.

When nothing can be done to satisfy (even approximately) the assumptions of the analysis of variance, or if the data are rankings, you should use a nonparametric testing and estimation procedure. When our assumptions are satisfied, these procedures, which rely on the comparative magnitudes of measurements (often ranks), are almost as powerful in detecting treatment differences as the tests presented in this chapter. When the assumptions are not satisfied, the nonparametric procedures may be more powerful.

13.17 SUMMARY AND COMMENTS

The completely randomized, the randomized block, and the Latin square designs are illustrations of experiments involving one, two, and three qualitative independent variables, respectively. The analysis of variance partitions the total sum of squares of deviations of the response measurements about their mean into portions associated with each independent variable and the experimental error. The former may be compared with the sum of squares for error, using mean squares and the F statistics, to see whether the mean squares for the independent variables are unusually large and, thereby, indicative of an effect on the response.

This chapter has presented a very brief introduction to the analysis of variance and its associated subject, the design of experiments. Experiments can be designed to investigate the effect of many quantitative and qualitative variables on a response. These may be variables of primary interest to the experimenter as well as nuisance variables, such as blocks, which we attempt to separate from the experimental error. They are subject to an analysis of variance when properly designed. A more extensive coverage of the basic concepts of experimental design and the analysis of experiments will be found in the references.

REFERENCES

Barr, A. J., J. H. Goodnight, J. P. Sall, and J. T. Helwig, *A User's Guide to SAS 76,* SAS Institute, P. O. Box 10066, Raleigh, N. C. 27605.

Brown, M. B., ed., *Biomedical Computer Programs.* Berkeley, Calif.: University of California, 1977.

Cochran, W. G., and G. M. Cox, *Experimental Designs,* 2nd ed. New York: John Wiley & Sons, Inc., 1957.

Guenther, W. C., *Analysis of Variance.* Englewood Cliffs, N.J.: Prentice-Hall, Inc., 1964.

Hicks, C. R., *Fundamental Concepts in the Design of Experiments.* New York: Holt, Rinehart and Winston, Inc., 1964.

Mendenhall, W., *An Introduction to Linear Models and the Design and Analysis of Experiments.* Belmont, Calif.: Wadsworth Publishing Company, Inc., 1968.

Ott, L., *An Introduction to Statistical Methods and Data Analysis.* North Scituate, Mass.: Duxbury Press, 1977.

Nie, N., C. H. Hull, J. G. Jenkins, K. Steinbrenner, and D. H. Bent, *Statistical Package for the Social Sciences,* 2nd ed. New York: McGraw-Hill Book Company, 1975.

Snedecor, G. W., and W. G. Cochran, *Statistical Methods,* 6th ed. Ames, Iowa: The Iowa State University Press, 1967.

Steel, R. G. D., and J. H. Torrie, *Principles and Procedures of Statistics.* New York: McGraw-Hill Book Company, 1960.

Winer, B. J., *Statistical Principles in Experimental Design,* 2nd ed. New York: McGraw-Hill Book Company, 1971.

SUPPLEMENTARY EXERCISES

13.26 State the assumptions underlying the analysis of variance of a completely randomized design.

13.27 Refer to Example 13.2. Calculate SSE by pooling the sums of squares of deviations within each of the four samples, and compare with the value obtained by subtraction. Note that this is an extension of the pooling procedure used in the two-sample case discussed in Section 13.2.

13.28 To compare the strengths of concrete produced by four experimental mixes, 3 specimens were prepared from each type of mix. Each of the 12 specimens was subjected to increasing compressive loads until breakdown. The following compressive loads in tons per square inch were attained at breakdown. Specimen numbers 1–12 are indicated in parentheses for identification purposes.

Mix	Comprehensive Load at Breakdown (tons/in.^{2})		
A	(1) 2.30	(5) 2.20	(9) 2.25
B	(2) 2.20	(6) 2.10	(10) 2.20
C	(3) 2.15	(7) 2.15	(11) 2.20
D	(4) 2.25	(8) 2.15	(12) 2.25

Assume that the requirements for a completely randomized design are met and analyze the data. State whether there is statistical support at the $\alpha = .05$ level of significance for the conclusion that the four types of concrete differ in average strength.

13.29 Refer to Exercise 13.28. Let μ_A and μ_B denote the mean strengths of concrete specimens prepared from mix *A* and *B*, respectively.
(a) Find a 90 percent confidence interval for μ_A.
(b) Find a 95 percent confidence interval for $\mu_A - \mu_B$.

13.30 Refer to Exercise 13.28. Suppose that the sand used in the mixes for samples 1, 2, 3, and 4 came from pit *A*, that the sand used for samples 5, 6, 7, and 8 came from pit *B*, and that sand for the other samples came from pit *C*. Analyze the data, assuming that the requirements for a randomized block are met with three blocks consisting respectively of samples 1, 2, 3, and 4; samples 5, 6, 7, and 8; and samples 9, 10, 11, and 12.
(a) At the 5 percent level, is there evidence of differences in concrete strength due to the sand used?
(b) Is there evidence at the 5 percent level that the four types of concrete differ in average strength?

(c) Does the conclusion of part (b) contradict the conclusion obtained in Exercise 13.28?

13.31 Refer to Exercise 13.30. Let μ_A and μ_B, respectively, denote the mean strengths of concrete specimens prepared from mix *A* and mix *B*.
(a) Find a 95 percent confidence interval for $\mu_A - \mu_B$.
(b) Is the interval found in part (a) the same interval found in Exercise 13.29(b)? Explain.

13.32 A study was initiated to investigate the effect of two drugs, administered simultaneously, in reducing human blood pressure. It was decided to utilize three levels of each drug and to include all nine combinations in the experiment. Nine high-blood-pressure patients were selected for the experiment, and one was randomly assigned to each of the nine drug combinations. The response observed was the drop in blood pressure over a fixed interval of time.
(a) Is this a randomized block design?
(b) Suppose that two patients were assigned to each of the nine drug combinations. What type of experimental design is this?

13.33 Refer to Exercise 13.32. Suppose that prior experimentation suggests that $\sigma = 20$.
(a) How many replications would be required to estimate any treatment (drug combination) mean correct to within ± 10?
(b) How many degrees of freedom will be available for estimating σ^2 when using the number of replications determined in part (a)?
(c) Give the approximate half-widths of a confidence interval for the difference in mean response for two treatments when using the number of replications determined in part (a).

13.34 A dealer has in stock three cars (car *A*, car *B*, and car *C*) of the same make and model. Wishing to compare these cars in gas consumption, a customer arranged to test each car with each of three brands of gasoline (brand *A*, brand *B*, and brand *C*). In each trial, a gallon of gasoline was added to an empty tank and the car was driven without stopping until it ran out of gasoline. The following table shows the number of miles covered in each of the nine trials.

Brand of Gasoline	Car *A*	Car *B*	Car *C*
A	22.4	17.0	19.2
B	20.8	19.4	20.2
C	21.5	18.7	21.2

(a) Should the customer conclude that the three cars differ in gas mileage? Test at the $\alpha = .05$ level.
(b) Do the data indicate that the brand of gasoline affects gas mileage?

13.35 Refer to Exercise 13.34. Suppose that gas mileage is unrelated to brand of gasoline; carry out an analysis of the data appropriate for a completely randomized design with three treatments.
(a) Should the customer conclude that the three cars differ in gas mileage? Test at the $\alpha = .05$ level.
(b) Compare your answer for part (a) in Exercise 13.34 with your answer for part (a) above; can you suggest a reason why blocking may be unwise in certain cases?

13.36 A portion of a questionnaire was constructed to enable judges to evaluate a certain aspect of observed classroom teaching. Four films portraying teaching performances, and differing markedly in the teaching characteristic under study, were viewed by each of eight judges. The order of viewing the four films was assigned in a random manner to each judge. The data obtained are shown below.

	Judges							
Films	1	2	3	4	5	6	7	8
1	9	10	7	5	12	7	8	6
2	4	9	3	0	6	8	2	4
3	12	16	10	9	11	10	10	14
4	9	11	7	8	12	7	7	8

(a) Give the type of design employed for this experiment and justify your diagnosis.
(b) How many degrees of freedom are available for estimating σ^2? Perform an analysis of variance on the data.
(c) Do the data provide sufficient evidence to indicate that the mean questionnaire score varies from film to film? Test by use of $\alpha = .05$.
(d) Suppose that the data did provide sufficient evidence to indicate differences among the mean questionnaire scores for the four films. Would this imply that the questionnaire was able to detect a difference in the teaching characteristic exhibited in the four films?

13.37 Refer to Exercise 13.28. About how many specimens per mix should be prepared to allow estimation of the difference in mean strengths for a preselected pair of specimen types to within .02 ton per square inch? Assume knowledge of the data given in Exercise 13.28.

13.38 Refer to Exercise 13.36. About how many judges should be used to allow estimation of the difference in mean questionnaire scores for a preselected pair of films to within 1 unit? Assume knowledge of the data given in Exercise 13.36.

Ψ 13.39 A completely randomized design was conducted to compare the effect of five different stimuli on reaction time. Twenty-seven people were employed in the experiment that was conducted using a completely randomized design. Regardless of the results of the analysis of variance, it is desired to compare stimuli *A* and *D*. The results of the experiment were as follows:

Stimulus	Reaction Time (sec)							Total	Mean
A	.8	.6	.6	.5				2.5	.625
B	.7	.8	.5	.5	.6	.9	.7	4.7	.671
C	1.2	1.0	.9	1.2	1.3	.8		6.4	1.067
D	1.0	.9	.9	1.1	.7			4.6	.920
E	.6	.4	.4	.7	.3			2.4	.48

(a) Conduct an analysis of variance and test for a difference in mean reaction time due to the five stimuli.
(b) Compare stimuli *A* and *D* to see if there is a difference in mean reaction time.

13.40 The experiment in Exercise 13.39 might have been more effectively conducted using a randomized block design with people as blocks since we would expect mean reaction time to vary from one person to another. Hence four people were used in a new experiment and each person was subjected to each of the five stimuli in a random order. The reaction times (sec) were as follows:

Subject	Stimulus				
	A	*B*	*C*	*D*	*E*
1	.7	.8	1.0	1.0	.5
2	.6	.6	1.1	1.0	.6
3	.9	1.0	1.2	1.1	.6
4	.6	.8	.9	1.0	.4

Conduct an analysis of variance and test for differences in treatments (stimuli).

13.41 The following measurements are on the thickness of cake icing for the Latin square design discussed in Section 13.11. The only difference between this exercise and the text discussion is that five (not three) mixes, *A, B, C, D,* and *E,* were employed in a 5×5 Latin square design. Thickness measurements are given in hundredths of an inch. Do the data provide sufficient evidence to indicate a difference in mean thickness of icing for the five cake mixtures? Estimate the difference in mean thickness for mixtures *A* and *B*.

Rows	Columns				
	1	2	3	4	5
1	*D* 9	*E* 18	*C* 6	*A* 8	*B* 11
2	*B* 12	*D* 17	*E* 10	*C* 4	*A* 5
3	*A* 6	*B* 16	*D* 10	*E* 9	*C* 4
4	*C* 4	*A* 13	*B* 11	*D* 8	*E* 13
5	*E* 14	*C* 11	*A* 7	*B* 10	*D* 15

13.42 Consider the following one-way classification consisting of three treatments, *A, B,* and *C,* where the number of observations per treatment varies from treatment to treatment.

A	*B*	*C*
24.2	24.5	26.0
27.5	22.7	
25.9		
24.7		

Do the data present sufficient evidence to indicate a difference between treatments?

13.43 Fifteen subjects with relatively equal reaction times and physical attributes were randomly grouped in five's. The first five were to react to a light stimulus, the second five to a sound stimulus, and the third five to a stimulus of both sound and light. The results are as follows:

Stimulus	Reaction Time (sec)				
Light	.36	.41	.45	.42	.31
Sound	.49	.42	.35	.48	.40
Both	.29	.37	.42	.41	.47

Is there sufficient evidence to indicate a difference in mean reaction times for the three stimuli? Use $\alpha = .05$.

13.44 A company wished to study the differences among four sales-training programs on the sales abilities of their sales personnel. Thirty-two people were randomly divided into four groups of equal size and the groups were then subjected to the different sales-training programs. Because there were some dropouts (illness, etc.) during the training programs, the number of trainees completing the programs varied from group to group. At the end of the training programs, each salesperson was randomly assigned a sales area from a group of sales areas that were judged to have equivalent sales potentials. The numbers of sales made by each of the four groups of salespeople during the first week after completing the training program are listed in the table.

(a) Do the data present sufficient evidence to indicate a difference in the mean achievement for the four training programs?
(b) Find a 90 percent confidence interval for the difference in the mean of sales that would be expected for persons subjected to training programs 1 and 4. Interpret the interval.
(c) Find a 90 percent confidence interval for the mean number of sales by persons subjected to training program 2.

	Training Program			
	1	2	3	4
	78	99	74	81
	84	86	87	63
	86	90	80	71
	92	93	83	65
	69	94	78	86
	73	85		79
		97		73
		91		70
Total	482	735	402	588

13.45 Four chemical plants, producing the same product and owned by the same company, discharge effluents into streams in the vicinity of their locations. To check on the extent of the pollution created by the effluents and to determine if this varies from plant to plant, the company collected random samples of liquid waste, five specimens for each of the four plants. The data are shown in the table. A SAS computer printout of the analysis of variance for the data is also shown. Use the information in the printout to answer the following questions:

Plant	Polluting Effluents (lb/gal of waste)				
A	1.65	1.72	1.50	1.37	1.60
B	1.70	1.85	1.46	2.05	1.80
C	1.40	1.75	1.38	1.65	1.55
D	2.10	1.95	1.65	1.88	2.00

STATISTICAL ANALYSIS SYSTEM

ANALYSIS OF VARIANCE PROCEDURE

DEPENDENT VARIABLE: Y

SOURCE	DF	SUM OF SQUARES	MEAN SQUARE	F VALUE	PR > F	R-SQUARE	C.V.
MODEL	3	0.46489500	0.15496500	5.20	0.0107	0.493679	10.1515
ERROR	16	0.47680000	0.02980000		STD DEV		Y MEAN
CORRECTED TOTAL	19	0.94169500			0.17262677		1.70050000

SOURCE	DF	ANOVA SS	F VALUE	PR > F
TRTMENTS	3	0.46489500	5.20	0.0107

(a) Do the data provide sufficient evidence to indicate a difference in the mean amount of effluents discharged by the four plants?

(b) If the maximum mean discharge of effluents is 1.5 lb/gal, do the data provide sufficient evidence to indicate that the limit is exceeded at plant *A*?

(c) Estimate the difference in the mean discharge of effluents between plants *A* and *D*, using a 95 percent confidence interval.

13.46 The yields of wheat in bushels per acre were compared for five different varieties, *A*, *B*, *C*, *D*, and *E*, at six different locations. Each variety was randomly assigned to a plot at each location. The results of the experiment are shown in the table. A SAS computer printout for the analysis of variance for the data is also shown below. Use the information in the printout to answer the following questions:

	Location					
Varieties	1	2	3	4	5	6
A	35.3	31.0	32.7	36.8	37.2	33.1
B	30.7	32.2	31.4	31.7	35.0	32.7
C	38.2	33.4	33.6	37.1	37.3	38.2
D	34.9	36.1	35.2	38.3	40.2	36.0
E	32.4	28.9	29.2	30.7	33.9	32.1

STATISTICAL ANALYSIS SYSTEM

ANALYSIS OF VARIANCE PROCEDURE

DEPENDENT VARIABLE: Y

SOURCE	DF	SUM OF SQUARES	MEAN SQUARE	F VALUE	PR > F	R-SQUARE	C.V.
MODEL	9	210.81166667	23.42351852	12.22	0.0001	0.846152	4.0499
ERROR	20	38.33000000	1.91650000		STD DEV		Y MEAN
CORRECTED TOTAL	29	249.14166667			1.38437712		34.18333333

SOURCE	DF	ANOVA SS	F VALUE	PR > F
BLOCKS	5	68.14166667	7.11	0.0006
TRTMENTS	4	142.67000000	18.61	0.0001

(a) Do the data provide sufficient evidence to indicate a difference in the mean yield for the five varieties of wheat?
(b) Do the data provide sufficient evidence to indicate that blocking on "location" increased the amount of information in the experiment?
(c) Find a 95 percent confidence interval for the difference in mean yields for varieties C and E.

EXPERIENCES WITH REAL DATA

It is interesting to note the difference in the values assigned to houses, automobiles, etc., by experienced appraisers. Some appraisers may tend to overvalue properties, others may undervalue them, and still others may vary up and down depending on the nature of the properties. Since you usually must pay for property evaluations, let us illustrate by comparing appraisers of automobiles. (*Note:* You could compare appraisers of antiques or other valuables using antique dealers, pawn dealers, etc.)

Most large used-car dealers have one person responsible for appraising the value of used cars. Select three of your local dealers for the comparison of appraisers and acquire five automobiles from persons in your class. Have each automobile appraised for purchase by each of the three dealers.

1. What type of experimental design has been suggested?
2. Use an analysis of variance to compare the mean appraisal price for the three auto dealers. Do the data provide sufficient evidence to indicate differences in mean appraisal values among the three dealers?
3. Estimate the difference in mean appraisal value for dealers 1 and 2.

The calculations for this experiment can be accomplished on an electronic desk calculator but you might wish to use packaged programs and an electronic computer to perform the calculations. Useful packaged programs are available in the Biomed (Biomedical Programs), the SAS (Statistical Analysis Systems), and the SPSS (Statistical Package for the Social Sciences) program libraries (see the references).

CHAPTER OBJECTIVES

GENERAL OBJECTIVE Chapters 8 and 9 presented statistical techniques for comparing two populations by comparing their respective population parameters. The techniques are applicable to data measured on a continuum and to data possessing normal population relative frequency distributions. The purpose of this chapter is to present a set of statistical test procedures to compare populations for the many types of data that do not satisfy these assumptions.

SPECIFIC OBJECTIVES

1. To explain the difference between parametric and nonparametric tests. *Section 14.1*
2. To explain why nonparametric tests are useful and, sometimes, essential. *Section 14.1*
3. To present a quick and easy method for comparing two populations, the sign test. *Section 14.2*
4. To show how statistical tests can be compared and to identify the advantages associated with nonparametric tests. *Sections 14.3, 14.9*
5. To present a nonparametric test for comparing two populations based on independent random samples, the Mann–Whitney U test. *Sections 14.4, 14.5*
6. To present a nonparametric test for comparing two populations for a paired-difference experiment, the Wilcoxon signed rank test. *Section 14.6*
7. To present a nonparametric test to detect nonrandomness in a sequence, the runs test. *Section 14.7*
8. To present a nonparametric test for correlation between two variables, Spearman's r_s *Section 14.8*

14 NONPARAMETRIC STATISTICS

14.1

INTRODUCTION

Some experiments yield response measurements that defy quantification. That is, they generate response measurements that can be ordered (ranked), but the location of the response on a scale of measurement is arbitrary. Although experiments of this type occur in almost all fields of study, they are particularly evident in social-science research and in studies of consumer preference. For example, suppose that a judge is employed to evaluate and rank the instructional abilities of four teachers or the edibility and taste characteristics of five brands of cornflakes. Since it is clearly impossible to give an exact measure of teacher competence or "tastability" of food, the response measurements are of a completely different character from those presented in preceding chapters. Nonparametric statistical methods are useful for analyzing this type of data.

Nonparametric statistical procedures apply not only to observations that are difficult to quantify. They are particularly useful in making inferences in situations where serious doubt exists about the assumptions that underlie standard methodology. For example, the *t* test for comparing a pair of means, Section 9.4, is based on the assumption that both populations are normally distributed with equal variances. The experimenter will never know whether these assumptions hold in a practical situation but will often be reasonably certain that departures from the assumptions will be small enough so that the properties of the statistical procedure will be undisturbed. That is, α and β will be approximately what the experimenter thinks they are. On the other hand, it is not uncommon for the experimenter seriously to question the assumptions and wonder whether he or she is using a valid statistical procedure. This difficulty may be circumvented by using a nonparametric statistical test and thereby avoiding reliance on a very uncertain set of assumptions.

Research has shown that nonparametric statistical tests are almost as capable of detecting differences among populations as the parametric methods of preceding chapters when normality and other assumptions are satisfied. They may be, and often are, more powerful in detecting population differences when the assumptions are not satisfied. For this reason, many statisticians advocate the use of nonparametric statistical procedures in preference to their parametric counterparts.

14.2

THE SIGN TEST FOR COMPARING TWO POPULATIONS

Without emphasizing the point, we employed a nonparametric statistical test as an alternative procedure for determining whether evidence existed to indicate a difference in the mean wear for the two types of tires in the paired-difference experiment, Section 9.5. Each pair of responses was compared and y (the number of times A exceeded B) was used as the test statistic. This nonparametric test is known as the sign test because y is the number of positive (or negative) signs associated with the differences. The implied null hypothesis is that the two population distributions are identical and the resulting technique is completely independent of the form of the distribution of differences. Thus, regardless of the distribution of differences, the probability that A exceeds B for a given pair will be $p = .5$ when the null hypothesis is true (that is, when the distributions for A and B are identical). Then y will possess a binomial probability distribution and a rejection region for y can be obtained using the binomial probability distribution of Chapter 6.

The following example will help you to recall how the sign test was constructed and how it is used in a practical situation.

EXAMPLE 14.1

The number of defective electrical fuses proceeding from each of two production lines, A and B, was recorded daily for a period of 10 days, with the following results:

	Production Lines	
Day	A	B
1	170	201
2	164	179
3	140	159
4	184	195
5	174	177
6	142	170
7	191	183
8	169	179
9	161	170
10	200	212

Assume that both production lines produced the same daily output. Compare the number, y_A, of defectives produced each day by production line A with the number, y_B, produced by production line B and let y equal the number of days when y_A exceeded y_B. Do the data present sufficient evidence to indicate that one production line tends to produce more defectives than the other or, equivalently, that $P(y_A > y_B) \neq 1/2$? State the null hypothesis to be tested and use y as a test statistic.

Solution Note that this is a paired-difference experiment. Let y be the number of times that the observation for production line A exceeds that for B in a given day. Under the null hypothesis that the two distributions of defectives are identical, the probability, p, that A exceeds B for a given pair is $p = .5$. Or, equivalently, we may wish to test an hypothesis that the binomial parameter, p, equals .5.

Very large or very small values of y are most contradictory to the null hypothesis. Therefore, the rejection region for the test will be located by

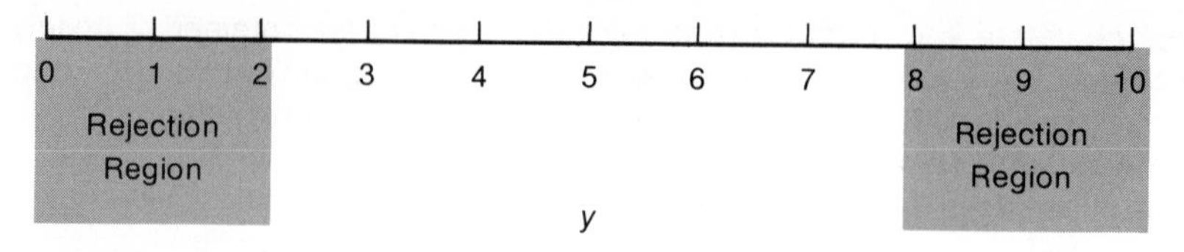

FIGURE 14.1

Rejection region for Example 14.1

including the most extreme values of y that at the same time provide an α that is feasible for the test.

Suppose that we would like α to be somewhere on the order of .05 or .10. We would commence the selection of the rejection region by including $y = 0$ and $y = 10$ and calculate the α associated with this region using $p(y)$ (the probability distribution for the binomial random variable, Chapter 6). Thus, with $n = 10$, $p = .5$,

$$\alpha = p(0) + p(10) = C_0^{10}(.5)^{10} + C_{10}^{10}(.5)^{10} = .002.$$

Since this value of α is too small, the region will be expanded by including the next pair of y values most contradictory to the null hypothesis, $y = 1$ and $y = 9$. The value of α for this region ($y = 0, 1, 9, 10$) may be obtained from Table 1, Appendix II:

$$\alpha = p(0) + p(1) + p(9) + p(10) = .022.$$

This also is too small, so we shall again expand the region to include $y = 0, 1, 2, 8, 9, 10$. You may verify that the corresponding value of α is .11. We will suppose that this value of α is acceptable to the experimenter and will employ $y = 0, 1, 2, 8, 9, 10$ as the rejection region for the test (see Figure 14.1).

From the data, we observe that $y = 2$ and therefore we reject the null hypothesis. Thus we conclude that sufficient evidence exists to indicate that the population distributions of number of defects are not identical. A quick glance at the data suggests that the mean number of defects for production line B tends to exceed that for A. The probability of rejecting the null hypothesis when true is only $\alpha = .11$, and therefore we are reasonably confident of our conclusion.

The experimenter in this example is using the test procedure as a rough tool for detecting faulty production lines. The rather large value of α is not likely to be disturbing because additional data can easily be collected if the experimenter is concerned about making a type I error in reaching a conclusion.

One problem that may occur when conducting a sign test is that the observations associated with one or more pairs may be equal and therefore result in ties. When this situation occurs, delete the tied pairs and reduce n, the total number of pairs.

You will also encounter situations where n, the number of pairs, is large.

Then the values of α associated with the sign test can be obtained by using the normal approximation to the binomial probability distribution discussed in Section 7.4. You can verify (by comparison of exact probabilities with their approximations) that these approximations will be quite adequate for n as small as 10. This is due to the symmetry of the binomial probability distribution for $p = .5$. For $n \geq 25$, the z test of Chapters 7 and 8 will be quite adequate, where

$$z = \frac{y - np}{\sqrt{npq}} = \frac{y - .5n}{.5\sqrt{n}}.$$

This would be testing the null hypothesis, $p = .5$, against the alternative, $p \neq .5$, for a two-tailed test or against the alternative, $p > .5$ (or $p < .5$), for a one-tailed test. The tests would utilize the familiar rejection regions of Chapter 8.

You will note that the data, Example 14.1, are the result of a paired-difference experiment. Suppose that the paired differences are normally distributed with a common variance, σ^2. Will the sign test detect a shift in location of the two populations as effectively as the Student's t test? Intuitively, we would suspect that the answer is "no," and this is correct, because the Student's t test utilizes comparatively more information. In addition to giving the sign of the difference, the t test uses the magnitudes of the observations to obtain a more accurate value of the sample means and variances. Thus one might say that the sign test is not as "efficient" as the Student's t test, but this statement is meaningful only if the populations conform to the assumption stated above, that is, the differences in paired observations are normally distributed with a common variance, σ_d^2. The sign test might be more efficient when these assumptions are not satisfied.

EXERCISES

14.1 What significance levels between $\alpha = .01$ and $\alpha = .15$ are available for a two-tailed sign test utilizing 25 paired observations? (Make use of tabulated values in Table 1, $n = 25$.) What are the corresponding rejection regions?

14.2 In Exercise 9.32, we compared property evaluations of two tax assessors, A and B. Their assessments for eight properties are shown in the table.

	Assessor	
Property	A	B
1	36.3	35.1
2	48.4	46.8
3	40.2	37.3
4	54.7	50.6
5	28.7	29.1
6	42.8	41.0
7	36.1	35.3
8	39.0	39.1

(a) Use the sign test to determine whether the data present sufficient evidence to indicate that one of the assessors tends to be consistently more conservative than the other, i.e., $P(y_A > y_B) \neq 1/2$. Test by using a value of α near .05

(b) Exercise 9.32 uses the t statistic to test the null hypothesis that there is no difference in the mean level of property assessments between assessors A and B. Check the answer (in the answer section) for Exercise 9.32 and compare it with your answer to part (a). Do the test results agree? Explain why the answers are (or are not) consistent.

14.3 Two gourmets rated 20 meals on a scale of 1 to 10. The data are shown in the table. Do the data provide sufficient evidence to indicate that one of the gourmets tends to give higher ratings than the other? Test by using the sign with a value of α near .05.

Meal	A	B	Meal	A	B
1	6	8	11	6	9
2	4	5	12	8	5
3	7	4	13	4	2
4	8	7	14	3	3
5	2	3	15	6	8
6	7	4	16	9	10
7	9	9	17	9	8
8	7	8	18	4	6
9	2	5	19	4	3
10	4	3	20	5	5

14.4 Clinical data concerning the effectiveness of two drugs in treating a particular disease were collected from 10 hospitals. The numbers of patients treated with the drugs varied from one hospital to another as well as between drugs within a given hospital. The data, in percent recovery, are shown below:

	Drug A			Drug B		
Hospital	Number in Group	Number Recovered	Percent Recovered	Number in Group	Number Recovered	Percent Recovered
1	84	63	75.0	96	82	85.4
2	63	44	69.8	83	69	83.1
3	56	48	85.7	91	73	80.2
4	77	57	74.0	47	35	74.5
5	29	20	69.0	60	42	70.0
6	48	40	83.3	27	22	81.5
7	61	42	68.9	69	52	75.4
8	45	35	77.8	72	57	79.2
9	79	57	72.2	89	76	85.4
10	62	48	77.4	46	37	80.4

Do the data present sufficient evidence to indicate a higher recovery rate for one of the two drugs?

(a) Test using the sign test. Choose your rejection region so that α is near .10.

(b) Why would it be inappropriate to use a Student's t test in analyzing the data?

14.5 An experiment was conducted to compare the tenderness of meat cuts subjected to two different meat tenderizers, A and B. To reduce the effect of extraneous variables, the data were paired by the specific meat cut, by applying the tenderizers to two cuts taken from the same steer, by cooking paired cuts together, and by using a single judge for each pair. After cooking, each cut was rated by a judge on a scale of 1 to 10, with 10 corresponding to the most tender condition of the cooked meat. The data, shown for a single judge, are given below. Do the data provide sufficient evidence to indicate that one of the two tenderizers tends to receive higher ratings than the other? Use a value of α near .05. [*Note:* $(1/2)^8 = .003906$.] Would a Student's t test be appropriate for analyzing these data? Explain.

	Tenderizer	
Cut	*A*	*B*
Shoulder roast	5	7
Chuck roast	6	5
Rib steak	8	9
Brisket	4	5
Club steak	9	9
Round steak	3	5
Rump roast	7	6
Sirloin steak	8	8
Sirloin tip steak	8	9
T-bone steak	9	10

14.3 A COMPARISON OF STATISTICAL TESTS

We have introduced two statistical test procedures utilizing given data for the same hypothesis, and we are now faced with the problem of comparison of tests. Which, if either, is better? One method would be to hold the sample size and α constant for both procedures and compare β, the probability of a type II error. Actually, statisticians prefer a comparison of the power of a test where

$$\text{power} = 1 - \beta.$$

Since β is the probability of failing to reject the null hypothesis when it is false, the power of the test is the probability of rejecting the null hypothesis when it is false and some specified alternative is true. It is the probability that the test will do what it was designed to do, that is, detect a departure from the null hypothesis when a departure exists.

Probably the most common method of comparing two test procedures is in terms of the relative efficiency of a pair of tests. Relative efficiency is the ratio of the sample sizes for the two test procedures required to achieve the same α and β for a given alternative to the null hypothesis.

In some situations, you may not be too concerned whether you are using the most powerful test. For example, you might choose to use the sign test over a more powerful competitor because of its ease of application. Thus you might view tests as microscopes that are utilized to detect departures from an hypothesized theory. One need not know the exact power of a microscope to use it in a biological investigation, and the same applies to statistical tests. If the test procedure detects a departure from the null hypothesis, we are delighted. If not, we can reanalyze the data by using a more powerful test, or we can increase the power of the microscope (test) by increasing the sample size.

14.4 THE USE OF RANKS FOR COMPARING TWO POPULATION DISTRIBUTIONS: INDEPENDENT RANDOM SAMPLES

A statistical test for comparing two populations based on independent random samples was proposed by F. Wilcoxon in 1945 (see the references). Suppose that you were to select independent random samples of n_1 and n_2 observations, each from two populations, call them I and II. Wilcoxon's idea was to combine the $n_1 + n_2 = n$ observations and to rank them, in order of magnitude, from 1 (the smallest) to n (the largest). Then if the observations were selected from identical populations, the rank sums for the samples should be more or less proportional to the sample sizes, n_1 and n_2. For example, if n_1 and n_2 were equal, you would expect the rank sums to be nearly equal. In contrast, if the observations in one population, say population I, tended to be larger than those in population II, then the observations in sample I would tend to receive the highest ranks and would have a larger than expected rank sum. Thus (sample sizes being equal) if one rank sum is very large (and, correspondingly, the other is very small), you may have evidence to indicate a difference between the two populations.

Mann and Whitney (see the references) proposed a statistical test in 1947 that also used the rank sums of the two samples, and their test can be shown to be equivalent to the Wilcoxon test. Because the Mann–Whitney U test and tables of critical values of U occur so often in the literature, we will explain its use in Section 14.5 and will give several examples of its application. In this section we illustrate the logic of the rank sum test and show you how the rejection region for the test and how α are determined.

EXAMPLE 14.2

The bacteria counts per unit volume are shown below for two types of cultures, *A* and *B*. Four observations were made for each culture:

Culture *A*	Culture *B*
27	32
31	29
26	35
25	28

Let n_1 and n_2 represent the number of observations in samples A and B, respectively. Then the rank sum procedure ranks each of the $(n_1 + n_2) = n$ measurements in order of magnitude and uses T, the sum of the ranks for A or B, as the test statistic. Thus, for the data given above, the corresponding ranks are

	Culture A	Culture B
	3	7
	6	5
	2	8
	1	4
Rank sum	12	24

Do these data present sufficient evidence to indicate a difference in the population distributions for A and B?

Solution Let T equal the rank sum for sample A (for this sample, $T = 12$). Certainly very small or very large values of T will provide evidence to indicate a difference between the two population distributions, and hence T, the rank sum, will be employed as a test statistic.

The rejection region for a given test will be obtained in the same manner as for the sign test. We will commence by selecting the most contradictory values of T as the rejection region and will add to these until α is of sufficient size.

The minimum rank sum would include the ranks 1, 2, 3, 4, or $T = 10$. Similarly, the maximum would include 5, 6, 7, 8, with $T = 26$. We will commence by including these two values of T in the rejection region. What is the corresponding value of α?

Finding the value of α is a probability problem that can be solved by using the methods of Chapter 4. If the populations are identical, every permutation of the eight ranks will represent a sample point and will be equally likely. Then α will represent the sum of the probabilities of the sample points (arrangements) which imply $T = 10$ or $T = 26$ in sample A. The total number of permutations of the eight ranks is 8!. The number of different arrangements of the ranks 1, 2, 3, 4 in sample A with the 5, 6, 7, 8 of sample B is (4!) (4!). Similarly, the number of arrangements that place the maximum value of T in sample A (ranks 5, 6, 7, 8) is (4!) (4!). Then the probability that $T = 10$ or $T = 26$ is

$$P(T = 10 \text{ or } 26) = p(10) + p(26) = \frac{2(4!)(4!)}{8!} = \frac{2}{C_4^8} = \frac{1}{35} = .029.$$

If this value of α is too small, the rejection region can be enlarged to include the next smallest and largest rank sums, $T = 11$ and $T = 25$. The rank sum $T = 11$ will include the ranks 1, 2, 3, 5, and

$$p(11) = \frac{4!4!}{8!} = \frac{1}{70}.$$

Similarly,

$$p(25) = \frac{1}{70}.$$

Then,

$$\alpha = p(10) + p(11) + p(25) + p(26) = \frac{2}{35} = .057.$$

Expansion of the rejection region to include 12 and 24 will substantially increase the value of α. The set of sample points giving a rank of 12 will be all sample points associated with rankings of (1, 2, 3, 6) and (1, 2, 4, 5). Thus

$$p(12) = \frac{2(4!)(4!)}{8!} = \frac{1}{35}$$

and

$$\alpha = p(10) + p(11) + p(12) + p(24) + p(25) + p(26)$$
$$= \frac{1}{70} + \frac{1}{70} + \frac{1}{35} + \frac{1}{35} + \frac{1}{70} + \frac{1}{70} = \frac{4}{35} = .114.$$

This value of α might be considered too large for practical purposes. In this case, we would be better satisfied with the rejection region $T = 10, 11, 25$, and 26.

The rank sum for the sample, $T = 12$, falls in the nonrejection region, and hence we do not have sufficient evidence to reject the hypothesis that the population distributions of bacteria counts for the two cultures are identical.

14.5

THE MANN–WHITNEY *U* TEST: INDEPENDENT RANDOM SAMPLES

The Mann–Whitney statistic, U, is obtained by ordering all $(n_1 + n_2)$ observations according to their magnitude and counting the number of observations in sample A that precede each observation in sample B. The U statistic is the sum of these counts.

For example, the eight observations of Example 14.2 are

25,	26,	27,	28,	29,	31,	32,	35.
A	*A*	*A*	*B*	*B*	*A*	*B*	*B*

The smallest B observation is 28, and $u_1 = 3$ observations from sample A precede it. Similarly, $u_2 = 3$ A observations precede the second B observation and $u_3 = 4$ and $u_4 = 4$ A observations precede the third and fourth B observations, respectively (32 and 35). Then,

$$U = u_1 + u_2 + u_3 + u_4 = 14.$$

Or you could count the number of B observations that precede each A observation and use the sum of the counts, U_B, as the U statistic. In either case, very large or small values of U will imply a separation of the ordered A and B observations and will provide evidence to indicate a difference (a shift in location) between the population distributions for A and B.

As noted in Section 14.4, it can be shown that the Mann–Whitney U statistic is related to Wilcoxon's rank sum. In fact, it can be shown (proof omitted) that

Formulas for the Mann–Whitney *U* Statistic

$$U_A = n_1 n_2 + \frac{n_1(n_1 + 1)}{2} - T_A,$$

$$U_B = n_1 n_2 + \frac{n_2(n_2 + 1)}{2} - T_B,$$

where $U_A + U_B = n_1 n_2$ and T_A and T_B are the rank sums for samples A and B, respectively.

As you can see from the formulas for U_A and U_B, U_A will be smaller when T_A is large, a situation that likely will occur when the population distribution of the A measurements is shifted to the right of the population distribution for the B measurements. Consequently, to conduct a one-tailed test to detect a shift in the A distribution to the right of the B distribution, you will reject the null hypothesis of "no difference in the population distributions" if U_A is less than some specified value, U_0. That is, you will reject H_0 for small values of U_A. Similarly, to conduct a one-tailed test to detect a shift of the B distribution to the right of the A distribution, you would reject H_0 if U_B is less than some specified value, say U_0. Consequently, the rejection region for the Mann–Whitney U test would appear as shown in Figure 14.2.

Table 8, Appendix II, gives the probability that an observed value of U will be less than some specified value, say U_0. This is the value of α for a one-tailed test. To conduct a two-tailed test, i.e., to detect a shift in the population distributions for the A and B measurements, in either direction, we will agree to always use U, the smaller of U_A or U_B, as the test statistic and reject H_0 for $U < U_0$ (see Figure 14.2). The value of α for the two-tailed test will be double the tabulated value given in Table 8, Appendix II.

To see how to locate the rejection region for the Mann–Whitney U test, suppose that $n_1 = 4$ and $n_2 = 5$. Then you would consult the third table in Table 8, Appendix II, the one corresponding to $n_2 = 5$. The first few lines of Table 8, $n_2 = 5$, are shown here in Table 14.1. Note that the table is constructed on the assumption that $n_1 \leq n_2$. That is, you will always identify the smaller sample as sample 1.

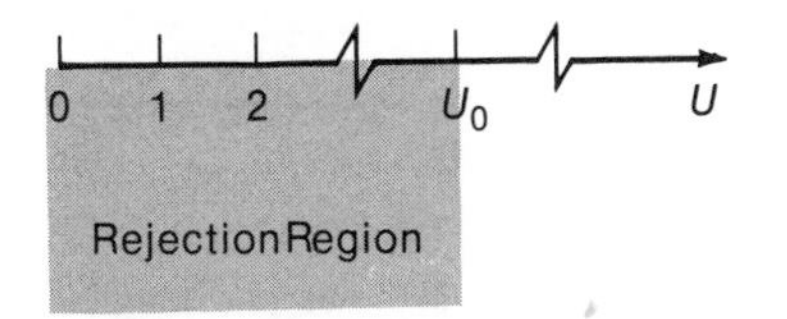

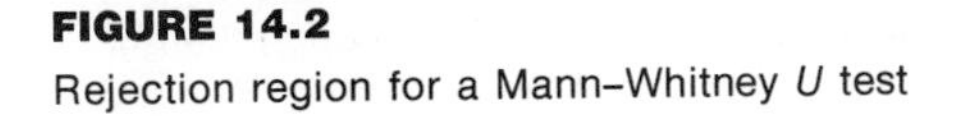

FIGURE 14.2

Rejection region for a Mann–Whitney U test

TABLE 14.1

An abbreviated version of Table 8, Appendix II; $P(U \leq U_0)$ for $n_2 = 5$

		$n_2 = 5$				
	n_1	1	2	3	4	5
	0	.1667	.0376	.0179	.0079	.0040
	1	.3333	.0952	.0357	.0159	.0079
	2	.5000	.1905	.0714	.0317	.0159
U_0	3		.2857	.1250	.0556	.0278
	4		.4286	.1964	.0952	.0476
	5		.5714	.2857	.1429	.0754
	⋮			⋮	⋮	⋮

Across the top of Table 14.1, you see values of n_1. Values of U_0 are shown down the left side of the table. The entries give the probability that U will assume a small value, namely, the probability that $U \leq U_0$. Since for our example, $n_1 = 4$, we will move across the top of the table to $n_1 = 4$. Move to the second row of the table corresponding to $U_0 = 2$ for $n_1 = 4$. Then you see the probability that U will be less than or equal to 2 is .0317. Similarly, moving across the row for $U_0 = 3$, you see that the probability that U is less than or equal to 3 is .0556. (This value is shaded in Table 14.1.) So, if you want to conduct a one-tailed Mann–Whitney U test with $n_1 = 4$ and $n_2 = 5$ and would like α to be near .05, you would reject the null hypothesis of equality of population relative frequency distributions when $U \leq 3$. The probability of a type I error for the test would be $\alpha = .0556$. If you use this same rejection region for a two-tailed test, i.e., $U \leq 3$, α would be double the tabulated value, or $\alpha = 2\,(.0556) = .1112$.

The Mann–Whitney *U* Test

Null Hypothesis, H_0: The population relative frequency distributions for A and B are identical.

Alternative Hypothesis, H_a: The two population relative frequency distributions are shifted in respect to their relative locations (a two-tailed test). Or H_a: The population relative frequency distribution for A is shifted to the right of the relative frequency distribution for population B (a one-tailed test).*

Test Statistic: For a two-tailed test, use U, the smaller of

$$U_A = n_1 n_2 + \frac{n_1(n_1 + 1)}{2} - T_A$$

and

$$U_B = n_1 n_2 + \frac{n_2(n_2 + 1)}{2} - T_B$$

* For the sake of convenience, we will describe the one-tailed test as one designed to detect a shift in the distribution of the A measurements to the right of the distribution of the B measurements. To detect a shift in the B distribution to the right of the A distribution, just interchange the letters A and B in the discussion.

where T_A and T_B are the rank sums for samples A and B, respectively. For a one-tailed test, use U_A.

Rejection Region:

1. For the two-tailed test and a given value of α, reject H_0 if $U \leq U_0$, where $P(U \leq U_0) = \alpha/2$. [*Note:* Observe that U_0 is the value such that $P(U \leq U_0)$ is equal to half of α.]
2. For a one-tailed test and a given value of α, reject H_0 if $U_A \leq U_0$, where $P(U_A \leq U_0) = \alpha$.

When applying the test to a set of data, you may find that some of the observations are of equal value. Ties in the observations can be handled by averaging the ranks that would have been assigned to the tied observations and assigning this average to each. Thus, if three observations are tied and are due to receive ranks 3, 4, 5, we would assign the rank of 4 to all three. The next observation in the sequence would receive the rank of 6, and ranks 3 and 5 would not appear. Similarly, if two observations are tied for ranks 3 and 4, each would receive a rank of 3.5, and ranks 3 and 4 would not appear.

EXAMPLE 14.3

Test the hypothesis that there is no difference in the population distributions for the bacteria-count data of Example 14.2.

Solution We have already noted that the Mann–Whitney U test and the Wilcoxon rank sum test are equivalent, so we should reach identically the same conclusions as for Example 14.2. Recall that the alternative hypothesis was that there was a difference in the distributions of bacteria counts for cultures A and B and that this implied a two-tailed test. Thus, since Table 8 gives values of $P(U \leq U_0)$ for specified sample sizes and values of U_0, we must double the tabulated value to find α. Suppose, as in Example 14.2, that we desire a value of α near .05. Checking Table 8 for $n_1 = n_2 = 4$, we find $P(U \leq 1) = .0286$. Using $U \leq 1$ as the rejection region (see Figure 14.3), α will equal $2(.0286) = .0572$ or, rounding to three decimal places, $\alpha = .057$ (the same value of α obtained for Example 14.2).

For the bacteria data, the rank sums are $T_A = 12$ and $T_B = 24$. Then

$$\begin{aligned} U_A &= n_1 n_2 + \frac{n_1(n_1 + 1)}{2} - T_A \\ &= (4)(4) + \frac{4(4 + 1)}{2} - 12 \\ &= 14, \end{aligned}$$

and

$$\begin{aligned} U_B &= n_1 n_2 + \frac{n_2(n_2 + 1)}{2} - T_B \\ &= (4)(4) + \frac{4(4 + 1)}{2} - 24 \\ &= 2. \end{aligned}$$

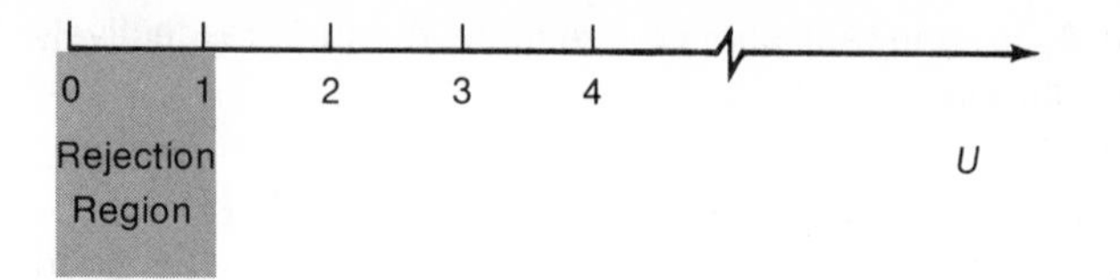

FIGURE 14.3
Rejection region for Example 14.3

Because the smaller observed value of U is 2 (calculated above), U does not fall in the rejection region. Hence there is not sufficient evidence to show a difference in the population distributions of bacteria counts for cultures A and B.

EXAMPLE 14.4

An experiment was conducted to compare the strength of two types of kraft papers, one a standard kraft paper of a specified weight and the other the same standard kraft paper treated with a chemical substance. Ten pieces of each type of paper, randomly selected from production, produced the strength measurements shown in the accompanying table. Test the hypothesis of "no difference in the distributions of strengths for the two types of paper" against the alternative hypothesis that the treated paper tends to be of greater strength.

	Standard A	Treated B
	1.21 (2)	1.49 (15)
	1.43 (12)	1.37 (7.5)
	1.35 (6)	1.67 (20)
	1.51 (17)	1.50 (16)
	1.39 (9)	1.31 (5)
	1.17 (1)	1.29 (3.5)
	1.48 (14)	1.52 (18)
	1.42 (11)	1.37 (7.5)
	1.28 (3.5)	1.44 (13)
	1.40 (10)	1.53 (19)
Rank sums	$T_A = 85.5$	$T_B = 124.5$

Solution The ranks are shown in parentheses alongside the $n_1 + n_2 = 10 + 10 = 20$ strength measurements and the rank sums, T_A and T_B, are shown below the columns. Since we wish to detect a shift in the distribution of the B measurements to the right of the distribution for the A measurements, we would reject the null hypothesis of "no difference in population strength distributions" when T_B is excessively large. Because this situation will occur when U_B is small, we shall conduct a one-tailed statistical test and reject the null hypothesis when $U_B \leq U_0$.

Suppose that we choose a value of α near .05. Then we can find U_0 by consulting the portion of Table 8, Appendix II, corresponding to $n_2 = 10$. The probability, $P(U \leq U_0)$, nearest .05 is .0526 and corresponds to $U_0 = 28$. Hence, we will reject if $U_B \leq 28$.

Calculating U_B, we have

$$\begin{aligned} U_B &= n_1 n_2 + \frac{n_2(n_2 + 1)}{2} - T_B \\ &= (10)(10) + \frac{(10)(11)}{2} - 124.5 \\ &= 30.5. \end{aligned}$$

As you can see, U_B is not less that $U_0 = 28$. Therefore, we cannot reject the null hypothesis. At the $\alpha = .05$ level of significance, there is not sufficient evidence to indicate that the treated kraft paper is stronger than the standard.

A simplified large-sample test ($n_1 \geq 10$ and $n_2 \geq 10$) can be obtained by using the familiar z statistic of Chapter 8. When the population distributions are identical, it can be shown that the U statistic has expected value and variance,

$$E(U) = \frac{n_1 n_2}{2}$$

and

$$V(U) = \frac{n_1 n_2 (n_1 + n_2 + 1)}{12},$$

and the distribution of

$$z = \frac{U - E(U)}{\sigma_U}$$

tends to normality with mean zero and variance equal to 1 as n_1 and n_2 become large. This approximation will be adequate when n_1 and n_2 are both greater than or equal to 10. Thus, for a two-tailed test with $\alpha = .05$, we would reject the null hypothesis if $|z| \geq 1.96$.

Observe that the z statistic will reach the same conclusion as the exact U test for Example 14.4. Thus

$$\begin{aligned} z &= \frac{30.5 - [(10)(10)/2]}{\sqrt{[(10)(10)(10 + 10 + 1)]/12}} = \frac{30.5 - 50}{\sqrt{2100/12}} = -\frac{19.5}{\sqrt{175}} \\ &= -\frac{19.5}{13.23} = -1.47. \end{aligned}$$

For a one-tailed test with $\alpha = .05$ located in the lower tail of the z distribution, we shall reject the null hypothesis if $z < -1.645$. You can see that $z = 1.47$ does not fall in the rejection region and that this test reaches the same conclusion as the exact U test of Example 14.4.

The Mann–Whitney *U* Test for Large Samples, $n_1 > 10$ and $n_2 > 10$

Null Hypothesis, H_0: The population relative frequency distributions for A and B are identical.

Alternative Hypothesis, H_a: The two population relative frequency distributions are not identical (a two-tailed test). Or H_a: The population relative frequency distribution for A is shifted to the right (or left) of the relative frequency distribution for population B (a one-tailed test).

Test Statistic:
$$z = \frac{U - (n_1 n_2/2)}{\sqrt{n_1 n_2 (n_1 + n_2 + 1)/12}}$$

Let $U = U_A$.

Rejection Region: Reject if $z \geq z_{\alpha/2}$ or $z < -z_{\alpha/2}$ for a two-tailed test. For a one-tailed test, place all of α in one tail of the z distribution. To detect a shift in the distribution of the A observations to the right of the distribution of the B observations, let $U = U_A$ and reject H_0 when $z < -z_\alpha$. To detect a shift in the opposite direction, let $U = U_A$ and reject H_0 when $z > z_\alpha$. Tabulated values of z are given in Table 3, Appendix II.

EXERCISES

14.6 Data collected on air pollution at 8 Ohio locations (presented earlier in Exercise 9.19) seem to suggest that air pollution, caused by pollution drifting from urban areas, is greater in rural areas than in urban areas. The data, reproduced from *Environmental Midwest* (December 1976), are shown in the table.

City	Designation	Maximum Concentration (ppm)	Days Exceeding Standard (%)	Number of Violations (Total)
Canton	urban	.14	44	148
Cincinnati	urban	.18	44	54
Cleveland	urban	.14	26	51
Columbus	urban	.15	27	113
Dayton	urban	.13	35	114
McConnelsville	rural	.16	56	239
Wilmington	rural	.18	58	259
Wooster	rural	.17	55	262

(a) In Exercise 9.19 we used a Student's t test to test the null hypothesis that there was no difference in the mean number of violations between urban and rural Ohio locations. Use the Mann–Whitney U test to test for a shift in distributions for the two locations. Are the conclusions of the t test and the Mann–Whitney U test in agreement? Explain why you think they should or should not be in agreement.

(b) Now use the Mann–Whitney U test to test a similar hypothesis concerning the maximum concentration of pollutants recorded over the time period.

14.7 We shall reanalyze the data of Exercise 9.22 by use of a nonparametric statistical test. The observations, which represent the dissolved oxygen content in water, measure the ability of a river, lake, or stream to support aquatic life. In this experiment, a pollution-control inspector suspected that a river community was releasing amounts of semitreated sewage into a river. To check his theory, he drew five randomly selected specimens of river water at a location above the town and another five below. The dissolved oxygen readings, in parts per million, are as follows:

Above Town	4.8	5.2	5.0	4.9	5.1
Below Town	5.0	4.7	4.9	4.8	4.9

(a) Use a one-tailed Mann–Whitney U test with $\alpha = .05$ to answer the question.
(b) Use a Student's t test (with $\alpha = .05$) to analyze the data (you may have done this in Exercise 9.22). Compare the conclusions reached in parts (a) and (b).

14.8 In an investigation of visual scanning behavior of deaf children, measurements of eye-movement rate were taken on nine deaf and nine hearing children. From the data given, does it appear that the distributions of eye-movement rates for deaf children (A) and hearing children (B) differ?

	Deaf Children (A)	Hearing Children (B)
	(15) 2.75	.89 (1)
	(11) 2.14	1.43 (7)
	(18) 3.23	1.06 (4)
	(10) 2.07	1.01 (3)
	(14) 2.49	.94 (2)
	(12) 2.18	1.79 (8)
	(17) 3.16	1.12 (5.5)
	(16) 2.93	2.01 (9)
	(13) 2.20	1.12 (5.5)
Rank sum	126	45

14.9 The life in months of service before failure of the color television picture tube in 8 television sets manufactured by firm A and 10 sets manufactured by firm B are as follows:

Firm	Life of Picture Tube (months)									
A	32	25	40	31	35	29	37	39		
B	41	39	36	47	45	34	48	44	43	33

Use the U test to analyze the data and test to see if the life in months of service before failure of the picture tube is the same for the picture tubes manufactured by each firm. (Use $\alpha = .10$.)

14.10 Is the Mann–Whitney U test appropriate for the data of Exercise 14.4? Explain.

14.11 A comparison of the weights of turtles caught in two different lakes was conducted to compare the effects of the two lake environments on turtle growth. All the turtles were of the same age and were tagged before being released in the lakes. The weight measurements for $n_1 = 10$ tagged turtles caught in lake 1 and $n_2 = 8$ caught in lake 2 are as follows:

Lake	Weight (oz)									
1	14.1	15.2	13.9	14.5	14.7	13.8	14.0	16.1	12.7	15.3
2	12.2	13.0	14.1	13.6	12.4	11.9	12.5	13.8		

Do the data provide sufficient evidence to indicate a difference in the distributions of weights for the tagged turtles exposed to the two lake environments? Use a Mann–Whitney U test with $\alpha = .05$ to answer the question.

14.12 Cancer treatment by means of chemicals, chemotherapy, utilizes chemicals which kill both cancer as well as normal cells. In some instances the toxicity of the cancer drug, i.e., its effect on normal cells, can be reduced by the simultaneous injection of a second drug. A study was conducted to determine whether a particular drug injection was beneficial in reducing the harmful effects of a chemotherapy treatment on the survival time for rats. Two randomly selected groups of rats, 12 rats in each group, were used for the experiment. Both groups, call them A and B, received the toxic drug in a dosage large enough to cause death, but in addition, group B received the antitoxin which was to reduce the toxic effect of the chemotherapy on normal cells. The test was terminated at the end of 20 days, or 480 hours. The lengths of survival time for the two groups of rats, to the nearest 4 hours, are shown in the table. Do the data provide sufficient evidence to indicate that rats receiving the antitoxin tend to survive longer after chemotherapy than those not receiving the toxin? Use the Mann–Whitney U test with a value of α near .05.

Only Chemotherapy A	Chemotherapy Plus Drug B
84	140
128	184
168	368
92	96
184	480
92	188
76	480
104	244
72	440
180	380
144	480
120	196

14.6

THE WILCOXON SIGNED RANK TEST FOR A PAIRED EXPERIMENT

A rank sum test proposed by Wilcoxon can be used to analyze the paired-difference experiment of Section 9.5 by considering the paired differences of the two treatments, A and B. Under the null hypothesis of "no differences in the distributions for A and B," you would expect (on the average) half of the differences in pairs to be negative and half to be positive. That is, the expected number of negative differences between pairs would be $n/2$ (where n is the number of pairs). Further, it would follow that positive and negative differences of equal absolute magnitude should occur with equal probability. If one were to order the differences according to their absolute values and rank them from smallest to largest, the expected rank sums for the negative and positive differences would be equal. Sizable differences in the sums of the ranks assigned to the positive and negative differences would provide evidence to indicate a shift in location between the distributions of responses for the two treatments, A and B.

To carry out the Wilcoxon test, calculate the differences $(y_A - y_B)$ for each of the n pairs. Differences equal to zero are eliminated and the number of pairs, n, is reduced accordingly. Rank the *absolute values* of the differences, assigning a 1 to the smallest, a 2 to the second smallest, and so on. Then calculate the rank sum for the negative differences and also calculate the rank sum for the positive differences. For a two-tailed test, we use the smaller of these two quantities, T, as a test statistic to test the null hypothesis that the two population relative frequency histograms are identical. The smaller the value of T, the greater will be the weight of evidence favoring rejection of the null hypothesis. Hence, we shall reject the null hypothesis if T is less than or equal to some value, say T_0.

To detect the one-sided alternative, that the distribution of the A observations is shifted to the right of the B observations, use the rank sum T^- of the negative differences and reject the null hypothesis for small values of T^-, say $T^- \leq T_0$. If we wish to detect a shift of the distribution of B observations to the right of the distribution of A observations, use the rank sum T^+ of the positive differences as a test statistic and reject for small values of T^+, say $T^+ \leq T_0$.

The probability that T is less than or equal to some value, T_0, has been calculated for a combination of sample sizes and values of T_0. These probabilities, given in Table 9, Appendix II, can be used to find the rejection region for the T test.

An abbreviated version of Table 9, Appendix II, is shown here in Table 14.2. Across the top of the table you see the number of differences (the number of pairs), n. Values of α (denoted by the symbol P in Wilcoxon's table) for a one-tailed test appear in the first column of the table. The second column gives values of α (P in Wilcoxon's notation) for a two-tailed test. Table entries are the critical values of T. You will recall that the critical value of a test statistic is the value that locates the upper (or lower) extreme of the rejection region.

For example, suppose you have $n = 7$ pairs and you are conducting a two-tailed test of the null hypothesis that the two population relative frequency distributions are identical. Checking the $n = 7$ column of Table 14.2 and using the second row (corresponding to $P = \alpha = .05$ for a two-tailed test), you see the entry, 2 (shaded in Table 14.2). This is T_0, the critical value of T. As noted earlier, the smaller the value of T, the greater will be the evidence to reject the null hypothesis. Therefore, you will reject the null hypothesis for all values of T less than or equal to 2. The rejection region for the Wilcoxon rank sum test for a paired experiment is always of the form: reject if $T \leq T_0$, where T_0 is the critical value of T. The rejection region is shown symbolically in Figure 14.4.

TABLE 14.2

An abbreviated version of Table 9, Appendix II; critical values of T

One-sided	Two-sided	$n = 5$	$n = 6$	$n = 7$	$n = 8$	$n = 9$	$n = 10$
$P = .05$	$P = .10$	1	2	4	6	8	11
$P = .025$	$P = .05$		1	2	4	6	8
$P = .01$	$P = .02$			0	2	3	5
$P = .005$	$P = .01$				0	2	3
One-sided	**Two-sided**	$n = 11$	$n = 12$	$n = 13$	$n = 14$	$n = 15$	$n = 16$
$P = .05$	$P = .10$	14	17	21	26	30	36
$P = .025$	$P = .05$	11	14	17	21	25	30
$P = .01$	$P = .02$	7	10	13	16	20	24
$P = .005$	$P = .01$	5	7	10	13	16	19
One-sided	**Two-sided**	$n = 17$					
$P = .05$	$P = .10$	41					
$P = .025$	$P = .05$	35					
$P = .01$	$P = .02$	28					
$P = .005$	$P = .01$	23					

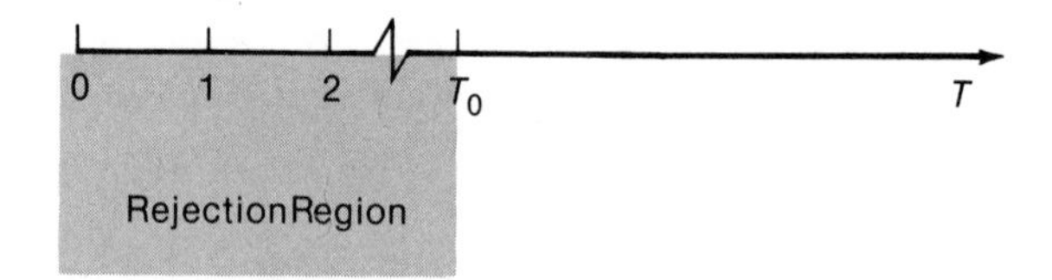

FIGURE 14.4

Rejection region for the Wilcoxon signed rank test for a paired experiment (reject if $T \leq T_0$)

Wilcoxon Signed Rank Test for a Paired Experiment

Null Hypothesis, H_0: The two population relative frequency distributions are identical.

Alternative Hypothesis, H_a: The two population relative frequency distributions differ in location (a two-tailed test). Or H_a: The population relative frequency distribution for A is shifted to the right* of the relative frequency distribution for population B (a one-tailed test).

Test Statistic:

1. For a two-tailed test, use T, the smaller of the rank sum for positive and the rank sum for negative differences.

*To detect a shift of the distribution of B observations to the right of the distribution of A observations, use the rank sum T^+ of the positive differences as the test statistic and reject H_0 if $T^+ \leq T_0$.

2. For a one-tailed test (to detect the alternative hypothesis described above), use the rank sum T^- of the negative differences.

Rejection Region:

1. For a two-tailed test, reject H_0 if $T \leq T_0$, where T_0 is the critical value given in Table 9, Appendix II.
2. For a one-tailed test (to detect the alternative hypothesis described above), use the rank sum T^- of the negative differences. Reject H_0 if $T^- \leq T_0$.

EXAMPLE 14.5

Test an hypothesis of no difference in population distributions of cake density for the paired-difference experiment, Exercise 9.70.

Solution The original data and differences in density for the six pairs of cakes are as follows:

	Density (oz/in.^{3})					
y_A	.135	.102	.098	.141	.131	.144
y_B	.129	.120	.112	.152	.135	.163
Difference $y_A - y_B$	.006	−.018	−.014	−.011	−.004	−.019
Rank	2	5	4	3	1	6

As with our other nonparametric tests, the null hypothesis to be tested is that the two population frequency distributions of cake densities are identical. The alternative hypothesis, which implies a two-tailed test, is that the distributions are different.

Because the amount of data is small, we will conduct our test using $\alpha = .10$. From Table 9, the critical value of T for a two-tailed test, $\alpha = .10$, is $T_0 = 2$. Hence we will reject H_0 if $T \leq 2$ (see Figure 14.5).

Now let us find the observed value of the test statistic, T. Checking the ranks of the differences, you can see that there is only one positive difference which has a rank of 2. Since we have agreed to always use the

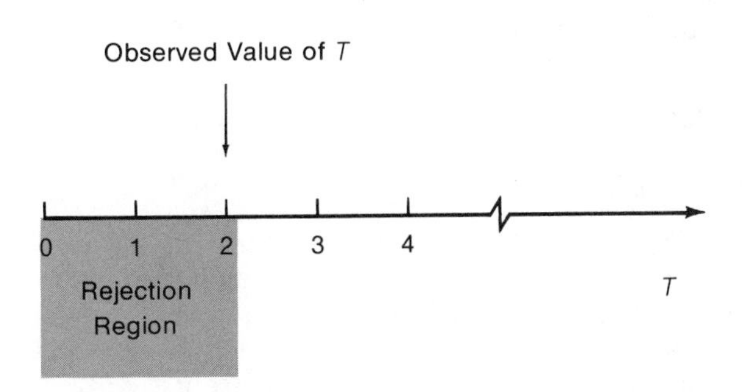

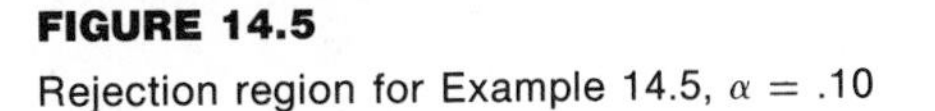

FIGURE 14.5
Rejection region for Example 14.5, $\alpha = .10$

smaller rank sum as the test statistic, the observed value of the test statistic is $T = 2$. Since $T = 2$ falls in the rejection region, we reject H_0 and conclude that the two population frequency distributions of cake densities differ.

Although Table 9, Appendix II, is applicable for values of n (number of data pairs) as large as $n = 50$, it is worth noting that T, like the Mann–Whitney U, will be approximately normally distributed when the null hypothesis is true and n is large (say 25 or more). This enables us to construct a large-sample z test, where

$$E(T) = \frac{n(n+1)}{4},$$

$$V(T) = \frac{n(n+1)(2n+1)}{24}.$$

Then the z statistic,

$$z = \frac{T - E(T)}{\sigma_T} = \frac{T - \dfrac{n(n+1)}{4}}{\sqrt{\dfrac{n(n+1)(2n+1)}{24}}},$$

can be used as a test statistic. Thus, for a two-tailed test and $\alpha = .05$, we would reject the hypothesis of "identical population distributions" when $|z| \geq 1.96$.

A Large-Sample Wilcoxon Signed Rank Test for a Paired Experiment: $n \geq 25$

Null Hypothesis, H_0: The population relative frequency distributions for A and B are identical.

Alternative Hypothesis, H_a: The two population relative frequency distributions differ in location (a two-tailed test). Or H_a: The population relative frequency distribution for A is shifted to the right (or left) of the relative frequency distribution for population B (a one-tailed test).

Test Statistic: $$z = \frac{T - [n(n+1)/4]}{\sqrt{[n(n+1)(2n+1)]/24}}$$

T can be either T^+ or T^-.

Rejection Region: Reject if $z \geq z_{\alpha/2}$ or $z < -z_{\alpha/2}$ for a two-tailed test. For a one-tailed test, place all of α in one tail of the z distribution. To detect a shift in the distribution of the A observations to the right of the distribution of the B observations, let $T = T^+$ and reject H_0 when $z > z_\alpha$. To detect a shift in the opposite direction, let $T = T^+$ and reject H_0 if $z < -z_\alpha$. Tabulated values of z are given in Table 3, Appendix II.

EXERCISES

14.13 In Exercise 14.2, we used the sign test to determine whether the data provided sufficient evidence to indicate a shift in the distributions of property assessments for assessors *A* and *B*.

(a) Use the Wilcoxon signed rank test for a paired experiment to test the null hypothesis that there is no difference in the distributions of property assessments between assessors *A* and *B*. Test by using a value of α near .05.

(b) Compare the conclusion of the test, part (a), with the conclusions derived from the *t* test, Exercise 9.32, and the sign test, Exercise 14.2. Explain why these test conclusions are (or are not) consistent.

14.14 The number of machine breakdowns per month was recorded for nine months on two identical machines used to make wire rope. The data are shown in the table.

	Machines	
Month	*A*	*B*
1	3	7
2	14	12
3	7	9
4	10	15
5	9	12
6	6	6
7	13	12
8	6	5
9	7	13

(a) Do the data provide sufficient evidence to indicate a difference in the monthly breakdown rates for the two machines? Test by using a value of α near .05.

(b) Can you think of a reason why the breakdown rates for the two machines might vary from month to month?

14.15 Refer to the comparison of gourmet mean ratings, Exercise 14.3, and use the Wilcoxon signed rank test to determine whether the data provide sufficient evidence to indicate a difference in the ratings of the two gourmets. Test by using a value of α near .05. Compare the results of this test with the results of the sign test, Exercise 14.3. Are the test conclusions consistent?

14.16 Two methods for controlling traffic, *A* and *B*, were employed at each of $n = 12$ intersections for a period of one week and the numbers of accidents occurring during this time period were recorded. The order of use (which method would be employed for the first week) was selected in a random manner.

Intersection	Method *A*	Method *B*	Intersection	Method *A*	Method *B*
1	5	4	7	2	3
2	6	4	8	4	1
3	8	9	9	7	9
4	3	2	10	5	2
5	6	3	11	6	5
6	1	0	12	1	1

Do the data provide sufficient evidence to indicate a shift in the distributions of accident rates for traffic control methods *A* and *B*?
(a) Analyze using a sign test.
(b) Analyze using the Wilcoxon rank sum test for a paired experiment.

14.17 Analyze the data of Exercise 14.5 using the Wilcoxon rank sum test for a paired experiment. Compare with your analysis, Exercise 14.5.

Ψ 14.18 Eight subjects were asked to perform a simple puzzle assembly task under normal conditions and under conditions of stress. During the stress condition, the subjects were told that a mild shock would be delivered 3 minutes after the start of the experiment and every 30 seconds thereafter until the task was completed. Blood-pressure readings were taken under both conditions. The following data represent the highest reading during the experiment:

Subject	Normal	Stress
1	126	130
2	117	118
3	115	125
4	118	120
5	118	121
6	128	125
7	125	130
8	120	120

Do the data present sufficient evidence to indicate higher blood-pressure readings during conditions of stress? Analyze the data using the Wilcoxon rank sum test for a paired experiment.

14.7

THE RUNS TEST: A TEST FOR RANDOMNESS

Consider a production process in which manufactured items emerge in sequence and each is classified as either defective (*D*) or nondefective (*N*). We have studied how one might compare the fraction defective over two equal time intervals using the normal deviate test, Chapter 8, and extended this to a test of an hypothesis of constant *p* over two or more time intervals using the chi-square test of Chapter 11. The purpose of these tests was to detect a change or trend in the fraction defective, *p*. Evidence to indicate increasing fraction defective might indicate the need for a process study to locate the source of difficulty. A decreasing value might suggest that a process quality-control program was having a beneficial effect in reducing the fraction defective.

Trends in fraction defective (or other quality measures) are not the only indication of lack of process control. A process may be causing periodic runs of defectives with the average fraction defective remaining constant, for all practical purposes, over long periods of time. For example, photoflash lamps are manufactured on a rotating machine with a fixed number of positions for bulbs. A bulb is placed on the machine at a given position, the air is removed, oxygen is pumped into the bulb, and the glass base is flame sealed. If a machine contains 20 positions, and several adjacent positions are faulty (too much heat in the sealing process), surges of defective lamps will emerge from the process in a periodic manner. Tests to compare the process fraction defective over equal intervals of time will not detect this periodic difficulty in the process. The periodicity, indicated by runs of defectives, is indicative of nonrandomness in the occurrence of defectives over time and can be detected by a test for randomness. The statistical test that we present, known as the runs test, is discussed in detail by A. Wald and J. Wolfowitz (see references). Other practical applications of the runs test will follow.

As the name implies, the runs test studies a sequence of events where each element in the sequence may assume one of two outcomes, say success (S) or failure (F). Thus the runs test is applied to a binary sequence of n_1 "successes" and n_2 "failures." If we think of the sequences of items emerging from a manufacturing process as defective (F) or nondefective (S), the observation of 20 items might yield

$$S\ S\ S\ S\ S\ F\ F\ S\ S\ S\ F\ F\ F\ S\ S\ S\ S\ S\ S\ S.$$

We notice the groupings of defectives and nondefectives and wonder whether this implies nonrandomness and, consequently, lack of process control.

DEFINITION

A run is defined to be a maximal subsequence of like elements.

For example, the first five successes is a subsequence of five like elements and it is maximal in the sense that it includes the maximum number of like elements before encountering an F. (The first four elements form a subsequence of like elements, but it is not maximal because the fifth element could also be included.) Consequently, the 20 elements shown above are arranged in five runs, the first containing five S's, the second containing two F's, and so on.

A very small or very large number of runs in a sequence would indicate nonrandomness. Therefore, let R (the number of runs in a sequence) be the test statistic, and let the rejection region be $R \leq k_1$ and $R \geq k_2$, as indicated in Figure 14.6. Let m denote the maximum possible number of runs. We must then find the probability distribution for R, $p(R)$, in order to calculate α and to locate a suitable rejection region for the test.

Suppose that the complete sequence contains n_1 S elements, and n_2 F elements, resulting in y_1 S runs and y_2 F runs, where $(y_1 + y_2) = R$. Then, for a given y_1, y_2 can equal y_1, $(y_1 - 1)$, or $(y_1 + 1)$, and $m = 2n_1$ if $n_1 = n_2$ and $m = 2n_1 + 1$ if $n_1 < n_2$. We will suppose that every distinguishable arrangement of the $(n_1 + n_2)$ elements in the sequence constitutes a simple event for the

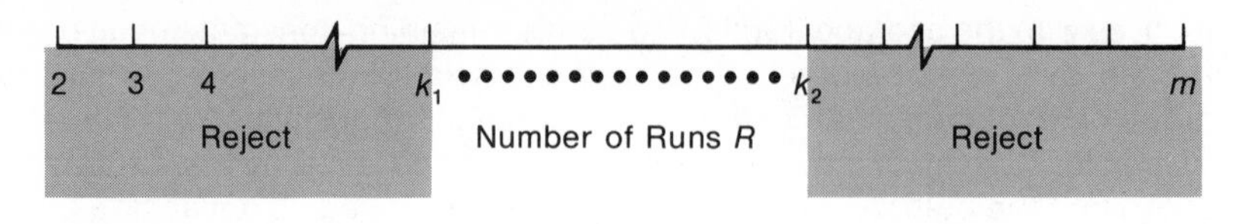

FIGURE 14.6
Rejection region for the runs test

experiment and that the sample points are equiprobable. It then remains for us to count the number of sample points that imply R runs.

The total number of distinguishable arrangements of n_1 S elements and n_2 F elements is $C_{n_1}^{n_1+n_2}$ and, therefore, the probability per sample point is $1/C_{n_1}^{n_1+n_2}$. The number of ways of achieving y_1 S runs is equal to the number of distinguishable arrangements of n_1 indistinguishable elements in y_1 cells, none of which is empty. The situation is represented in Figure 14.7. You will note that this is equal to the number of ways of distributing the $(y_1 - 1)$ identical inner bars in the $(n_1 - 1)$ spaces between the S elements (the outer two bars remain fixed). Consequently, it is equal to the number of ways of selecting $(y_1 - 1)$ spaces (for the bars) out of the $(n_1 - 1)$ spaces available. This will equal $C_{y_1-1}^{n_1-1}$.

The number of ways of observing y_1 S runs and y_2 F runs, obtained by applying the *mn* rule, is

$$(C_{y_1-1}^{n_1-1})(C_{y_2-1}^{n_2-1}).$$

This gives the number of sample points in the event "y_1 S runs and y_2 F runs." Then multiplying this number by the probability per sample point, we obtain the probability of exactly y_1 S runs and y_2 F runs to be

$$\begin{aligned} p(y_1, y_2) &= \frac{(C_{y_1-1}^{n_1-1})(C_{y_2-1}^{n_2-1})}{C_{n_1}^{n_1+n_2}} \qquad \text{for } y_1 = y_2 + 1 \text{ or } y_1 = y_2 - 1, \\ &= 2\,\frac{(C_{y_1-1}^{n_1-1})(C_{y_2-1}^{n_2-1})}{C_{n_1}^{n_1+n_2}} \qquad \text{for } y_1 = y_2. \end{aligned}$$

Then $p(R)$ equals the sum of $p(y_1, y_2)$ over all values of y_1 and y_2 such that $(y_1 + y_2) = R$.

To illustrate the use of the formula, the event $R = 4$ could occur when $y_1 = 2$ and $y_2 = 2$ with either the S or F elements commencing the sequences. For example, you could have $y_1 = 2$, $y_2 = 2$ with either of the sequences, $S\ F\ F\ S\ S\ F$ or $F\ S\ S\ F\ F\ S$. Consequently,

$$p(4) = 2P(y_1 = 2, y_2 = 2).$$

On the other hand, $R = 5$ could occur when $y_1 = 2$ and $y_2 = 3$, or $y_1 = 3$ and $y_2 = 2$, and these points are mutually exclusive. Then

$$p(5) = P(y_1 = 3, y_2 = 2) + P(y_1 = 2, y_2 = 3).$$

$$|S|SSSS|SS| \ldots\ldots |SS|SSS|S|$$

FIGURE 14.7
Distribution of n_1 S elements in y_1 cells (none empty)

EXAMPLE 14.6

Suppose that a sequence consists of $n_1 = 5$ S elements and $n_2 = 3$ F elements. Calculate the probability of observing $R = 3$ runs. Also, calculate $P(R \leq 3)$.

Solution Three runs could occur when $y_1 = 2$ and $y_2 = 1$ or $y_1 = 1$ and $y_2 = 2$. Then,

$$\begin{aligned} p(3) &= P(y_1 = 2, y_2 = 1) + P(y_1 = 1, y_2 = 2) \\ &= \frac{(C_1^4)(C_0^2)}{C_5^8} + \frac{(C_0^4)(C_1^2)}{C_5^8} \\ &= \frac{4}{56} + \frac{2}{56} = .107. \end{aligned}$$

Next, we require $P(R \leq 3) = p(2) + p(3)$. Accordingly, note that

$$\begin{aligned} p(2) &= 2P(y_1 = 1, y_2 = 1) \\ &= 2\frac{(C_0^4)(C_0^2)}{C_5^8} = \frac{2}{56} = .036. \end{aligned}$$

Thus, the probability of 3 or less runs is .143.

The values of $P(R \leq a)$ have already been calculated and are given in Table 10, Appendix II, for all combinations of n_1 and n_2, where n_1 and n_2 are less than or equal to 10. These can be used to locate the rejection regions for a one- or two-tailed test. An abbreviated version of Table 10, Appendix II, is shown here in Table 14.3.

For example, suppose that you have $n_1 = 4$ elements of one type and $n_2 = 5$ of another. Further, suppose that you wish to conduct a one-tailed test of the hypothesis of randomness and will reject if the number of runs is too small. Then the probability that you will observe $R = 3$ or less runs is shown in Table 14.3 as $\alpha = .071$ (shaded in the table). Thus, the sample sizes (n_1, n_2) are shown at the left of the table and the number a is recorded across the top of the table. The table entries give the probability that $R \leq a$.

On the other hand, suppose you wish to detect an overly large number of runs, rejecting the null hypothesis of randomness when R is large. Then for $n_1 = 4$ and $n_2 = 5$, the table gives $P(R \leq 7) = .929$. Therefore, it follows that $P(R \geq 8) = 1 - P(R \leq 7) = 1 - .929 = .071$. Then $\alpha = .071$ for this one-tailed test.

The two preceding paragraphs show how to find α if you conduct either a one-tailed test for undermixing or a one-tailed test for overmixing. A two-tailed test of nonrandomness to detect either undermixing or overmixing can be conducted by locating the rejection regions as shown in Figure 14.8 and determining the critical values in the same manner as used for the one-tailed tests.

To illustrate the selection of the alternative hypothesis for the runs test, consider a test of the null hypothesis that the sequence of defective and nondefective photoflash bulbs emerging from a production line is random. If specific heat-sealing positions on the rotating sealing machine were operating improperly, defectives would be produced in a spaced and systematic manner, a condition

TABLE 14.3

An abbreviated version of Table 10, Appendix II, for the runs test

	$P(R \leq a)$							
(n_1, n_2)	2	3	4	5	6	7	8	9
(2,3)	.200	.500	.900	1.000				
(2,4)	.133	.400	.800	1.000				
(2,5)	.095	.333	.714	1.000				
(2,6)	.071	.286	.643	1.000				
(2,7)	.056	.250	.583	1.000				
(2,8)	.044	.222	.533	1.000				
⋮	⋮	⋮	⋮	⋮				
(3,7)	.017	.083	.283	.583	.833	1.000		
(3,8)	.012	.067	.236	.533	.788	1.000		
(3,9)	.009	.055	.200	.491	.745	1.000		
(3,10)	.007	.045	.171	.455	.706	1.000		
(4,4)	.029	.114	.371	.629	.886	.971	1.000	
(4,5)	.016	.071	.262	.500	.786	.929	.992	1.000
(4,6)	.010	.048	.190	.405	.690	.881	.976	1.000
(4,7)	.006	.033	.142	.333	.606	.833	.954	1.000
(4,8)	.004	.024	.109	.279	.533	.788	.929	1.000
⋮	⋮	⋮	⋮	⋮	⋮	⋮	⋮	⋮

that could result in either a larger or a smaller than expected number of runs (overmixing or undermixing of defectives and nondefectives). To detect this situation, you would want to use a two-tailed test, rejecting the null hypothesis of randomness for either very large or very small values of R. Then you would place part of α in the upper tail of the R distribution and part in the lower tail and select a two-tailed rejection region, as shown in Figure 14.8.

A very small number of runs (undermixing) could occur if the heat supply (natural gas) to all sealing positions was inadvertently reduced. Then all sealing positions would produce defectives. The machine would produce alternatively surges of defectives and nondefectives (depending on whether the gas supply was deficient or adequate), which would yield a smaller number of runs than expected. To detect this situation, you would run a one-tailed test, rejecting the null hypothesis of randomness for small values of R.

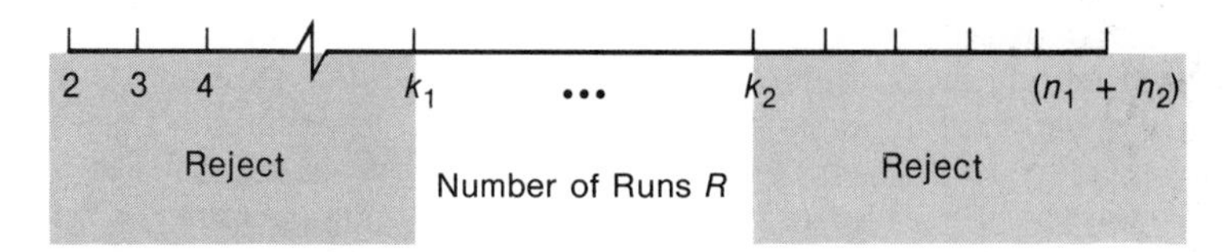

FIGURE 14.8

Rejection region for the runs test

The Runs Test

Null Hypothesis, H_0: The sequence of elements, call them S's and F's, have been produced in a random manner.

Alternative Hypothesis, H_a: The elements have been produced in a non-random sequence (a two-tailed test). Or H_a: The process is nonrandom due solely to overmixing (an upper one-tailed test) or due solely to under-mixing (a lower one-tailed test).

Test Statistic: R, the number of runs.

Rejection Region: For a two-tailed test, reject if $R \leq k_1$ or $R \geq k_2$ (see Figure 14.8), where $P(R \leq k_1) + P(R \geq k_2) = \alpha$ and k_1 and k_2 are obtained from Table 10, Appendix II. For a lower one-tailed test, reject if $R \leq k_1$, where $P(R \leq k_1) = \alpha$ and k_1 is obtained from Table 10. Similarly, for an upper one-tailed test, reject if $R \geq k_2$, where $P(R \geq k_2) = \alpha$ and k_2 is obtained from Table 10.

EXAMPLE 14.7

A true-false examination was constructed with the answers running in the following sequence:

$$T\ F\ F\ T\ F\ T\ F\ T\ T\ F\ T\ F\ F\ T\ F\ T\ F\ T\ T\ F.$$

Does this sequence indicate a departure from randomness in the arrangement of T and F answers?

Solution The sequence contains $n_1 = 10$ T and $n_2 = 10$ F answers with $y = 16$ runs. Nonrandomness can be indicated by either an unusually small or an unusually large number of runs, and, consequently, we will be concerned with a two-tailed test.

Suppose that we wish to use α approximately equal to .05 with .025 or less in each tail of the rejection region. Then from Table 10 with $n_1 = n_2 = 10$, we note that $P(R \leq 6) = .019$ and $P(R \leq 15) = .981$. Then $P(R \geq 16) = .019$, and we would reject the hypothesis of randomness if $R \leq 6$ or $R \geq 16$. Since $R = 16$ for the observed data, we conclude that evidence exists to indicate nonrandomness in the professor's arrangement of answers. His attempt to mix the answers was overdone.

A second application of the runs test is in detecting nonrandomness of a sequence of quantitative measurements over time. These sequences, known as time series, occur in many fields. For example, the measurement of a quality characteristic of an industrial product, blood pressure of a human, and the price of a stock on the stock market all vary over time. Departures in randomness in a series, caused either by trends or periodicities, can be detected by examining the deviations of the time-series measurements from their average. Negative and positive deviations could be denoted by S and F, respectively, and we could then test this time sequence of deviations for nonrandomness. We will illustrate with an example.

EXAMPLE 14.8

Paper is produced in a continuous process. Suppose that a brightness measurement, y, is made on the paper once every hour and that the results appear as shown in Figure 14.9.

The average for the 15 sample measurements, $\bar{y}$, appears as shown. Note the deviations about $\bar{y}$. Do these indicate a lack of randomness and thereby suggest periodicity in the process and lack of control?

Solution The sequence of negative (S) and positive (F) deviations as indicated in Figure 14.9 is

$$S\ S\ S\ S\ F\ F\ S\ F\ F\ S\ F\ S\ S\ S\ S.$$

Then $n_1 = 10$, $n_2 = 5$, and $R = 7$. Consulting Table 10, $P(R \leq 7) = .455$. This value of R is not improbable, assuming the hypothesis of randomness to be true. Consequently, there is not sufficient evidence to indicate nonrandomness in the sequence of brightness measurements.

The runs test can also be used to compare two population frequency distributions for a two-sample unpaired experiment. Thus it provides an alternative to the Mann–Whitney U test, Section 14.5. If the measurements for the two samples are arranged in order of magnitude, they will form a sequence. The measurements for samples 1 and 2 can be denoted as S and F, respectively, and we are once again concerned with a test for randomness. If all measurements for sample 1 are smaller than those for sample 2, the sequence will result in $S\ S\ S\ S \ldots S\ F\ F\ F \ldots F$, or $R = 2$ runs. Thus a small value of R will provide evidence of a difference in population frequency distributions, and the rejection region chosen would be $R \leq a$. This rejection region would imply a one-tailed statistical test. An illustration of the application of the runs test to compare two population frequency distributions will be left as an exercise for the reader.

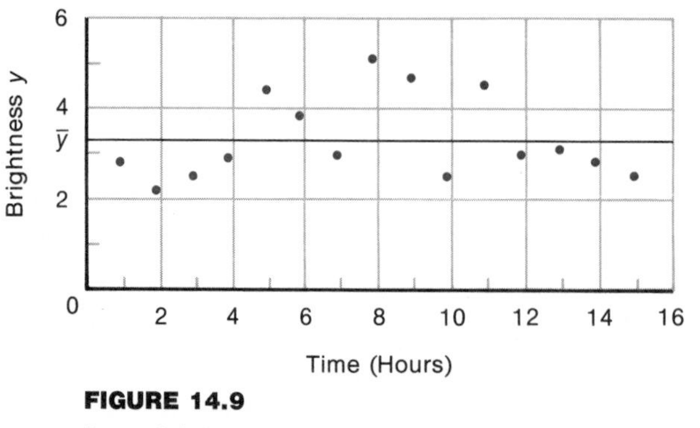

FIGURE 14.9

Paper brightness versus time

Note that most practical applications of the runs test result in one-tailed tests. Usually we wish to detect a departure from randomness that implies a smaller than expected number of runs.

As in the case of the other nonparametric test statistics studied in earlier sections of this chapter, the probability distribution for R tends to normality as n_1 and n_2 become large. The approximation is good when n_1 and n_2 are both greater than 10. Consequently, we may use the z statistic as a large-sample test statistic, where

$$z = \frac{R - E(R)}{\sqrt{V(R)}}$$

and

$$E(R) = \frac{2n_1n_2}{n_1 + n_2} + 1,$$

$$V(R) = \frac{2n_1n_2(2n_1n_2 - n_1 - n_2)}{(n_1 + n_2)^2(n_1 + n_2 - 1)}$$

are the expected value and variance of R, respectively.

Large-Sample Runs Test: $n_1 > 10$ and $n_2 > 10$

Null Hypothesis, H_0: The sequence of elements, call them S's and F's, have been produced in a random manner.

Alternative Hypothesis, H_a: The elements have been produced in a nonrandom sequence (a two-tailed test). Or H_a: The process is nonrandom due solely to overmixing (an upper one-tailed test) or due solely to undermixing (a lower one-tailed test).

Test Statistic:

$$z = \frac{R - \dfrac{2n_1n_2}{n_1 + n_2} + 1}{\sqrt{\dfrac{2n_1n_2(2n_1n_2 - n_1 - n_2)}{(n_1 + n_2)^2(n_1 + n_2 - 1)}}}.$$

Rejection Region: Reject if $z \geq z_{\alpha/2}$ or $z < -z_{\alpha/2}$ for a two-tailed test. For a one-tailed test, place all of α in one tail of the z distribution and reject if $z > z_\alpha$ (for an upper one-tailed test) or reject if $z < -z_\alpha$ (for a lower one-tailed test). Tabulated values of z are given in Table 3, Appendix II.

EXERCISES

14.19 A filler machine in a food-processing plant is set so that the mean fill is 16 ounces. When the process is in control, the actual load per can should vary about this mean in a random manner with a variance that remains stable over time. Suppose a process that has been operating for a period of time shows the following fill readings:

15.87, 16.14, 15.98, 16.03, 16.05, 16.01, 16.08, 15.94

(a) Do the data present sufficient evidence to indicate that the process mean differs from 16.0 ounces?

(b) Do the data suggest that the process is behaving in a nonrandom manner? (*Hint:* Use the runs test to answer this question.)

14.20 A union supervisor claims that applicants for jobs are selected without regard to race. An examination of the hiring records of the local, one that contains all male members, gave the following sequence of white and black hirings:

W, W, W, W, B, W, W, W, B, B, W, B, B

Do these data suggest a nonrandom racial selection in the hiring of the union's members?

14.21 Fifteen experimental batteries were selected at random from a lot at pilot plant *A*, and 15 standard batteries were selected at random from production at plant *B*. All 30 batteries were simultaneously placed under an electrical load of the same magnitude. The first battery to fail was an *A*, and second a *B*, the third a *B*, etc. The following sequence shows the order of failure for the 30 batteries:

A B B B A B A A B B B B A B A B B B B A A B A A A B A A A A

(a) Use the large-sample theory for the *U* test to determine (use $\alpha = .05$) if there is sufficient evidence to conclude that the mean life for the experimental batteries is greater than the mean life for the standard batteries.

(b) If, indeed, the experimental batteries have the greater mean life, what would be the effect on the expected number of runs? Use the large-sample theory for the runs test to test (using $\alpha = .05$) whether there is a difference in the distribution of battery life for the two populations.

14.8

RANK CORRELATION COEFFICIENT

In the preceding sections we have used ranks to indicate the relative magnitude of observations in nonparametric tests for comparison of treatments. We will now employ the same technique in testing for a relation between two ranked variables. Two common rank correlation coefficients are the Spearman r_s and the Kendall τ. We will present the Spearman r_s because its computation is identical to that for the sample correlation coefficient, r, of Chapter 10. Kendall's rank correlation coefficient is discussed in detail in Kendall (1961).

Suppose that eight elementary science teachers have been ranked by a judge according to their teaching ability, and all have taken a "national teachers' examination." The data are as follows:

Teacher	Judge's Rank	Examination Score
1	7	44
2	4	72
3	2	69
4	6	70
5	1	93
6	3	82
7	8	67
8	5	80

Do the data suggest an agreement between the judge's ranking and the examination score? Or one might express this question by asking whether a correlation exists between ranks and test scores.

The two variables of interest are rank and test score. The former is already in rank form, and the test scores may be ranked similarly, as shown below. The ranks for tied observations are obtained by averaging the ranks that the tied observations would occupy as for the Mann–Whitney U statistic.

Teacher	Judge's Rank (x_i)	Test Rank (y_i)
1	7	1
2	4	5
3	2	3
4	6	4
5	1	8
6	3	7
7	8	2
8	5	6

The Spearman rank correlation coefficient, r_s, is calculated using the ranks as the paired measurements on the two variables, x and y, in the formula for r, Chapter 10. Thus,

Spearman's Rank Correlation Coefficient

$$r_s = \frac{SS_{xy}}{\sqrt{SS_x SS_y}},$$

where x_i and y_i represent the ranks of the ith pair of observations and

$$SS_{xy} = \sum_{i=1}^{n} (x_i - \bar{x})(y_i - \bar{y}) = \sum_{i=1}^{n} x_i y_i - \frac{\left(\sum_{i=1}^{n} x_i\right)\left(\sum_{i=1}^{n} y_i\right)}{n},$$

$$SS_x = \sum_{i=1}^{n} (x_i - \bar{x})^2 = \sum_{i=1}^{n} x_i^2 - \frac{\left(\sum_{i=1}^{n} x_i\right)^2}{n},$$

$$SS_y = \sum_{i=1}^{n} (y_i - \bar{y})^2 = \sum_{i=1}^{n} y_i^2 - \frac{\left(\sum_{i=1}^{n} y_i\right)^2}{n}.$$

When there are no ties in either the x observations or the y observations, the above expression for r_s algebraically reduces to the simpler expression

$$r_s = 1 - \frac{6\sum_{i=1}^{n} d_i^2}{n(n^2 - 1)}, \qquad \text{where } d_i = x_i - y_i.$$

If the number of ties is small in comparison with the number of data pairs, little error will result in using this shortcut formula. We shall illustrate the use of the formula by an example.

EXAMPLE 14.9

Calculate r_s for the teacher–judge test-score data.

Solution The differences and squares of differences between the two rankings are as follows:

Teacher	x_i	y_i	d_i	d_i^2
1	7	1	6	36
2	4	5	−1	1
3	2	3	−1	1
4	6	4	2	4
5	1	8	−7	49
6	3	7	−4	16
7	8	2	6	36
8	5	6	−1	1
Total				144

Substituting into the formula for r_s,

$$r_s = 1 - \frac{6\sum_{i=1}^{n} d_i^2}{n(n^2 - 1)} = 1 - \frac{6(144)}{8(64 - 1)}$$

$$= -.714.$$

The Spearman rank correlation coefficient may be employed as a test statistic to test an hypothesis of "no association" between two populations. We assume that the n pairs of observations, (x_i, y_i), have been randomly selected and therefore "no association between the populations" would imply a random assignment of the n ranks within each sample. Each random assignment (for the two samples) would represent a sample point associated with the experiment and a value of r_s could be calculated for each. Thus it is possible to calculate the probability that r_s assumes a large absolute value due solely to chance and thereby suggests an association between populations when none exists.

The rejection region for a two-tailed test is shown in Figure 14.10. If the alternative hypothesis is that the correlation between x and y is negative, you would reject H_0 for negative values of r_s that are close to -1 (in the lower tail of Figure 14.10). Similarly, if the alternative hypothesis is that the correlation between x and y is positive, you would reject H_0 for large positive values of r_s (in the upper tail of Figure 14.10).

The critical values of r_s are given in Table 11, Appendix II. An abbreviated version of Table 11 is shown here in Table 14.4.

Across the top of Table 14.4 (and Table 11, Appendix II) are recorded values of α that you might wish to use for a one-tailed test of the null hypothesis of "no association" between x and y. The number of rank pairs, n, appears at the left side of the table. The table entries give the critical value r_0 for a one-tailed test. Thus, $P(r_s \geq r_0) = \alpha$. For example, suppose you have $n = 8$ rank pairs and

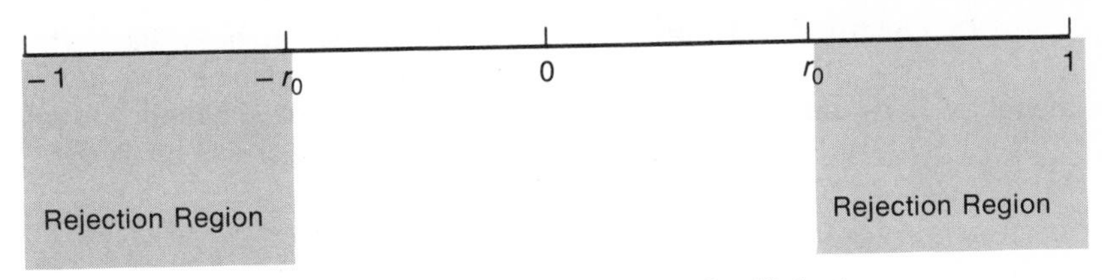

FIGURE 14.10

Rejection region for a two-tailed test of the null hypothesis of "no association," using Spearman's rank correlation test

the research hypothesis is that the correlation between the ranks is positive. Then you would want to reject the null hypothesis of "no association" only for large positive values of r_s and would use a one-tailed test. Referring to Table 14.4 and using the row corresponding to $n = 8$ and the column for $\alpha = .05$, you read $r_0 = 0.643$. Therefore, you would reject H_0 for all values of r_s greater than or equal to 0.643.

The test is conducted in exactly the same manner if you wish to test only the alternative hypothesis that the ranks are negatively correlated. The only difference is that you would reject the null hypothesis if $r_s \leq -0.643$. That is, you just place a minus sign in front of the tabulated value of r_0 to get the lower-tail critical value.

TABLE 14.4

An abbreviated version of Table 11, Appendix II, for Spearman's rank correlation test

n	$\alpha = 0.05$	$\alpha = 0.025$	$\alpha = 0.01$	$\alpha = 0.005$
5	0.900	—	—	—
6	0.829	0.886	0.943	—
7	0.714	0.786	0.893	—
8	0.643	0.738	0.833	0.881
9	0.600	0.683	0.783	0.833
10	0.564	0.648	0.745	0.794
11	0.523	0.623	0.736	0.818
12	0.497	0.591	0.703	0.780
13	0.475	0.566	0.673	0.745
14	0.457	0.545		
15	0.441	0.525		
16	0.425			
17	0.412			
18	0.399	⋮	⋮	⋮
19	0.388			
20	0.377			
⋮	⋮			

To conduct a two-tailed test, reject the null hypothesis if $r_s \geq r_0$ or $r_s \leq -r_0$. The value of α for the test will be double the value shown at the top of the table. For example, if $n = 8$ and you choose the 0.025 column, you will reject H_0 if $r_s \geq .738$ or $r_s \leq -.738$. The α value for the test will be $2(.025) = .05$.

Spearman's Rank Correlation Test

Null Hypothesis, H_0: There is no association between the rank pairs.

Alternative Hypothesis, H_a: There is an association between the rank pairs (a two-tailed test). Or H_a: The correlation between the rank pairs is positive (or negative) (a one-tailed test).

Test Statistic:

$$r_s = \frac{SS_{xy}}{\sqrt{SS_x SS_y}},$$

where x_i and y_i represent the ranks of the ith pair of observations.

Rejection Region: For a two-tailed test, reject H_0 if $r_s \geq r_0$ or $r_s \leq -r_0$, where r_0 is given in Table 11, Appendix II. Double the tabulated probability to obtain the value of α for the two-tailed test. For a one-tailed test, reject H_0 if $r_s \geq r_0$ (for an upper-tailed test) or $r_s \leq -r_0$ (for a lower-tailed test). The α value for a one-tailed test is the value shown in Table 11, Appendix II.

EXAMPLE 14.10

Test an hypothesis of "no association" between the populations for Example 14.9.

Solution The critical value of r_s for a one-tailed test with $\alpha = .05$ and $n = 8$ is .643. Let us assume that a correlation between judge's rank and the teachers' test scores could not possibly be positive. (Low rank means good teaching and should be associated with a high test score if the judge and test measure teaching ability.) The alternative hypothesis would be that the population rank correlation coefficient, ρ_s, is less than zero, and we would be concerned with a one-tailed statistical test. Thus α for the test would be the tabulated value .05, and we would reject the null hypothesis if $r_s \leq -.643$.

The calculated value of the test statistic, $r_s = -.714$, is less than the critical value for $\alpha = .05$. Hence the null hypothesis would be rejected at the $\alpha = .05$ level of significance. It appears that some agreement does exist between the judge's rankings and the test scores. However, it should be noted that this agreement could exist when *neither* provides an adequate yardstick for measuring teaching ability. For example, the association could exist if both the judge and those who constructed the teacher's examination possessed a completely erroneous, but identical, concept of the characteristics of good teaching.

EXERCISES

14.22 In Exercise 10.90, we provided data on the bending stiffness and flex of tennis rackets. In addition to engineering tests, *Tennis Industry Magazine* in Florida, a leading tennis business magazine, obtained a subjective evaluation of racket vibration, stiffness, and torque (twist) by 29 male and female tennis professionals. A summary of the subjective ratings of the professionals for each racket is shown in the table.* Calculate r_s for the following pairs of variables:

Racket	Vibration Center (Sweet Spot) Hits	Vibration Off-Center Hits	Flex	Torque
Wilson T-4000	55	71	46	71
Davis Classic	59	71	43	65
Yamaha YFG-30	55	61	59	65
P.D.P. Fiberstaff	46	55	74	54
Aldila Cannon	50	61	66	53
Dunlop-Fort	56	65	60	60
Garcia "Pro" Royal	59	68	51	66
Head Comp. II	54	66	63	59
Yonex T-7500	51	53	69	58
F. Willys-Devastator	76	75	73	73
Wilson Kramer	58	66	55	65
Garcia 240	60	65	70	63

(a) Flex and torque.
(b) Vibration from sweet-spot hits and vibration from off-center hits.
(c) Torque and vibration from off-center hits.
Test to determine whether the data provide sufficient evidence to indicate a correlation between each of the pairs of variables, (a), (b), and (c). Test using a value of α near .05.

14.23 A school principal suspected that a teacher's attitude toward a first grader depended on her original judgment of the child's ability. He also suspected that much of that judgment was based on the first grader's IQ score—which was usually known to the teacher. After 3 weeks of teaching, a teacher was asked to rank the nine children in her class from 1 (highest) to 9 (lowest) as to her opinion of their ability. Calculate r_s for the following teacher-IQ ranks:

Teacher	1	2	3	4	5	6	7	8	9
IQ	3	1	2	4	5	7	9	6	8

14.24 Refer to Exercise 14.23. Do the data provide sufficient evidence to indicate a positive correlation between the teacher's ranks and the ranks of the IQs? Use $\alpha = .05$.

14.25 Two art critics each ranked 12 paintings by contemporary (but anonymous) artists in accordance with their appeal to the respective critics. The ratings are shown in the

* *Source:* R. Gillen, ed., "Equipment Preview 1976: How to Pick the Right Racquet for Your Game," *Tennis USA,* February 1976, pp. 87–91, 99.

table. Do the critics seem to agree on their ratings of contemporary art? That is, do the data provide sufficient evidence to indicate a positive correlation between critics *A* and *B*? Test by using a value of α near .05.

Paintings	Critic *A*	Critic *B*
1	6	5
2	4	6
3	9	10
4	1	2
5	2	3
6	7	8
7	3	1
8	8	7
9	5	4
10	10	9

14.26 Refer to Exercise 14.3, which gives the ratings of two gourmets, *A* and *B*, for 20 different meals.

(a) Are the ratings of the two gourmets positively correlated? Test by using Spearman's r_s and a value of α near .05.

(b) Explain why it would or would not be reasonable to test for a correlation between the two gourmets' ratings using the *t* test of Chapter 10.

14.27 An experiment was conducted to study the relationship between the ratings of a tobacco leaf grader and the moisture content of the corresponding tobacco leaves. Twelve leaves were rated by the grader on a scale of 1 to 10 and corresponding readings of moisture content were made. The data are as follows:

Leaf	Grader's Rating	Moisture Content
1	9	.22
2	6	.16
3	7	.17
4	7	.14
5	5	.12
6	8	.19
7	2	.10
8	6	.12
9	1	.05
10	10	.20
11	9	.16
12	3	.09

Calculate r_s. Do the data provide sufficient evidence to indicate an association between the grader's ratings and the moisture content of the leaves?

14.9

SOME GENERAL COMMENTS ON NONPARAMETRIC STATISTICAL TESTS

The nonparametric statistical tests presented in the preceding pages represent only a few of the many nonparametric statistical methods of inference available. A much larger collection of nonparametric test procedures, along with worked examples, are given by Siegel (1956) and Conover (1971).

We have indicated that nonparametric statistical procedures are particularly useful when the experimental observations are susceptible to ordering but cannot be measured on a quantitative scale. Parametric statistical procedures usually cannot be applied to this type of data, hence all inferential procedures must be based on nonparametric methods.

A second application of nonparametric statistical methods is in testing hypotheses associated with populations of quantitative data when uncertainty exists concerning the satisfaction of assumptions about the form of the population distributions. Just how useful are nonparametric methods for this situation?

It is well known that many statistical test and estimation procedures are only slightly affected by moderate departures from the assumed form of the population frequency distribution and that parametric test procedures are more efficient than their nonparametric equivalents when the assumptions underlying parametric tests are true. However, this gain in efficiency is often very slight and, when the assumptions are not true, nonparametric methods may be more efficient than their parametric counterparts. For this reason, and because of their ease of application, nonparametric methods are often recommended.

REFERENCES

Conover, W. J., *Practical Nonparametric Statistics*. New York: John Wiley & Sons, Inc., 1971.

Kendall, M. G., and A. Stuart, *The Advanced Theory of Statistics*, Vol. 2. New York: Hafner Press, 1961.

Mann, H. B., and D. R. Whitney, "On a Test of Whether One of Two Random Variables Is Stochastically Larger Than the Other," *Ann. Math. Stat.*, *18* (1947), pp. 50–60.

Noether, G. E., *Elements of Nonparametric Statistics*. New York: John Wiley & Sons, Inc., 1967.

Siegel, S., *Nonparametric Statistics for the Behavioral Sciences*. New York: McGraw-Hill Book Company, 1956.

Wald, A., and J. Wolfowitz, "On a Test of Whether Two Samples Are from the Same Population," *Ann. Math. Stat.*, *2* (1940), pp. 147–162.

Wilcoxon, F., "Individual Comparisons by Ranking Methods," *Biometrics*, *1* (1945), pp. 80–83.

SUPPLEMENTARY EXERCISES

14.28 A psychological experiment was conducted to compare the lengths of response time (in seconds) for two different stimuli. In order to remove natural person-to-person variability in the responses, both stimuli were applied to each of nine subjects, thus permitting an analysis of the difference between stimuli *within* each person.

Subject	Stimulus 1	Stimulus 2
1	9.4	10.3
2	7.8	8.9
3	5.6	4.1
4	12.1	14.7
5	6.9	8.7
6	4.2	7.1
7	8.8	11.3
8	7.7	5.2
9	6.4	7.8

(a) Use the sign test to determine whether sufficient evidence exists to indicate a difference in mean response for the two stimuli. Use a rejection region for which $\alpha \leq .05$.

(b) Test the hypothesis of no difference in mean response using Student's t test.

14.29 Refer to Exercise 14.28. Test the hypothesis that no difference exists in the distributions of responses for the two stimuli, using the Wilcoxon signed rank test. Use a rejection region for which α is as near as possible to the α achieved in Exercise 14.28(a).

14.30 To compare two junior high schools, A and B, in academic effectiveness, an experiment was designed requiring the use of 10 sets of identical twins, each twin having just completed the sixth grade. In each case, the twins in the same set had obtained their schooling in the same classrooms at each grade level. One child was selected at random from each pair of twins and assigned to school A. The remaining children were sent to school B. Near the end of the ninth grade, a certain achievement test was given to each child in the experiment. The results are shown in the following table.

	Twin Pair									
School	1	2	3	4	5	6	7	8	9	10
A	67	80	65	70	86	50	63	81	86	60
B	39	75	69	55	74	52	56	72	89	47

(a) Test (using the sign test) the hypothesis that the two schools are the same in academic effectiveness, as measured by scores on the achievement test, against the alternative that the schools are not equally effective. Use a level of significance as near as possible to $\alpha = .05$.

(b) Suppose it were known that junior high school A had a superior faculty and better learning facilities. Test the hypothesis of equal academic effectiveness against the alternative that school A is superior. Use a level of significance as near as possible to $\alpha = .05$.

14.31 Refer to Exercise 14.30. What answers are obtained if Wilcoxon's signed rank test is used in analyzing the data? Compare with the answers to Exercise 14.30.

14.32 The coded values for a measure of brightness in paper (light reflectivity), prepared by two different processes, are given below for samples of size nine drawn randomly from each of the two processes:

Process	Brightness								
A	6.1	9.2	8.7	8.9	7.6	7.1	9.5	8.3	9.0
B	9.1	8.2	8.6	6.9	7.5	7.9	8.3	7.8	8.9

Do the data present sufficient evidence ($\alpha = .10$) to indicate a difference in the populations of brightness measurements for the two processes?
(a) Use the Mann–Whitney U test.
(b) Use Student's t test.

14.33 Refer to Exercise 14.32. What answer is obtained if the runs test is used in analyzing the data? Compare with the answers to Exercise 14.32.

14.34 Four Republican and four Democratic politicians attended a civic banquet and selected their seats in the following order (left to right): R, D, D, D, D, R, R, R. Does this seating sequence provide sufficient evidence to imply nonrandomness in the seating selection and a consequent grouping by party affiliation?

14.35 The conditions (D for diseased, S for sound) of the individual trees in a row of 10 poplars were found to be from left to right: $S, S, D, D, S, D, D, D, S, S$. Is there sufficient evidence to indicate nonrandomness in the sequence and therefore the possibility of contagion?

14.36 Items emerging from a continuous production process were classified as defective or nondefective. A sequence of items, observed over time, was as follows: $D, N, N, N, N, N, N, D, D, N, N, N, N, N, N, D, D, D, N, N, N, N, N, D, N, N, N, D, D, N, N, N, D, D$.
(a) Give the approximate probability that $R \leq 11$ where $n_1 = 11$ and $n_2 = 23$.
(b) Do these data suggest lack of randomness in the occurrence of defectives (D) and nondefectives (N)? Use the large-sample approximation for the runs test.

14.37 A quality-control chart has been maintained for a certain measurable characteristic of items taken from a conveyor belt at a certain point in a production line. The measurements obtained today in order of time are:

68.2, 71.6, 69.3, 71.6, 70.4, 65.0, 63.6, 64.7,
65.3, 64.2, 67.6, 68.6, 66.8, 68.9, 66.8, 70.1.

(a) Classify the measurements in this time series as above or below the sample mean and determine (use the runs test) whether consecutive observations suggest lack of stability in the production process.
(b) Divide the time period into two equal parts and compare the means, using Student's t test. Do the data provide evidence of a shift in the mean level of the quality characteristic?

14.38 If (as in the case of measurements produced by two well-calibrated measuring instruments) the means of two populations are equal, it is possible to use the Mann–Whitney U statistic for testing hypotheses concerning the population variances as follows:
(1) Rank the combined sample.

(2) Number the ranked observations "from the outside in"; that is, number the smallest observation 1; the largest, 2; the next-to-smallest, 3; the next-to-largest, 4; etc. This final sequence of numbers induces an ordering on the symbols A (population A items) and B (population B items). If $\sigma_A^2 > \sigma_B^2$, one would expect to find a preponderance of A's near the first of the sequences, and thus a relatively small "sum of ranks" for the A observations.

(a) Given the following measurements produced by well-calibrated precision instruments A and B, test at near the $\alpha = .05$ level to determine whether the more expensive instrument, B, is more precise than A. (Note that this would imply a one-tailed test.) Use the Mann–Whitney U test.

Instrument A	Instrument B
1,060.21	1,060.24
1,060.34	1,060.28
1,060.27	1,060.32
1,060.36	1,060.30
1,060.40	

(b) Test, using the F statistic of Section 9.7.

14.39 A large corporation selects college graduates for employment, using both interviews and a psychological-achievement test. Interviews conducted at the home office of the company were far more expensive than the tests which could be conducted on campus. Consequently, the personnel office was interested in determining whether the test scores were correlated with interview ratings and whether tests could be substituted for interviews. The idea was not to eliminate interviews but to reduce their number. To determine whether correlation was present, 10 prospects were ranked during interviews, and tested. The paired scores are as follows:

Subject	Interview Rank	Test Score
1	8	74
2	5	81
3	10	66
4	3	83
5	6	66
6	1	94
7	4	96
8	7	70
9	9	61
10	2	86

Calculate the Spearman rank correlation coefficient, r_s. Rank 1 is assigned to the candidate judged to be the best.

14.40 Refer to Exercise 14.39. Do the data present sufficient evidence to indicate that the correlation between interview rankings and test scores is less than zero? If this evidence does exist, can we say that tests could be used to reduce the number of interviews?

14.41 A political scientist wished to examine the relationship of the voter image of a conservative political candidate and the distance between the residences of the voter and the candidate. Each of 12 voters rated the candidate on a scale of 1 to 20. The data are as follows:

Voter	Rating	Distance
1	12	75
2	7	165
3	5	300
4	19	15
5	17	180
6	12	240
7	9	120
8	18	60
9	3	230
10	8	200
11	15	130
12	4	130

Calculate the Spearman rank correlation coefficient, r_s.

14.42 Refer to Exercise 14.41. Do these data provide sufficient evidence to indicate a negative correlation between rating and distance?

14.43 A comparison of reaction times for two different stimuli in a psychological word-association experiment produced the following results when applied to a random sample of 16 people:

Stimulus	Reaction Time (sec)							
1	1	3	2	1	2	1	3	2
2	4	2	3	3	1	2	3	3

Do the data present sufficient evidence to indicate a difference in mean reaction time for the two stimuli? Use the Mann–Whitney U statistic and test using $\alpha = .05$. (*Note:* This test was conducted using Student's t in the exercises for Chapter 9. Compare your results.)

14.44 The following table gives the scores of a group of 15 students in mathematics and art.

Student	Math	Art	Student	Math	Art
1	22	53	8	60	71
2	37	68	9	62	55
3	36	42	10	65	74
4	38	49	11	66	68
5	42	51	12	56	64
6	58	65	13	66	67
7	58	51	14	67	73
			15	62	65

Use Wilcoxon's signed rank test to determine if the median scores for these students differ significantly for the two subjects.

14.45 Refer to Exercise 14.44. Compute Spearman's rank correlation coefficient for these data and test H_0: $\rho = 0$ at the 10 percent level of significance.

14.46 Calculate the probability that $U \leq 2$ for $n_1 = n_2 = 5$. Assume that no ties will be present.

14.47 Calculate $P(R \leq 6)$ for the runs test where $n_1 = n_2 = 8$.

EXPERIENCES WITH REAL DATA

Students in humanities courses are frequently called upon to interpret the thoughts of an artist or the abstract writings of an author. Particularly, the student may be expected to detect the outstanding characteristics of the art as seen by the eye of the professor or by other art critics. Or, the situation may be reversed and the artistic creations (term papers, essays, paintings, etc.) of the students may be rated by the professor. This brings to mind a question, whether art is "in the eye of the beholder" or whether some aspect of a piece of art will appear favorably to all (or almost all) viewers. To test the theory that experts may have very different criteria for rating art objects, select 10 paintings, prints, essays, or books. Ask two "art critics" who might be considered expert in that art medium to rank the objects. In order that the rankings truly test our theory, the tests should be conducted independently of one another (the fact that ranking is being done by another person should be kept confidential). Collect the data and calculate r_s. Test the hypothesis of "no association" between the rankings of the critics.

15

A SUMMARY AND CONCLUSION

The preceding fourteen chapters construct a picture of statistics centered about the dominant feature of the subject, statistical inference. Inference, the objective of statistics, runs as a thread through the entire book, from the phrasing of the inference through the discussion of the probabilistic mechanism and the presentation of the reasoning involved in making the inference, to the formal elementary discussion of the theory of statistical inference presented in Chapter 8. What is statistics, what is its purpose, and how does it accomplish its objective? If we have answered these questions to your satisfaction, if each chapter and section seems to fulfill a purpose and to complete a portion of the picture, we have in some measure accomplished our instructional objective.

Chapter 1 presented statistics as a scientific tool utilized in making inferences, a prediction, or a decision concerning a population of measurements based upon information contained in a sample. Thus statistics is employed in the evolutionary process known as the scientific method—which in essence is the observation of nature—in order that we may form inferences or theories concerning the structure of nature and test the theories against repeated observations. Inherent in this objective is the sampling and experimentation that purchase a quantity of information which, hopefully, will be employed to provide the best inferences concerning the population from which the sample was drawn.

The method of phrasing the inference, that is, describing a set of measurements, was presented in Chapter 3 in terms of a frequency distribution and associated numerical descriptive measures. In particular we noted that the frequency distribution is subject to a probabilistic interpretation and that the numerical descriptive measures are more suitable for inferential purposes because we can more easily associate with them a measure of their goodness. Finally, a secondary but extremely important result of our study of numerical descriptive measures involved the notion of variation, its measurement in terms of a standard deviation, and its interpretation by using Tchebysheff's Theorem and the Empirical Rule. Thus, while concerned with describing a set of measurements—the population—we provided the basis for a description of the sampling distributions of estimators and the z test statistic to be considered as we progressed in our study.

The mechanism involved in making an inference—a decision—concerning the parameters of a population of die throws was introduced in Chapter 4. We hypothesized the population of die throws to be known, that is, that the die was perfectly balanced, and then drew a sample of $n = 10$ tosses from the population. Observing 10 ones, we concluded that the observed sample was highly improbable, assuming our hypothesis to be true, and we therefore rejected the hypothesis and concluded that the die was unbalanced. Thus, we noted that the theory of probability assumes the population known, and reasons from the population to the sample. Statistical inference, using probability, observes the sample and attempts to make inferences concerning the population. Fundamental to this procedure is a probabilistic model for the frequency distribution of the population, the acquisition of which was considered in Chapters 4 and 5.

The methodology of Chapters 4 and 5 was employed in Chapter 6 in the construction of a probability distribution, that is, a model for the frequency distribution, of a discrete random variable generated by the binomial experiment. The binomial experiment was chosen as an example because the acquisition of the probabilities, $p(y)$, is a task easily handled by the beginner and thus gives him an opportunity to utilize his probabilistic tools. In addition, it was chosen because of its utility, which was exemplified by the cold vaccine and lot acceptance sampling problems. Particularly, the inferential aspects of these problems were noted, with emphasis upon the reasoning involved.

Study of a useful continuous random variable in Chapter 7 centered about the Central Limit Theorem, its suggested support of the Empirical Rule, and its use in describing the probability distribution of sample means and sums. Indeed, we used the Central Limit Theorem to justify the use of the Empirical Rule and the normal probability distribution as an approximation to the binomial probability distribution when the number of trials, n, is large. Through examples we attempted to reinforce the probabilistic concept of statistical inference introduced in preceding chapters and induce you, as a matter of intuition, to employ statistical reasoning in making inferences.

Chapter 8 formally discussed statistical inference, estimation and tests of hypotheses, the methods of measuring the goodness of the inference, and presented a number of estimators and test statistics which, because of the Central Limit Theorem, possess approximately a normal probability distribution in repeated sampling. These notions were carried over to the discussion of the small-sample tests and estimators in Chapter 9.

Chapters 10 and 11 attempted, primarily, to broaden the view of the beginner, presenting two rather interesting and unique inferential problems. While stress was placed upon the relation between two variables, y and x, in Chapter 10, we were primarily interested in studying the use of the method of least squares in obtaining an equation for predicting y as a function of many variables. Thus, while the methodology appropriate to a single variable is useful, it was not the sole objective of our discussion. The analysis of enumerative data presents a methodology that is interesting and extremely useful in the social sciences as well as in many other areas.

Chapters 1 through 11 dealt with the philosophy and the methodology for making inferences and for measuring the quantity of information in a sample. Thus they form a necessary foundation for understanding how and why noise and volume affect the quantity of information in an experiment and set the stage for a discussion of experimental design in Chapter 12.

The analysis of variance, Chapter 13, extended the comparison of treatment means in the completely random and blocked designs (Chapter 9) to more than two treatments. Simultaneously, it introduced a completely new way to view the analysis of data through the analysis (or partitioning) of variance. Nonparametric statistical methods, Chapter 14, provided a discussion of some very useful statistical test procedures that are particularly appropriate for ordinal data.

Finally, we note that the methodology presented in this introduction to statistics represents a very small sample of the population of statistical methodology that is available to the researcher. It is a bare introduction and nothing more. The design of experiments—an extremely useful topic—was barely touched. Indeed, the methodology presented, although very useful, is intended primarily to serve as a vehicle suitable for conveying to you the philosophy involved.

APPENDIX I

USEFUL STATISTICAL TESTS AND CONFIDENCE INTERVALS

I. Inferences concerning the mean of a population

A. Sample size, n, is large ($n > 30$)

1. Test:

Null hypothesis: $\mu = \mu_0$.
Alternative hypothesis: $\mu \neq \mu_0$.
Test statistic:

$$z = \frac{\bar{y} - \mu_0}{\sigma / \sqrt{n}}.$$

If σ is unknown and the sample is large, use

$$s = \sqrt{\frac{\sum_{i=1}^{n} (y_i - \bar{y})^2}{n - 1}}$$

as an estimate of σ.

Rejection region:

$$\left.\begin{array}{l}\text{Reject if } z \text{ is greater than } 1.96 \\ \text{Reject if } z \text{ is less than } -1.96\end{array}\right\} \text{for } \alpha = .05.$$

2. $100(1 - \alpha)$ percent confidence interval:

$$\bar{y} \pm \frac{z_{\alpha/2}\, \sigma}{\sqrt{n}}.$$

B. Small samples, $n < 30$, and the observations are nearly normally distributed

1. Test:

Null hypothesis: $\mu = \mu_0$.
Alternative hypothesis: $\mu \neq \mu_0$.
Test statistic:

$$t = \frac{\bar{y} - \mu_0}{s / \sqrt{n}}.$$

Rejection region: See t tables.

2. $100(1 - \alpha)$ percent confidence interval:

$$\bar{y} \pm \frac{t_{\alpha/2}\, s}{\sqrt{n}}.$$

II. Inferences concerning the difference between the means of two populations

A. Large samples

1. Assumptions:
   (a) Population I has mean equal to μ_1 and variance equal to σ_1^2.
   (b) Population II has mean equal to μ_2 and variance equal to σ_2^2.
2. Some results:
   (a) Let $\bar{y}_1$ be the mean of a random sample of n_1 observations from population I, and $\bar{y}_2$ be the mean of an independent and random sample of n_2 observations from population II. Consider the difference, $(\bar{y}_1 - \bar{y}_2)$.
   (b) It can be shown that the mean of $(\bar{y}_1 - \bar{y}_2)$ is $(\mu_1 - \mu_2)$ and its

variance is $\frac{\sigma_1^2}{n_1} + \frac{\sigma_2^2}{n_2}$. Furthermore, for large samples, $(\bar{y}_1 - \bar{y}_2)$ will be approximately normally distributed.

3. Test:
   Null hypothesis: $(\mu_1 - \mu_2) = D_0$. [*Note:* We are usually testing the hypothesis that $(\mu_1 - \mu_2) = 0$, i.e., $(\mu_1 = \mu_2)$.]
   Alternative hypothesis: $(\mu_1 - \mu_2) \neq D_0$.
   Test statistic:

$$z = \frac{(\bar{y}_1 - \bar{y}_2) - D_0}{\sqrt{\sigma_1^2/n_1 + \sigma_2^2/n_2}}.$$

   If the null hypothesis is that $\mu_1 = \mu_2$, then $D_0 = 0$ and

$$z = \frac{\bar{y}_1 - \bar{y}_2}{\sqrt{\sigma_1^2/n_1 + \sigma_2^2/n_2}}.$$

   If σ_1^2 and σ_2^2 are unknown and n is large, use s_1^2 and s_2^2 as estimates.

   Rejection region:

$$\left.\begin{array}{r}\text{Reject if } z > 1.96 \\ \text{or } z < -1.96\end{array}\right\} \text{ for } \alpha = .05.$$

4. $100(1 - \alpha)$ percent confidence interval:

$$(\bar{y}_1 - \bar{y}_2) \pm z_{\alpha/2}\sqrt{\frac{\sigma_1^2}{n_1} + \frac{\sigma_2^2}{n_2}}.$$

B. Small samples
   1. Assumptions: Both populations approximately normally distributed and

$$\sigma_1^2 = \sigma_2^2.$$

   2. Test
      Null hypothesis: $(\mu_1 - \mu_2) = D_0$.
      Alternative hypothesis: $(\mu_1 - \mu_2) \neq D_0$,
      Test statistic:

$$t = \frac{(\bar{y}_1 - \bar{y}_2) - D_0}{s\sqrt{\frac{1}{n_1} + \frac{1}{n_2}}},$$

      where s is a pooled estimate of σ:

$$s = \sqrt{\frac{(n_1 - 1)s_1^2 + (n_2 - 1)s_2^2}{n_1 + n_2 - 2}}.$$

      Rejection region: See t tables.
   3. $100(1 - \alpha)$ percent confidence interval:

$$(\bar{y}_1 - \bar{y}_2) \pm t_{\alpha/2}\, s\sqrt{\frac{1}{n_1} + \frac{1}{n_2}}.$$

III. Inferences concerning a probability, p

A. Assumptions for a "binomial experiment":

1. Experiment consists of n identical trials each resulting in one of two outcomes, say success and failure.
2. The probability of success is equal to p and remains the same from trial to trial.
3. The trials are independent of each other.
4. The variable measured is y = number of successes observed during the n trials.

B. Results:

1. The estimator of p is $\hat{p} = \frac{y}{n}$.
2. The average value of $\hat{p}$ is p.
3. The variance of $\hat{p}$ is equal to $\frac{pq}{n}$.

C. Test (n large):

Null hypothesis: $p = p_0$.

Alternative hypothesis: $p \neq p_0$ (two-tailed test).

Test statistic:

$$z = \frac{\frac{y}{n} - p_0}{\sqrt{\frac{p_0 q_0}{n}}}.$$

Rejection region: Reject if $|z| \geq 1.96$. (*Note:* $\alpha = .05$.)

D. 100 $(1 - \alpha)$ percent confidence interval (n large):

$$\hat{p} \pm z_{\alpha/2}\sqrt{\frac{\hat{p}\hat{q}}{n}}.$$

IV. Inferences comparing two probabilities, p_1 and p_2

A. Assumption: Independent random samples are drawn from each of two binomial populations.

	Population I	Population II
Probability of success	p_1	p_2
Sample size	n_1	n_2
Observed successes	y_1	y_2

B. Results:

1. The estimated difference between p_1 and p_2 is

$$\hat{p}_1 - \hat{p}_2 = \frac{y_1}{n_1} - \frac{y_2}{n_2}.$$

2. The average value of $(\hat{p}_1 - \hat{p}_2)$ is $(p_1 - p_2)$.
3. The variance of $(\hat{p}_1 - \hat{p}_2)$ is

$$\frac{p_1 q_1}{n_1} + \frac{p_2 q_2}{n_2}.$$

C. Test (n_1 and n_2 large):
Null hypothesis: $p_1 = p_2 = p$.
Alternative hypothesis: $p_1 \neq p_2$ (two-tailed test).
Test statistic:

$$z = \frac{\frac{y_1}{n_1} - \frac{y_2}{n_2}}{\sqrt{\widehat{p}\widehat{q}\left(\frac{1}{n_1} + \frac{1}{n_2}\right)}}, \qquad \text{where } \widehat{p} = \frac{y_1 + y_2}{n_1 + n_2}.$$

Rejection region: Reject if $|z| \geq 1.96$. (*Note:* $\alpha = .05$.)

D. $100(1 - \alpha)$ percent confidence interval (n_1 and n_2 large):

$$(\widehat{p}_1 - \widehat{p}_2) \pm z_{\alpha/2}\sqrt{\frac{\widehat{p}_1\widehat{q}_1}{n_1} + \frac{\widehat{p}_2\widehat{q}_2}{n_2}}.$$

V. Inferences concerning the variance of a population

A. Assumption: Population measurements are normally distributed.

B. Test:
Null hypothesis: $\sigma^2 = \sigma_0^2$.
Alternative hypothesis: $\sigma^2 > \sigma_0^2$. (*Note:* This implies a one-tailed test.)
Test statistic:

$$\chi^2 = \frac{(n-1)s^2}{\sigma_0^2}.$$

Rejection region: Reject if χ^2 is greater than or equal to χ_α^2 (see table of χ^2 values). For example, if $\alpha = .05$ and $n = 10$, reject if χ^2 is greater than 16.919.

C. $100(1 - \alpha)$ percent confidence interval:

$$\frac{(n-1)s^2}{\chi_U^2} < \sigma^2 < \frac{(n-1)s^2}{\chi_L^2}.$$

VI. Tests for comparing the equality of two variances

A. Assumptions:

1. Population I has a normal distribution with mean μ_1 and variance σ_1^2.
2. Population II has a normal distribution with mean μ_2 and variance σ_2^2.
3. Two independent random samples are drawn, n_1 measurements from population I, n_2 from population II.

B. Test:
Null hypothesis: $\sigma_1^2 = \sigma_2^2$.
Alternative hypothesis: $\sigma_1^2 > \sigma_2^2$ (one-tailed test).
Test statistic:

$$F = \frac{s_1^2}{s_2^2}.$$

Rejection region: Reject if F is greater than or equal to $F_\alpha(n_1 - 1, n_2 - 1)$.

C. $100(1 - \alpha)$ percent confidence interval:

$$\frac{s_1^2}{s_2^2} \cdot \frac{1}{F_{v_1,v_2}} < \frac{\sigma_1^2}{\sigma_2^2} < \frac{s_1^2}{s_2^2} \cdot F_{v_2,v_1},$$

where

$\nu_1 = n_1 - 1$ and $\nu_2 = n_2 - 1$.

APPENDIX II
TABLES

Table 1

Binomial probability tables

Tabulated values are $\sum_{y=0}^{a} p(y)$.

(Computations are rounded at third decimal place.)

(a) $n = 5$

	P													
a	0.01	0.05	0.10	0.20	0.30	0.40	0.50	0.60	0.70	0.80	0.90	0.95	0.99	*a*
0	.951	.774	.590	.328	.168	.078	.031	.010	.002	.000	.000	.000	.000	0
1	.999	.977	.919	.737	.528	.337	.188	.087	.031	.007	.000	.000	.000	1
2	1.000	.999	.991	.942	.837	.683	.500	.317	.163	.058	.009	.001	.000	2
3	1.000	1.000	1.000	.993	.969	.913	.812	.663	.472	.263	.081	.023	.001	3
4	1.000	1.000	1.000	1.000	.998	.990	.969	.922	.832	.672	.410	.226	.049	4

(b) $n = 10$

	P													
a	0.01	0.05	0.10	0.20	0.30	0.40	0.50	0.60	0.70	0.80	0.90	0.95	0.99	*a*
0	.904	.599	.349	.107	.028	.006	.001	.000	.000	.000	.000	.000	.000	0
1	.996	.914	.736	.376	.149	.046	.011	.002	.000	.000	.000	.000	.000	1
2	1.000	.988	.930	.678	.383	.167	.055	.012	.002	.000	.000	.000	.000	2
3	1.000	.999	.987	.879	.650	.382	.172	.055	.011	.001	.000	.000	.000	3
4	1.000	1.000	.998	.967	.850	.633	.377	.166	.047	.006	.000	.000	.000	4
5	1.000	1.000	1.000	.994	.953	.834	.623	.367	.150	.033	.002	.000	.000	5
6	1.000	1.000	1.000	.999	.989	.945	.828	.618	.350	.121	.013	.001	.000	6
7	1.000	1.000	1.000	1.000	.998	.988	.945	.833	.617	.322	.070	.012	.000	7
8	1.000	1.000	1.000	1.000	1.000	.998	.989	.954	.851	.624	.264	.086	.004	8
9	1.000	1.000	1.000	1.000	1.000	1.000	.999	.994	.972	.893	.651	.401	.096	9

(c) $n = 15$

	P													
a	0.01	0.05	0.10	0.20	0.30	0.40	0.50	0.60	0.70	0.80	0.90	0.95	0.99	a
0	.860	.463	.206	.035	.005	.000	.000	.000	.000	.000	.000	.000	.000	0
1	.990	.829	.549	.167	.035	.005	.000	.000	.000	.000	.000	.000	.000	1
2	1.000	.964	.816	.398	.127	.027	.004	.000	.000	.000	.000	.000	.000	2
3	1.000	.995	.944	.648	.297	.091	.018	.002	.000	.000	.000	.000	.000	3
4	1.000	.999	.987	.836	.515	.217	.059	.009	.001	.000	.000	.000	.000	4
5	1.000	1.000	.998	.939	.722	.403	.151	.034	.004	.000	.000	.000	.000	5
6	1.000	1.000	1.000	.982	.869	.610	.304	.095	.015	.001	.000	.000	.000	6
7	1.000	1.000	1.000	.996	.950	.787	.500	.213	.050	.004	.000	.000	.000	7
8	1.000	1.000	1.000	.999	.985	.905	.696	.390	.131	.018	.000	.000	.000	8
9	1.000	1.000	1.000	1.000	.996	.966	.849	.597	.278	.061	.002	.000	.000	9
10	1.000	1.000	1.000	1.000	.999	.991	.941	.783	.485	.164	.013	.001	.000	10
11	1.000	1.000	1.000	1.000	1.000	.998	.982	.909	.703	.352	.056	.005	.000	11
12	1.000	1.000	1.000	1.000	1.000	1.000	.996	.973	.873	.602	.184	.036	.000	12
13	1.000	1.000	1.000	1.000	1.000	1.000	1.000	.995	.965	.833	.451	.171	.010	13
14	1.000	1.000	1.000	1.000	1.000	1.000	1.000	1.000	.995	.965	.794	.537	.140	14

(d) $n = 20$

a	0.01	0.05	0.10	0.20	0.30	0.40	0.50	0.60	0.70	0.80	0.90	0.95	0.99	a
0	.818	.358	.122	.012	.001	.000	.000	.000	.000	.000	.000	.000	.000	0
1	.983	.736	.392	.069	.008	.001	.000	.000	.000	.000	.000	.000	.000	1
2	.999	.925	.677	.206	.035	.004	.000	.000	.000	.000	.000	.000	.000	2
3	1.000	.984	.867	.411	.107	.016	.001	.000	.000	.000	.000	.000	.000	3
4	1.000	.997	.957	.630	.238	.051	.006	.000	.000	.000	.000	.000	.000	4
5	1.000	1.000	.989	.804	.416	.126	.021	.002	.000	.000	.000	.000	.000	5
6	1.000	1.000	.998	.913	.608	.250	.058	.006	.000	.000	.000	.000	.000	6
7	1.000	1.000	1.000	.968	.772	.416	.132	.021	.001	.000	.000	.000	.000	7
8	1.000	1.000	1.000	.990	.887	.596	.252	.057	.005	.000	.000	.000	.000	8
9	1.000	1.000	1.000	.997	.952	.755	.412	.128	.017	.001	.000	.000	.000	9
10	1.000	1.000	1.000	.999	.983	.872	.588	.245	.048	.003	.000	.000	.000	10
11	1.000	1.000	1.000	1.000	.995	.943	.748	.404	.113	.010	.000	.000	.000	11
12	1.000	1.000	1.000	1.000	.999	.979	.868	.584	.228	.032	.000	.000	.000	12
13	1.000	1.000	1.000	1.000	1.000	.994	.942	.750	.392	.087	.002	.000	.000	13
14	1.000	1.000	1.000	1.000	1.000	.998	.979	.874	.584	.196	.011	.000	.000	14
15	1.000	1.000	1.000	1.000	1.000	1.000	.994	.949	.762	.370	.043	.003	.000	15
16	1.000	1.000	1.000	1.000	1.000	1.000	.999	.984	.893	.589	.133	.016	.000	16
17	1.000	1.000	1.000	1.000	1.000	1.000	1.000	.996	.965	.794	.323	.075	.001	17
18	1.000	1.000	1.000	1.000	1.000	1.000	1.000	.999	.992	.931	.608	.264	.017	18
19	1.000	1.000	1.000	1.000	1.000	1.000	1.000	1.000	.999	.988	.878	.642	.182	19

(Column headings give P.)

(e) $n = 25$

	P													
a	0.01	0.05	0.10	0.20	0.30	0.40	0.50	0.60	0.70	0.80	0.90	0.95	0.99	*a*
0	.778	.277	.072	.004	.000	.000	.000	.000	.000	.000	.000	.000	.000	0
1	.974	.642	.271	.027	.002	.000	.000	.000	.000	.000	.000	.000	.000	1
2	.998	.873	.537	.098	.009	.000	.000	.000	.000	.000	.000	.000	.000	2
3	1.000	.966	.764	.234	.033	.002	.000	.000	.000	.000	.000	.000	.000	3
4	1.000	.993	.902	.421	.090	.009	.000	.000	.000	.000	.000	.000	.000	4
5	1.000	.999	.967	.617	.193	.029	.002	.000	.000	.000	.000	.000	.000	5
6	1.000	1.000	.991	.780	.341	.074	.007	.000	.000	.000	.000	.000	.000	6
7	1.000	1.000	.998	.891	.512	.154	.022	.001	.000	.000	.000	.000	.000	7
8	1.000	1.000	1.000	.953	.677	.274	.054	.004	.000	.000	.000	.000	.000	8
9	1.000	1.000	1.000	.983	.811	.425	.115	.013	.000	.000	.000	.000	.000	9
10	1.000	1.000	1.000	.994	.902	.586	.212	.034	.002	.000	.000	.000	.000	10
11	1.000	1.000	1.000	.998	.956	.732	.345	.078	.006	.000	.000	.000	.000	11
12	1.000	1.000	1.000	1.000	.983	.846	.500	.154	.017	.000	.000	.000	.000	12
13	1.000	1.000	1.000	1.000	.994	.922	.655	.268	.044	.002	.000	.000	.000	13
14	1.000	1.000	1.000	1.000	.998	.966	.788	.414	.098	.006	.000	.000	.000	14
15	1.000	1.000	1.000	1.000	1.000	.987	.885	.575	.189	.017	.000	.000	.000	15
16	1.000	1.000	1.000	1.000	1.000	.996	.946	.726	.323	.047	.000	.000	.000	16
17	1.000	1.000	1.000	1.000	1.000	.999	.978	.846	.488	.109	.002	.000	.000	17
18	1.000	1.000	1.000	1.000	1.000	1.000	.993	.926	.659	.220	.009	.000	.000	18
19	1.000	1.000	1.000	1.000	1.000	1.000	.998	.971	.807	.383	.033	.001	.000	19
20	1.000	1.000	1.000	1.000	1.000	1.000	1.000	.991	.910	.579	.098	.007	.000	20
21	1.000	1.000	1.000	1.000	1.000	1.000	1.000	.998	.967	.766	.236	.034	.000	21
22	1.000	1.000	1.000	1.000	1.000	1.000	1.000	1.000	.991	.902	.463	.127	.002	22
23	1.000	1.000	1.000	1.000	1.000	1.000	1.000	1.000	.998	.973	.729	.358	.026	23
24	1.000	1.000	1.000	1.000	1.000	1.000	1.000	1.000	1.000	.996	.928	.723	.222	24

Table 2
Table of e^{-x}

X	e^{-x}	X	e^{-x}	X	e^{-x}	X	e^{-x}
0.00	1.000000	2.60	.074274	5.10	.006097	7.[illegible]	[illegible]
0.10	.904837	2.70	.067206	5.20	.005517	7.[illegible]	[illegible]
0.20	.818731	2.80	.060810	5.30	.004992	7.[illegible]	.000410
0.30	.740818	2.90	.055023	5.40	.004517	7.90	.000371
0.40	.670320	3.00	.049787	5.50	.004087	8.00	.000336
0.50	.606531	3.10	.045049	5.60	.003698	8.10	.000304
0.60	.548812	3.20	.040762	5.70	.003346	8.20	.000275
0.70	.496585	3.30	.036883	5.80	.003028	8.30	.000249
0.80	.449329	3.40	.033373	5.90	.002739	8.40	.000225
0.90	.406570	3.50	.030197	6.00	.002479	8.50	.000204
1.00	.367879	3.60	.027324	6.10	.002243	8.60	.000184
1.10	.332871	3.70	.024724	6.20	.002029	8.70	.000167
1.20	.301194	3.80	.022371	6.30	.001836	8.80	.000151
1.30	.272532	3.90	.020242	6.40	.001661	8.90	.000136
1.40	.246597	4.00	.018316	6.50	.001503	9.00	.000123
1.50	.223130	4.10	.016573	6.60	.001360	9.10	.000112
1.60	.201897	4.20	.014996	6.70	.001231	9.20	.000101
1.70	.182684	4.30	.013569	6.80	.001114	9.30	.000091
1.80	.165299	4.40	.012277	6.90	.001008	9.40	.000083
1.90	.149569	4.50	.011109	7.00	.000912	9.50	.000075
2.00	.135335	4.60	.010052	7.10	.000825	9.60	.000068
2.10	.122456	4.70	.009095	7.20	.000747	9.70	.000061
2.20	.110803	4.80	.008230	7.30	.000676	9.80	.000056
2.30	.100259	4.90	.007447	7.40	.000611	9.90	.000050
2.40	.090718	5.00	.006738	7.50	.000553	10.00	.000045
2.50	.082085						

Table 3: Normal curve areas

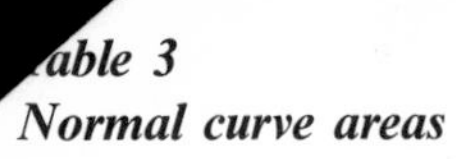

able 3
Normal curve areas

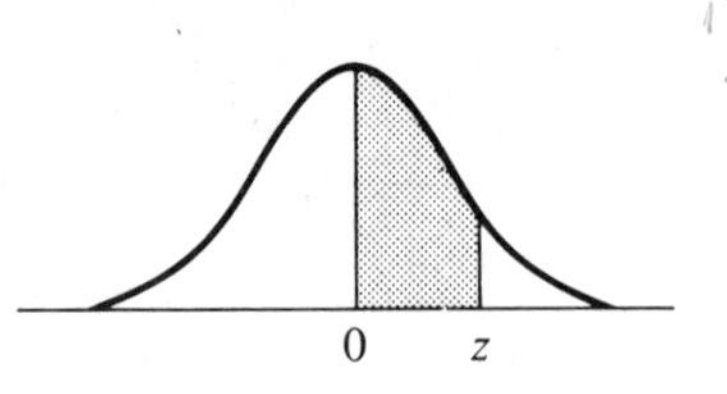

z	.00	.01	.02	.03	.04	.05	.06	.07	.08	.09
0.0	.0000	.0040	.0080	.0120	.0160	.0199	.0239	.0279	.0319	.0359
0.1	.0398	.0438	.0478	.0517	.0557	.0596	.0636	.0675	.0714	.0753
0.2	.0793	.0832	.0871	.0910	.0948	.0987	.1026	.1064	.1103	.1141
0.3	.1179	.1217	.1255	.1293	.1331	.1368	.1406	.1443	.1480	.1517
0.4	.1554	.1591	.1628	.1664	.1700	.1736	.1772	.1808	.1844	.1879
0.5	.1915	.1950	.1985	.2019	.2054	.2088	.2123	.2157	.2190	.2224
0.6	.2257	.2291	.2324	.2357	.2389	.2422	.2454	.2486	.2517	.2549
0.7	.2580	.2611	.2642	.2673	.2704	.2734	.2764	.2794	.2823	.2852
0.8	.2881	.2910	.2939	.2967	.2995	.3023	.3051	.3078	.3106	.3133
0.9	.3159	.3186	.3212	.3238	.3264	.3289	.3315	.3340	.3365	.3389
1.0	.3413	.3438	.3461	.3485	.3508	.3531	.3554	.3577	.3599	.3621
1.1	.3643	.3665	.3686	.3708	.3729	.3749	.3770	.3790	.3810	.3830
1.2	.3849	.3869	.3888	.3907	.3925	.3944	.3962	.3980	.3997	.4015
1.3	.4032	.4049	.4066	.4082	.4099	.4115	.4131	.4147	.4162	.4177
1.4	.4192	.4207	.4222	.4236	.4251	.4265	.4279	.4292	.4306	.4319
1.5	.4332	.4345	.4357	.4370	.4382	.4394	.4406	.4418	.4429	.4441
1.6	.4452	.4463	.4474	.4484	.4495	.4505	.4515	.4525	.4535	.4545
1.7	.4554	.4564	.4573	.4582	.4591	.4599	.4608	.4616	.4625	.4633
1.8	.4641	.4649	.4656	.4664	.4671	.4678	.4686	.4693	.4699	.4706
1.9	.4713	.4719	.4726	.4732	.4738	.4744	.4750	.4756	.4761	.4767
2.0	.4772	.4778	.4783	.4788	.4793	.4798	.4803	.4808	.4812	.4817
2.1	.4821	.4826	.4830	.4834	.4838	.4842	.4846	.4850	.4854	.4857
2.2	.4861	.4864	.4868	.4871	.4875	.4878	.4881	.4884	.4887	.4890
2.3	.4893	.4896	.4898	.4901	.4904	.4906	.4909	.4911	.4913	.4916
2.4	.4918	.4920	.4922	.4925	.4927	.4929	.4931	.4932	.4934	.4936
2.5	.4938	.4940	.4941	.4943	.4945	.4946	.4948	.4949	.4951	.4952
2.6	.4953	.4955	.4956	.4957	.4959	.4960	.4961	.4962	.4963	.4964
2.7	.4965	.4966	.4967	.4968	.4969	.4970	.4971	.4972	.4973	.4974
2.8	.4974	.4975	.4976	.4977	.4977	.4978	.4979	.4979	.4980	.4981
2.9	.4981	.4982	.4982	.4983	.4984	.4984	.4985	.4985	.4986	.4986
3.0	.4987	.4987	.4987	.4988	.4988	.4989	.4989	.4989	.4990	.4990

This table is abridged from Table I of *Statistical Tables and Formulas*, by A. Hald (New York: John Wiley & Sons, Inc., 1952). Reproduced by permission of A. Hald and the publishers, John Wiley & Sons, Inc.

Table 4
Critical values of t

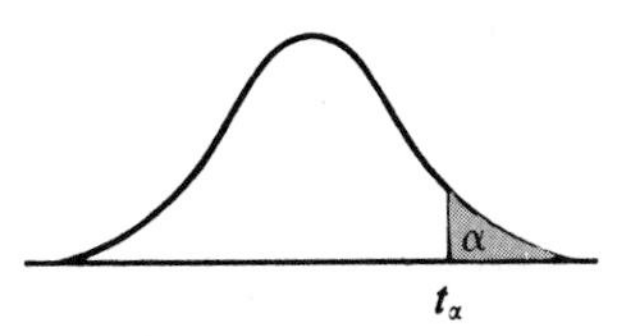

d.f.	$t_{.100}$	$t_{.050}$	$t_{.025}$	$t_{.010}$	$t_{.005}$	d.f.
1	3.078	6.314	12.706	31.821	63.657	1
2	1.886	2.920	4.303	6.965	9.925	2
3	1.638	2.353	3.182	4.541	5.841	3
4	1.533	2.132	2.776	3.747	4.604	4
5	1.476	2.015	2.571	3.365	4.032	5
6	1.440	1.943	2.447	3.143	3.707	6
7	1.415	1.895	2.365	2.998	3.499	7
8	1.397	1.860	2.306	2.896	3.355	8
9	1.383	1.833	2.262	2.821	3.250	9
10	1.372	1.812	2.228	2.764	3.169	10
11	1.363	1.796	2.201	2.718	3.106	11
12	1.356	1.782	2.179	2.681	3.055	12
13	1.350	1.771	2.160	2.650	3.012	13
14	1.345	1.761	2.145	2.624	2.977	14
15	1.341	1.753	2.131	2.602	2.947	15
16	1.337	1.746	2.120	2.583	2.921	16
17	1.333	1.740	2.110	2.567	2.898	17
18	1.330	1.734	2.101	2.552	2.878	18
19	1.328	1.729	2.093	2.539	2.861	19
20	1.325	1.725	2.086	2.528	2.845	20
21	1.323	1.721	2.080	2.518	2.831	21
22	1.321	1.717	2.074	2.508	2.819	22
23	1.319	1.714	2.069	2.500	2.807	23
24	1.318	1.711	2.064	2.492	2.797	24
25	1.316	1.708	2.060	2.485	2.787	25
26	1.315	1.706	2.056	2.479	2.779	26
27	1.314	1.703	2.052	2.473	2.771	27
28	1.313	1.701	2.048	2.467	2.763	28
29	1.311	1.699	2.045	2.462	2.756	29
inf.	1.282	1.645	1.960	2.326	2.576	inf.

From "Table of Percentage Points of the *t*-Distribution." Computed by Maxine Merrington, *Biometrika*, Vol. 32 (1941), p. 300. Reproduced by permission of Professor E. S. Pearson.

Table 5
Critical values of chi square

d.f.	$\chi^2_{0.995}$	$\chi^2_{0.990}$	$\chi^2_{0.975}$	$\chi^2_{0.950}$	$\chi^2_{0.900}$
1	0.0000393	0.0001571	0.0009821	0.0039321	0.0157908
2	0.0100251	0.0201007	0.0506356	0.102587	0.210720
3	0.0717212	0.114832	0.215795	0.351846	0.584375
4	0.206990	0.297110	0.484419	0.710721	1.063623
5	0.411740	0.554300	0.831211	1.145476	1.61031
6	0.675727	0.872085	1.237347	1.63539	2.20413
7	0.989265	1.239043	1.68987	2.16735	2.83311
8	1.344419	1.646482	2.17973	2.73264	3.48954
9	1.734926	2.087912	2.70039	3.32511	4.16816
10	2.15585	2.55821	3.24697	3.94030	4.86518
11	2.60321	3.05347	3.81575	4.57481	5.57779
12	3.07382	3.57056	4.40379	5.22603	6.30380
13	3.56503	4.10691	5.00874	5.89186	7.04150
14	4.07468	4.66043	5.62872	6.57063	7.78953
15	4.60094	5.22935	6.26214	7.26094	8.54675
16	5.14224	5.81221	6.90766	7.96164	9.31223
17	5.69724	6.40776	7.56418	8.67176	10.0852
18	6.26481	7.01491	8.23075	9.39046	10.8649
19	6.84398	7.63273	8.90655	10.1170	11.6509
20	7.43386	8.26040	9.59083	10.8508	12.4426
21	8.03366	8.89720	10.28293	11.5913	13.2396
22	8.64272	9.54249	10.9823	12.3380	14.0415
23	9.26042	10.19567	11.6885	13.0905	14.8479
24	9.88623	10.8564	12.4011	13.8484	15.6587
25	10.5197	11.5240	13.1197	14.6114	16.4734
26	11.1603	12.1981	13.8439	15.3791	17.2919
27	11.8076	12.8786	14.5733	16.1513	18.1138
28	12.4613	13.5648	15.3079	16.9279	18.9392
29	13.1211	14.2565	16.0471	17.7083	19.7677
30	13.7867	14.9535	16.7908	18.4926	20.5992
40	20.7065	22.1643	24.4331	26.5093	29.0505
50	27.9907	29.7067	32.3574	34.7642	37.6886
60	35.5346	37.4848	40.4817	43.1879	46.4589
70	43.2752	45.4418	48.7576	51.7393	55.3290
80	51.1720	53.5400	57.1532	60.3915	64.2778
90	59.1963	61.7541	65.6466	69.1260	73.2912
100	67.3276	70.0648	74.2219	77.9295	82.3581

From "Tables of the Percentage Points of the χ^2-Distribution." *Biometrika*, Vol. 32 (1941), pp. 188–189, by Catherine M. Thompson. Reproduced by permission of Professor E. S. Pearson.

$\chi^2_{0.100}$	$\chi^2_{0.050}$	$\chi^2_{0.025}$	$\chi^2_{0.010}$	$\chi^2_{0.005}$	d.f.
2.70554	3.84146	5.02389	6.63490	7.87944	1
4.60517	5.99147	7.37776	9.21034	10.5966	2
6.25139	7.81473	9.34840	11.3449	12.8381	3
7.77944	9.48773	11.1433	13.2767	14.8602	4
9.23635	11.0705	12.8325	15.0863	16.7496	5
10.6446	12.5916	14.4494	16.8119	18.5476	6
12.0170	14.0671	16.0128	18.4753	20.2777	7
13.3616	15.5073	17.5346	20.0902	21.9550	8
14.6837	16.9190	19.0228	21.6660	23.5893	9
15.9871	18.3070	20.4831	23.2093	25.1882	10
17.2750	19.6751	21.9200	24.7250	26.7569	11
18.5494	21.0261	23.3367	26.2170	28.2995	12
19.8119	22.3621	24.7356	27.6883	29.8194	13
21.0642	23.6848	26.1190	29.1413	31.3193	14
22.3072	24.9958	27.4884	30.5779	32.8013	15
23.5418	26.2962	28.8454	31.9999	34.2672	16
24.7690	27.5871	30.1910	33.4087	35.7185	17
25.9894	28.8693	31.5264	34.8053	37.1564	18
27.2036	30.1435	32.8523	36.1908	38.5822	19
28.4120	31.4104	34.1696	37.5662	39.9968	20
29.6151	32.6705	35.4789	38.9321	41.4010	21
30.8133	33.9244	36.7807	40.2894	42.7956	22
32.0069	35.1725	38.0757	41.6384	44.1813	23
33.1963	36.4151	39.3641	42.9798	45.5585	24
34.3816	37.6525	40.6465	44.3141	46.9278	25
35.5631	38.8852	41.9232	45.6417	48.2899	26
36.7412	40.1133	43.1944	46.9630	49.6449	27
37.9159	41.3372	44.4607	48.2782	50.9933	28
39.0875	42.5569	45.7222	49.5879	52.3356	29
40.2560	43.7729	46.9792	50.8922	53.6720	30
51.8050	55.7585	59.3417	63.6907	66.7659	40
63.1671	67.5048	71.4202	76.1539	79.4900	50
74.3970	79.0819	83.2976	88.3794	91.9517	60
85.5271	90.5312	95.0231	100.425	104.215	70
96.5782	101.879	106.629	112.329	116.321	80
107.565	113.145	118.136	124.116	128.299	90
118.498	124.342	129.561	135.807	140.169	100

Table 6
Percentage points of the F distribution

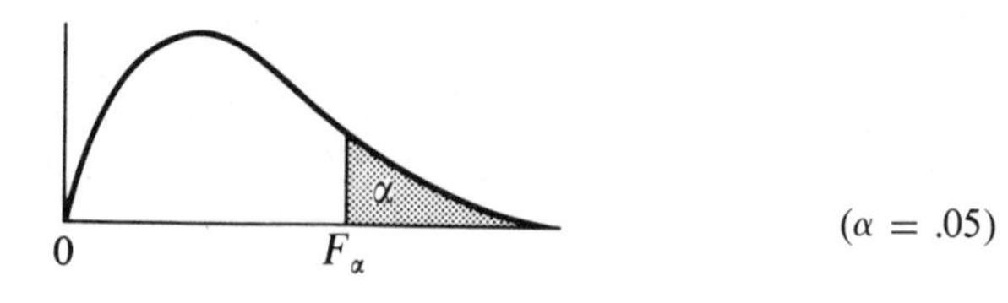

($\alpha = .05$)

Numerator Degrees of Freedom

ν_2 \ ν_1 (Denominator Degrees of Freedom)	1	2	3	4	5	6	7	8	9
1	161.4	199.5	215.7	224.6	230.2	234.0	236.8	238.9	240.5
2	18.51	19.00	19.16	19.25	19.30	19.33	19.35	19.37	19.38
3	10.13	9.55	9.28	9.12	9.01	8.94	8.89	8.85	8.81
4	7.71	6.94	6.59	6.39	6.26	6.16	6.09	6.04	6.00
5	6.61	5.79	5.41	5.19	5.05	4.95	4.88	4.82	4.77
6	5.99	5.14	4.76	4.53	4.39	4.28	4.21	4.15	4.10
7	5.59	4.74	4.35	4.12	3.97	3.87	3.79	3.73	3.68
8	5.32	4.46	4.07	3.84	3.69	3.58	3.50	3.44	3.39
9	5.12	4.26	3.86	3.63	3.48	3.37	3.29	3.23	3.18
10	4.96	4.10	3.71	3.48	3.33	3.22	3.14	3.07	3.02
11	4.84	3.98	3.59	3.36	3.20	3.09	3.01	2.95	2.90
12	4.75	3.89	3.49	3.26	3.11	3.00	2.91	2.85	2.80
13	4.67	3.81	3.41	3.18	3.03	2.92	2.83	2.77	2.71
14	4.60	3.74	3.34	3.11	2.96	2.85	2.76	2.70	2.65
15	4.54	3.68	3.29	3.06	2.90	2.79	2.71	2.64	2.59
16	4.49	3.63	3.24	3.01	2.85	2.74	2.66	2.59	2.54
17	4.45	3.59	3.20	2.96	2.81	2.70	2.61	2.55	2.49
18	4.41	3.55	3.16	2.93	2.77	2.66	2.58	2.51	2.46
19	4.38	3.52	3.13	2.90	2.74	2.63	2.54	2.48	2.42
20	4.35	3.49	3.10	2.87	2.71	2.60	2.51	2.45	2.39
21	4.32	3.47	3.07	2.84	2.68	2.57	2.49	2.42	2.37
22	4.30	3.44	3.05	2.82	2.66	2.55	2.46	2.40	2.34
23	4.28	3.42	3.03	2.80	2.64	2.53	2.44	2.37	2.32
24	4.26	3.40	3.01	2.78	2.62	2.51	2.42	2.36	2.30
25	4.24	3.39	2.99	2.76	2.60	2.49	2.40	2.34	2.28
26	4.23	3.37	2.98	2.74	2.59	2.47	2.39	2.32	2.27
27	4.21	3.35	2.96	2.73	2.57	2.46	2.37	2.31	2.25
28	4.20	3.34	2.95	2.71	2.56	2.45	2.36	2.29	2.24
29	4.18	3.33	2.93	2.70	2.55	2.43	2.35	2.28	2.22
30	4.17	3.32	2.92	2.69	2.53	2.42	2.33	2.27	2.21
40	4.08	3.23	2.84	2.61	2.45	2.34	2.25	2.18	2.12
60	4.00	3.15	2.76	2.53	2.37	2.25	2.17	2.10	2.04
120	3.92	3.07	2.68	2.45	2.29	2.17	2.09	2.02	1.96
∞	3.84	3.00	2.60	2.37	2.21	2.10	2.01	1.94	1.88

From "Tables of Percentage Points of the Inverted Beta (*F*)-Distribution," *Biometrika*, Vol. 33 (1943), pp. 73–88, by Maxine Merrington and Catherine M. Thompson. Reproduced by permission of the *Biometrika* Trustees.

10	12	15	20	24	30	40	60	120	∞	ν_1 / ν_2
241.9	243.9	245.9	248.0	249.1	250.1	251.1	252.2	253.3	254.3	1
19.40	19.41	19.43	19.45	19.45	19.46	19.47	19.48	19.49	19.50	2
8.79	8.74	8.70	8.66	8.64	8.62	8.59	8.57	8.55	8.53	3
5.96	5.91	5.86	5.80	5.77	5.75	5.72	5.69	5.66	5.63	4
4.74	4.68	4.62	4.56	4.53	4.50	4.46	4.43	4.40	4.36	5
4.06	4.00	3.94	3.87	3.84	3.81	3.77	3.74	3.70	3.67	6
3.64	3.57	3.51	3.44	3.41	3.38	3.34	3.30	3.27	3.23	7
3.35	3.28	3.22	3.15	3.12	3.08	3.04	3.01	2.97	2.93	8
3.14	3.07	3.01	2.94	2.90	2.86	2.83	2.79	2.75	2.71	9
2.98	2.91	2.85	2.77	2.74	2.70	2.66	2.62	2.58	2.54	10
2.85	2.79	2.72	2.65	2.61	2.57	2.53	2.49	2.45	2.40	11
2.75	2.69	2.62	2.54	2.51	2.47	2.43	2.38	2.34	2.30	12
2.67	2.60	2.53	2.46	2.42	2.38	2.34	2.30	2.25	2.21	13
2.60	2.53	2.46	2.39	2.35	2.31	2.27	2.22	2.18	2.13	14
2.54	2.48	2.40	2.33	2.29	2.25	2.20	2.16	2.11	2.07	15
2.49	2.42	2.35	2.28	2.24	2.19	2.15	2.11	2.06	2.01	16
2.45	2.38	2.31	2.23	2.19	2.15	2.10	2.06	2.01	1.96	17
2.41	2.34	2.27	2.19	2.15	2.11	2.06	2.02	1.97	1.92	18
2.38	2.31	2.23	2.16	2.11	2.07	2.03	1.98	1.93	1.88	19
2.35	2.28	2.20	2.12	2.08	2.04	1.99	1.95	1.90	1.84	20
2.32	2.25	2.18	2.10	2.05	2.01	1.96	1.92	1.87	1.81	21
2.30	2.23	2.15	2.07	2.03	1.98	1.94	1.89	1.84	1.78	22
2.27	2.20	2.13	2.05	2.01	1.96	1.91	1.86	1.81	1.76	23
2.25	2.18	2.11	2.03	1.98	1.94	1.89	1.84	1.79	1.73	24
2.24	2.16	2.09	2.01	1.96	1.92	1.87	1.82	1.77	1.71	25
2.22	2.15	2.07	1.99	1.95	1.90	1.85	1.80	1.75	1.69	26
2.20	2.13	2.06	1.97	1.93	1.88	1.84	1.79	1.73	1.67	27
2.19	2.12	2.04	1.96	1.91	1.87	1.82	1.77	1.71	1.65	28
2.18	2.10	2.03	1.94	1.90	1.85	1.81	1.75	1.70	1.64	29
2.16	2.09	2.01	1.93	1.89	1.84	1.79	1.74	1.68	1.62	30
2.08	2.00	1.92	1.84	1.79	1.74	1.69	1.64	1.58	1.51	40
1.99	1.92	1.84	1.75	1.70	1.65	1.59	1.53	1.47	1.39	60
1.91	1.83	1.75	1.66	1.61	1.55	1.50	1.43	1.35	1.25	120
1.83	1.75	1.67	1.57	1.52	1.46	1.39	1.32	1.22	1.00	∞

Table 7
Percentage points of the F distribution

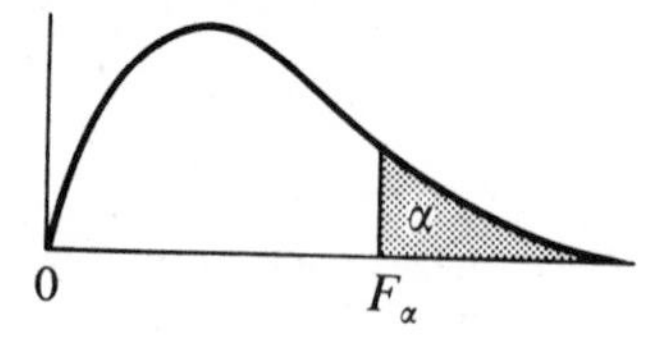

($\alpha = .01$)

Numerator Degrees of Freedom

Denominator Degrees of Freedom ν_2 \ ν_1	1	2	3	4	5	6	7	8	9
1	4052	4999.5	5403	5625	5764	5859	5928	5982	6022
2	98.50	99.00	99.17	99.25	99.30	99.33	99.36	99.37	99.39
3	34.12	30.82	29.46	28.71	28.24	27.91	27.67	27.49	27.35
4	21.20	18.00	16.69	15.98	15.52	15.21	14.98	14.80	14.66
5	16.26	13.27	12.06	11.39	10.97	10.67	10.46	10.29	10.16
6	13.75	10.92	9.78	9.15	8.75	8.47	8.26	8.10	7.98
7	12.25	9.55	8.45	7.85	7.46	7.19	6.99	6.84	6.72
8	11.26	8.65	7.59	7.01	6.63	6.37	6.18	6.03	5.91
9	10.56	8.02	6.99	6.42	6.06	5.80	5.61	5.47	5.35
10	10.04	7.56	6.55	5.99	5.64	5.39	5.20	5.06	4.94
11	9.65	7.21	6.22	5.67	5.32	5.07	4.89	4.74	4.63
12	9.33	6.93	5.95	5.41	5.06	4.82	4.64	4.50	4.39
13	9.07	6.70	5.74	5.21	4.86	4.62	4.44	4.30	4.19
14	8.86	6.51	5.56	5.04	4.69	4.46	4.28	4.14	4.03
15	8.68	6.36	5.42	4.89	4.56	4.32	4.14	4.00	3.89
16	8.53	6.23	5.29	4.77	4.44	4.20	4.03	3.89	3.78
17	8.40	6.11	5.18	4.67	4.34	4.10	3.93	3.79	3.68
18	8.29	6.01	5.09	4.58	4.25	4.01	3.84	3.71	3.60
19	8.18	5.93	5.01	4.50	4.17	3.94	3.77	3.63	3.52
20	8.10	5.85	4.94	4.43	4.10	3.87	3.70	3.56	3.46
21	8.02	5.78	4.87	4.37	4.04	3.81	3.64	3.51	3.40
22	7.95	5.72	4.82	4.31	3.99	3.76	3.59	3.45	3.35
23	7.88	5.66	4.76	4.26	3.94	3.71	3.54	3.41	3.30
24	7.82	5.61	4.72	4.22	3.90	3.67	3.50	3.36	3.26
25	7.77	5.57	4.68	4.18	3.85	3.63	3.46	3.32	3.22
26	7.72	5.53	4.64	4.14	3.82	3.59	3.42	3.29	3.18
27	7.68	5.49	4.60	4.11	3.78	3.56	3.39	3.26	3.15
28	7.64	5.45	4.57	4.07	3.75	3.53	3.36	3.23	3.12
29	7.60	5.42	4.54	4.04	3.73	3.50	3.33	3.20	3.09
30	7.56	5.39	4.51	4.02	3.70	3.47	3.30	3.17	3.07
40	7.31	5.18	4.31	3.83	3.51	3.29	3.12	2.99	2.89
60	7.08	4.98	4.13	3.65	3.34	3.12	2.95	2.82	2.72
120	6.85	4.79	3.95	3.48	3.17	2.96	2.79	2.66	2.56
∞	6.63	4.61	3.78	3.32	3.02	2.80	2.64	2.51	2.41

From "Tables of Percentage Points of the Inverted Beta (*F*)-Distribution," *Biometrika*, Vol. 33 (1943), pp. 73–88, by Maxine Merrington and Catherine M. Thompson. Reproduced by permission of the *Biometrika* Trustees.

10	12	15	20	24	30	40	60	120	∞	ν_1 / ν_2
6056	6106	6157	6209	6235	6261	6287	6313	6339	6366	1
99.40	99.42	99.43	99.45	99.46	99.47	99.47	99.48	99.49	99.50	2
27.23	27.05	26.87	26.69	26.60	26.50	26.41	26.32	26.22	26.13	3
14.55	14.37	14.20	14.02	13.93	13.84	13.75	13.65	13.56	13.46	4
10.05	9.89	9.72	9.55	9.47	9.38	9.29	9.20	9.11	9.02	5
7.87	7.72	7.56	7.40	7.31	7.23	7.14	7.06	6.97	6.88	6
6.62	6.47	6.31	6.16	6.07	5.99	5.91	5.82	5.74	5.65	7
5.81	5.67	5.52	5.36	5.28	5.20	5.12	5.03	4.95	4.86	8
5.26	5.11	4.96	4.81	4.73	4.65	4.57	4.48	4.40	4.31	9
4.85	4.71	4.56	4.41	4.33	4.25	4.17	4.08	4.00	3.91	10
4.54	4.40	4.25	4.10	4.02	3.94	3.86	3.78	3.69	3.60	11
4.30	4.16	4.01	3.86	3.78	3.70	3.62	3.54	3.45	3.36	12
4.10	3.96	3.82	3.66	3.59	3.51	3.43	3.34	3.25	3.17	13
3.94	3.80	3.66	3.51	3.43	3.35	3.27	3.18	3.09	3.00	14
3.80	3.67	3.52	3.37	3.29	3.21	3.13	3.05	2.96	2.87	15
3.69	3.55	3.41	3.26	3.18	3.10	3.02	2.93	2.84	2.75	16
3.59	3.46	3.31	3.16	3.08	3.00	2.92	2.83	2.75	2.65	17
3.51	3.37	3.23	3.08	3.00	2.92	2.84	2.75	2.66	2.57	18
3.43	3.30	3.15	3.00	2.92	2.84	2.76	2.67	2.58	2.49	19
3.37	3.23	3.09	2.94	2.86	2.78	2.69	2.61	2.52	2.42	20
3.31	3.17	3.03	2.88	2.80	2.72	2.64	2.55	2.46	2.36	21
3.26	3.12	2.98	2.83	2.75	2.67	2.58	2.50	2.40	2.31	22
3.21	3.07	2.93	2.78	2.70	2.62	2.54	2.45	2.35	2.26	23
3.17	3.03	2.89	2.74	2.66	2.58	2.49	2.40	2.31	2.21	24
3.13	2.99	2.85	2.70	2.62	2.54	2.45	2.36	2.27	2.17	25
3.09	2.96	2.81	2.66	2.58	2.50	2.42	2.33	2.23	2.13	26
3.06	2.93	2.78	2.63	2.55	2.47	2.38	2.29	2.20	2.10	27
3.03	2.90	2.75	2.60	2.52	2.44	2.35	2.26	2.17	2.06	28
3.00	2.87	2.73	2.57	2.49	2.41	2.33	2.23	2.14	2.03	29
2.98	2.84	2.70	2.55	2.47	2.39	2.30	2.21	2.11	2.01	30
2.80	2.66	2.52	2.37	2.29	2.20	2.11	2.02	1.92	1.80	40
2.63	2.50	2.35	2.20	2.12	2.03	1.94	1.84	1.73	1.60	60
2.47	2.34	2.19	2.03	1.95	1.86	1.76	1.66	1.53	1.38	120
2.32	2.18	2.04	1.88	1.79	1.70	1.59	1.47	1.32	1.00	∞

Table 8
Distribution function of U

$P(U \leq U_0)$; U_0 is the argument
$n_1 \leq n_2$; $3 \leq n_2 \leq 10$

$n_2 = 3$

U_0 \ n_1	1	2	3
0	.25	.10	.05
1	.50	.20	.10
2		.40	.20
3		.60	.35
4			.50

$n_2 = 4$

U_0 \ n_1	1	2	3	4
0	.2000	.0667	.0286	.0143
1	.4000	.1333	.0571	.0286
2	.6000	.2667	.1143	.0571
3		.4000	.2000	.1000
4		.6000	.3143	.1714
5			.4286	.2429
6			.5714	.3429
7				.4429
8				.5571

$n_2 = 5$

U_0 \ n_1	1	2	3	4	5
0	.1667	.0476	.0179	.0079	.0040
1	.3333	.0952	.0357	.0159	.0079
2	.5000	.1905	.0714	.0317	.0159
3		.2857	.1250	.0556	.0278
4		.4286	.1964	.0952	.0476
5		.5714	.2857	.1429	.0754
6			.3929	.2063	.1111
7			.5000	.2778	.1548
8				.3651	.2103
9				.4524	.2738
10				.5476	.3452
11					.4206
12					.5000

$n_2 = 6$

U_0 \ n_1	1	2	3	4	5	6
0	.1429	.0357	.0119	.0048	.0022	.0011
1	.2857	.0714	.0238	.0095	.0043	.0022
2	.4286	.1429	.0476	.0190	.0087	.0043
3	.5714	.2143	.0833	.0333	.0152	.0076
4		.3214	.1310	.0571	.0260	.0130
5		.4286	.1905	.0857	.0411	.0206
6		.5714	.2738	.1286	.0628	.0325
7			.3571	.1762	.0887	.0465
8			.4524	.2381	.1234	.0660
9			.5476	.3048	.1645	.0898
10				.3810	.2143	.1201
11				.4571	.2684	.1548
12				.5429	.3312	.1970
13					.3961	.2424
14					.4654	.2944
15					.5346	.3496
16						.4091
17						.4686
18						.5314

$n_2 = 7$

U_0 \ n_1	1	2	3	4	5	6	7
0	.1250	.0278	.0083	.0030	.0013	.0006	.0003
1	.2500	.0556	.0167	.0061	.0025	.0012	.0006
2	.3750	.1111	.0333	.0121	.0051	.0023	.0012
3	.5000	.1667	.0583	.0212	.0088	.0041	.0020
4		.2500	.0917	.0364	.0152	.0070	.0035
5		.3333	.1333	.0545	.0240	.0111	.0055
6		.4444	.1917	.0818	.0366	.0175	.0087
7		.5556	.2583	.1152	.0530	.0256	.0131
8			.3333	.1576	.0745	.0367	.0189
9			.4167	.2061	.1010	.0507	.0265
10			.5000	.2636	.1338	.0688	.0364
11				.3242	.1717	.0903	.0487
12				.3939	.2159	.1171	.0641
13				.4636	.2652	.1474	.0825
14				.5364	.3194	.1830	.1043
15					.3775	.2226	.1297
16					.4381	.2669	.1588
17					.5000	.3141	.1914
18						.3654	.2279
19						.4178	.2675
20						.4726	.3100
21						.5274	.3552
22							.4024
23							.4508
24							.5000

$n_2 = 8$

U_0 \ n_1	1	2	3	4	5	6	7	8
0	.1111	.0222	.0061	.0020	.0008	.0003	.0002	.0001
1	.2222	.0444	.0121	.0040	.0016	.0007	.0003	.0002
2	.3333	.0889	.0242	.0081	.0031	.0013	.0006	.0003
3	.4444	.1333	.0424	.0141	.0054	.0023	.0011	.0005
4	.5556	.2000	.0667	.0242	.0093	.0040	.0019	.0009
5		.2667	.0970	.0364	.0148	.0063	.0030	.0015
6		.3556	.1394	.0545	.0225	.0100	.0047	.0023
7		.4444	.1879	.0768	.0326	.0147	.0070	.0035
8		.5556	.2485	.1071	.0466	.0213	.0103	.0052
9			.3152	.1414	.0637	.0296	.0145	.0074
10			.3879	.1838	.0855	.0406	.0200	.0103
11			.4606	.2303	.1111	.0539	.0270	.0141
12			.5394	.2848	.1422	.0709	.0361	.0190
13				.3414	.1772	.0906	.0469	.0249
14				.4040	.2176	.1142	.0603	.0325
15				.4667	.2618	.1412	.0760	.0415
16				.5333	.3108	.1725	.0946	.0524
17					.3621	.2068	.1159	.0652
18					.4165	.2454	.1405	.0803
19					.4716	.2864	.1678	.0974
20					.5284	.3310	.1984	.1172
21						.3773	.2317	.1393
22						.4259	.2679	.1641
23						.4749	.3063	.1911
24						.5251	.3472	.2209
25							.3894	.2527
26							.4333	.2869
27							.4775	.3227
28							.5225	.3605
29								.3992
30								.4392
31								.4796
32								.5204

$n_2 = 9$

U_0 \ n_1	1	2	3	4	5	6	7	8	9
0	.1000	.0182	.0045	.0014	.0005	.0002	.0001	.0000	.0000
1	.2000	.0364	.0091	.0028	.0010	.0004	.0002	.0001	.0000
2	.3000	.0727	.0182	.0056	.0020	.0008	.0003	.0002	.0001
3	.4000	.1091	.0318	.0098	.0035	.0014	.0006	.0003	.0001
4	.5000	.1636	.0500	.0168	.0060	.0024	.0010	.0005	.0002
5		.2182	.0727	.0252	.0095	.0038	.0017	.0008	.0004
6		.2909	.1045	.0378	.0145	.0060	.0026	.0012	.0006
7		.3636	.1409	.0531	.0210	.0088	.0039	.0019	.0009
8		.4545	.1864	.0741	.0300	.0128	.0058	.0028	.0014
9		.5455	.2409	.0993	.0415	.0180	.0082	.0039	.0020
10			.3000	.1301	.0559	.0248	.0115	.0056	.0028
11			.3636	.1650	.0734	.0332	.0156	.0076	.0039
12			.4318	.2070	.0949	.0440	.0209	.0103	.0053
13			.5000	.2517	.1199	.0567	.0274	.0137	.0071
14				.3021	.1489	.0723	.0356	.0180	.0094
15				.3552	.1818	.0905	.0454	.0232	.0122
16				.4126	.2188	.1119	.0571	.0296	.0157
17				.4699	.2592	.1361	.0708	.0372	.0200
18				.5301	.3032	.1638	.0869	.0464	.0252
19					.3497	.1942	.1052	.0570	.0313
20					.3986	.2280	.1261	.0694	.0385
21					.4491	.2643	.1496	.0836	.0470
22					.5000	.3035	.1755	.0998	.0567
23						.3445	.2039	.1179	.0680
24						.3878	.2349	.1383	.0807
25						.4320	.2680	.1606	.0951
26						.4773	.3032	.1852	.1112
27						.5227	.3403	.2117	.1290
28							.3788	.2404	.1487
29							.4185	.2707	.1701
30							.4591	.3029	.1933
31							.5000	.3365	.2181
32								.3715	.2447
33								.4074	.2729
34								.4442	.3024
35								.4813	.3332
36								.5187	.3652
37									.3981
38									.4317
39									.4657
40									.5000

$n_2 = 10$

U_0 \ n_1	1	2	3	4	5	6	7	8	9	10
0	.0909	.0152	.0035	.0010	.0003	.0001	.0001	.0000	.0000	.0000
1	.1818	.0303	.0070	.0020	.0007	.0002	.0001	.0000	.0000	.0000
2	.2727	.0606	.0140	.0040	.0013	.0005	.0002	.0001	.0000	.0000
3	.3636	.0909	.0245	.0070	.0023	.0009	.0004	.0002	.0001	.0000
4	.4545	.1364	.0385	.0120	.0040	.0015	.0006	.0003	.0001	.0001
5	.5455	.1818	.0559	.0180	.0063	.0024	.0010	.0004	.0002	.0001
6		.2424	.0804	.0270	.0097	.0037	.0015	.0007	.0003	.0002
7		.3030	.1084	.0380	.0140	.0055	.0023	.0010	.0005	.0002
8		.3788	.1434	.0529	.0200	.0080	.0034	.0015	.0007	.0004
9		.4545	.1853	.0709	.0276	.0112	.0048	.0022	.0011	.0005
10		.5455	.2343	.0939	.0376	.0156	.0068	.0031	.0015	.0008
11			.2867	.1199	.0496	.0210	.0093	.0043	.0021	.0010
12			.3462	.1518	.0646	.0280	.0125	.0058	.0028	.0014
13			.4056	.1868	.0823	.0363	.0165	.0078	.0038	.0019
14			.4685	.2268	.1032	.0467	.0215	.0103	.0051	.0026
15			.5315	.2697	.1272	.0589	.0277	.0133	.0066	.0034
16				.3177	.1548	.0736	.0351	.0171	.0086	.0045
17				.3666	.1855	.0903	.0439	.0217	.0110	.0057
18				.4196	.2198	.1099	.0544	.0273	.0140	.0073
19				.4725	.2567	.1317	.0665	.0338	.0175	.0093
20				.5275	.2970	.1566	.0806	.0416	.0217	.0116
21					.3393	.1838	.0966	.0506	.0267	.0144
22					.3839	.2139	.1148	.0610	.0326	.0177
23					.4296	.2461	.1349	.0729	.0394	.0216
24					.4765	.2811	.1574	.0864	.0474	.0262
25					.5235	.3177	.1819	.1015	.0564	.0315
26						.3564	.2087	.1185	.0667	.0376
27						.3962	.2374	.1371	.0782	.0446
28						.4374	.2681	.1577	.0912	.0526
29						.4789	.3004	.1800	.1055	.0615
30						.5211	.3345	.2041	.1214	.0716
31							.3698	.2299	.1388	.0827
32							.4063	.2574	.1577	.0952
33							.4434	.2863	.1781	.1088
34							.4811	.3167	.2001	.1237
35							.5189	.3482	.2235	.1399
36								.3809	.2483	.1575
37								.4143	.2745	.1763
38								.4484	.3019	.1965
39								.4827	.3304	.2179
40								.5173	.3598	.2406
41									.3901	.2644
42									.4211	.2894
43									.4524	.3153
44									.4841	.3421
45									.5159	.3697
46										.3980
47										.4267
48										.4559
49										.4853
50										.5147

Computed by M. Pagano at the Department of Statistics, University of Florida.

Table 9
Critical values of T in the Wilcoxon matched-pairs signed ranks test

$n = 5(1)50$

One-sided	Two-sided	$n = 5$	$n = 6$	$n = 7$	$n = 8$	$n = 9$	$n = 10$
$\alpha = .05$	$\alpha = .10$	1	2	4	6	8	11
$\alpha = .025$	$\alpha = .05$		1	2	4	6	8
$\alpha = .01$	$\alpha = .02$			0	2	3	5
$\alpha = .005$	$\alpha = .01$				0	2	3
One-sided	Two-sided	$n = 11$	$n = 12$	$n = 13$	$n = 14$	$n = 15$	$n = 16$
$\alpha = .05$	$\alpha = .10$	14	17	21	26	30	36
$\alpha = .025$	$\alpha = .05$	11	14	17	21	25	30
$\alpha = .01$	$\alpha = .02$	7	10	13	16	20	24
$\alpha = .005$	$\alpha = .01$	5	7	10	13	16	19
One-sided	Two-sided	$n = 17$	$n = 18$	$n = 19$	$n = 20$	$n = 21$	$n = 22$
$\alpha = .05$	$\alpha = .10$	41	47	54	60	68	75
$\alpha = .025$	$\alpha = .05$	35	40	46	52	59	66
$\alpha = .01$	$\alpha = .02$	28	33	38	43	49	56
$\alpha = .005$	$\alpha = .01$	23	28	32	37	43	49
One-sided	Two-sided	$n = 23$	$n = 24$	$n = 25$	$n = 26$	$n = 27$	$n = 28$
$\alpha = .05$	$\alpha = .10$	83	92	101	110	120	130
$\alpha = .025$	$\alpha = .05$	73	81	90	98	107	117
$\alpha = .01$	$\alpha = .02$	62	69	77	85	93	102
$\alpha = .005$	$\alpha = .01$	55	68	68	76	84	92
One-sided	Two-sided	$n = 29$	$n = 30$	$n = 31$	$n = 32$	$n = 33$	$n = 34$
$\alpha = .05$	$\alpha = .10$	141	152	163	175	188	201
$\alpha = .025$	$\alpha = .05$	127	137	148	159	171	183
$\alpha = .01$	$\alpha = .02$	111	120	130	141	151	162
$\alpha = .005$	$\alpha = .01$	100	109	118	128	138	149
One-sided	Two-sided	$n = 35$	$n = 36$	$n = 37$	$n = 38$	$n = 39$	
$\alpha = .05$	$\alpha = .10$	214	228	242	256	271	
$\alpha = .025$	$\alpha = .05$	195	208	222	235	250	
$\alpha = .01$	$\alpha = .02$	174	186	198	211	224	
$\alpha = .005$	$\alpha = .01$	160	171	183	195	208	
One-sided	Two-sided	$n = 40$	$n = 41$	$n = 42$	$n = 43$	$n = 44$	$n = 45$
$\alpha = .05$	$\alpha = .10$	287	303	319	336	353	371
$\alpha = .025$	$\alpha = .05$	264	279	295	311	327	344
$\alpha = .01$	$\alpha = .02$	238	252	267	281	297	313
$\alpha = .005$	$\alpha = .01$	221	234	248	262	277	292
One-sided	Two-sided	$n = 46$	$n = 47$	$n = 48$	$n = 49$	$n = 50$	
$\alpha = .05$	$\alpha = .10$	389	408	427	446	466	
$\alpha = .025$	$\alpha = .05$	361	379	397	415	434	
$\alpha = .01$	$\alpha = .02$	329	345	362	380	398	
$\alpha = .005$	$\alpha = .01$	307	323	339	356	373	

From "Some Rapid Approximate Statistical Procedures" (1964), 28, F. Wilcoxon and R. A. Wilcox. Reproduced with the kind permission of R. A. Wilcox and the Lederle Laboratories.

Table 10
Distribution of the total number of runs R in samples of size (n_1, n_2); $P(R \leq a)$

	a								
(n_1, n_2)	2	3	4	5	6	7	8	9	10
(2,3)	.200	.500	.900	1.000					
(2,4)	.133	.400	.800	1.000					
(2,5)	.095	.333	.714	1.000					
(2,6)	.071	.286	.643	1.000					
(2,7)	.056	.250	.583	1.000					
(2,8)	.044	.222	.533	1.000					
(2,9)	.036	.200	.491	1.000					
(2,10)	.030	.182	.455	1.000					
(3,3)	.100	.300	.700	.900	1.000				
(3,4)	.057	.200	.543	.800	.971	1.000			
(3,5)	.036	.143	.429	.714	.929	1.000			
(3,6)	.024	.107	.345	.643	.881	1.000			
(3,7)	.017	.083	.283	.583	.833	1.000			
(3,8)	.012	.067	.236	.533	.788	1.000			
(3,9)	.009	.055	.200	.491	.745	1.000			
(3,10)	.007	.045	.171	.455	.706	1.000			
(4,4)	.029	.114	.371	.629	.886	.971	1.000		
(4,5)	.016	.071	.262	.500	.786	.929	.992	1.000	
(4,6)	.010	.048	.190	.405	.690	.881	.976	1.000	
(4,7)	.006	.033	.142	.333	.606	.833	.954	1.000	
(4,8)	.004	.024	.109	.279	.533	.788	.929	1.000	
(4,9)	.003	.018	.085	.236	.471	.745	.902	1.000	
(4,10)	.002	.014	.068	.203	.419	.706	.874	1.000	
(5,5)	.008	.040	.167	.357	.643	.833	.960	.992	1.000
(5,6)	.004	.024	.110	.262	.522	.738	.911	.976	.998
(5,7)	.003	.015	.076	.197	.424	.652	.854	.955	.992
(5,8)	.002	.010	.054	.152	.347	.576	.793	.929	.984
(5,9)	.001	.007	.039	.119	.287	.510	.734	.902	.972
(5,10)	.001	.005	.029	.095	.239	.455	.678	.874	.958
(6,6)	.002	.013	.067	.175	.392	.608	.825	.933	.987
(6,7)	.001	.008	.043	.121	.296	.500	.733	.879	.966
(6,8)	.001	.005	.028	.086	.226	.413	.646	.821	.937
(6,9)	.000	.003	.019	.063	.175	.343	.566	.762	.902
(6,10)	.000	.002	.013	.047	.137	.288	.497	.706	.864
(7,7)	.001	.004	.025	.078	.209	.383	.617	.791	.922
(7,8)	.000	.002	.015	.051	.149	.296	.514	.704	.867
(7,9)	.000	.001	.010	.035	.108	.231	.427	.622	.806
(7,10)	.000	.001	.006	.024	.080	.182	.355	.549	.743
(8,8)	.000	.001	.009	.032	.100	.214	.405	.595	.786
(8,9)	000	.001	.005	.020	.069	.157	.319	.500	.702
(8,10)	.000	.000	.003	.013	.048	.117	.251	.419	.621
(9,9)	.000	.000	.003	.012	.044	.109	.238	.399	.601
(9,10)	.000	.000	.002	.008	.029	.077	.179	.319	.510
(10,10)	.000	.000	.001	.004	.019	.051	.128	.242	.414

	α									
(n_1, n_2)	11	12	13	14	15	16	17	18	19	20
(2,3)										
(2,4)										
(2,5)										
(2,6)										
(2,7)										
(2,8)										
(2,9)										
(2,10)										
(3,3)										
(3,4)										
(3,5)										
(3,6)										
(3,7)										
(3,8)										
(3,9)										
(3,10)										
(4,4)										
(4,5)										
(4,6)										
(4,7)										
(4,8)										
(4,9)										
(4,10)										
(5,5)										
(5,6)	1.000									
(5,7)	1.000									
(5,8)	1.000									
(5,9)	1.000									
(5,10)	1.000									
(6,6)	.998	1.000								
(6,7)	.992	.999	1.000							
(6,8)	.984	.998	1.000							
(6,9)	.972	.994	1.000							
(6,10)	.958	.990	1.000							
(7,7)	.975	.996	.999	1.000						
(7,8)	.949	.988	.998	1.000	1.000					
(7,9)	.916	.975	.994	.999	1.000					
(7,10)	.879	.957	.990	.998	1.000					
(8,8)	.900	.968	.991	.999	1.000	1.000				
(8,9)	.843	.939	.980	.996	.999	1.000	1.000			
(8,10)	.782	.903	.964	.990	.998	1.000	1.000			
(9,9)	.762	.891	.956	.988	.997	1.000	1.000	1.000		
(9,10)	.681	.834	.923	.974	.992	.999	1.000	1.000	1.000	
(10,10)	.586	.758	.872	.949	.981	.996	.999	1.000	1.000	1.000

From "Tables for Testing Randomness of Grouping in a Sequence of Alternatives," C. Eisenhart and F. Swed, *Annals of Mathematical Statistics*, Volume 14 (1943). Reproduced with the kind permission of the Editor, *Annals of Mathematical Statistics*.

Table 11
Critical values of Spearman's rank correlation coefficient

n	$\alpha = 0.05$	$\alpha = 0.025$	$\alpha = 0.01$	$\alpha = 0.005$
5	0.900	—	—	—
6	0.829	0.886	0.943	—
7	0.714	0.786	0.893	—
8	0.643	0.738	0.833	0.881
9	0.600	0.683	0.783	0.833
10	0.564	0.648	0.745	0.794
11	0.523	0.623	0.736	0.818
12	0.497	0.591	0.703	0.780
13	0.475	0.566	0.673	0.745
14	0.457	0.545	0.646	0.716
15	0.441	0.525	0.623	0.689
16	0.425	0.507	0.601	0.666
17	0.412	0.490	0.582	0.645
18	0.399	0.476	0.564	0.625
19	0.388	0.462	0.549	0.608
20	0.377	0.450	0.534	0.591
21	0.368	0.438	0.521	0.576
22	0.359	0.428	0.508	0.562
23	0.351	0.418	0.496	0.549
24	0.343	0.409	0.485	0.537
25	0.336	0.400	0.475	0.526
26	0.329	0.392	0.465	0.515
27	0.323	0.385	0.456	0.505
28	0.317	0.377	0.448	0.496
29	0.311	0.370	0.440	0.487
30	0.305	0.364	0.432	0.478

From "Distribution of Sums of Squares of Rank Differences for Small Samples," E. G. Olds, *Annals of Mathematical Statistics*, Volume 9 (1938). Reproduced with the kind permission of the Editor, *Annals of Mathematical Statistics*.

Table 12
Squares, cubes, and roots

Roots of numbers other than those given directly may be found by the following relations:

$$\sqrt{100n} = 10\sqrt{n}; \quad \sqrt{1000n} = 10\sqrt{10n}; \quad \sqrt{\tfrac{1}{10}n} = \tfrac{1}{10}\sqrt{10n};$$
$$\sqrt{\tfrac{1}{100}n} = \tfrac{1}{10}\sqrt{n}; \quad \sqrt{\tfrac{1}{1000}n} = \tfrac{1}{100}\sqrt{10n};$$

n	n^2	$\sqrt{n}$	$\sqrt{10n}$	n	n^2	$\sqrt{n}$	$\sqrt{10n}$
				30	900	5.477 226	17.32051
1	1	1.000 000	3.162 278	31	961	5.567 764	17.60682
2	4	1.414 214	4.472 136	32	1 024	5.656 854	17.88854
3	9	1.732 051	5.477 226	33	1 089	5.744 563	18.16590
4	16	2.000 000	6.324 555	34	1 156	5.830 952	18.43909
5	25	2.236 068	7.071 068	35	1 225	5.916 080	18.70829
6	36	2.449 490	7.745 967	36	1 296	6.000 000	18.97367
7	49	2.645 751	8.366 600	37	1 369	6.082 763	19.23538
8	64	2.828 427	8.944 272	38	1 444	6.164 414	19.49359
9	81	3.000 000	9.486 833	39	1 521	6.244 998	19.74842
10	100	3.162 278	10.00000	**40**	1 600	6.324 555	20.00000
11	121	3.316 625	10.48809	41	1 681	6.403 124	20.24846
12	144	3.464 102	10.95445	42	1 764	6.480 741	20.49390
13	169	3.605 551	11.40175	43	1 849	6.557 439	20.73644
14	196	3.741 657	11.83216	44	1 936	6.633 250	20.97618
15	225	3.872 983	12.24745	45	2 025	6.708 204	21.21320
16	256	4.000 000	12.64911	46	2 116	6.782 330	21.44761
17	289	4.123 106	13.03840	47	2 209	6.855 655	21.67948
18	324	4.242 641	13.41641	48	2 304	6.928 203	21.90890
19	361	4.358 899	13.78405	49	2 401	7.000 000	22.13594
20	400	4.472 136	14.14214	**50**	2 500	7.071 068	22.36068
21	441	4.582 576	14.49138	51	2 601	7.141 428	22.58318
22	484	4.690 416	14.83240	52	2 704	7.211 103	22.80351
23	529	4.795 832	15.16575	53	2 809	7.280 110	23.02173
24	576	4.898 979	15.49193	54	2 916	7.348 469	23.23790
25	625	5.000 000	15.81139	55	3 025	7.416 198	23.45208
26	676	5.099 020	16.12452	56	3 136	7.483 315	23.66432
27	729	5.196 152	16.43168	57	3 249	7.549 834	23.87467
28	784	5.291 503	16.73320	58	3 364	7.615 773	24.08319
29	841	5.385 165	17.02939	59	3 481	7.618 146	24.28992

n	n^2	$\sqrt{n}$	$\sqrt{10n}$	n	n^2	$\sqrt{n}$	$\sqrt{10n}$
60	3 600	7.745 967	24.49490	**100**	10 000	10.00000	31.62278
61	3 721	7.810 250	24.69818	101	10 201	10.04998	31.78050
62	3 844	7.874 008	24.89980	102	10 404	10.09950	31.93744
63	3 969	7.937 254	25.09980	103	10 609	10.14889	32.09361
64	4 096	8.000 000	25.29822	104	10 816	10.19804	32.24903
65	4 225	8.062 258	25.49510	105	11 025	10.24695	32.40370
66	4 356	8.124 038	25.69047	106	11 236	10.29563	32.55764
67	4 489	8.185 353	25.88436	107	11 449	10.34408	32.71085
68	4 624	8.246 211	26.07681	108	11 664	10.39230	32.86335
69	4 761	8.306 624	26.26785	109	11 881	10.44031	33.01515
70	4 900	8.366 600	26.45751	**110**	12 100	10.48809	33.16625
71	5 041	8.426 150	26.64583	111	12 321	10.53565	33.31666
72	5 184	8.485 281	26.83282	112	12 544	10.58301	33.46640
73	5 329	8.544 004	27.01851	113	12 769	10.63015	33.61547
74	5 476	8.602 325	27.20294	114	12 996	10.67708	33.76389
75	5 625	8.660 254	27.38613	115	13 225	10.72381	33.91165
76	5 776	8.717 798	27.56810	116	13 456	10.77033	34.05877
77	5 929	8.774 964	27.74887	117	13 689	10.81665	34.20526
78	6 084	8.831 761	27.92848	118	13 924	10.86278	34.35113
79	6 241	8.888 194	28.10694	119	14 161	10.90871	34.49638
80	6 400	8.944 272	28.28427	**120**	14 400	10.95445	34.64102
81	6 561	9.000 000	28.46050	121	14 641	11.00000	34.78505
82	6 724	9.055 385	28.63564	122	14 884	11.04536	34.92850
83	6 889	9.110 434	28.80972	123	15 129	11.09054	35 07136
84	7 056	9.165 151	28.98275	124	15 376	11.13553	35 21363
85	7 225	9.219 544	29.15476	125	15 625	11.18034	35.35534
86	7 396	9.273 618	29.32576	126	15 876	11.22497	35.49648
87	7 569	9.327 379	29.49576	127	16 129	11.26943	35.63706
88	7 744	9.380 832	29.66479	128	16 384	11.31371	35.77709
89	7 921	9.433 981	29.83287	129	16 641	11.35782	35.91657
90	8 100	9.486 833	30.00000	**130**	16 900	11.40175	36.05551
91	8 281	9.539 392	30.16621	131	17 161	11.44552	36.19392
92	8 464	9.591 663	30.33150	132	17 424	11.48913	36.33180
93	8 649	9.643 651	30.49590	133	17 689	11.53256	36.46917
94	8 836	9.695 360	30.65942	134	17 956	11.57584	36.60601
95	9 025	9.746 794	30.82207	135	18 225	11.61895	36.74235
96	9 216	9.797 959	30.98387	136	18 496	11.66190	36.87818
97	9 409	9.848 858	31.14482	137	18 769	11.70470	37.01351
98	9 604	9.899 495	31.30495	138	19 044	11.74734	37.14835
99	9 801	9.949 874	31.46427	139	19 321	11.78983	37.28270

n	n^2	$\sqrt{n}$	$\sqrt{10n}$	n	n^2	$\sqrt{n}$	$\sqrt{10n}$
140	19 600	11.83216	37.41657	**180**	32 400	13.41641	42.42641
141	19 881	11.87434	37.54997	181	32 761	13.45362	42.54409
142	20 164	11.91638	37.68289	182	33 124	13.49074	42.66146
143	20 449	11.95826	37.81534	183	33 489	13.52775	42.77850
144	20 736	12.00000	37.94733	184	33 856	13.56466	42.89522
145	21 025	12.04159	38.07887	185	34 225	13.60147	43.01163
146	21 316	12.08305	38.20995	186	34 596	13.63818	43.12772
147	21 609	12.12436	38.34058	187	34 969	13.67479	43.24350
148	21 904	12.16553	38.47077	188	35 344	13.71131	43.35897
149	22 201	12.20656	38.60052	189	35 721	13.74773	43.47413
150	22 500	12.24745	38.72983	**190**	36 100	13.78405	43.58899
151	22 801	12.28821	38.85872	191	36 481	13.82027	43.70355
152	23 104	12.32883	38.98718	192	36 864	13.85641	43.81780
153	23 409	12.36932	39.11521	193	37 249	13.89244	43.93177
154	23 716	12.40967	39.24283	194	37 636	13.92839	44.04543
155	24 025	12.44990	39.37004	195	38 025	13.96424	44.15880
156	24 336	12.49000	39.49684	196	38 416	14.00000	44.27189
157	24 649	12.52996	39.62323	197	38 809	14.03567	44.38468
158	24 964	12.56981	39.74921	198	39 204	14.07125	44.49719
159	25 281	12.60952	39.87480	199	39 601	14.10674	44.60942
160	25 600	12.64911	40.00000	**200**	40 000	14.14214	44.72136
161	25 921	12.68858	40.12481	201	40 401	14.17745	44.83302
162	26 244	12.72792	40.24922	202	40 804	14.21267	44.94441
163	26 569	12.76715	40.37326	203	41 209	14.24781	45.05552
164	26 806	12.80625	40.49691	204	41 616	14.28286	45.16636
165	27 225	12.84523	40.62019	205	42 025	14.31782	45.27693
166	27 556	12.88410	40.74310	206	42 436	14.35270	45.38722
167	27 889	12.92285	40.86563	207	42 849	14.38749	45.49725
168	28 224	12.96148	40.98780	208	43 264	14.42221	45.60702
169	28 561	13.00000	41.10961	209	43 681	14.45683	45.71652
170	28 900	13.03840	41.23106	**210**	44 100	14.49138	45.82576
171	29 241	13.07670	41.35215	211	44 521	14.52584	45.93474
172	29 584	13.11488	41.47288	212	44 944	14.56022	46.04346
173	29 929	13.15295	41.59327	213	45 369	14.59452	46.15192
174	30 276	13.19091	41.71331	214	45 796	14.62874	46.26013
175	30 625	13.22876	41.83300	215	46 225	14.66288	46.36809
176	30 976	13.26650	41.95235	216	46 656	14.69694	46.47580
177	31 329	13.30413	42.07137	217	47 089	14.73092	46.58326
178	31 684	13.34166	42.19005	218	47 524	14.76482	46.69047
179	32 041	13.37909	42.30829	219	47 961	14.79865	46.79744

n	n^2	$\sqrt{n}$	$\sqrt{10n}$	n	n^2	$\sqrt{n}$	$\sqrt{10n}$
220	48 400	14.83240	46.90416	**260**	67 600	16.12452	50.99020
221	48 841	14.86607	47.01064	261	68 121	16.15549	51.08816
222	49 284	14.89966	47.11688	262	68 644	16.18641	51.18594
223	49 729	14.93318	47.22288	263	69 169	16.21727	51.28353
224	50 176	14.96663	47.32864	264	69 696	16.24808	51.38093
225	50 625	15.00000	47.43416	265	70 225	16.27882	51.47815
226	51 076	15.03330	47.53946	266	70 756	16.30951	51.57519
227	51 529	15.06652	47.64452	267	71 289	16.34013	51.67204
228	51 984	15.09967	47.74935	268	71 824	16.37071	51.76872
229	52 441	15.13275	47.85394	269	72 361	16.40122	51.86521
230	52 900	15.16575	47.95832	**270**	72 900	16.43168	51.96152
231	53 361	15.19868	48.06246	271	73 441	16.46208	52.05766
232	53 824	15.23155	48.16638	272	73 984	16.49242	52.15362
233	54 289	15.26434	48.27007	273	74 529	16.52271	52.24940
234	54 756	15.29706	48.37355	274	75 076	16.55295	52.34501
235	55 225	15.32971	48.47680	275	75 625	16.58312	52.44044
236	55 696	15.36229	48.57983	276	76 176	16.61235	52.53570
237	56 169	15.39480	48.68265	277	76 729	16.64332	52.63079
238	56 644	15.42725	48.78524	278	77 284	16.67333	52.72571
239	57 121	15.45962	48.88763	279	77 841	16.70329	52.82045
240	57 600	15.49193	48.98979	**280**	78 400	16.73320	52.91503
241	58 081	15.52417	49.09175	281	78 961	16.76305	53.00943
242	58 564	15.55635	49.19350	282	79 524	16.79286	53.10367
243	59 049	15.58846	49.29503	283	80 089	16.82260	53.19774
244	59 536	15.62050	49.39636	284	80 656	16.85230	53.29165
245	60 025	15.65248	49.49747	285	81 225	16.88194	53.38539
246	60 516	15.68439	49.59839	286	81 796	16.91153	53.47897
247	61 009	15.71623	49.69909	287	82 369	16.94107	53.57238
248	61 504	15.74902	49.79960	288	82 944	16.97056	53.66563
249	62 001	15.77973	49.89990	289	83 521	17.00000	53.75872
250	62 500	15.81139	50.00000	**290**	84 100	17.02939	53.85165
251	63 001	15.84298	50.09990	291	84 681	17.05872	53.94442
252	63 504	15.87451	50.19960	292	85 264	17.08801	54.03702
253	64 009	15.90597	50.29911	293	85 849	17.11724	54.12947
254	64 516	15.93738	50.39841	294	86 436	17.14643	54.22177
255	65 025	15.96872	50.49752	295	87 025	17.17556	54.31390
256	65 536	16.00000	50.59644	296	87 616	17.20465	54.40588
257	66 049	16.03122	50.69517	297	88 209	17.23369	54.49771
258	66 564	16.06238	50.79370	298	88 804	17.26268	54.58938
259	67 081	16.09348	50.89204	299	89 401	17.29162	54.68089

n	n^2	$\sqrt{n}$	$\sqrt{10n}$	n	n^2	$\sqrt{n}$	$\sqrt{10n}$
300	90 000	17.32051	54.77226	**340**	115 600	18.43909	58.30952
301	90 601	17.34935	54.86347	341	116 281	18.46619	58.39521
302	91 204	17.37815	54.95453	342	116 964	18.49324	58.48077
303	91 809	17.40690	55.04544	343	117 649	18.52026	58.56620
304	92 416	17.43560	55.13620	344	118 336	18.54724	58.65151
305	93 025	17.46425	55.22681	345	119 025	18.57418	58.73670
306	93 636	17.49286	55.31727	346	119 716	18.60108	58.82176
307	94 249	17.52142	55.40758	347	120 409	18.62794	58.90671
308	94 864	17.54993	55.49775	348	121 104	18.65476	58.99152
309	95 481	17.57840	55.58777	349	121 801	18.68154	59.07622
310	96 100	17.60682	55.67764	**350**	122 500	18.70829	59.16080
311	96 721	17.63519	55.76737	351	123 201	18.73499	59.24525
312	97 344	17.66352	55.85696	352	123 904	18.76166	59.32959
313	97 969	17.69181	55.94640	353	124 609	18.78829	59.41380
314	98 596	17.72005	56.03570	354	125 316	18.81489	59.49790
315	99 225	17.74824	56.12486	355	126 025	18.84144	59.58188
316	99 856	17.77639	56.21388	356	126 736	18.86796	59.66574
317	100 489	17.80449	56.30275	357	127 449	18.89444	59.74948
318	101 124	17.83255	56.39149	358	128 164	18.92089	59.83310
319	101 761	17.86057	56.48008	359	128 881	18.94730	59.91661
320	102 400	17.88854	56.56854	**360**	129 600	18.97367	60.00000
321	103 041	17.91647	56.65686	361	130 321	19.00000	60.08328
322	103 684	17.94436	56.74504	362	131 044	19.02630	60.16644
323	104 329	17.97220	56.83309	363	131 769	19.05256	60.24948
324	104 976	18.00000	56.92100	364	132 496	19.07878	60.33241
325	105 625	18.02776	57.00877	365	133 225	19.10497	60.41523
326	106 276	18.05547	57.09641	366	133 956	19.13113	60.49793
327	106 929	18.08314	57.18391	367	134 689	19.15724	60.58052
328	107 584	18.11077	57.27128	368	135 424	19.18333	60.66300
329	108 241	18.13836	57.35852	369	136 161	19.20937	60.74537
330	108 900	18.16590	57.44563	**370**	136 900	19.23538	60.82763
331	109 561	18.19341	57.53260	371	137 641	19.26136	60.90977
332	110 224	18.22087	57.61944	372	138 384	19.28730	60.99180
333	110 889	18.24829	57.70615	373	139 129	19.31321	61.07373
334	111 556	18.27567	57.79273	374	139 876	19.33908	61.15554
335	112 225	18.30301	57.87918	375	140 625	19.36492	61.23724
336	112 896	18.33030	57.96551	376	141 376	19.39072	61.31884
337	113 569	18.35756	58.05170	377	142 129	19.41649	61.40033
338	114 244	18.38478	58.13777	378	142 884	19.44222	61.48170
339	114 921	18.41195	58.22371	379	143 641	19.46792	61.56298

n	n^2	$\sqrt{n}$	$\sqrt{10n}$	n	n^2	$\sqrt{n}$	$\sqrt{10n}$
380	144 400	19.49359	61.64414	**420**	176 400	20.49390	64.80741
381	145 161	19.51922	61.72520	421	177 241	20.51828	64.88451
382	145 924	19.54482	61.80615	422	178 084	20.54264	64.96153
383	146 689	19.57039	61.88699	423	178 929	20.56696	65.03845
384	147 456	19.59592	61.96773	424	179 776	20.59126	65.11528
385	148 225	19.62142	62.04837	425	180 625	20.61553	65.19202
386	148 996	19.64688	62.12890	426	181 476	20.63977	65.26868
387	149 769	19.67232	62.20932	427	182 329	20.66398	65.34524
388	150 544	19.69772	62.28965	428	183 184	20.68816	65.42171
389	151 321	19.72308	62.36986	429	184 041	20.71232	65.49809
390	152 100	19.74842	62.44998	**430**	184 900	20.73644	65.57439
391	152 881	19.77372	62.52999	431	185 761	20.76054	65.65059
392	153 664	19.79899	62.60990	432	186 624	20.78461	65.72671
393	154 449	19.82423	62.68971	433	187 489	20.80865	65.80274
394	155 236	19.84943	62.76942	434	188 356	20.83267	65.87868
395	156 025	19.87461	62.84903	435	189 225	20.85665	65.95453
396	156 816	19.89975	62.92853	436	190 096	20.88061	66.03030
397	157 609	19.92486	63.00794	437	190 969	20.90454	66.10598
398	158 404	19.94994	63.08724	438	191 844	20.92845	66.18157
399	159 201	19.97498	63.16645	439	192 721	20.95233	66.25708
400	160 000	20.00000	63.24555	**440**	193 600	20.97618	66.33250
401	160 801	20.02498	63.32456	441	194 481	21.00000	66.40783
402	161 604	20.04994	63.40347	442	195 364	21.02380	66.48308
403	162 409	20.07486	63.48228	443	196 249	21.04757	66.55825
404	163 216	20.09975	63.56099	444	197 136	21.07131	66.63332
405	164 025	20.12461	63.63961	445	198 025	21.09502	66.70832
406	164 836	20.14944	63.71813	446	198 916	21.11871	66.78323
407	165 649	20.17424	63.79655	447	199 809	21.14237	66.85806
408	166 464	20.19901	63.87488	448	200 704	21.16601	66.93280
409	167 281	20.22375	63.95311	449	201 601	21.18962	67.00746
410	168 100	20.24864	64.03124	**450**	202 500	21.21320	67.08204
411	168 921	20.27313	64.10928	451	203 401	21.23676	67.15653
412	169 744	20.29778	64.18723	452	204 304	21.26029	67.23095
413	170 569	20.32240	54.26508	453	205 209	21.28380	67.30527
414	171 396	20.34699	64.34283	454	206 116	21.30728	67.37952
415	172 225	20.37155	64.42049	455	207 025	21.33073	67.45369
416	173 056	20.39608	64.49806	456	207 936	21.35416	67.52777
417	173 889	20.42058	64.57554	457	208 849	21.37756	67.60178
418	174 724	20.44505	64.65292	458	209 764	21.40093	67.67570
419	175 561	20.46949	64.73021	459	210 681	21.42429	67.74954

n	n^2	$\sqrt{n}$	$\sqrt{10n}$	n	n^2	$\sqrt{n}$	$\sqrt{10n}$
460	211 600	21.44761	67.82330	**500**	250 000	22.36068	70.71068
461	212 521	21.47091	67.89698	501	251 001	22.38303	70.78135
462	213 444	21.49419	67.97058	502	252 004	22.40536	70.85196
463	214 369	21.51743	68.04410	503	253 009	22.42766	70.92249
464	215 296	21.54066	68.11755	504	254 016	22.44994	70.99296
465	216 225	21.56386	68.19091	505	255 025	22.47221	71.06335
466	217 156	21.58703	68.26419	506	256 036	22.49444	71.13368
467	218 089	21.61018	68.33740	507	257 049	22.51666	71.20393
468	219 024	21.63331	68.41053	508	258 064	22.53886	71.27412
469	219 961	21.65641	68.48957	509	259 081	22.56103	71.34424
470	220 900	21.67948	68.55655	**510**	260 100	22.58318	71.41428
471	221 841	21.70253	68.62944	511	261 121	22.60531	71.48426
472	222 784	21.72556	68.70226	512	262 144	22.62742	71.55418
473	223 729	21.74856	68.77500	513	263 169	22.64950	71.62402
474	224 676	21.77154	68.84766	514	264 196	22.67157	71.69379
475	225 625	21.79449	68.92024	515	265 225	22.69361	71.76350
476	226 576	21.81742	68.99275	516	266 256	22.71563	71.83314
477	227 529	21.84033	69.06519	517	267 289	22.73763	71.90271
478	228 484	21.86321	69.13754	518	268 324	22.75961	71.97222
479	229 441	21.88607	69.20983	519	269 361	22.78157	72.04165
480	230 400	21.90890	69.28203	**520**	270 400	22.80351	72.11103
481	231 361	21.93171	69.35416	521	271 441	22.82542	72.18033
482	232 324	21.95450	69.42622	522	272 484	22.84732	72.24957
483	233 289	21.97726	69.49820	523	273 529	22.86919	72.31874
484	234 256	22.00000	69.57011	524	274 576	22.89105	72.38784
485	235 225	22.02272	69.64194	525	275 625	22.91288	72.45688
486	236 196	22.04541	69.71370	526	276 676	22.93469	72.52586
487	237 169	22.06808	69.78539	527	277 729	22.95648	72.59477
488	238 144	22.09072	69.85700	528	278 784	22.97825	72.66361
489	239 121	22.11334	69.92853	529	279 841	23.00000	72.73239
490	240 100	22.13594	70.00000	**530**	280 900	23.02173	72.80110
491	241 081	22.15852	70.07139	531	281 961	23.04344	72.86975
492	242 064	22.18107	70.14271	532	283 024	23.06513	72.93833
493	243 049	22.20360	70.21396	533	284 089	23.08679	73.00685
494	244 036	22.22611	70.28513	534	285 156	23.10844	73.07530
495	245 025	22.24860	70.35624	535	286 225	23.13007	73.14369
496	246 016	22.27106	70.42727	536	287 296	23.15167	73.21202
497	247 009	22.29350	70.49823	537	288 369	23.17326	73.28028
498	248 004	22.31591	70.56912	538	289 444	23.19483	73.34848
499	249 001	22.33831	70.63993	539	290 521	23.21637	73.41662

n	n^2	$\sqrt{n}$	$\sqrt{10n}$	n	n^2	$\sqrt{n}$	$\sqrt{10n}$
540	291 600	23.23790	73.48469	**580**	336 400	24.08319	76.15773
541	292 681	23.25941	73.55270	581	337 561	24.10394	76.22336
542	293 764	23.28089	73.62065	582	338 724	24.12468	76.28892
543	294 849	23.30236	73.68853	583	339 889	24.14539	76.35444
544	295 936	23.32381	73.75636	584	341 056	24.16609	76.41989
545	297 025	23.34524	73.82412	585	342 225	24.18677	76.48529
546	298 116	23.36664	73.89181	586	343 396	24.20744	76.55064
547	299 209	23.38803	73.95945	587	344 569	24.22808	76.61593
548	300 304	23.40940	74.02702	588	345 744	24.24871	76.68116
549	301 401	23.43075	74.09453	589	346 921	24.26932	76.74634
550	302 500	23.45208	74.16198	**590**	348 100	24.28992	76.81146
551	303 601	23.47339	74.22937	591	349 281	24.31049	76.87652
552	304 704	23.49468	74.29670	592	350 464	24.33105	76.94154
553	305 809	23.51595	74.36397	593	351 649	24.35159	77.00649
554	306 916	23.53720	74.43118	594	352 836	24.37212	77.07140
555	308 025	23.55844	74.49832	595	354 025	24.39262	77.13624
556	309 136	23.57965	74.56541	596	355 216	24.41311	77.20104
557	310 249	23.60085	74.63243	597	356 409	24.43358	77.26578
558	311 364	23.62202	74.69940	598	357 604	24.45404	77.33046
559	312 481	23.64318	74.76630	599	358 801	24.47448	77.39509
560	313 600	23.66432	74.83315	**600**	360 000	24.49490	77.45967
561	314 721	23.68544	74.89993	601	361 201	24.51530	77.52419
562	315 844	23.70654	74.96666	602	362 404	24.53569	77.58866
563	316 969	23.72762	75.03333	603	363 609	24.55606	77.65307
564	318 096	23.74868	75.09993	604	364 816	24.57641	77.71744
565	319 225	23.76973	75.16648	605	366 025	24.59675	77.78175
566	320 356	23.79075	75.23297	606	367 236	24.61707	77.84600
567	321 489	23.81176	75.29940	607	368 449	24.63737	77.91020
568	322 624	23.83275	75.36577	608	369 664	24.65766	77.97435
569	323 761	23.85372	75.43209	609	370 881	24.67793	78.03845
570	324 900	23.87467	75.49834	**610**	372 100	24.69818	78.10250
571	326 041	23.89561	75.56454	611	373 321	24.71841	78.16649
572	327 184	23.91652	75.63068	612	374 544	24.73863	78.23043
573	328 329	23.93742	75.69676	613	375 769	24.75884	78.29432
574	329 476	23.95830	75.76279	614	376 996	24.77902	78.35815
575	330 625	23.97916	75.82875	615	378 225	24.79919	78.42194
576	331 776	24.00000	75.89466	616	379 456	24.81935	78.48567
577	332 929	24.02082	75.96052	617	380 689	24.83948	78.54935
578	334 084	24.04163	76.02631	618	381 924	24.85961	78.61298
579	335 241	24.06242	76.09205	619	383 161	24.87971	78.67655

n	n^2	$\sqrt{n}$	$\sqrt{10n}$	n	n^2	$\sqrt{n}$	$\sqrt{10n}$
620	384 400	24.89980	78.74008	**660**	435 600	25.69047	81.24038
621	385 641	24.91987	78.80355	661	436 921	25.70992	81.30191
622	386 884	24.93993	78.86698	662	438 244	25.72936	81.36338
623	388 129	24.95997	78.93035	663	439 569	25.74879	81.42481
624	389 376	24.97999	78.99367	664	440 896	25.76820	81.48620
625	390 625	25.00000	79.05694	665	442 225	25.78759	81.54753
626	391 876	25.01999	79.12016	666	443 556	25.80698	81.60882
627	393 129	25.03997	79.18333	667	444 889	25.82634	81.67007
628	394 384	25.05993	79.24645	668	446 224	25.84570	81.73127
629	395 641	25.07987	79.30952	669	447 561	25.86503	81.79242
630	396 900	25.09980	79.37254	**670**	448 900	25.88436	81.85353
631	398 161	25.11971	79.43551	671	450 241	25.90367	81.91459
632	399 424	25.13961	79.49843	672	451 584	25.92296	81.97561
633	400 689	25.15949	79.56130	673	452 929	25.94224	82.03658
634	401 956	25.17936	79.62412	674	454 276	25.96151	82.09750
635	403 225	25.19921	79.68689	675	455 625	25.98076	82.15838
636	404 496	25.21904	79.74961	676	456 976	26.00000	82.21922
637	405 769	25.23886	79.81228	677	458 329	26.01922	82.28001
638	407 044	25.25866	79.87490	678	459 684	26.03843	82.34076
639	408 321	25.27845	79.93748	679	461 041	26.05763	82.40146
640	409 600	25.29822	80.00000	**680**	462 400	26.07681	82.46211
641	410 881	25.31798	80.06248	681	463 761	26.09598	82.52272
642	412 164	25.33772	80.12490	682	465 124	26.11513	82.58329
643	413 449	25.35744	80.18728	683	466 489	26.13427	82.64381
644	414 736	25.37716	80.24961	684	467 856	26.15339	82.70429
645	416 025	25.39685	80.31189	685	469 225	26.17250	82.76473
646	417 316	25.41653	80.37413	686	470 596	26.19160	82.82512
647	418 609	25.43619	80.43631	687	471 969	26.21068	82.88546
648	419 904	25.45584	80.49845	688	473 344	26.22975	82.94577
649	421 201	25.47548	80.56054	689	474 721	26.24881	83.00602
650	422 500	25.49510	80.62258	**690**	476 100	26.26785	83.06624
651	423 801	25.51470	80.68457	691	477 481	26.28688	83.12641
652	425 104	25.53429	80.74652	692	478 864	26.30589	83.18654
653	426 409	25.55386	80.80842	693	480 249	26.32489	83.24662
654	427 716	25.57342	80.87027	694	481 636	26.34388	83.30666
655	429 025	25.59297	80.93207	695	483 025	26.36285	83.36666
656	430 336	25.61250	80.99383	696	484 416	26.38181	83.42661
657	431 649	25.63201	81.05554	697	485 809	26.40076	83.48653
658	432 964	25.65151	81.11720	698	487 204	26.41969	83.54639
659	434 281	25.67100	81.17881	699	488 601	26.43861	83.60622

n	n^2	$\sqrt{n}$	$\sqrt{10n}$	n	n^2	$\sqrt{n}$	$\sqrt{10n}$
700	490 000	26.45751	83.66600	**740**	547 600	27.20294	86.02325
701	491 401	26.47640	83.72574	741	549 081	27.22132	86.08136
702	492 804	26.49528	83.78544	742	550 564	27.23968	86.13942
703	494 209	26.51415	83.84510	743	552 049	27.25803	86.19745
704	495 616	26.53300	83.90471	744	553 536	27.27636	86.25543
705	497 025	26.55184	83.96428	745	555 025	27.29469	86.31338
706	498 436	26.57066	84.02381	746	556 516	27.31300	86.37129
707	499 849	26.58947	84.08329	747	558 009	27.33130	86.42916
708	501 264	26.60827	84.14274	748	559 504	27.34959	86.48699
709	502 681	26.62705	84.20214	749	561 001	27.36786	86.54479
710	504 100	26.64583	84.26150	**750**	562 500	27.38613	86.60254
711	505 521	26.66458	84.32082	751	564 001	27.40438	86.66026
712	506 944	26.68333	84.38009	752	565 504	27.42262	86.71793
713	508 369	26.70206	84.43933	753	567 009	27.44085	86.77557
714	509 796	26.72078	84.49852	754	568 516	27.45906	86.83317
715	511 225	26.73948	84.55767	755	570 025	27.47726	86.89074
716	512 656	26.75818	84.61678	756	571 536	27.49545	86.94826
717	514 089	26.77686	84.67585	757	573 049	27.51363	87.00575
718	515 524	26.79552	84.73488	758	574 564	27.53180	87.06320
719	516 961	26.81418	84.79387	759	576 081	27.54995	87.12061
720	518 400	26.83282	84.85281	**760**	577 600	27.56810	87.17798
721	519 841	26.85144	84.91172	761	579 121	27.58623	87.23531
722	521 284	26.87006	84.97058	762	580 644	27.60435	87.29261
723	522 729	26.88866	85.02941	763	582 169	27.62245	87.34987
724	524 176	26.90725	85.08819	764	583 696	27.64055	87.40709
725	525 625	26.92582	85.14693	765	585 225	27.65863	87.46428
726	527 076	26.94439	85.20563	766	586 756	27.67671	87.52143
727	528 529	26.96294	85.26429	767	588 289	27.69476	87.57854
728	529 984	26.98148	85.32292	768	589 824	27.71281	87.63561
729	531 441	27.00000	85.38150	769	591 361	27.73085	87.69265
730	532 900	27.01851	85.44004	**770**	592 900	27.74887	87.74964
731	534 361	27.03701	85.49854	771	594 441	27.76689	87.80661
732	535 824	27.05550	85.55700	772	595 984	27.78489	87.86353
733	537 289	27.07397	85.61542	773	597 529	27.80288	87.92042
734	538 756	27.09243	85.67380	774	599 076	27.82086	87.97727
735	540 225	27.11088	85.73214	775	600 625	27.83882	88.03408
736	541 696	27.12932	85.79044	776	602 176	27.85678	88.09086
737	543 169	27.14774	85.84870	777	603 729	27.87472	88.14760
738	544 644	27.16616	85.90693	778	605 284	27.89265	88.20431
739	546 121	27.18455	85.96511	779	606 841	27.91057	88.26098

n	n^2	$\sqrt{n}$	$\sqrt{10n}$	n	n^2	$\sqrt{n}$	$\sqrt{10n}$
780	608 400	27.92848	88.31761	**820**	672 400	28.63564	90.55385
781	609 961	27.94638	88.37420	821	674 041	28.65310	90.60905
782	611 524	27.96426	88.43076	822	675 684	28.67054	90.66422
783	613 089	27.98214	88.48729	823	677 329	28.68798	90.71935
784	614 656	28.00000	88.54377	824	678 976	28.70540	90.77445
785	616 225	28.01785	88.60023	825	680 625	28.72281	90.82951
786	617 796	28.03569	88.65664	826	682 726	28.74022	90.88454
787	619 369	28.05352	88.71302	827	683 929	28.75761	90.93954
788	620 944	28.07134	88.76936	828	685 584	28.77499	90.99451
789	622 521	28.08914	88.82567	829	687 241	28.79236	91.04944
790	624 100	28.10694	88.88194	**830**	688 900	28.80972	91.10434
791	625 681	28.12472	88.93818	831	690 561	28.82707	91.15920
792	627 264	28.14249	88.99428	832	692 224	28.84441	91.21403
793	628 849	28.16026	89.05055	833	693 889	28.86174	91.26883
794	630 436	28.17801	89.10668	834	695 556	28.87906	91.32360
795	632 025	28.19574	89.16277	835	697 225	28.89637	91.37833
796	633 616	28.21347	89.21883	836	698 896	28.91366	91.43304
797	635 209	28.23119	89.27486	837	700 569	28.93095	91.48770
798	636 804	28.24889	89.33085	838	702 244	28.94823	91.54234
799	638 401	28.26659	89.38680	839	703 921	28.96550	91.59694
800	640 000	28.28472	89.44272	**840**	705 600	28.98275	91.65151
801	641 601	28.30194	89.49860	841	707 281	29.00000	91.70605
802	643 204	28.31960	89.55445	842	708 964	29.01724	91.76056
803	644 809	28.33725	89.61027	843	710 649	29.03446	91.81503
804	646 416	28.35489	89.66605	844	712 336	29.05168	91.86947
805	648 025	28.37252	89.72179	845	714 025	29.06888	91.92388
806	649 636	28.39014	89.77750	846	715 716	29.08608	91.97826
807	651 249	28.40775	89.83318	847	717 409	29.10326	92.03260
808	652 864	28.42534	89.88882	848	719 104	29.12044	92.08692
809	654 481	28.44293	89.94443	849	720 801	29.13760	92.14120
810	656 100	28.46050	90.00000	**850**	722 500	29.15476	92.19544
811	657 721	28.47806	90.05554	851	724 201	29.17190	92.24966
812	659 344	28.49561	90.11104	852	725 904	29.18904	92.30385
813	660 969	28.51315	90.16651	853	727 609	29.20616	92.35800
814	662 596	28.53069	90.22195	854	729 316	29.22328	92.41212
815	664 225	28.54820	90.27735	855	731 025	29.24038	92.46621
816	665 856	28.56571	90.33272	856	732 736	29.25748	92.52027
817	667 489	28.58321	90.38805	857	734 449	29.27456	92.57429
818	669 124	28.60070	90.44335	858	736 164	29.29164	92.62829
819	670 761	28.61818	90.49862	859	737 881	29.30870	92.68225

n	n^2	$\sqrt{n}$	$\sqrt{10n}$	n	n^2	$\sqrt{n}$	$\sqrt{10n}$
860	739 600	29.32576	92.73618	**900**	810 000	30.00000	94.86833
861	741 321	29.34280	92.79009	901	811 801	30.01666	94.92102
862	743 044	29.35984	92.84396	902	813 604	30.03331	94.97368
863	744 769	29.37686	92.89779	903	815 409	30.04996	95.02631
864	746 496	29.39388	92.95160	904	817 216	30.06659	95.07891
865	748 225	29.41088	93.00538	905	819 025	30.08322	95.13149
866	749 956	29.42788	93.05912	906	820 836	30.09983	95.18403
867	751 689	29.44486	93.11283	907	822 649	30.11644	95.23655
868	753 424	29.46184	93.16652	908	824 464	30.13304	95.28903
869	755 161	29.47881	93.22017	909	826 281	30.14963	95.34149
870	756 900	29.49576	93.27379	**910**	828 100	30.16621	95.39392
871	758 641	29.51271	93.32738	911	829 921	30.18278	95.44632
872	760 384	29.52965	93.38094	912	831 744	30.19934	95.49869
873	762 129	29.54657	93.43447	913	833 569	30.21589	95.55103
874	763 876	29.56349	93.48797	914	835 396	30.23243	95.60335
875	765 625	29.58040	93.54143	915	837 225	30.24897	95.65563
876	767 376	29.59730	93.59487	916	839 056	30.26549	95.70789
877	769 129	29.61419	93.64828	917	840 889	30.28201	95.76012
878	770 884	29.63106	93.70165	918	842 724	30.29851	95.81232
879	772 641	29.64793	93.75500	919	844 561	30.31501	95.86449
880	774 400	29.66479	93.80832	**920**	846 400	30.33150	95.91663
881	776 161	29.68164	93.86160	921	848 241	30.34798	95.96874
882	777 924	29.69848	93.91486	922	850 084	30.36445	96.02083
883	779 689	29.71532	93.96808	923	851 929	30.38092	96.07289
884	781 456	29.73214	94.02127	924	853 776	30.39737	96.12492
885	783 225	29.74895	94.07444	925	855 625	30.41381	96.17692
886	784 996	29.76575	94.12757	926	857 476	30.43025	96.22889
887	786 769	29.78255	94.18068	927	859 329	30.44667	96.28084
888	788 544	29.79933	94.23375	928	861 184	30.46309	96.33276
889	790 321	29.81610	94.28680	929	863 041	30.47950	96.38465
890	792 100	29.83287	94.33981	**930**	864 900	30.49590	96.43651
891	793 881	29.84962	94.39280	931	866 761	30.51229	96.48834
892	795 664	29.86637	94.44575	932	868 624	30.52868	96.54015
893	797 449	29.88311	94.49868	933	870 489	30.54505	96.59193
894	799 236	29.89983	94.55157	934	872 356	30.56141	96.64368
895	801 025	29.91655	94.60444	935	874 225	30.57777	96.69540
896	802 816	29.93326	94.65728	936	876 096	30.59412	96.74709
897	804 609	29.94996	94.71008	937	877 969	30.61046	96.79876
898	806 404	29.96665	94.76286	938	879 844	30.62679	96.85040
899	808 201	29.98333	94.81561	939	881 721	30.64311	96.90201

n	n^2	$\sqrt{n}$	$\sqrt{10n}$	n	n^2	$\sqrt{n}$	$\sqrt{10n}$
940	883 600	30.65942	96.95360	**970**	940 900	31.14482	98.48858
941	885 481	30.67572	97.00515	971	942 841	31.16087	98.53933
942	887 364	30.69202	97.05668	972	944 784	31.17691	98.59006
943	889 249	30.70831	97.10819	973	946 729	31.19295	98.64076
944	891 136	30.72458	97.15966	974	948 676	31.20897	98.69144
945	893 025	30.74085	97.21111	975	950 625	31.22499	98.74209
946	894 916	30.75711	97.26253	976	952 576	31.24100	98.79271
947	896 809	30.77337	97.31393	977	954 529	31.25700	98.84331
948	898 704	30.78961	97.36529	978	956 484	31.27299	98.89388
949	900 601	30.80584	97.41663	979	958 441	31.28898	98.94443
950	902 500	30.82207	97.46794	**980**	960 400	31.30495	98.99495
951	904 401	30.83829	97.51923	981	962 361	31.32092	99.04544
952	906 304	30.85450	97.57049	982	964 324	31.33688	99.09591
953	908 209	30.87070	97.62172	983	966 289	31.35283	99.14636
954	910 116	30.88689	97.67292	984	968 256	31.36877	99.19677
955	912 025	30.90307	97.72410	985	970 225	31.38471	99.24717
956	913 936	30.91925	97.77525	986	972 196	31.40064	99.29753
957	915 849	30.93542	97.82638	987	974 169	31.41656	99.34787
958	917 764	30.95158	97.87747	988	976 144	31.43247	99.39819
959	919 681	30.96773	97.92855	989	978 121	31.44837	99.44848
960	921 600	30.98387	97.97959	**990**	980 100	31.46427	99.49874
961	923 521	31.00000	98.03061	991	982 081	31.48015	99.54898
962	925 444	31.01612	98.08160	992	984 064	31.49603	99.59920
963	927 369	31.03224	98.13256	993	986 049	31.51190	99.64939
964	929 296	31.04835	98.18350	994	988 036	31.52777	99.69955
965	931 225	31.06445	98.23441	995	990 025	31.54362	99.74969
966	933 156	31.08054	98.28530	996	992 016	31.55947	99.79980
967	935 089	31.09662	98.33616	997	994 009	31.57531	99.84989
968	937 024	31.11270	98.38699	998	996 004	31.59114	99.89995
969	938 961	31.12876	98.43780	999	998 001	31.60696	99.94999
				1000	1000 000	31.62278	100.00000

From *Handbook of Tables for Probability and Statistics*, 2nd ed. Edited by William H. Beyer (Cleveland: The Chemical Rubber Company, 1968). Reproduced by permission of the publishers.

Table 13
Random numbers

Line/Col.	(1)	(2)	(3)	(4)	(5)	(6)	(7)	(8)	(9)	(10)	(11)	(12)	(13)	(14)
1	10480	15011	01536	02011	81647	91646	69179	14194	62590	36207	20969	99570	91291	90700
2	22368	46573	25595	85393	30995	89198	27982	53402	93965	34095	52666	19174	39615	99505
3	24130	48360	22527	97265	76393	64809	15179	24830	49340	32081	30680	19655	63348	58629
4	42167	93093	06243	61680	07856	16376	39440	53537	71341	57004	00849	74917	97758	16379
5	37570	39975	81837	16656	06121	91782	60468	81305	49684	60672	14110	06927	01263	54613
6	77921	06907	11008	42751	27756	53498	18602	70659	90655	15053	21916	81825	44394	42880
7	99562	72905	56420	69994	98872	31016	71194	18738	44013	48840	63213	21069	10634	12952
8	96301	91977	05463	07972	18876	20922	94595	56869	69014	60045	18425	84903	42508	32307
9	89579	14342	63661	10281	17453	18103	57740	84378	25331	12566	58678	44947	05585	56941
10	85475	36857	53342	53988	53060	59533	38867	62300	08158	17983	16439	11458	18593	64952
11	28918	69578	88231	33276	70997	79936	56865	05859	90106	31595	01547	85590	91610	78188
12	63553	40961	48235	03427	49626	69445	18663	72695	52180	20847	12234	90511	33703	90322
13	09429	93969	52636	92737	88974	33488	36320	17617	30015	08272	84115	27156	30613	74952
14	10365	61129	87529	85689	48237	52267	67689	93394	01511	26358	85104	20285	29975	89868
15	07119	97336	71048	08178	77233	13916	47564	81056	97735	85977	29372	74461	28551	90707
16	51085	12765	51821	51259	77452	16308	60756	92144	49442	53900	70960	63990	75601	40719
17	02368	21382	52404	60268	89368	19885	55322	44819	01188	65255	64835	44919	05944	55157
18	01011	54092	33362	94904	31273	04146	18594	29852	71585	85030	51132	01915	92747	64951
19	52162	53916	46369	58586	23216	14513	83149	98736	23495	64350	94738	17752	35156	35749
20	07056	97628	33787	09998	42698	06691	76988	13602	51851	46104	88916	19509	25625	58104
21	48663	91245	85828	14346	09172	30168	90229	04734	59193	22178	30421	61666	99904	32812
22	54164	58492	22421	74103	47070	25306	76468	26384	58151	06646	21524	15227	96909	44592
23	32639	32363	05597	24200	13363	38005	94342	28728	35806	06912	17012	64161	18296	22851
24	29334	27001	87637	87308	58731	00256	45834	15398	46557	41135	10367	07684	36188	18510
25	02488	33062	28834	07351	19731	92420	60952	61280	50001	67658	32586	86679	50720	94953

Abridged from *Handbook of Tables for Probability and Statistics*, Second Edition, edited by William H. Beyer (Cleveland: The Chemical Rubber Company, 1968). Reproduced by permission of the publishers, The Chemical Rubber Company.

26	81525	72295	04839	96423	24878	82651	66566	14778	76797	14780	13300	87074	79666	95725
27	29676	20591	68086	26432	46901	20849	89768	81536	86645	12659	92259	57102	80428	25280
28	00742	57392	39064	66432	84673	40027	32832	61362	98947	96067	64760	64584	96096	98253
29	05366	04213	25669	26422	44407	44048	37937	63904	45766	66134	75470	66520	34693	90449
30	91921	26418	64117	94305	26766	25940	39972	22209	71500	64568	91402	42416	07844	69618
31	00582	04711	87917	77341	42206	35126	74087	99547	81817	42607	43808	76655	62028	76630
32	00725	69884	62797	56170	86324	88072	76222	36086	84637	93161	76038	65855	77919	88006
33	69011	65795	95876	55293	18988	27354	26575	08625	40801	59920	29841	80150	12777	48501
34	25976	57948	29888	88604	67917	48708	18912	82271	65424	69774	33611	54262	85963	03547
35	09763	83473	73577	12908	30883	18317	28290	35797	05998	41688	34952	37888	38917	88050
36	91567	42595	27958	30134	04024	86385	29880	99730	55536	84855	29080	09250	79656	73211
37	17955	56349	90999	49127	20044	59931	06115	20542	18059	02008	73708	83517	36103	42791
38	46503	18584	18845	49618	02304	51038	20655	58727	28168	15475	56942	53389	20562	87338
39	92157	89634	94824	78171	84610	82834	09922	25417	44137	48413	25555	21246	35509	20468
40	14577	62765	35605	81263	39667	47358	56873	56307	61607	49518	89656	20103	77490	18062
41	98427	07523	33362	64270	01638	92477	66969	98420	04880	45585	46565	04102	46880	45709
42	34914	63976	88720	82765	34476	17032	87589	40836	32427	70002	70663	88863	77775	69348
43	70060	28277	39475	46473	23219	53416	94970	25832	69975	94884	19661	72828	00102	66794
44	53976	54914	06990	67245	68350	82948	11398	42878	80287	88267	47363	46634	06541	97809
45	76072	29515	40980	07391	58745	25774	22987	80059	39911	96189	41151	14222	60697	59583
46	90725	52210	83974	29992	65831	38857	50490	83765	55657	14361	31720	57375	56228	41546
47	64364	67412	33339	31926	14883	24413	59744	92351	97473	89286	35931	04110	23726	51900
48	08962	00358	31662	25388	61642	34072	81249	35648	56891	69352	48373	45578	78547	81788
49	95012	68379	93526	70765	10592	04542	76463	54328	02349	17247	28865	14777	62730	92277
50	15664	10493	20492	38391	91132	21999	59516	81652	27195	48223	46751	22923	32261	85653
51	16408	81899	04153	53381	79401	21438	83035	92350	36693	31238	59649	91754	72772	02338
52	18629	81953	05520	91962	04739	13092	97662	24822	94730	06496	35090	04822	86774	98289
53	73115	35101	47498	87637	99016	71060	88824	71013	18735	20286	23153	72924	35165	43040
54	57491	16703	23167	49323	45021	33132	12544	41035	80780	45393	44812	12515	98931	91202
55	30405	83946	23792	14422	15059	45799	22716	19792	09983	74353	68668	30429	70735	25490
56	16631	35006	85900	98275	32388	52390	16815	69298	82732	38480	73817	32523	41961	44437
57	96773	20206	42559	78985	05300	22164	24369	54224	35083	19687	11052	91491	60383	19746
58	38935	64202	14349	82674	66523	44133	00697	35552	35970	19124	63318	29686	03387	59846
59	31624	76384	17403	53363	44167	64486	64758	75366	76554	31601	12614	33072	60332	92325
60	78919	19474	23632	27889	47914	02584	37680	20801	72152	39339	34806	08930	85001	87820
61	03931	33309	57047	74211	63445	17361	62825	39908	05607	91284	68833	25570	38818	46920
62	74426	33278	43972	10119	89917	15665	52872	73823	73144	88662	88970	74492	51805	99378
63	09066	00903	20795	95452	92648	45454	09552	88815	16553	51125	79375	97596	16296	66092
64	42238	12426	87025	14267	20979	04508	64535	31355	86064	29472	47689	05974	52468	16834
65	16153	08002	26504	41744	81959	65642	74240	56302	00033	67107	77510	70625	28725	34191

Line/Col.	(1)	(2)	(3)	(4)	(5)	(6)	(7)	(8)	(9)	(10)	(11)	(12)	(13)	(14)
66	21457	40742	29820	96783	29400	21840	15035	34537	33310	06116	95240	15957	16572	06004
67	21581	57802	02050	89728	17937	37621	47075	42080	97403	48626	68995	43805	33386	21597
68	55612	78095	83197	33732	05810	24813	86902	60397	16489	03264	88525	42786	05269	92532
69	44657	66999	99324	51281	84463	60563	79312	93454	68876	25471	93911	25650	12682	73572
70	91340	84979	46949	81973	37949	61023	43997	15263	80644	43942	89203	71795	99533	50501
71	91227	21199	31935	27022	84067	05462	35216	14486	29891	68607	41867	14951	91696	85065
72	50001	38140	66321	19924	72163	09538	12151	06878	91903	18749	34405	56087	82790	70925
73	65390	05224	72958	28609	81406	39147	25549	48542	42627	45233	57202	94617	23772	07896
74	27504	96131	83944	41575	10573	08619	64482	73923	36152	05184	94142	25299	84387	34925
75	37169	94851	39117	89632	00959	16487	65536	49071	39782	17095	02330	74301	00275	48280
76	11508	70225	51111	38351	19444	66499	71945	05422	13442	78675	84081	66938	93654	59894
77	37449	30362	06694	54690	04052	53115	62757	95348	78662	11163	81651	50245	34971	52924
78	46515	70331	85922	38329	57015	15765	97161	17869	45349	61796	66345	81073	49106	79860
79	30986	81223	42416	58353	21532	30502	32305	86482	05174	07901	54339	58861	74818	46942
80	63798	64995	46583	09785	44160	78128	83991	42865	92520	83531	80377	35909	81250	54238
81	82486	84846	99254	67632	43218	50076	21361	64816	51202	88124	41870	52689	51275	83556
82	21885	32906	92431	09060	64297	51674	64126	62570	26123	05155	59194	52799	28225	85762
83	60336	98782	07408	53458	13564	59089	26445	29789	85205	41001	12535	12133	14645	23541
84	43937	46891	24010	25560	86355	33941	25786	54990	71899	15475	95434	98227	21824	19585
85	97656	63175	89303	16275	07100	92063	21942	18611	47348	20203	18534	03862	78095	50136
86	03299	01221	05418	38982	55758	92237	26759	86367	21216	98442	08303	56613	91511	75928
87	79626	06486	03574	17668	07785	76020	79924	25651	83325	88428	85076	72811	22717	50585
88	85636	68335	47539	03129	65651	11977	02510	26113	99447	68645	34327	15152	55230	93448
89	18039	14367	61337	06177	12143	46609	32989	74014	64708	00533	35398	58408	13261	47908
90	08362	15656	60627	36478	65648	16764	53412	09013	07832	41574	17639	82163	60859	75567
91	79556	29068	04142	16268	15387	12856	66227	38358	22478	73373	88732	09443	82558	05250
92	92608	82674	27072	32534	17075	27698	98204	63863	11951	34648	88022	56148	34925	57031
93	23982	25835	40055	67006	12293	02753	14827	23235	35071	99704	37543	11601	35503	85171
94	09915	96306	05908	97901	28395	14186	00821	80703	70426	75647	76310	88717	37890	40129
95	59037	33300	26695	62247	69927	76123	50842	43834	86654	70959	79725	93872	28117	19233
96	42488	78077	69882	61657	34136	79180	97526	43092	04098	73571	80799	76536	71255	64239
97	46764	86273	63003	93017	31204	36692	40202	35275	57306	55543	53203	18098	47625	88684
98	03237	45430	55417	63282	90816	17349	88298	90183	36600	78406	06216	95787	42579	90730
99	86591	81482	52667	61582	14972	90053	89534	76036	49199	43716	97548	04379	46370	28672
100	38534	01715	94964	87288	65680	43772	39560	12918	86537	62738	19636	51132	25739	56947

ANSWERS

CHAPTER 2

2.1 1, −1, −1
2.2 1, 1/2, .0625 = 1/16
2.3 3, 12
2.4 1, 8, 27
2.5 2, −3, −1
2.6 0, 1, 4
2.7 2, 5, 10
2.8 $y + 1$
2.9 (a) −9
(b) 65
(c) $y_1 + y_2 + y_3 + y_4 - 8$
(d) $3y + 12$
2.10 (a) 0, 3, 8, 15
(b) $\sum_{y=1}^{4} (y^2 - 1)$
(c) 26
2.11 (a) 4, 1, 0, 1, 4
(b) $\sum_{y=1}^{5} (y - 3)^2$
(c) 10
2.12 (b) $\sum_{i=1}^{5} y_i$, $\sum_{i=1}^{5} y_i^2$
(c) 3, 3
(d) 73, 73, 5329
2.13 (b) 3220
(c) 960; 331, 400; 1,000,000
2.14 (a) 105
(b) 1300
(c) 3136
(d) $105x$
2.15 2
2.16 25
2.17 −6
2.18 16
2.19 $1 + p + p^2 + p^3$
2.20 $4y + 7$
2.21 77
2.22 (a) 1, 5, 11, 19
(b) $\sum_{y=1}^{4} y^2 + \sum_{y=1}^{4} y - 4$
(c) 36
2.23 (a) $(y - 1)^2$
(b) $\sum_{y=1}^{5} (y - 1)^2$
(c) 30
2.24 (a) $c + 1$, $c + 2$, $c + 3$
(b) $\sum_{y=1}^{3} (c + y)$
(c) $3c + 6$
2.25 (a) 3
(b) 7
(c) 11
(d) −1
(e) −5
(f) $4a^2 + 3$
(g) $-4a + 3$
(h) $7 - 4y$
2.26 (a) 0
(b) 25
(c) 9
(d) $(x - 2)^2$
(e) $(a - 3)^2$
2.27 (a) 7
(b) $(a + b)^2 - (a + b) + 1$
2.28 (a) 1
(b) undefined, division by 0
(c) 17/5
(d) $(a^2 + 1)/(1 - a)$
2.29 $p(0) = 1$, $p(1) = 1 - a$
2.30 (a) 1/4
(b) −6
(c) $\frac{3}{x^2} - \frac{3}{x} + 1$
2.31 (a) 2
(b) 2
(c) 2
2.32 14.1, 43.63, 198.81
2.33 yields 5(3.1) = 15.5 less than original sum; −1.4
2.34 36
2.35 21
2.36 25
2.37 $5x^2 + 30$
2.38 $6x^2 + 55$
2.39 $9 + 6i$
2.40 $10 + 10y^2$
2.41 $y_1 + y_2 - 3$
2.42 $\sum_{i=1}^{n} y_i - na$
2.43 $\sum_{i=1}^{n} y_i^2 - 2a \sum_{i=1}^{n} y_i - na^2$
2.44 36
2.45 −29

2.46 431
2.47 610
2.48 654
2.49 649.077
2.54 (a) 18
(b) 76
2.55 22
2.56 (a) 27.7
(b) 110.79
2.57 .196

CHAPTER 3

3.1 (b) 25/60
3.2 (b) 36/50
3.4 (a) 2
(b) 1.581
3.5 (a) .68
(b) .95
(c) $\approx$0.00
3.6 (a) at least 3/4 of measurements in interval .15–.19; at least 8/9 of measurements in interval .14–.20
(b) approximately 68% of measurements in interval .16–.18; approximately 95% of measurements in interval .15–.19; almost all of measurements in interval .14–.20
(c) no, sample is too small for the data to possess a mound-shaped distribution
3.7 at least 3/4 of measurements in interval 0–140; at least 8/9 of measurements in interval 0–193
3.8 (a) 82% of all families have incomes less than $25,000
3.9 yes, too many standard deviations away from mean
3.10 (a) 2.8
(b) 3.2, 1.789
3.11 (a) 1.375
(b) .839, .916
3.12 (a) $s \approx .15$
(b) $\bar{y} = .76$, $s = .165$
3.13 (b) 23.65
(c) 1; 1.368
3.14 (b) 7.729
(c) 1.985

Interval	Actual	Tchebysheff's Theorem	Empirical Rule
$\bar{y} \pm s$	.714	≥ 0	$\approx$.68
$\bar{y} \pm 2s$	.957	$\geq 3/4$	$\approx$.95
$\bar{y} \pm 3s$	1.000	$\geq 8/9$	$\approx$1.00

3.15 (a) 32.1
(b) 8.025
(c) 7.671

3.16

Interval	Actual	Tchebysheff's Theorem	Empirical Rule
$\bar{y} \pm s$	.74	≥ 0	$\approx$.68
$\bar{y} \pm 2s$	.94	$\geq 3/4$	$\approx$.95
$\bar{y} \pm 3s$	.98	$\geq 8/9$	$\approx$1.00

3.17 (a) 1.4, 19.6
(b) 1.4, 19.6
3.18 (a) 2.04, 2.806
(b) and (c)

k	$\bar{y} \pm ks$	Actual	Tchebysheff's Theorem	Empirical Rule
1	−.77, 4.85	.84	≥ 0	$\approx$.68
2	−3.57, 7.65	.92	$\geq 3/4$	$\approx$.95
3	−6.38, 10.46	1.00	$\geq 8/9$	$\approx$1.00

3.19 3,001; 8.5
3.20 .01983, .0000013667
3.21 .01983, .0000013667
3.22 15, 1.714
3.23 7.729, $s^2 = 3.9398$, $s = 1.985$
3.26 3, 6.8, 2.608
3.27 2.286, 3.905, 1.976
3.32 .68; .96; yes; yes
3.33 .68; .95
3.34 7.12, .017, .130
3.35 .8; 1.0; yes; not mound-shaped
3.36 .475 vs. .502
3.37 (a) 7.75
(b) 59.2, 10.369 vs. 7.75

3.39 .815
3.40 (2, 26)
3.41 .95
3.42 .16
3.44 80.367, 189.206, 13.755
3.45 14.75 vs. 13.755
3.46

k	$\bar{y} \pm ks$	Actual	Tchebysheff's Theorem	Empirical Rule
1	66.61, 94.12	.77	≥ 0	$\approx .68$
2	52.86, 107.88	.93	$\geq 3/4$	$\approx .95$

3.47 4.29, 4
3.49 $5/2.5 = 2$ vs. 1.976
3.50 $7/2.5 = 2.8$ vs. 2.608;
$.3/2.5 = .12$ vs. .130
3.51 $.5/4 = .125$ vs. .132
3.52 no; at least 8/9, probably more
3.53 1.333; 1.467
3.54 .815
3.55 45
3.57 (a) 101.594, 13.327
(b) 13.75
(c)

Interval	Actual	Tchebysheff's Theorem	Empirical Rule
$\bar{y} \pm s$	.719	≥ 0	$\approx .68$
$\bar{y} \pm 2s$	.938	$\geq 3/4$	$\approx .95$
$\bar{y} \pm 3s$	1.000	$\geq 8/9$	≈ 1.00

3.58

k	$\bar{y} \pm ks$	Approximate Fraction in Interval
1	415, 425	$\approx .68$
2	410, 430	$\approx .95$
3	405, 435	≈ 1.00

3.59 (a) .025
(b) .84
3.60 (b) .66, 1.387
(c) .95; .96; yes; yes
3.61 (a) at least 3/4 have between 145 and 205 teachers
(b) .16
3.62 $R/4 = 4.25$; yes
3.63 (a)

k	$\bar{y} \pm ks$	Approximate Fraction in Interval
1	420, 570	$\approx .68$
2	345, 645	$\approx .95$
3	270, 720	≈ 1.00

(b) .16
3.64 .12, .017 yield 7.12 and .017
3.65 1.2, 1.7 yield 7.12 and .017
3.66 2.652, .2518
3.67 1.52 and 25.177 yield 2.652 and .2518
3.68 17.105, .0173
3.69 1.05, 1.73 yield 17.105, .0173

CHAPTER 4

4.1 A: {4}, B: {2, 4, 6}, C: {1, 2}, D: {4}, E: {2, 4, 6}, F contains no sample points; 1/6; 1/2; 0
4.2 (a) choose 2 of 4 cans without replacement
(b) $\{E_1E_2\}$, $\{E_1W_1\}$, $\{E_1W_2\}$, $\{E_2W_1\}$, $\{E_2W_2\}$, $\{W_1W_2\}$
(c) 1/6
4.3 (a) choose 2 of 5 cans without replacement
(b) $\{E_1E_2\}$, $\{E_1E_3\}$, $\{E_1W_1\}$, $\{E_1W_2\}$, $\{E_2E_3\}$, $\{E_2W_1\}$, $\{E_2W_2\}$, $\{E_3W_1\}$, $\{E_3W_2\}$, $\{W_1W_2\}$
(c) 1/10
4.4 (a) {choice of points 1, 2, 3}: {111}, {112}, {113}, {121}, {122}, {123}, {131}, {132}, {133}, {211}, {212}, {213}, {221}, {222}, {223}, {231}, {232}, {233}, {311}, {312}, {313}, {321}, {322}, {323}, {331}, {332}, {333}
(b) {123}, {132}, {213}, {231}, {312}, {321}
(c) $P(E_i) = 1/27$, $P(A) = 2/9$
4.5 (a) rank A, B, C

(b) $\{ABC\}$, $\{ACB\}$, $\{BAC\}$, $\{BCA\}$, $\{CAB\}$, $\{CBA\}$
(c) 1/3, 1/3

4.6 (a) 2/27
(b) 20/27

4.7 3/10

4.8 1/6, 1/18

4.9 (a) assign four men, 1 to each of four jobs
(b) (1234), (1243), (1324), (1342), (1423), (1432), (2134), (2143), (2314), (2341), (2413), (2431), (3124), (3142), (3214), (3241), (3412), (3421), (4123), (4132), (4213), (4231), (4312), (4321); (1 and 2 minority)
(c) 1/6

4.10 (a) select 2 from 5 without replacement
(b) (12), (13), (14), (15), (23), (24), (25), (34), (35), (45)
(c) 1/10

4.11 (a) (1111), (1112), (1113), (1121), (1122), (1123), (1131), . . . , (3333); 81 points in all
(b) (3331), (3332), (3333): 3 from latest batch
(c) (3331), (3332), (3313), (3323), (3133), (3233), (1333), (2333)
(d) all of (c) plus (3333)
(e) 1/81, 1/27, 8/81, 1/9

4.12 .008

4.13 (a) 3/8
(b) 5/8
(c) 3/4
(d) 1/4
(e) 2/5
(f) no, E_4 and E_6 are in A and B
(g) $P(A|B) \neq P(A)$ and $P(B|A) \neq P(B)$; no

4.14 (a) .45
(b) .27
(c) .78

4.15 (a) .125
(b) .875

4.16 (a) .095
(b) .855
(c) .005
(d) .955

4.17 (a) .000125
(b) .1354
(c) .1426
(d) $p > .05$

4.18 .0214

4.19 .5, no

4.20 (a) 5/6
(b) 25/36
(c) 11/36

4.21 .5952

4.22 (a) sample points correspond to a pair of birth dates (d_1, d_2), where d_i, $i = 1, 2$, is the birth date for person i; d_i can assume values 1, 2, 3, . . . , 365
(b) $365^2 = 133{,}225$
(c) A contains sample points (1, 1), (2, 2), (3, 3), . . . , (365, 365)
(d) 1/365
(e) 364/365

4.23 (a) $P(A) = .9918, P(B) = .0082$
(b) $P(A) = .9836, P(B) = .0164$

4.24 .25

4.25 .3130

4.26 .9412

4.27 .6667

4.28 .6585

4.29 720

4.30 12,144

4.31 3,628,800

4.32 5040

4.33 5.720645×10^{12}

4.34 3; 9; 59,049

4.35 380,204,032

4.36 2,598,960

4.37 1/56

4.38 (a) 49,995,000
(b) .00000012

4.39 (a) $5^{20} \approx 9.536743164 \times 10^{13}$
(b) $5^{-20} \approx 1.048576 \times 10^{-14}$

4.40 (a) 25
(b) 4/25 = .16
4.41 (a) .0001049
(b) learning occurred
4.42 1/120; 7/24
4.43 (a) 1/14 = .0714
(b) .2381
(c) .4396
(d) .4696
(e) .4506
4.44 4/7 = .571
4.45 (a) (C_1C_2), (C_1A_1), (C_1A_2), (C_2C_1), (C_2A_1), (C_2A_2), (A_1C_1), (A_1C_2), (A_1A_2), (A_2C_1), (A_2C_2), (A_2A_1)
(b) (C_1C_2), (C_1A_1), (C_1A_2), (C_2C_1), (C_2A_1), (C_2A_2)
(c) (C_1A_1), (C_1A_2), (C_2A_1), (C_2A_2), (A_1C_1), (A_1C_2), (A_2C_1), (A_2C_2)
(d) (A_1A_2), (A_2A_1)
4.46 1/2, 2/3, 1/3, 5/6
4.47 (a) (*TTTT*), (*TTTH*), (*TTHT*), (*THTT*), (*HTTT*), (*TTHH*), (*THTH*), (*HTTH*), (*HTHT*), (*HHTT*), (*THHT*), (*HHHT*), (*HHTH*), (*HTHH*), (*THHH*), (*HHHH*)
(b) (*HHHT*), (*HHTH*), (*HTHH*), (*THHH*)
(c) 1/16, 1/4
4.48 identical to coin-tossing problem, Exercise 4.47; *H* represents customer preference for style 1, *T* corresponds to style 2
(c) (*HHHH*), (*TTTT*)
(d) 1/8
4.49 (a) *S* contains 21 sample points
(b) *A* contains 6 sample points
(c) 2/7
4.50 1/6
4.51 .375, .5
4.52 1/15, 3/5
4.53 1/36, 1/6
4.54 5/21
4.55 1/4, 1/12, 0, 1/3
4.56 1/2, 1/4, 1/3, 3/4, 5/6, 1/6, 11/12
4.57 .3913
4.58 1/2, 1/5, no, no, 2/5
4.59 1/2, 2/3, 1/3, 5/6, 1/6, 0, 2/3
4.60 1/27, 8/81, 2/81, 1/9, 1/9, 1/27, 1/9
4.61 $P(AB) = P(A)P(B) > 0$
4.62 1/2, 2/3, 1/3, 5/6
4.63 0, 1/3
4.64 (a) .73
(b) .27
4.65 36
4.66 8
4.67 12
4.68 20,000
4.70 21 sample points, 5/21
4.71 (a) 6
(b) 15
4.72 15, 1/15, 3/5
4.73 12
4.74 120, 625
4.75 120, 180
4.76 18!/9!
4.77 42
4.78 294
4.79 1/28
4.80 720
4.81 (a) spade royal flush
(b) royal flush or spade flush
(c) .0004952
(d) .000001539
(e) .0000003848
(f) .0004964
4.82 .05
4.83 (a) 56
(b) 6
(c) 6/56
4.84 (a) .81
(b) .01
(c) .99
4.85 .999999
4.86 (a) .8
(b) .64
(c) .36
4.87 .432
4.88 1, .4783, .00781
4.89 15
4.90 .0256, .1296
4.91 .6
4.92 1/81

4.93 yes, $4!(3!)^4/12!$
4.94 .0625, yes
4.95 .000778, .0000003848, .0004964
4.96 .63
4.97 .2, .1
4.98 6/10
4.99 1/42
4.100 (a) 15/28
(b) 55/56
4.101 (a) 1/4
(b) 1/12
(c) 1/2
4.102 5/6
4.103 (a) 45
(b) 10
(c) 10/45 = 2/9
4.104 .8704
4.105 4/25
4.106 44/105
4.107 .0625
4.108 4
4.109 (a) .128
(b) .488
4.110 (a) 1/5
(b) 2/5
(c) 1/2
4.111 (a) 1/8
(b) 1/64
(c) not necessarily; they could have studied together, etc.
4.112 .903
4.113 8/15, 16/31

CHAPTER 5

5.1 (a) C
(b) D
(c) D
(d) D
(e) D
5.2 (a) D
(b) C
(c) C
(d) D
(e) D
5.3 (a) C
(b) C
(c) D
(d) D
(e) C
5.4 $p(0) = .64$, $p(1) = .32$, $p(2) = .04$
5.5 $p(1) = 1/4$, $p(2) = 1/4$, $p(3) = 1/4$, $p(4) = 1/4$
5.6 $p(0) = 1/5$, $p(1) = 3/5$, $p(2) = 1/5$
5.8 (a) $p(0) = .857375$, $p(1) = .135375$, $p(2) = .007125$, $p(3) = .000125$
(c) .00725
5.9 (a) .1, .09, .081
(b) $p(y) = (.9)^{y-1}(.1)$
5.10 1.625; .734; yes
5.11 1.4
5.12 $2050
5.13 2960
5.14 13,800.388
5.15 (a) 2.15
(b) 1.5275
(d) .95
5.16 (a) 120
(b) 20
(c) 1
5.17 10; (12), (13), (14), (15), (23), (24), (25), (34), (35), (45)
5.18 1.731×10^{13}
5.19 $p(0) = .125$, $p(1) = .375$, $p(2) = .375$, $p(3) = .125$; 1.5; .75
5.20 (a) 2.5
(b) 7.5
(c) 1.25, 1.118
5.21 (a) D
(b) C
(c) D
(d) D
(e) C
5.22 (a) 1/10
(b) 2.6
(c) 1.84
5.23 1.0 vs. $\geq .75$
5.24 $\mu = 1$
5.25 .4

5.26 2.917; 1.0
5.28 1.0 vs. $\geq 3/4$
5.29 $p(1) = 3/10$, $p(2) = 6/10$, $p(3) = 1/10$; .7
5.30 1.8, .36
5.31 8333.333
5.32 $p(0) = .0256$, $p(1) = .1536$, $p(2) = .3456$, $p(3) = .3456$, $p(4) = .1296$; .4752
5.33 $p(0) = .008$, $p(1) = .096$, $p(2) = .384$, $p(3) = .512$; .992
5.34 2.4; .48
5.35 (a) $p(2) = 1/36$, $p(3) = 2/36$, $p(4) = 3/36$, $p(5) = 4/36$, $p(6) = 5/36$, $p(7) = 6/36$, $p(8) = 5/36$, $p(9) = 4/36$, $p(10) = 3/36$, $p(11) = 2/36$, $p(12) = 1/36$
(b) 7
(c) 5.8333
(e) 34/36 = .944
5.36 $p(y) = C_y^2 C_{3-y}^4 / C_3^6$, $y = 0, 1, 2$
5.37 120
5.38 (a) $p(y) = C_y^4 (1/3)^y (2/3)^{4-y}$
(b) 1/9 (c) 4/3
(d) 8/9
5.39 (a) $p(y) = C_y^3 (1/3)^y (2/3)^{3-y}$, $y = 0, 1, 2, 3$
(b) 7/27
(c) 1
(d) 2/3
5.40 (a) $p(0) = 125/216$, $p(1) = 75/216$, $p(2) = 15/216$, $p(3) = 1/216$
(b) 16/216
(c) 108/216
(d) .4167
5.41 (a) 2.13
(b) 1.8131; 1.347
(c) .95
(d) .027
5.42 (a) $p(0) = 1/42$, $p(1) = 10/42$, $p(2) = 20/42$, $p(3) = 10/42$, $p(4) = 1/42$
(b) 11/42
(c) 2
(d) 2/3
5.43 2150
5.44 187.20
5.45 (b) .135
(c) .947
5.46 .368, .368, .184, .632
5.47 .0498
5.48 (a) .084, .125
(b) no
5.49 no
5.50 yes
5.51 1023.7, 40.9%
5.54 .5
5.55 1.25
5.56 (a) 2/7
(b) 5/7

CHAPTER 6

6.1 (a) .0081
(b) .4116
(c) .2401
6.3 $p(2) = .1240$; $P(y \geq 2) = .1497$; yes
6.4

$p = .1$		$p = .5$		$p = .9$	
y	$p(y)$	y	$p(y)$	y	$p(y)$
0	.5314	0	.0156	0	.000001
1	.3543	1	.0938	1	.0001
2	.0984	2	.2344	2	.0012
3	.0146	3	.3125	3	.0146
4	.0012	4	.2344	4	.0984
5	.0001	5	.0938	5	.3543
6	.000001	6	.0156	6	.5314

6.5 (a) .9606
(b) .9941
6.6 .0037, .3439, $p < .9$ because $P(y \geq 2) = .0523$
6.7 (a) .3277
(b) .4096
(c) .0067
6.8 Their performance would seem to be below the 80% rate of success.
6.9 (a) .1681
(b) .5282

6.10 (a) .0523
(b) .0486

6.11 600, 420, 20.4939,
$P(y > 700) \leq .04$

6.12 Observed 71% is more than 16 standard deviations away from 50%; consequently, probability is very small.

6.13 No, the probability of observing 62% is very small because the observed 62% lies more than three standard deviations away from 50%; yes.

6.14 $\mu = 1875$; $\sigma^2 = 468.75$,
$\sigma = 21.651$

k	$\mu \pm k\sigma$	Tchebysheff's Theorem
2	1832, 1918	$\geq 3/4$
3	1810, 1940	$\geq 8/9$

6.15 (a) 800
(b) 12.649
(c) No, $y = 900$ is nearly 8 standard deviations above the mean.

6.16 (a) 5000
(b) 50
(c) no

6.17 $\mu = 100$, $\sigma^2 = 90$; $\mu \pm 2\sigma$:
$81 \leq y \leq 119$

6.18 (a) .590
(b) .168
(c) .031
(d) 0
(e) 1

6.19 (a) .919
(b) .528
(c) .188
(d) 0
(e) 1

6.20 (a) .349
(b) .028
(c) .001
(d) 0
(e) 1

6.21 (a) .736
(b) .149
(c) .010
(d) 0
(e) 1

6.22 for a given value of n and p, P(accept) increases as a increases; for a given value of a and p, P(accept) decreases as n increases

6.23 (b) 3
(c) 1

6.24 $a = 1$

6.25 $n = 25$, $a = 1$

6.26 (a) .125
(b) .7518

6.27 No; the probability of this event is .343 when $p = .30$.

6.28 (a) .151
(b) .302

6.29 (a) .036
(b) .352

6.30 (a) .118, larger
(b) .164

6.31 yes, $P(8, 9, 10 \mid p = .5) = .055$; H_0: $p = .5$; H_a: $p > .5$; yes; for rejection region $y \leq 8$, $\alpha = .045$

6.32 (a) H_0: $p = .5$; H_a: $p < .5$
(b) rejection region: $y = 0, 1, 2, 3, 4$
(c) program was ineffective

6.33 (a) H_0: $p = .5$; H_a: $p \neq .5$
(b) .03125
(c) .7378

6.34 (a) $P(0, 1, 6, 7 \mid p = .5) = .125$
(b) .1497
(c) not sufficient evidence to reject H_0

6.35 (a) .077
(b) $y = 10$ is not in the rejection region; therefore, there is not sufficient evidence to reject H_0.

6.36 .922, .078

6.37 (a) H_0: $p = .5$; H_a: $p \neq .5$
(b) .063
(c) .410

6.39 (a) $p(0) = .125$, $p(1) = .375$, $p(2) = .375$, $p(3) = .125$

(c) 1.5, .866

(d)

k	$\mu \pm k\sigma$	Probability
1	.63, 2.37	.75
2	−.23, 3.23	1.0

6.40 (a) $p(0) = .729$, $p(1) = .243$, $p(2) = .027$, $p(3) = .001$
(c) .3, .520
(d) .729, .972
6.41 .32805; .99999; if you use Table 1, Appendix II, the answers will differ slightly due to rounding error; .329; 1.0
6.42 (a) .9606
(b) .9994
6.43 .03125, .1875
6.44 .608
6.45 $\mu = 2$, $\sigma = 1.342$; 0 to 4.684 $(\mu \pm 2\sigma)$
6.46 (a) .234
(b) .136
6.47 .151
6.48 .608
6.49 (a) $p(y) = C_y^{20}(1/5)^y(4/5)^{20-y}$, $y = 0, 1, \ldots, 20$
(b) ≈ 0
6.50 (a) .387
(b) .651
(c) .264
6.51 (a) .012
(b) .630
(c) trials could be dependent
6.52 $P(y = 8) = .115$; $P(y \geq 8) = .228$
6.53 3000, no
6.54 no; $y = 1373$ lies more than 16 standard deviations away from μ when split is 50–50
6.55 (a) (25, 5)
(b) (25, 5)
6.56 (a) .983
(b) .736
(c) .392
(d) .069
6.57 (a) H_0: $p = 1/3$; H_a: $p > 1/3$
(b) .0453
(c) .7734
(d) less
6.58 (a) 80
(b) 64
(c) (64, 96)
6.59 (a) $p(y) = C_y^3(.7)^y(.3)^{3-y}$
(b) .343
(c) .657
6.60 using Table 1, Appendix II:
(a) .919
(b) .357
6.61 (a) 20
(b) 4
(c) .006
(d) theory is incorrect
6.62 11
6.63 reject H_0: $p = .5$; $\alpha = .042$; $\beta = .196$
6.64 .017; do not support
6.65 (a) H_0: $p = .5$; H_a: $p \neq .5$
(b) .116
(c) .392
6.66 9
6.71 .04
6.72 .987

CHAPTER 7

7.1 (a) .4192
(b) .4599
7.2 (a) .2734
(b) .2734
7.3 (a) .8301
(b) .1452
7.4 (a) .6826
(b) .9544
(c) .9974
7.5 1.96
7.6 .86
7.7 −1.12
7.8 .85
7.9 −1.645
7.10 1.645
7.11 2.575
7.12 (a) 1.273
(b) .1020
7.13 .1562, .0012

7.14 .0505
7.15 .0668
7.16 (a) .0778
(b) .0274
7.17 .0401
7.18 .0475
7.20 3.5, .765
7.21 3.5, .541
7.22 approximately normal, 1500, 35.355
7.23 (a) ≈ 0
(b) .0764
(c) .9236
7.24 (a) normal
(b) .0571
(c) $\mu = 21.948$
7.26 (a) .546
(b) .5468
7.27 (a) .245
(b) .2483
7.28 .1251
7.29 ≈ 0; yes
7.30 $z = 1.461$
7.31 .9441
7.32 near 0
7.33 $z = 2.882$; yes
7.34 (a) .3849
(b) .3159
7.35 (a) .4452
(b) .2734
7.36 (a) .4279
(b) .1628
7.37 (a) .4251
(b) .4778
7.38 (a) .3227
(b) .1586
7.39 (a) .1572
(b) .5701
7.40 (a) .0730
(b) .8623
7.41 .7734
7.42 .9115
7.43 $z_0 = 0$
7.44 $z_0 = 1.10$
7.45 .1904
7.46 .1596
7.47 .5457
7.48 .8612
7.49 $z_0 = .6745$
7.50 no
7.51 .0668
7.52 .2266
7.53 .2643
7.54 .16
7.55 (a) .578
(b) .5752
7.56 (a) .421
(b) .4013
7.57 85.36
7.58 $z = -1.26$; no
7.59 .9554
7.60 yes, $z = 1.89$
7.61 .0901
7.62 7.301
7.63 yes, $z = -3.27$
7.64 $z = -1.96$; $y \leq 28$ or $y \geq 52$
7.65 .117
7.66 .9929
7.67 (a) .0071
(b) .0571
7.68 every 383.5 hours
7.69 (a) 141
(b) .0401
7.70 (a) 40
(b) .4
(c) .0062
7.71 .898
7.72 (a) .0228
(b) .00052
7.73 .3307
7.74 1072 days
7.75 .9474
7.76 .9599
7.77 (a) .178
(b) .392
7.78 (a) .1725
(b) .4052
7.79 (a) 20
(b) 4
(c) .9913
7.80 .1922
7.81 .3557
7.82 .1314
7.83 (a) .0228
(b) .00052
7.84 limit total number of passengers to $n = 11$

CHAPTER 8

8.1 (a) range/4 = 2.65
(b) 1.7 ± .62
8.2 39.8° ± 4.865°
8.3 7.2% ± .792%
8.4 1280 ± 28.4
8.5 .1587
8.6 (a) .8098
(b) .9356
8.7 64.3 ± 5.239
8.8 .145 ± .0015
8.9 3.7 ± .204; random sampling
8.10 81 ± 3.326
8.11 7.9 ± 1.152
8.12 (a) 7.2 ± .751
(b) 2.5 ± .738
8.13 2.8 ± .443
8.14 .27 ± .035
8.15 .63 ± .113
8.16 (a) No; this is not a random sample.
(b) .145 ± .035
8.17 .7 ± .045
8.18 .2 ± .017
8.19 .92 ± .017
8.20 .15 ± .064
8.21 .11 ± .047
8.22 .11 ± .0897
8.23 .16 ± .099
8.24 .08 ± .0505
8.25 approximately 40,000
8.26 approximately 100
8.27 approximately 666
8.28 approximately 72
8.29 approximately 2400
8.30 approximately 100; no
8.31 approximately 136
8.32 approximately 358
8.33 (a) H_0: $\mu = 80$
(b) H_a: $\mu \neq 80$
(c) $z = -3.75$; reject H_0
8.34 yes, $z = -25.298$; reject H_0
8.35 yes, $z = -3.926$; reject H_0
8.36 no, $z = -1.176$; do not reject H_0
8.37 .1342
8.38 $z = 7.013$; reject H_0: $\mu_1 = \mu_2$
8.39 yes, $z = 2.858$
8.40 approximately 0; yes, $z = 4.96$; reject H_0; yes
8.41 H_0: $p_1 = p_2$; yes, $z = 14.017$; reject H_0
8.43 9.7 ± 2.32
8.44 9.7 ± 1.908
8.45 .1 ± .025
8.46 1.7 ± .493
8.47 1.7 ± .406
8.48 26.3 ± .675
8.49 approximately 2400
8.50 approximately 400; 6800 ± 49
8.51 .04 ± .1055
8.52 approximately 768
8.53 .6170
8.54 approximately 100
8.55 approximately 44
8.58 yes, $z = -2.811$
8.59 $z = 1.577$; do not reject H_0
8.60 yes, $z = 4.472$
8.61 no, $z = 1.000$
8.62 random sample; either σ known or sample size 30 or more
8.63 $z = -3.122$; reject H_0: $p = .95$
8.64 $z = 4.052$; reject H_0: $\mu = 5$
8.65 $z = 6.894$; reject H_0: $\mu_1 = \mu_2$
8.66 no, $z = 1.684$
8.67 approximately 68
8.68 $z = -2.5$; reject H_0: $\mu = 1600$
8.69 .1151, .0228
8.70 no, $z = 1.5$
8.71 309
8.72 $z = -1.667$, yes
8.73 .8 ± .047
8.74 (a) H_0: $p = .2$; H_a: $p > .2$
(b) .0749
8.75 yes, $z = 4.00$
8.76 (a) .0625 ± .0242
(b) approximately 586
8.77 22 ± 1.552
8.78 approximately 10,000
8.79 .69 ± .082
8.80 approximately 21
8.81 (a) H_0: $\mu = 1100$; H_a: $\mu < 1100$
(b) $z < -1.645$
(c) yes, $z = -1.897$
8.82 (a) .2 ± .0784
(b) approximately 400
8.83 approximately 200
8.84 yes, $z = 4.00$
8.85 .07 ± .096
8.86 yes, $z = 5.237$

8.87 $34 \pm .588$
8.88 approximately 64
8.89 6147
8.90 (a) yes, $z = 4.333$
(b) 13 ± 5.88
8.91 (a) H_0: $p = .2$; H_a: $p > .2$
(c) .0351
(d) reject H_0
8.92 $.1 \pm .041$
8.93 yes, $z = 8.667$
8.94 .0456
8.95 .6700
8.96 .4681
8.97 .387, .651

8.98

μ	β
873	.3446
875	.6103
877	.8274

8.99

μ	β
873	.6141
875	.7794
877	.8906

8.100 yes, $z = 5.44$
8.101 no, $z = 1.31$
8.102 $.42 \pm .0608$
8.103 $.58 \pm .018$
8.104 (a) yes, $z = 11.236$
(b) $.14 \pm .027$
8.105 $5.4 \pm .277$
8.106 (a) no, $z = .915$
(c) $1.37 \pm .111$
(d) $.07 \pm .126$
8.107 yes, $z = 2.465$

CHAPTER 9

9.1 $85.733 \pm .744$
9.2 no, $t = -.647$
9.3 28.935 ± 3.676
9.4 32.333 ± 30.220
9.5 no, population distribution not normal
9.6 yes, $t = 2.80$; reject H_0
9.7 $26.4 \pm .877$
9.8 yes, $t = 5.985$; reject H_0
9.9 31 ± 2.625
9.10 $3.782 \pm .1296$
9.11 $4.26 \pm .715$
9.12 no, $t = -1.436$; do not reject H_0
9.13 \$368.33 ± \$156.723
9.14 no, $t = -1.195$; do not reject H_0
9.15 no, $t = -1.950$; do not reject H_0
9.16 $2.433 \pm .849$
9.17 approximately 20
9.18 no, $t = 1.567$; do not reject H_0
9.19 yes, $t = 6.127$; reject H_0
9.20 no, population distribution not normal
9.21 yes, $t = 4.535$; reject H_0
9.22 (a) no, $t = 1.606$; do not reject H_0
(b) $.14 \pm .201$
9.23 yes, $t = -3.354$; reject H_0
9.24 -6.3 ± 3.946
9.25 (a) approximately 142
9.26 (a) no, sample sizes permit use of z test
(b) yes, $t = 3.801$; reject H_0
9.27 yes, $t = 3.038$; reject H_0
9.28 (a) yes, $t = 4.326$; reject H_0
(b) 1.58 ± 1.014
9.30 approximately 67
9.31 no, $t = 1.862$; do not reject H_0
9.32 (a) yes, $t = 2.821$; reject H_0
(b) 1.4875 ± 1.247 (a 95% confidence interval)
(d) yes
9.33 42.125 ± 6.280
9.34 (.227, 2.194)
9.35 $\chi^2 = 22.449$, reject H_0: $\sigma^2 = .49$
9.36 (1.408, 31.264)
9.37 (a) no, $t = -.232$; do not reject H_0
(b) yes, $\chi^2 = 19.98$; reject H_0
9.38 (.00476, .00980)
9.39 (a) no, 1000 is 5 standard deviations away from 800

(b) yes, $z = 3.262$; reject H_0
(c) yes; $\chi^2 = 57.281$; for $\alpha = .05$, reject H_0

9.40 (24.582, 73.243)
9.41 yes, $F = 2.486$; reject H_0
9.42 (1.544, 4.003) (using $F_{49,49} \approx 1.6$)
9.43 (a) $\sigma_1^2 = \sigma_2^2$
(b) no, $F = 5.897$
9.44 (a) no, $F = 2.904$
(b) (.050, .254)
9.45 (a) yes, $F = 2.197$
9.47 no, $t = -1.341$
9.48 795 ± 7.949
9.49 approximately 31
9.50 24.7 ± 1.054
9.51 for $\alpha = .05$, yes; $t = 2.108$
9.52 yes, $t = 2.635$
9.53 $7.1 \pm .07$
9.54 11.3 ± 1.44
9.55 77.763 ± 6.337
9.57 yes, $t = 1.861$
9.58 4.9 ± 4.541
9.59 no, $F = 2.123$
9.60 yes, $F = 3.268$
9.61 yes, $t = 2.2$
9.62 no, $t = 1.712$
9.63 yes, $t = 2.497$
9.64 $-.75 \pm .772$
9.65 $.875 \pm .829$
9.66 $.875 \pm .967$; yes
9.67 $t = 9.568$; reject H_0: $\mu_1 = \mu_2$
9.68 $.259 \pm .047$ unpaired;
$.259 \pm .050$ paired
9.69 when within-group variation is less than between-group variation
9.70 no, $t = 2.571$
9.71 $.01 \pm .010$
9.72 (a) yes, $t = -3.408$
(b) $1.385 \pm .068$
9.73 no, $t = -1.583$
9.74 -3.033 ± 3.472
9.75 approximately 239
9.76 (29.301, 391.152)
9.77 $\chi^2 = 12.6$; do not reject H_0: $\sigma^2 = .01$
9.78 (.00896, .05814)
9.80 (a) two-tailed; H_a: $\sigma_1^2 \neq \sigma_2^2$
(b) lower-tailed; H_a: $\sigma_1^2 < \sigma_2^2$
(c) upper-tailed; H_a: $\sigma_1^2 > \sigma_2^2$
9.81 $F = 1.922$
9.82 (.781, 4.728)
9.83 (20.939, 935.680)
9.84 yes, $F = 2.407$
9.85 no, $\chi^2 = 7.008$
9.86 (.185, 2.465)
9.87 (a) yes, $t = 6.0$
(b) $1.2 \pm .564$
9.88 no, $t = -1.8$
9.89 yes; $t = 2.425$
9.90 17.75 ± 7.276
9.91 (a) yes, $t = 4.968$
(b) 15.667 ± 6.941
9.92 15.967 ± 7.141
9.93 (a) $t = 2.657$
(b) 4.542 ± 3.046
9.94 -10.186 ± 1.228
9.95 yes, $t = 2.945$

CHAPTER 10

10.1 y intercept = 2, slope = 3
10.2 y intercept = 2, slope = 1/2
10.4 (a) $\hat{y} = 2.6 + .9x$
10.5 (a) $\hat{y} = 5.16 - .531x$
10.6 (a) $\hat{y} = 3 + 4.286x$
10.7 (b) $\hat{y} = 48.795 - .0393x$
(d) 29.145
10.8 SSE = 1.1, $s^2 = .367$
10.9 SSE = .0777, $s^2 = .0194$
10.10 SSE = 2.5714, $s^2 = .6429$
10.11 SSE = 60.3184, $s^2 = 6.702$
10.12 yes, $t = 4.70$
10.13 $.9 \pm .451$
10.14 yes, $t = -15.949$
10.15 $-.531 \pm .092$
10.16 yes, $t = 22.361$
10.17 $4.286 \pm .409$
10.18 yes, $t = -4.769$
10.19 $-.0393 \pm .0187$
10.20 (a) $\hat{y} = 3 + .475x$
(c) 5.025
10.21 yes, $t = 3.791$; no
10.22 $.475 \pm .289$
10.23 (a) $\hat{y} = 4.3 + 1.5x$

(c) 1.531
(d) yes, $t = 3.834$
10.24 (a) $\hat{y} = .9552 + .0462x$
(c) .00274
(d) yes, $t = 10.55$
(e) $.0462 \pm .0079$
(f) .0462 pounds
10.25 (a) $\hat{y} = .1067 + .0016x$
(c) .0003239
(d) no, $t = 1.092$
(e) $.0016 \pm .00273$
(f) .0016 pounds
10.26 $3.5 \pm .780$
10.27 29.145 ± 1.943
10.28 12.5 ± 2.002
10.29 (a) 46.343 ± 1.646
10.30 (a) 27.406 ± 3.313
(b) $56{,}017.87 \pm 1{,}341.879$
10.31 3.5 ± 1.625
10.32 46.343 ± 4.205
10.33 $1.279 \pm .137$
10.34 $49{,}988.605 \pm 3{,}385.344$
10.35 (a) $\hat{y} = -.627 + 1.080x$
(b) 10.169 ± 2.326
10.36 (a) $\hat{y} = .72 + .118x$
(c) $.118 \pm .008$
(d) reject H_0; $t = 4.587$
(e) $7.230 \pm .342$
10.40 (a) $\hat{y} = 307.917 + 34.583x$
(b) .874; .764
(c) 515.415 ± 27.010
(d) 515.415 ± 97.384
(e) 76.4
10.41 (a) $\hat{y} = 8.267 - 1.314x$
(c) $-.982$
(d) 96.5%
10.42 (a) $\hat{y} = -.933 + 1.314x$
(c) .982
(d) 96.5%
10.43 (a) $\hat{y} = 198.925 + .0385x$
(c) .3297
(d) 10.9%
10.45 y intercept $= -2$, slope $= -2/3$
10.46 y intercept $= -5/3$, slope $= 2/3$
10.47 y intercept $= 0$, slope $= 2$
10.48 $\hat{y} = 1.00 + .70x$
10.49 (a) $\hat{y} = 2.643 + .554x$
10.50 .3667
10.51 .6554
10.52 yes, $t = 3.656$
10.53 yes, $t = 3.618$
10.54 all points fall on a straight line
10.55 σ^2
10.56 $.7 \pm .451$
10.57 $.554 \pm .393$
10.58 $1.7 \pm .78$
10.59 $2.089 \pm .880$
10.60 3.75 ± 1.850
10.61 .904
10.62 .851
10.63 72.36
10.64 (a) $\hat{y} = 46 - .317x$
(c) 19.033
10.65 $-.317 \pm .102$
10.66 30.167 ± 2.500
10.67 30.167 ± 8.291
10.68 $-.872$
10.69 (a) $\hat{y} = -.333 + 1.4x$
(c) 1.2410
10.70 yes, $t = 3.974$
10.71 (a) $\hat{y} = -5.497 + .066x$
(c) .1800
10.72 yes, $t = 3.38$
10.73 .730
10.74 $2.382 \pm .274$
10.75 $2.382 \pm .816$
10.76 (a) $\hat{y} = .067 + .517x$
(b) $2.133 \pm .282$
10.77 No, observations are not independent.
10.78 (a) $\hat{y} = 2.6 - .7x$
(c) .30, .316
10.79 yes, $t = -7.00$
10.80 $-.70 \pm .318$
10.81 (a) $1.9 \pm .848$
(b) smaller
10.82 (a) .9707
(b) 94.23
10.83 (a) $\hat{y} = .8 + 1.4x$
(c) .4, .1333
10.84 6.4 ± 1.325
10.85 $1.4 \pm .272$
10.86 do not reject H_0: $\beta_0 = 0$; $t = 2.089$
10.87 (a) .9899
(b) 98%

10.88 (a) .7738
(b) .5988
(c) $\hat{y} = 11.238 + 1.309x$
10.89 (a) .9803
(b) .9610
(c) $\hat{y} = 21.867 + 14.967x$
(d) yes
10.90 (a) no; $t = 2.066$
(b) .2992
10.91 (a) $\hat{y} = 9.590 + .653x$
(b) 21.346 ± 2.291
10.92 (a) $\hat{y} = .025 + .821x$
(c) yes, $t = 8.958$
(d) $.034 \pm .021$

CHAPTER 11

11.2 yes, $X^2 = 24.48$ ($\chi^2_{.05} = 7.81$)
11.3 no, $X^2 = .658$ ($\chi^2_{.05} = 7.81$)
11.4 yes, $X^2 = 31.77$ ($\chi^2_{.05} = 9.49$)
11.5 no, $X^2 = 13.58$ ($\chi^2_{.10} = 17.275$)
11.6 yes, $X^2 = 173.64$ ($\chi^2_{.05} = 12.59$)
11.7 yes, $X^2 = 141.03$ ($\chi^2_{.01} = 6.63$)
11.8 yes, $X^2 = 6.75$ ($\chi^2_{.01} = 6.63$)
11.9 yes, $X^2 = 52.73$ ($\chi^2_{.05} = 5.99$)
11.10 no, $X^2 = 1.89$ ($\chi^2_{.10} = 4.61$)
11.11 no, $X^2 = 2.87$ ($\chi^2_{.10} = 4.61$)
11.12 yes, $X^2 = 8.22$ ($\chi^2_{.10} = 2.71$)
11.13 (a) yes, $X^2 = 18.53$
($\chi^2_{.05} = 3.84$)
(b) yes, $z = 4.30$ ($z_{.05} = 1.645$)
11.14 yes, $X^2 = 8.75$ ($\chi^2_{.05} = 5.99$)
11.15 yes, $X^2 = 38.43$ ($\chi^2_{.01} = 16.81$)
11.16 yes for $\alpha = .10$, $X^2 = 5.49$
($\chi^2_{.10} = 4.61$)
11.17 (a) yes, $X^2 = 14.19$
($\chi^2_{.01} = 11.34$)
11.18 no, $X^2 = 10.4$ ($\chi^2_{.05} = 11.07$)
11.19 (a) reject H_0: $p = 1/6$,
$z = -2.30$
11.20 yes, $X^2 = 7.19$ ($\chi^2_{.05} = 5.99$)
11.21 no, $X^2 = 12.92$ ($\chi^2_{.10} = 13.36$)
11.22 no, $X^2 = 4.4$ ($\chi^2_{.10} = 7.78$)
11.23 yes, $X^2 = 54.11$ ($\chi^2_{.01} = 20.09$)
11.24 yes, $X^2 = 21.51$ ($\chi^2_{.01} = 11.34$)
11.25 no, $X^2 = 3.97$ ($\chi^2_{.05} = 5.99$)
11.26 yes, $X^2 = 6.49$ ($\chi^2_{.05} = 5.99$)
11.27 $.0906 \pm .1035$
11.28 yes, $X^2 = 6.18$ ($\chi^2_{.05} = 5.99$)
11.30 $\hat{p} = .5$; 6.25, 25, 37.5, 25, 6.25; reject H_0: $X^2 = 8.56$
($\chi^2_{.05} = 7.81$)
11.31 yes, $X^2 = 11.63$ ($\chi^2_{.05} = 9.49$)
11.32 no, $X^2 = 5.02$ ($\chi^2_{.05} = 5.99$)
11.33 yes, $X^2 = 27.17$ ($\chi^2_{.10} = 6.25$)
11.34 yes, $X^2 = 12.5$ ($\chi^2_{.10} = 4.61$)
11.35 yes, $X^2 = 7.48$ ($\chi^2_{.10} = 2.71$)
11.36 (a) yes, $X^2 = 10.27$
($\chi^2_{.05} = 3.84$)
(b) yes, $z = 3.205$
($z_{.05} = 1.645$)
11.37 yes, $X^2 = 7.62$ ($\chi^2_{.01} = 6.63$)
11.38 yes, $X^2 = 6.19$ ($\chi^2_{.05} = 5.99$);
$.415 \pm .057$
11.39 yes, $X^2 = 15.85$ ($\chi^2_{.10} = 14.68$)
11.40 yes, $X^2 = 233.36$ ($\chi^2_{.01} = 15.09$)
11.41 yes, $X^2 = 18.82$ ($\chi^2_{.05} = 15.51$)

CHAPTER 12

12.6 No, people who work during these hours will not be included.
12.7 $n_1 = n_2 = n/2$, signal amplification
12.8 $n_1 = 34$, $n_2 = 56$
12.9 $n_1 = n_2 = 48$
12.10 For clue to answer, see Figure 12.3 and its discussion.
12.12 yes
12.13 (a) randomized block design
12.14 (a) plots
(b) six fertilizer combinations
(c) randomized block design
(d) noise reduction
12.29 *BAC, CBA, ACB*
12.30 three observations each at $x = 2$ and $x = 5$; signal amplification
12.31 (a) 1.46 times as large
(b) slightly more than twice (2.14) as many observations
12.32 test for curvature
12.33 noise reduction (this is a randomized block design)

CHAPTER 13

13.1 SSE = 10.75, $F = 2.93$; do not reject H_0

13.2 (a)

Source	d.f.	SS	MS	F
Method	2	641.88	320.94	5.15
Error	8	498.67	62.33	
Total	10	1140.55		

(b) yes, $F = 5.15$ ($F_{.05} = 4.46$)

13.3 (a) 76 ± 8.14
(b) 66.33 ± 10.51
(c) 9.67 ± 13.30
(d) No, they are not independent.

13.4 (a)

Source	d.f.	SS	MS
Regimens	2	64.31	32.155
Error	25	338.02	13.5208
Total	27	402.33	

(b) no, $F = 2.38$ ($F_{.05} = 3.39$)
(c) 5.03 ± 3.48
(d) 3.69 ± 2.525

13.5 (a) yes, $F = 13.03$ ($F_{.05} = 3.71$)
(b) 7.5 ± 4.67

13.6 (a) yes, $F = 5.7$ ($F_{.05} = 4.26$)
(b) 5.83 ± 5.41
(c) 60.5 ± 3.54
(d) yes (note that the observations are the means of four assembly times)

13.7 (a) Each observation is the mean length of 10 leaves.
(b) yes, $F = 57.38$ ($F_{.05} = 3.10$)
(c) reject H_0, $t = 12.105$
(d) $2.367 \pm .337$

13.8 (a) yes, $F = 63.66$ ($F_{.05} = 3.29$)
(b) $1.66 \pm .273$

13.9 (a) no, $F = .87$ ($F_{.05} = 2.88$)
(b) 2.7 ± 3.75
(c) 27.5 ± 2.65
(d) 26

13.10 (a) yes, $F = 6.46$ ($F_{.05} = 5.14$)
(b) yes, $F = 3.75$ ($F_{.05} = 4.76$)
(c) $1.2 \pm .65$

13.11 (a) yes, $F = 19.44$ ($F_{.05} = 4.76$)
(b) yes, $F = 40.21$ ($F_{.05} = 5.14$)
(c) $1.4 \pm .597$

13.12 (a)

Source	d.f.	SS	MS	F
Treatments	2	38	19	10.06
Blocks	3	61.667	20.556	10.88
Error	6	11.333	1.889	
Total	11	111.000		

yes, $F = 10.06$ ($F_{.05} = 5.14$)
(b) yes, $F = 10.88$ ($F_{.05} = 4.76$)
(c) 3.5 ± 1.888

13.13 (a)

Source	d.f.	SS	MS	F
Treatments	2	524,177.167	262,088.58	258.19
Blocks	3	173,415	57,805	56.95
Error	6	6,090.5	1,015.083	
Total	11	703,681.67		

(b) 6
(c) yes, $F = 258.19$ ($F_{.05} = 5.14$)
(d) yes, $F = 56.95$ ($F_{.05} = 4.76$)
(e) 22.529
(f) 237.25 ± 55.13

13.14 (a) yes, $F = 5.77$ ($F_{.05} = 3.74$)
(b) 1.25 ± 1.501

13.15 (a) yes, $F = 7.20$ ($F_{.05} = 5.14$)
(b) 2.275 ± 1.194
(c) yes, $F = 16.61$ ($F_{.05} = 4.76$)

13.16

Source	d.f.	SS	MS	F
Rows	2	46.89	43.44	11.11
Columns	2	22.89	11.44	5.42
Treatments	2	91.56	45.78	21.68
Error	2	4.22	2.11	
Total	8	165.56		

7.33 ± 5.105

13.17 (a) yes, $F = 103.62$ ($F_{.05} = 4.76$)

(b) $2.55 \pm .366$
(c) no, $F = .16$ ($F_{.05} = 4.76$)
(d) no, $F = 2.02$ ($F_{.05} = 4.76$)

13.18 (a)

Source	d.f.	SS	MS	F
Rows	3	80.00	26.67	.41
Columns	3	91.50	30.50	.46
Treatments	3	36.50	12.17	.19
Error	6	394.00	65.67	
Total	15	602.00		

(b) no, $F = .19$ ($F_{.05} = 4.76$)
(c) 1.75 ± 11.133

13.19 (a) no, $F = 3.02$ ($F_{.05} = 4.76$)
(b) no, $F = .62$ ($F_{.05} = 4.76$)
(c) no, $F = .60$ ($F_{.05} = 4.76$)
(d) no
(e) $.975 \pm 1.4903$

13.20 $n_i = 11$
13.21 $b = 16$
13.22 $b = 21$
13.23 (a) randomized block design
(b) 32
13.24 $b = 18$
13.25 assume $\sigma^2 \approx .03$; $b = 48$
13.27 SSE = 1196.6

13.28

Source	d.f.	SS	MS	F
Treatments	3	.015	.005	2.00
Error	8	.020	.0025	
Total	11	.035		

no, $F = 2.00$ ($F_{.05} = 4.07$)

13.29 (a) $2.25 \pm .054$
(b) $.083 \pm .094$

13.30

Source	d.f.	SS	MS	F
Treatments	3	.015	.005	6.00
Blocks	2	.015	.0075	9.00
Error	6	.005	.00083	
Total	11	.035		

(a) yes, $F = 9.00$ ($F_{.05} = 5.14$)
(b) yes, $F = 6.00$ ($F_{.05} = 4.76$)
(c) no

13.31 (a) $.0833 \pm .0577$
(b) no
13.32 (a) no
(b) completely randomized design
13.33 (a) 16
(b) 135
(c) for a 95% confidence interval and $z \approx 2$, half width is 14.14
13.34 (a) no, $F = 6.20$ ($F_{.05} = 6.94$)
(b) no, $F = .54$ ($F_{.05} = 6.94$)

13.35

Source	d.f.	SS	MS	F
Cars	2	15.469	7.734	7.33
Error	6	6.333	1.056	
Total	8	21.802		

(a) yes, $F = 7.33$ ($F_{.05} = 5.14$)

13.36 (a) randomized block design

(b)

Source	d.f.	SS	MS	F
Films	3	198.34	66.11	23.57
Judges	7	106.97	15.28	5.45
Error	21	58.91	2.81	
Total	31	364.22		

(c) yes, $F = 23.57$ ($F_{.05} = 3.07$)
(d) no

13.37 50
13.38 23

13.39 (a)

Source	d.f.	SS	MS	F
Treatments	4	1.212	.303	11.67
Error	22	.571	.026	
Total	26	1.783		

reject H_0, $F = 11.67$ ($F_{.05} = 2.82$)
(b) $t = -2.73$, reject H_0: $\mu_D - \mu_A = 0$

13.40

Source	d.f.	SS	MS	F
Treatments	4	.787	.197	27.78
Blocks	3	.140	.047	6.59
Error	12	.085	.007	
Total	19	1.012		

$F = 27.78$ ($F_{.05} = 3.26$); reject H_0: $\mu_A = \mu_B = \mu_C = \mu_D = \mu_E$

13.41 yes, $F = 17.34$ ($F_{.05} = 3.26$); 4.2 ± 2.27 (95% confidence interval)

13.42 no, $F = 1.53$ ($F_{.05} = 6.94$)

13.43 no, $F = .63$ ($F_{.05} = 3.89$)

13.44 (a) yes, $F = 9.84$ ($F_{.05} = 3.03$)
(b) 6.83 ± 6.341
(c) 91.875 ± 4.151

13.45 (a) yes, $F = 5.20$ ($F_{.05} = 3.24$)
(b) no, $t = .88$ ($t_{.05} = 1.746$)
(c) $.348 \pm .231$

13.46 (a) yes, $F = 18.61$ ($F_{.05} = 2.87$)
(b) yes, $F = 7.11$ ($F_{.05} = 2.71$)
(c) 5.1 ± 1.67

CHAPTER 14

14.1 $\{0, 1, \ldots, 6, 19, 20, \ldots, 25\}$ for $\alpha = .014$; $\{0, 1, \ldots, 7, 18, 19, \ldots, 25\}$ for $\alpha = .044$; $\{0, 1, \ldots, 8, 17, 18, \ldots, 25\}$ for $\alpha = .108$

14.2 (a) H_0: $p = 1/2$; H_a: $p \neq 1/2$; rejection region: $\{0, 1, 7, 8\}$; $y = 6$; do not reject H_0 at $\alpha = .07$

14.3 rejection region: $\{0, 1, \ldots, 4, 13, 14, \ldots, 17\}$ for $\alpha = .049$; $y = 8$; no

14.4 (a) yes, $y = 2$; rejection region: $\{0, 1, 2, 8, 9, 10\}$ for $\alpha = .11$
(b) variance not constant

14.5 no, $y = 2$; rejection region: $\{0, 1, 7, 8\}$ for $\alpha = .0703$

14.6 (a) $U_B = 0$; reject H_0; rejection region: $\{0, 1\}$ for $\alpha = .0714$
(b) $U_B = 2.5$; do not reject H_0; rejection region: $\{0, 1\}$ for $\alpha = .0714$

14.7 (a) $U_A = 6$; do not reject H_0; rejection region: $\{0, 1, \ldots, 4\}$ (one-tailed) for $\alpha = .0476$
(b) do not reject H_0: $\mu_A - \mu_B = 0$; $t = 1.606$ ($t_{.05} = 1.86$)

14.8 yes, $U_A = 0$; rejection region: $\{0, 1, \ldots, 18\}$ for $\alpha = .0504$

14.9 $U_B = 12.5$; reject H_0; rejection region: $\{0, 1, \ldots, 21\}$ for $\alpha = .1012$

14.10 No, samples are not independent.

14.11 yes, $U_1 = 8$; rejection region: $\{0, 1, \ldots, 18\}$ for $\alpha = .0546$

14.12 yes, $z = -3.49$

14.13 (a) $T = 3$; rejection region: $\{T \leq 4\}$; reject H_0; consistent with results of Exercise 9.32

14.14 (a) no, $T = 6.5$; rejection region: $\{T \leq 4\}$

14.15 no, $T = 73.5$; rejection region: $\{T \leq 35\}$; consistent with results of Exercise 14.3

14.16 (a) no, $y = 8$, rejection region: $\{0, 1, 2, 9, 10, 11\}$ for $\alpha = .0654$
(b) no, $T = 14.5$; rejection region: $\{T \leq 11\}$ for $\alpha = .05$

14.17 $T = 7$; do not reject H_0; rejection region: $\{T \leq 4\}$ for $\alpha = .05$

14.18 yes, $T = 3.5$; rejection region: $\{T \leq 4\}$ for $\alpha = .05$

14.19 (a) no, $t = -.422$
(b) no, $R = 5$; rejection region: $\{2, 7\}$ for $\alpha = .107$

14.20 no, $R = 6$; rejection region: $\{2, 3, 10, 11\}$ for $\alpha = .081$

14.21 (a) reject H_0: $\mu_B - \mu_A = 0$; $z = -1.804$
(b) would be small number; do not reject H_0; $z = -.37$

14.22 (a) $-.465$; do not reject H_0: $\rho_s = 0$
(b) .708; reject H_0: $\rho_s = 0$
(c) .819; reject H_0: $\rho_s = 0$
14.23 .867
14.24 yes
14.25 $r_s = .903$; yes
14.26 (a) yes, $r_s = .660$
14.27 yes, $r_s = .9118$
14.28 (a) do not reject H_0; $y = 2$; rejection region: $\{0, 1, 8, 9\}$ for $\alpha = .0390$
(b) do not reject H_0; $t = 1.65$; rejection region: $|t| > 2.306$ for $\alpha = .05$
14.29 do not reject H_0; $T = 10.5$; rejection region: $\{T \leq 6\}$ for $\alpha = .05$
14.30 (a) do not reject H_0; $y = 7$; rejection region: $\{0, 1, 9, 10\}$ for $\alpha = .022$
(b) do not reject H_0; $y = 7$; rejection region: $\{8, 9, 10\}$ for $\alpha = .055$
14.31 (a) reject H_0; $T = 6$; rejection region: $T \leq 8$ for $\alpha = .05$
(b) reject H_0; $T = 6$; rejection region: $T \leq 11$ for $\alpha = .05$
14.32 (a) no, $U = 32$; rejection region: $U \leq 21$ for $\alpha = .094$
(b) no, $t = .30$
14.33 no, $R = 13$; rejection region: $\{2, 3, 4, 5, 6, 7\}$ for $\alpha = .109$
14.34 no, $R = 3$; rejection region: $\{2\}$ for $\alpha = .029$
14.35 no, $R = 5$; rejection region: $\{2, 3\}$ for $\alpha = .04$
14.36 (a) near 0, $z = -1.95$
(b) yes, $z = -1.95$
14.37 (a) AAAAABBBBBBABABA; $R = 7$; no evidence of lack of stability; rejection region: $\{2, 3, 4, 5, 13, 14, 15, 16\}$ for $\alpha = .06$
(b) $t = .574$; do not reject H_0
14.38 (a) $U_B = 3$; reject H_0; rejection region: $U \leq 3$ for $\alpha = .0556$
(b) $F = 4.91$; do not reject H_0; rejection region: $F \geq 9.12$
14.39 $-.845$
14.40 yes; yes
14.41 $-.593$
14.42 yes
14.43 $U = 17.5$; no; rejection region: $U \leq 13$ for $\alpha = .0498$
14.44 $T = 14$; reject H_0; rejection region: $T \leq 16$ for $\alpha = .01$ or $T \leq 25$ for $\alpha = .05$
14.45 .67684, reject H_0; rejection region: $|r_s| \geq .441$ for $\alpha = .10$
14.46 .0159
14.47 .100

INDEX